ELECTROMAGNETIC DEVICES

ELECTROMAGNETIC DEVICES

BY

HERBERT C. ROTERS

Director of Research, Fairchild Aviation Corporation
Formerly Assistant Professor of Electrical Engineering
Stevens Institute of Technology

NEW YORK
JOHN WILEY & SONS, Inc.
LONDON: CHAPMAN & HALL, Limited
1941

COPYRIGHT, 1941, BY

HERBERT C. ROTERS

All Rights Reserved

This book or any part thereof must not be reproduced in any form without the written permission of the publisher.

Copyright, Canada, 1941, International Copyright, 1941
HERBERT C. ROTERS, Proprietor

All Foreign Rights Reserved
Reproduction in whole or in part forbidden

PRINTED IN U. S. A.

PRESS OF
BRAUNWORTH & CO., INC.
BUILDERS OF BOOKS
BRIDGEPORT, CONN

PREFACE

This work is intended primarily as a textbook for graduate students in electrical engineering and as a reference book for engineers concerned with the development or design of new magnetic devices. A familiar knowledge of the subject matter generally taught in undergraduate courses in electrical engineering is assumed.

The book is divided into two parts: the first eight chapters in which the fundamental background theory and methods applicable to all types of magnetic circuits and non-rotary electromagnetic devices are developed, and the last six chapters in which these principles are applied to the solution of a variety of problems. These solutions have been developed in general terms, and each is followed by the detailed numerical solution for a particular set of data. It has been my experience that such specific and detailed solutions are invaluable to students who must study by themselves or away from an instructor.

The accurate solution of practical electromagnetic problems has always been difficult owing to the effects of magnetic leakage and the non-linear relationship between magnetic flux and magnetic intensity in ferromagnetic materials. Special methods have been developed to handle magnetic leakage, and great emphasis has been placed on graphical and step-by-step methods for the practical solution of problems involving the non-linearity caused by hysteresis and saturation in iron, and the variation of magnetic leakage with motion.

The material of this book has been presented to graduate students at the Stevens Institute of Technology for several years, both in the day school and in an evening lecture course covering a school year (one two-hour lecture and four hours of home preparation and problems per week).

A large amount of quantitative data is presented, both on magnetic materials and electromagnets, many of which are original and have been obtained by painstaking laboratory work done at Stevens by graduate and undergraduate students working on these projects, and also by me. Thanks are due Stevens for providing the apparatus, facilities, and special magnetic equipment required.

This work was undertaken in 1932 at the suggestion of Dr. Alan Hazeltine, formerly head of the Department of Electrical Engineering and at the present time Professor of Physical Mathematics at Stevens. Many of the methods presented represent an outgrowth from some original work on rational methods of magnet design done by Dr. Hazeltine. I wish to acknowledge the scheme of "estimating permeance" of Chapter V, and the design methods of Chapter X, as coming from this source. My thanks are also due Dr. Hazeltine for the many helpful suggestions and frequent time spent with me in consultation and checking some of the more difficult parts of the manuscript.

Many of the original problems and methods used in the latter chapters are an outgrowth of my consulting practice. I am indebted to the Fairchild Aviation Corporation for the origin of many of the problems on dynamic characteristics and the models built and tested to check the computed results.

My generous thanks are due the many instructors in the Electrical Engineering Department at Stevens who from time to time have assisted by making drawings, computations, and suggestions; to my colleague, Professor William L. Sullivan, for his criticism and helpful suggestions in connection with the development of the step-by-step methods of Chapter XII; to Mr. Harry W. Phair and Mr. Frank W. Stellwagen of the Fairchild Aviation Corporation for their assistance in correcting and criticizing the manuscript; and to my wife, Mary D. Roters, for her assistance in correcting the proof and preparing the index.

HERBERT C. ROTERS

JAMAICA, N. Y.
January, 1941

CONTENTS

CHAPTER	PAGE
I. Introduction	1
II. Magnetic Properties of Iron and Some of Its Alloys	14
III. The Theory of Operation of Electromagnets and the Factors Entering into Their Efficient Design	73
IV. Calculation of Magnetic Circuits Containing Iron and Air Gaps of Known Permeance	84
V. Calculation of the Permeance of Flux Paths through Air between Surfaces of High-Permeability Material	116
VI. Coils	151
VII. Heating of Magnet Coils	178
VIII. Magnetic Forces	196
IX. Characteristics of Tractive Magnets: Selection of Best Type for a Specific Duty	228
X. Design of Tractive Magnets: Design Procedure: Illustrative Designs	245
XI. Time-Delayed Magnets	336
XII. High-Speed Magnets	363
XIII. Alternating-Current Magnets	419
XIV. Relays	483
Appendix	531
Index	533

ELECTROMAGNETIC DEVICES

CHAPTER I

INTRODUCTION

1. General

This book is intended primarily for those who will be responsible for the design and development of new magnetic equipment. Developments in this field have been very rapid, and the present indications are that they will continue to remain so. The airplane alone, particularly the military plane, has required the development of many new electromagnetic devices for instruments and control. Two reasons for the necessity of control devices are the general inaccessibility of the equipment, which makes remote control desirable, and the severe requirements of functional operation of this type of equipment.

Although the average electrical engineer is not necessarily interested in the design of electromagnetic devices, he is interested in the theory of their design. This is said because an intelligent understanding of the operation and limitations of electromagnetic machinery and devices must be based upon a knowledge of the fundamental laws of electromagnetism, electromagnetic force relations, and the limitations of commercially available materials.

Successful development work in any field cannot be predicated upon superficial knowledge, but in every instance must depend upon a sound understanding of the fundamentals of the field, coupled to an imagination toned down by practical experience. Though it is true that experience is of inestimable value, it can sometimes be overrated. When dealing with the mathematically unpredictable, such as human emotions and relations, experience, mellowed by maturity, is the essential requisite. When dealing with an inanimate object which functions in direct response to the laws of mechanics and electricity, an exact, quantitative knowledge is a requisite if an analysis or design from only a functional point of view is required. If the particular object is to be practically and economically applied, experience does become an essential factor.

An exact quantitative knowledge of a subject can be gained only by breaking it down into its essential component parts, studying and mastering these, and then putting them together in their proper relation and perspective. This book has been written with this in mind. In the following articles the subject matter has been broken down by chapters to show the essential parts and their correlation.

2. Properties of Magnetic Materials

On this basis the first important subdivision is a knowledge of the materials available for building magnetic devices. Tremendous advances have been made in the last few years both in the number of types of magnetic materials available and in the improvement of their properties. These recent advances are discussed in the latter part of this chapter. A full discussion of all the properties of magnetic materials which are important in predetermining the complete magnetic performance of direct- and alternating-current electromagnets, transformers, polarized chokes and transformers, permanent magnets, etc., is presented in Chapter II. A very complete set of data, many of which were taken especially for the purpose at the Stevens' Laboratory, covering all the materials commercially available, is also presented.

3. Factors Entering into the Efficient Design of an Electromagnet

In actual operation, because of the transient and unsymmetrical nature of the magnetic cycle, the electromagnet is very complex. The energy changes occurring during a complete cycle include a storage of magnetic energy, mechanical work done, hysteresis energy loss, disgorgement of magnetic energy in the form of a spark, and finally the energy associated with the residual force. The efficiency of a magnetic device in any particular application depends on how skillfully the designer can control those energy changes which are desirable and mitigate those which are undesirable. Chapter III presents a complete discussion of all these energy changes, based on theoretical considerations and actual experimental data.

4. Magnetic Circuit Calculations

Predetermination of the magnetic saturation occurring in an electromagnetic device, or the available magnetic potential across an air gap, depends upon an accurate analysis of the magnetic circuit. Magnetic circuit calculations range from a simple series magnetic circuit without leakage, to the more involved circuits with distributed leakage, unsymmetrical parallel circuits, polarized cores involving anhysteretic magnetization and incremental permeabilities, and residual-flux predetermination for both soft and permanent magnet cores. All the various types of magnetic circuit calculations are covered in Chapter IV.

5. Permeance of Air Paths

The calculation of the permeance of flux paths through air, or the more general problem of the determination of solenoidal or lamellar fields, is a problem which, except for the more simple configurations, has defied the efforts of mathematicians for centuries. A two-dimensional field emanating from continuous cylindrical surfaces can be evaluated analytically by a sufficiently astute mathematician. A two-dimensional field with discontinuities has been evaluated for the special case of intersecting plane surfaces. In practice, however, simple two-dimensional fields do not occur. The actual field is always three-dimensional with several discontinuities in the form of corners and edges. With patience, two-dimensional fields can be handled by the method of field mapping; but three-dimensional fields, unless they can be considered to vary in only two dimensions, are difficult, if not impossible, to handle. Accuracy in the predetermination of the force of a magnet is, to a very great measure, dependent upon the accurate evaluation of air-path permeances. In Chapter V, a heuristic method of evaluating these permeances is developed. The efficacy of any method of computation must ultimately depend on the accuracy of the results obtained and the relative effort required. Numerous examples of computed results checked by experiment are given.

6. Coils for Magnets

The heart of any electromagnetic device is the exciting coil. The construction of a coil, aside from the choice of wire size, is purely a mechanical matter. Sufficient space must be provided for the required number of turns, sufficient insulation to withstand the highest probable surge voltage, and a mechanical construction which will insure ruggedness, strength, high space factor, and low cost. Commercial winding practice dictates coil construction, winding tolerances, insulation, etc. Chapter VI gives complete data on winding practice, and also the method of computing coil performance.

7. Temperature Rise

Heating, as applied to average-sized electromagnets or other small non-moving electromagnetic devices, is discussed in Chapter VII. The actual temperature rise of a body as a function of time due to the internal evolution of heat is complicated, depending upon its thermal capacity, thermal conductivity, and the heat-dissipating capacity of its external surfaces. The latter depends upon the nature of the radiating surface, its temperature, and whether or not the air surrounding it is still. The problem is first treated from a theoretical point of view under ideal conditions, and the results so derived are correlated with experimental data. In this manner, not only have the proper empirical constants been derived, but also the limits of validity of the various theoretical formulas have been determined.

8. Electromagnetic Force Formulas

The actual manner in which the work of a magnet will be obtained, that is, a large force through a short stroke or a small force through a long stroke, depends on the shape of the working pole faces. Force formulas, with all their limitations and corrections, and for all manner of pole-face shapes, and types of action, are derived in Chapter VIII.

9. Magnet-Pole-Face Types

The proper type of pole face for any particular magnet depends on the relative values of the force and the stroke. The use of the right pole face results in the least weight for a given work. Chapter IX discusses the force-stroke characteristics of the various types of pole faces, and develops complete data for determining the most economical pole-face type for a given force-stroke characteristic.

10. Design of Direct-Current Electromagnets

The actual procedure of designing a magnet to meet given specifications is, like all machine design problems, an essentially heuristic procedure. All types of magnet, while essentially the same in fundamentals, differ in the detailed procedure necessary to get a working design. The amount of labor necessary to get a satisfactory and economical design depends on how close the first choice of basic design factors is to the optimum values. These optimum values, for any particular type, vary with the ratio of the force to the stroke. In Chapter X the criteria for an optimum design, the procedure of design, and basic design factors are detailed in such a manner that an optimum design can be obtained with rapidity for any one of six basic types of electromagnet.

11. Time-Delayed Magnets

In many instances it is necessary to provide for a time-delay action in an electromagnet. Thus, it is sometimes desired to delay the execution of the work of the magnet an appreciable time after the energizing impulse has been initiated. When the time delay is short this can be incorporated in the electrical design of the magnet itself. When long time delays are necessary the delay mechanism must be external to the magnet. The problem of the predetermination of time delay can be handled mathematically with good results, provided that the inductance of the magnet remains substantially constant when the current varies, and there is no motion of the armature during the time-delay period. When these conditions do not occur, the solution must be obtained by a step-by-step method applied to the differential equation for the circuit. Complete data on time-delay methods, with their mathematical solution, and the correlation of the mathematical solution with experimental results are given in Chapter XI.

12. Quick-Acting or High-Speed Magnets

The problem of predetermining the time of action of a magnet, though probably one of the most difficult problems in magnet design, is also the most fascinating. It is of particular interest in high-speed magnets, where the shortness of the time required for the magnet to complete its work is the yardstick of merit. Such magnets are used in many modern control devices. The problem is one of electrodynamics, involving two variables, space and time. The partial differential equations that result, because of their non-constant coefficients, must be solved by a step-by-step integration. Complete details of the method of solution, and the correlation of the computed and experimental results for an actual magnet, are given in Chapter XII. In addition, a rational method of developing a preliminary design for short-stroke high-speed magnets to meet required specifications is developed.

13. Alternating-Current Magnets

The alternating-current magnet differs from the direct-current magnet, principally because the current through the coil is not determined by the resistance of the coil, as it is in a direct-current magnet, but instead is limited by the self-induced voltage produced by the alternating flux. As the flux linkage does not vary with the position of the plunger, the current depends on plunger position. This causes the alternating-current magnet to have an odd-shaped force-stroke characteristic. Aside from this, the fact that the flux is alternating produces other effects. Thus the force pulsates from zero to a maximum twice during each cycle of supply current. The average force, however, bears the same relation to the current as with direct current. In some applications where it is undesirable to allow the force to fall to zero it is necessary to split the working flux into two parts which are out of phase. This is generally accomplished by means of a shading coil.

Alternating-current power magnets are, in general, undesirable. Their volt-ampere consumption, which bears a definite minimum relationship to the frequency and the work, is high compared to that of the corresponding direct-current one. The iron circuit must be thoroughly laminated to prevent excessive eddy-current losses. Their one redeeming feature is that they are inherently high-speed magnets. They are extensively used as relays and for applications, such as electromagnetic hammers, which depend on the alternating-current characteristics. For power magnets it is common to use a direct-current magnet powered from a rectifier. Chapter XIII presents a complete discussion of the alternating-current magnet from the above point of view and complete details regarding their design.

14. Relays

The relay, as its name implies, is a device intended to repeat or relay. In an electromagnetic sense it is a small-work magnet of low power consumption for closing contacts. However, from the point of view of design, the relay is very general and embraces all the problems of special magnets. Thus a relay must always be designed with respect to its residual force which determines the current at which it will release; in addition, it must often be designed with regard to its speed of action, or for a definite time delay. When operating on single-phase alternating current, shading coils must be used to prevent chattering. Power consumption, rather than heating, is the general design limitation which distinguishes it from a tractive magnet. Relays are discussed in Chapter XIV.

RECENT ADVANCES IN FERROMAGNETIC MATERIALS

It is safe to say that the only advance in the design of magnets in the last few years has been made because of new ferromagnetic materials [1]

---

[1] Extremely interesting discussions of new developments in this field and their commercial applications may be found in the following references:

V. E. LEGG, "Survey of Magnetic Materials and Applications in the Telephone System," *Bell System Technical Journal*, July, 1939.

which have been developed. In some instances, as with Permalloy and Alnico, the effect has been so great as to change the whole course of design.

15. Soft Magnetic Materials

In the field of the "soft magnetic" materials a complete exploration of all the possible alloys of iron, nickel, and cobalt has resulted in the production of many useful alloys. These alloys may be divided into three groups:

1. Ferronickels iron-nickel alloys which have very high initial and maximum permeabilities and very low hysteresis loss.
2. Ferrocobalts: iron-cobalt alloys which have a high permeability that endures well beyond the saturation limit of ordinary iron.
3. Constant-permeability alloys: iron-nickel-cobalt alloys which have permeabilities that are invariant within limited ranges of flux density, and a hysteresis loss which is zero or negligible.

In the field of electromagnet design the first two of these are of special interest. Ferronickel (47 per cent) is of particular use in the design of sensitive relays. In this application it affords two important properties: a relatively high flux density with a very small magnetizing force, and a very small residual effect. In other applications the low hysteresis loss and high resistivity of ferronickel are very useful as they allow the construction of special devices, such as voltage generators and transformers, which must have negligible losses. Ferrocobalt is chiefly useful in tractive magnets where a high density of force is desirable and in other applications where its high saturation density results in a substantial reduction in size and weight. It also has a high incremental permeability at high polarizing flux densities. Constant-permeability alloys are particularly suitable for use in circuit elements in which distortion and energy loss must be a minimum. In the telephone industry, these materials are known as Permalloy, Permendur, and Perminvar, respectively.

T. D. YENSEN, "Magnetic Materials and Preparation," McGraw-Hill Book Co., 1937. This is the fourth chapter of a book entitled "Introduction to Ferromagnetism" by FRANCIS BITTER.
C. E. WEBB, "Recent Developments in Magnetic Materials," *Journal of the Institution of Electrical Engineers*, March, 1938.
These articles are very extensive and have complete bibliographies of all the important references in the literature.

16. Tractive and Residual Effects in Commercial Soft Magnetic Materials

As tractive force is a very important consideration in the design of electromagnets and relays, Fig. 1 is presented in order to give a definite picture of the relative tractive effort which can be obtained from the various special alloys commercially obtainable.[2]

From these curves it is apparent that ferrocobalt is superior to ordinary iron in producing a tractive effect. Unfortunately, except for special applications, the price of ferrocobalt is prohibitive. The 34.5 per cent variety is much less expensive than the vanadium variety and is

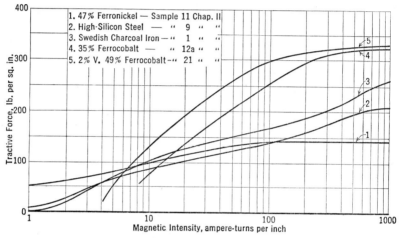

FIG. 1. Curves of tractive effort as a function of magnetic intensity for various magnetic materials commercially available. Data for Curve 5 taken from *Bell System Technical Journal*, Vol. XVIII, No. 3, page 438.

much easier to fabricate. The addition of vanadium is necessary if the product is to be rolled into sheets and also is advantageous as it increases the resistivity. On the other hand, where it is desired to produce a large force with very low magnetizing currents the ferronickel is shown to be superior.

In relays and sometimes in tractive magnets, the residual force produced by the coercive intensity of the material is important. This effect can be compared only by an examination of the demagnetization curves of the materials. Figure 2 shows the demagnetization curves of the alloys of Fig. 1,[2] except that for sample 12a which has been replaced by sample 12b.

[2] The samples referred to in Figs. 1 and 2 are described in Art. 33, Chapter II.

In order to show the effect of the coercive intensity in producing residual tractive force, the force curves, A and B, have been drawn to the right of the axis of ordinates. Curve A shows the residual produced when there is an ideal gap of zero length in the magnetic circuit. The force values are obtained by projecting, horizontally from the residual flux densities, on the axis of ordinates to Curve A as shown by the dashed lines. In actual practice, however, there is always an air gap in the magnetic circuit. This air-gap length being taken as one-one thousandth part [3] of the iron length, the air-gap permeance line OC may be drawn. The intersection of this line with the demagnetization curves gives the actual air-gap flux densities with the air gap present. Projecting

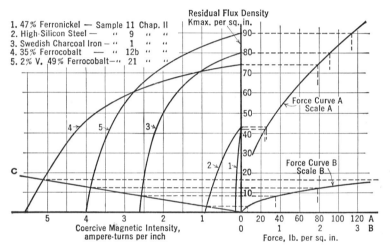

Fig. 2. Demagnetization curves and residual tractive effort for various magnetic materials commercially available.

these intersection points on the force curve B, by the horizontal lines shown, the residual forces are obtained. To make the comparison more vivid these forces are tabulated below. The advantage of ferronickel and high-silicon steel in producing a small residual effect is strikingly shown.

17. Importance of Impurities in Soft Magnetic Materials

The beginning of the development of magnetic materials may be dated at approximately the year 1900 when investigators attempted to find the cause of aging in transformer iron. In those days the best iron

---

[3] This is about the smallest gap length that is commercially practicable in the usual design.

RESIDUAL FORCE FOLLOWING EXCITATION TO SATURATION

Material	Residual Force, lb. per sq. in.	
	No Air Gap	0.1% Air Gap
47% ferronickel	25	0.0035
High-silicon steel	24	0.12
Swedish charcoal iron	89	0.9
35% ferrocobalt	77	3.5
2% V–49% ferrocobalt	112	1.9

available was Swedish charcoal iron. Owing to the low resistivity of this iron, the eddy-current loss was high, which when combined with the hysteresis loss produced a relatively high total core loss. It was found that the temperature rise of the iron due to its losses when used in transformers caused the core loss to increase rapidly, sometimes doubling it in the course of a few months. The only cure was to disassemble the transformers in order that the iron might be reannealed.

Attempts were made to discover the cause of this effect, known as aging, and while there were conflicting theories, the cause has remained obscure until recently. However, practical progress was made, and in 1900 Barrett, Brown, and Hadfield brought forth their silicon steel. By alloying the charcoal iron with $2\frac{1}{2}$ per cent silicon they found that aging disappeared, hysteresis loss decreased 25 per cent, and the resistivity was increased fourfold, greatly reducing the total core loss. Since that time considerable study has been devoted to the iron-silicon alloys. The commercial product of today is far better than Hadfield's best laboratory alloy.

The interesting feature of the reduction of iron loss by alloying with silicon is that the entire effect is now definitely traced to the removal of minute quantities of such impurities as oxygen and carbon by a process of absorption, which, though not actually removing the impurities in the physical sense, renders them impotent.

Research work carried out during the last few years has indicated that absolutely pure iron is an ideal magnetic material, having zero hysteresis loss and infinite permeability. Actual experiments on single crystals of very pure iron have shown relative permeabilities of several hundred thousand and extremely low hysteresis loss.

Iron of this exceptional quality has been produced only in the form of single crystals. The final removal of the last traces of impurities is accomplished by long annealing at high temperatures in an atmosphere of hydrogen. This not only reduces the ordinary oxides but also eliminates the residual amounts of carbon, sulphur, and phosphorus as gases (CH_4, H_2S, PH_3). Iron specially prepared and treated in this way has shown total impurities of the order of only 0.01 per cent.

Commercially such material is at present impractical, owing to its high cost and the difficulty of producing it in the form of rods and sheets. Even if it could be produced it would not be desirable for alternating-current apparatus because of its low resistivity and consequently high eddy-current loss. For these reasons alloys using silicons, or some similar material such as aluminum or vanadium, which mitigate the effect of the residual impurities of commercial iron and at the same time produce a high resistivity, must still be used.

Other very important factors in the production of iron of supermagnetic qualities are grain size and grain orientation. It has been definitely proved that as the grain size is increased the hysteresis loss is decreased. Furthermore, if the iron is rolled so that the axes of the grains have a parallel orientation as compared to a random orientation, a material of exceedingly high permeability and low hysteresis loss is obtained. This latter process, known as "fibering," has the effect of giving the sheet magnetic properties resembling those of a single crystal. Such materials are commercially available.[4]

In the case of ferronickel and ferrocobalt, prolonged annealing at high temperatures in an atmosphere of hydrogen has shown a remarkable improvement in their magnetic properties. Fifty per cent ferronickel when treated in this way has shown maximum permeabilities over 150,000 as compared to about 40,000 for the usual product. This product is commercially available under the name of "Hipernik." Likewise, 2 per cent vanadium Permendur when specially hydrogen-annealed shows a large increase in permeability and a slight increase in saturation density.

18. Structure and Grain Orientation of Magnetic Materials

The accepted explanation of many of the effects just described is based upon the existence of a definite lattice structure of the crystals of the material which imparts to the crystal an axis of easy magnetizability. Any portion of the matter having a continuous lattice structure and a definite orientation throughout is called a grain. An ideal magnet material will have regularity of the lattice structure throughout. As the regularity of the lattice structure is interrupted at the boundaries of the grains, anything which increases the number of grains or their random orientation is undesirable. Likewise, anything which increases the size of the grains and decreases the random orientation of these grains is desirable.

Impurities in the metal tend to make the grain size small, as they constitute the focal points from which crystallization starts. Thus small particles of oxides and carbides, when present in solid solution in the metal, will cause the crystallization to start at many points simultaneously, producing small grains. Consequently, the total absence of impurities is essential if large grains are to be produced.

Another factor controlling the grain size is the manner in which the metal is worked and the subsequent heat treatment. Considerable success in producing large grains has been obtained with silicon steel by cold rolling followed by annealing. If this is done in a special way, a preferred orientation of the lattice structure of the grains can be obtained which results in a magnetic material having properties approaching that of a single crystal.[4]

Cold working of a material will introduce internal strains or lattice distortions and tend to make it hard. Annealing by heating to a high temperature with very slow cooling produces a homogeneous structure free from lattice distortion. Thus magnetic materials, which in the process of fabrication have been stressed beyond the elastic limit, should be carefully annealed.

19. Hard Magnetic Materials

Advances in this field also have been astounding in the last decade. In contrast to what has just been said about soft magnetic materials, the problem in producing a hard magnetic material is to introduce lattice distortion in the crystal deliberately. Such a distortion or strain in the material can be produced by cold working, but where it is intended to produce a strong permanent magnet material it is best done by quench hardening or by dispersion hardening. Either process depends on the existence of a heterogeneous structure involving the presence of two components, one of which is dispersed in a finely divided form throughout the matrix causing severe lattice distortion. In all the earlier types of permanent-magnet steels, such as carbon, tungsten, chromium, and cobalt, the hardness depends on the formation of a martensitic structure on quenching. These steels have the inherent disadvantage that the martensitic structure producing the lattice distortion is unstable and subject to deterioration from vibration or high temperature.

[4] See Art. 30, Chapter II.

In the last few years new permanent-magnet materials have been developed in which the lattice distortion is produced by a dispersion of the second component in finely divided form throughout the matrix during a carefully controlled cooling process. Such steels are known as dispersion-hardened alloys. The best-known materials in this group are the nickel-iron-aluminum and nickel-iron-aluminum-cobalt alloys which appear under the names of Nipermag and Alnico III, for the former group, and Alnico I, II, and IV for the latter group.[5]

These materials are fabricated by casting or sintering. Sintering is a process whereby a properly proportioned mixture of the ingredients of the alloy in a finely divided form are molded into a solid metal by a process involving the use of temperatures a few hundred degrees below the melting point in combination with very high pressures. When the material is cast it must be finished by grinding, and when sintered it is formed into molds to the finished size.

These dispersion-hardened alloys have an exceedingly large hysteresis loss and hence form powerful permanent magnets. The commercial product averages about twice the available magnetic energy of the best grade of cobalt steel and several times that of the best grade of tungsten steel. This improvement is so great that it has revolutionized the design of many pieces of electrical equipment. Thus many types of small apparatus which formerly depended upon an electromagnetic winding to produce a constant flux can now be more economically designed utilizing these permanent-magnet materials.

[5] See Art. 32, Chapter II.

CHAPTER II

MAGNETIC PROPERTIES OF IRON AND SOME OF ITS ALLOYS

20. Normal Hysteresis Loop and Magnetization Curve

The magnetic performance of a piece of iron is dependent on its previous magnetic history, so much so in fact that certain apparatus, such as some sensitive relays, etc., must be operated in a definite magnetic sequence or cycle if they are to give reliable results.[1] This effect is due to a magnetic property of the iron termed *hysteresis*, meaning a lag; that is, the effect lags behind the cause, and hence an immediate effect (magnetic flux) may be due to an immediate cause (magnetomotive force) modified by some previous cause of which the observer has no knowledge. It is for this reason that, before making any magnetic measurements on iron, it must first be put through a very definite magnetic cycle so that the previous history is known. For ease of comparison, all curves giving the relation between the flux and magnetomotive force of a piece of iron are usually plotted with flux density and magnetic intensity as axes. Such a curve is called a *magnetization curve*.

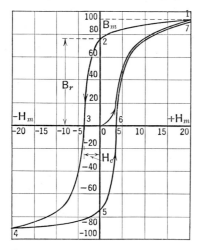

FIG. 1. Hysteresis cycle in soft iron having no previous magnetic history (sample 3).

Consider an iron ring sample which has been thoroughly demagnetized,[2] Fig. 1. The ring sample is chosen merely for its simplicity and

---

[1] This is particularly true at flux densities well below the knee of the magnetization curve, in which range most sensitive relays operate.

[2] A piece of iron is demagnetized by subjecting it to a slowly reversing magnetic intensity, which has an initial value at least as great as any previously applied value, and gradually decreases to zero. This process effectively removes the previous magnetic history of the iron.

convenience; for the sake of definiteness and proportion Fig. 1 has been drawn from actual test data taken on sample 3, the data for which are given on pages 47 and 69.

As the exciting magnetomotive force on the ring is gradually increased the flux density rises on the curve shown by 0–1 of Fig. 1, reaching the flux density B_m at a magnetic intensity of H_m. Such a curve is called a *rising magnetization curve*. If, after reaching point 1, the magnetic intensity is gradually decreased, the flux density will fall along the

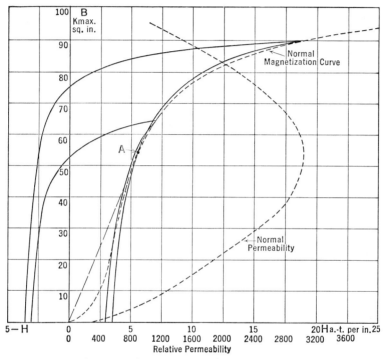

Fig. 2. Normal hysteresis loops and normal magnetization curve for soft iron (sample 3).

curve 1–2, reaching the value B_r at $H = 0$. This flux density B_r retained by the piece of iron after being magnetized is called the *residual flux density*, or retentivity. The cycle is continued by increasing H in the negative direction, causing the flux to decrease along the curve 2–3. The value of magnetic intensity H_c required to decrease the flux density to zero is called the *coercive magnetic intensity* or the coercive force. The cycle is completed by allowing the magnetic intensity to decrease to $-H_m$ and thence increase to $+H_m$ causing the flux density to fall along the curve 3–4, and finally to rise along the curve 4–5–6–7. Point 7 will not in general coincide with point 1, being usually slightly lower. If the magnetic intensity is now varied from $+H$ to $-H$ back to $+H$ several times, points 1 and 7 of these succeeding loops will gradually coincide, forming what is termed a *normal* or *symmetrical hysteresis loop*. When the magnetic intensity has been alternated sufficiently between the same positive and negative values to obtain a closed hysteresis loop, whether it be a symmetrical loop or not, the iron is said to be in the *cyclic state*.

The locus of the extremities of the normal hysteresis loops of a material is called its *normal magnetization curve*. Such a curve is shown in Fig. 2, for the same sample (3) as Fig. 1, along with several of the normal hysteresis loops from which it was derived. It will be noticed that at the higher magnetizations the normal magnetization curve lies outside the hysteresis loop.

If the rising magnetization curve of Fig. 1 is plotted on Fig. 2 it will be almost indistinguishable from the normal magnetization curve except in the region of steep slope, where it is slightly lower. For all practical purposes we shall consider these two curves to coincide for mild steel.

Figures 11 and 15 give normal magnetization curves for many commercial materials.

21. Permeability

The quotient of the flux density by the magnetic intensity as read from the normal magnetization curve is known as the absolute value of the *normal* permeability or merely the permeability of the material. If this permeability is divided by that of a vacuum the relative permeability is obtained. In references to ferromagnetic materials the permeability is almost always given relative to that of a vacuum, and hence, when we refer to the permeability hereafter, we shall mean the relative permeability unless specifically stated otherwise.

The permeability curve derived from the normal magnetization curve is shown in Fig. 2. The value of the permeability at $B = 0$, which is the slope of the tangent to the magnetization curve at $B = 0$ divided by the permeability of air, is known as the *initial permeability*. It is equal to 300 for this sample, and generally is of the order of several hundred for the ordinary commercial material.

Figure 15 shows normal permeability curves for several samples of commercial silicon steels.

The *maximum permeability*, shown as 3040 in Fig. 2, occurs at that point on the normal magnetization curve where the line OA through the origin is tangent to it. Maximum permeabilities of commercial materials range from a few hundred to several thousand.

22. Incremental Permeability

It is very convenient, when dealing with certain types of apparatus where a small alternating magnetic intensity is superposed on a constant magnetic intensity, to speak of the *incremental permeability*, which is the quotient of ΔB by ΔH, ΔH being in the reverse direction from the

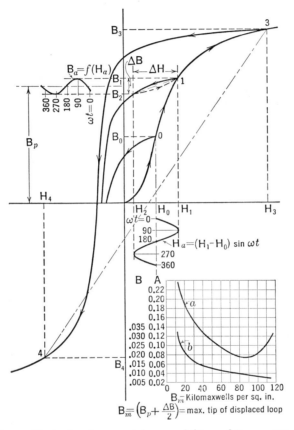

FIG. 3a. Curves illustrating incremental permeability and its manner of variation.

FIG. 3b. Empirical constants for Spooner's formula for incremental permeability. Reprinted from Spooner's "Properties and Testing of Magnetic Materials," by courtesy of the McGraw-Hill Book Co.

change in H immediately preceding. There is no restriction as to the magnitude of ΔB or ΔH, or as to the position on the hysteresis loop or magnetization curve at which they are taken.

Suppose that in Fig. 3a a direct polarizing magnetic intensity of

H_0 is impressed producing a constant flux density B_0 as shown on the normal magnetization curve. Let there now be superposed upon H_0 a sinusoidal magnetic intensity H_a such that the resultant magnetic intensity will vary between the limits of H_1 and H_2 as shown. This will cause the flux density to rise along the normal magnetization curve (same as rising magnetization curve) from 0 until point 1 is reached, where the magnetic intensity is at its maximum value H_1. At 1, H_a commences to decrease, causing the flux density to fall along a normal hysteresis loop until point 2 is reached, where the magnetic intensity is at its minimum value H_2. At 2, H_a increases and the flux density rises along the dashed curve to 1 again, after which the *minor* hysteresis loop 1–2 is traced over and over. It is therefore apparent that as far as the change in flux density caused by H_a is concerned the apparent absolute permeability of the iron is given by the slope $\Delta B/\Delta H$ of the axis of the minor loop 1–2. This permeability is known as the *incremental permeability*, and it is usually given relative to a vacuum. It is of chief interest in the design of chokes and transformers carrying direct current.

If H_a is increased so that the resultant magnetic intensity varies between H_3 and H_4 the incremental permeability will be proportional to the slope of the line 3–4 as shown. In general for a given biasing magnetic intensity below the knee of the normal magnetization curve the incremental permeability first increases and then decreases as the magnitude of H_a is increased.

If H_a is kept constant at some value which is not very small, the incremental permeability will decrease as the polarizing flux density increases. This is because the tops of the hysteresis loops become flatter with increasing magnetizations, as can be seen from Fig. 3a.

If H_a is allowed to approach zero the incremental permeability is proportional to the initial slope of the descending branch of the normal hysteresis loop corresponding to H_0. This particular incremental permeability is called the *reversible permeability*. Because the descending branch of the hysteresis loop has less slope as the magnetization is increased the reversible permeability is always less than the initial permeability. The reversible permeability is of particular interest where the superposed magnetic intensity is very small, as occurs in audio-frequency transformers. As ΔB is usually the independent variable in most applications dealing with incremental permeability, the values of incremental permeability are always plotted as a function of ΔB instead of as a function of ΔH.

When dealing with incremental permeabilities it makes considerable difference whether the variation in flux density or the variation in magnetic intensity is the independent variable. In Fig. 3a, H has been made the independent variable, and it will be noticed that the impressed sinusoidal variation H_a produces a flux density variation B_a which is not sinusoidal. Furthermore, B_a is not superposed upon the flux density B_0 that would be produced by the polarizing magnetic intensity H_0, but upon B_p, which is greater than B_0, depending upon the magnitude of H_a. Thus, when H_a is the independent variable the value of the polarizing flux density B_p is not easily determinable. Likewise, the wave form of the induced voltage E_a will be very complex compared to that of H_a. H_a will be the independent variable in constant current circuits and in some resonant circuits. Thus if an inductance is connected in the plate circuit of a screen-grid tube, the alternating component of plate current is determined almost entirely by the applied grid voltage.

More often, however, the flux-density variation is the independent variable. This is true in all constant- or substantially constant-voltage circuits; thus in a transformer supplying a half-wave rectifier the variation in flux density is forced to be sinusoidal because the supply voltage is sinusoidal; in a filter choke operating from a low-impedance tube the wave form of voltage across the choke, and hence the flux variation, is also determined entirely by the wave form of the supply voltage.

The discussion so far might lead one to believe that the incremental permeability is defined if the magnitude of ΔB or ΔH is specified, together with either the polarizing magnetic intensity H_0, or the polarizing flux density B_p. This, however, is not true, and in order that the incremental permeability be determinate it is necessary that the wave shape of ΔB or ΔH be defined in addition.

Referring to Fig. 3c, if a sinusoidal variation in magnetic intensity, equal to ΔH and polarized by H_0, is impressed on the magnetization curve shown it will produce the wave of flux density variation designated by B, having a total pulsation of ΔB as shown. The incremental permeability will then be designated as $\Delta B/\Delta H$. If, however, the same magnitude of pulsation ΔH, alternating about the same polarizing intensity H_0, but not sinusoidal in shape, is impressed on the magnetization curve it will produce the wave of flux-density variation designated as B' having a total pulsation of $\Delta B'$ as shown. The incremental permeability of the sample will be designated as $\Delta B'/\Delta H$. It is quite obvious from the figure that these two values of incremental permeability obtained for the same specified value of ΔH and H_0 are not the same. The same conclusion would be reached if ΔB and B_p were specified in magnitude. It therefore follows that if the incremental permeability is to be determinate the wave shape of either ΔB or ΔH must be specified besides the value of polarizing flux density or magnetic intensity.

The most useful definition [3] of incremental permeability is that where the pulsation in flux density is maintained sinusoidal, and where H_0 is specified, thus:

$$\mu_\Delta = \frac{\Delta B}{\Delta H} \quad (\Delta B \text{ sinusoidal})$$

The sinusoidal wave shape for ΔB is the most common one, as most apparatus operates on an impressed voltage of sinusoidal wave form. This will produce a sinusoidal flux wave if the circuit resistance is small compared to the reactance, as is usual. Likewise the value of polarizing current, and hence H_0, are generally known, whereas the value of the polarizing flux density is seldom known.

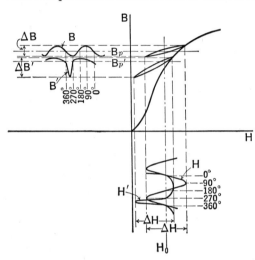

FIG. 3c. Curves illustrating the effect of wave form on incremental permeability.

Figure 3c also shows that the polarizing flux density is dependent on the wave shape. Thus B_p and B'_p, for the sinusoidal and distorted waves, respectively, are not the same even though ΔH and H_0 are the same.

Wave-shape distortion is appreciable only for the higher values of ΔB. Thus from 0 to 3 kilomaxwells per square inch, which covers the operating range of audio-frequency transformers, the incremental permeability is independent of wave form.

In Fig. 19 are shown curves of the incremental permeability plotted as a function of the maximum superposed sinusoidal alternating flux density $\Delta B/2$, for various values of H_0. The data are given for samples 7, 8, and 9 of low-, medium-, and high-silicon steel, respectively.[4]

These curves show that the incremental permeability increases with

---

[3] L. G. A. Sim, "Incremental Permeability and Inductance: The Role of Waveform in Measurement," *Wireless Engineer*, Vol. 12, 1935, Nos. 136, 137.

[4] The data shown in Fig. 19 were taken by Charles Rouault, and checked by Igor Bensin, as a thesis project at Stevens Institute under the supervision of the author. For a complete description of the experimental method and a critical discussion of the results see "An Investigation of Incremental Permeability," by Charles Rouault, Stevens Institute, June, 1939, and "An Improved Method of Measuring Incremental

superposed alternating flux density for any constant value of polarizing magnetic intensity until the maximum flux density reached in the cycle passes the knee of the saturation curve. For any constant value of superposed alternating flux density the incremental permeability decreases as the polarizing magnetic intensity increases.

Where no experimental data are available, the incremental permeability can be approximated by an empirical method if the normal magnetization or permeability curve for the material is known. This method, developed by Spooner,[5] is based on the experimental fact that, if the incremental permeability is plotted as a function of the flux pulsation ΔB (the maximum flux density of the tip of the minor loop being held constant), the result will be very close to a straight line which can be represented by the simple equation

$$\mu_\Delta \propto a + b \times \Delta B$$

The incremental permeability is further found to be a function of the flux density of the top of the minor loop farthest removed from zero. The resulting formula for incremental permeability is:

$$\mu_\Delta = \mu_{B_m}(a + b \times \Delta B), \qquad (1)$$

where μ_{B_m} is the normal permeability corresponding to the top of the minor loop farthest removed from zero.

ΔB is the amplitude of the total flux pulsation of the minor loop expressed in kilomaxwells per square inch.

a and b are constants which have been determined experimentally from tests of various magnetic materials. The values of these constants are given in Fig. 3b.

Permeability," by Igor Bensin, Stevens Institute, June, 1940. Briefly, Mr. Rouault excited a thoroughly laminated ring sample with a variable sinusoidal voltage derived from a commercial power line through a variable auto-transformer. In series with the exciting circuit, a low-resistance source of direct current for polarizing and a standard resistance were inserted. The total resistance drop to alternating current was never allowed to exceed 10 per cent of the impressed voltage in order to avoid distortion of the impressed sinusoidal wave form. The magnitude of ΔB was computed from the measured r.m.s. value of voltage induced in an insulated coil wound on the sample, and the cross section of the sample. ΔH was determined by comparing the current wave, obtained by impressing the voltage drop across the standard resistance on a cathode-ray oscillograph, with a standard sine wave shape of known r.m.s. value. This comparison was based on making the crest-to-trough value of the standard wave the same as that of the distorted exciting wave, and thence computing ΔH as $2\sqrt{2}\, E_{r.m.s.} N/R_s l_i$, where $E_{r.m.s.}$ is the root mean square value of the standard comparison wave, N the turns on the exciting winding, R_s the resistance of the standard, and l_i the mean length of the iron ring sample.

[5] Thomas Spooner, "Permeability," *Jour. A.I.E.E.*, Vol. 42, p. 42; January, 1923.

Spooner states that this method gives results with fair accuracy for all classes of ferromagnetic materials if the amplitude of ΔB is considerable, but that for very small values of ΔB (reversible permeability) it is unreliable.[5a] One difficulty sometimes encountered in applying this method is finding the flux density B_m at the tip of the minor loop if the polarizing magnetic intensity H_0 is given instead of the polarizing flux density B_p. If anhysteretic magnetization curves, like those of Figs. 18a and b, are available it may be calculated as $B_p + (\Delta B/2)$ if the even harmonic distortion of flux pulsation ΔB is small. If the maximum value of the alternating magnetic intensity H_a is known, B_m may be approximated from the normal magnetization curve corresponding to a magnetic intensity equal to $H_0 + H_a$, provided that B_m so located is above the point of maximum permeability. For B_m below the point of maximum permeability Spooner states that the effective magnetization curve may be taken as a straight line between the origin and the point of the normal magnetization curve corresponding to maximum permeability.

It will be noticed that another effect of superposing an alternating field on that produced by a direct current is to change the apparent permeability of the iron for direct current. Referring to Fig. 3a, the permeability for the direct magnetizing force H_0 alone is B_0/H_0, while with the alternating flux density B_a superposed it becomes B_p/H_0. Spooner[6] states that when ΔB is small the apparent permeability over the entire range of the magnetization curve up to fairly high densities is increased over the normal value, but when ΔB is large it is reduced, and at high densities is less than the normal permeability.

If the incremental permeability of the iron is known, the value of the polarizing flux density B_p can be approximately [7] determined for any value of ΔB and H_0 as follows: From ΔB and μ_Δ calculate ΔH. Then B_1, the maximum minor loop density, can be determined from the normal magnetization curve from H_1 equal to $H_0 + (\Delta H/2)$. Subtracting $\Delta B/2$ from B_1, the polarizing flux density B_p can be found. The value of B_p plotted against H_0 for constant values of ΔB or ΔH is sometimes called an anhysteretic curve (apparent magnetization curve with superposed alternating flux). Such curves, obtained experimentally, are

---

[5a] Data taken by Bensin (*op. cit.*) indicate that ΔB should be not less than about 20 kilomaxwells per square inch.

[6] T. Spooner, "Effect of a Superposed Alternating Field on Apparent Magnetic Permeability and Hysteresis Loss," *Physical Review*, 1925.

[7] Owing to the even harmonic distortion present in either the alternating flux density or magnetic intensity variation the axis of either cannot be determined exactly by taking the mean of the maximum and minimum values.

given in Fig. 18a and b for samples 8 and 9 of medium- and high-silicon steel, respectively.

23. Saturation

As the flux density of a piece of iron increases, the slope of the magnetization curve equal to dB/dH, sometimes called the differential permeability, first increases and then decreases. When the differential permeability is only a few times greater than that of a vacuum the iron is said to be saturated. The term saturation as ordinarily used is broad, including almost the entire range of the magnetization curve above the sharp bend called the knee. However, there is a very definite flux density, called the saturation density, designated by B_s, that the iron itself will carry. Any flux density in excess of this value is considered as being carried by the void which would be left if the iron were removed. Thus the flux density in any piece of iron can be considered as consisting of two parts: B_f, the so-called ferric flux density carried by the iron itself; and μH, the flux density that would exist in the absence of the iron if the magnetic intensity were maintained constant.

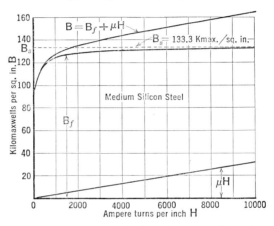

Fig. 4a. Separation of the flux of a piece of iron into its ferric component and air component.

The actual flux density in the iron can then be expressed by the following equation:

$$B = B_f + \mu H$$

This is illustrated in Fig. 4a for a sample (8) of medium-silicon steel:

The saturation density B_s can be extrapolated from data at lower densities by means of the following well-known relation: [8]

$$\mu_f = k(B_s - B_f)$$

---

[8] This relation, first discovered by Fröhlich, was later modified by Kennelly so as to be in a more useful form. Kennelly, *Trans. A.I.E.E.*, Vol. 8, p. 485, 1891; or see Steinmetz, "Theory and Calculation of Electric Circuits," First Edition, p. 43, McGraw-Hill Book Co.

which states that the permeability is proportional to the magnetizability. If we introduce for ferric permeability μ_f its equivalent B_f/H and solve for B_f we will obtain:

$$B_f = \frac{H}{\dfrac{1}{kB_s} + \dfrac{H}{B_s}}$$

Letting $1/B_s = \sigma$, and $1/kB_s = \alpha$, we have:

$$B_f = \frac{H}{\alpha + \sigma H}$$

$$\nu_f = \frac{H}{B_f} = \alpha + \sigma H$$

where ν_f is the ferric reluctivity. In other words, the law may be stated that the ferric reluctivity is a linear function of the magnetic intensity; if ν_f is plotted against H for any sample of steel the points should fall on a straight line. Actually, however, it is found that this law does not hold until the knee of the saturation curve has been passed.

In Fig. 4b, the ferric reluctivity as computed from the normal magnetization curve 1 of Fig. 11a, for a sample of annealed Swedish charcoal

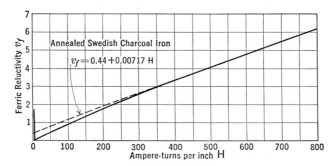

Fig. 4b. Ferric reluctivity curves for extrapolating for the saturation density by Kennelly's method.

iron, is shown plotted against the magnetic intensity. It will be noticed that the straight part of the curve which can be extrapolated, begins at about 400 ampere-turns per inch; this point will vary with different materials, some such as cobalt steel requiring a much higher value of magnetic intensity before straightening out.

The slope of this curve is equal to 0.00717, corresponding to a value of B_s equal to 139.4 kilomaxwells per square inch. The intercept α is considered to be a measure of the magnetic hardness of the material; that is, the greater the α, the harder the material.

In Table II there are given the saturation densities B_s for various magnetic materials. Those for samples 1, 7, 8, 9, 11, 12a, 12b, and 20 have been extrapolated from their normal magnetization curves by the above method.

24. Residual Flux Density and Coercive Intensity

It is often desirable, when dealing with the force produced by the residual flux of a magnet, or with permanent magnets, to know the value of the residual flux density B_r, and the coercive intensity H_c, that will be produced by various maximum-values of magnetic intensity. If the coercive intensity and the residual flux density are plotted as functions of the maximum magnetizing intensity it will be found that these curves have the same shape as normal magnetization curves (see Fig. 13). Sanford and Cheney[9] have found that these curves, because of their similarity to ordinary magnetization curves, may be extrapolated in an exactly similar manner. Thus if H_m/B_r or H_m/H_c are plotted against H_m as the independent variable, the resulting curves have the same shape as those of ν_f plotted against H, discussed in Art. 23. The following analytical expressions, which they have checked to a high order of precision, result:

$$\frac{H_m}{B_r} = a_1 + b_1 H_m \tag{5}$$

$$\frac{H_m}{H_c} = a_2 + b_2 H_m \tag{6}$$

a_1 and a_2 are intercepts on the axis of ordinates, and b_1 and b_2 are the reciprocals of the saturation values of B_r and H_c, respectively. These expressions, like that of Art. 23, cannot be used at very low values of H_m, where the reluctivity curve ν_f is not a straight line.

In Fig. 5a are shown curves of H_m/B_r and H_m/H_c plotted against H_m for sample 1 of annealed Swedish charcoal iron. From these curves the values of $b_1 = 0.01275$ and $b_2 = 0.3775$ are obtained, giving saturation values of $B_r = 78.4$ and $H_c = 2.65$, respectively. In Fig. 5b are shown similar curves taken from data by Sanford and Cheney[9] for 36 per cent cobalt permanent-magnet steel. These curves give values of $b_1 = 0.01736$ and $b_2 = 0.002295$, corresponding to saturation values of $B_r = 57.6$ and $H_c = 436$, respectively.

[9] "The Variation of Residual Induction and Coercive Force with Magnetizing Force," *Bur. Standards Sci. Paper*, 384, 1920.

Figures 5c and 5d show H_c and B_r plotted against H_m for both samples. Note the exact resemblance to ordinary magnetization curves. In

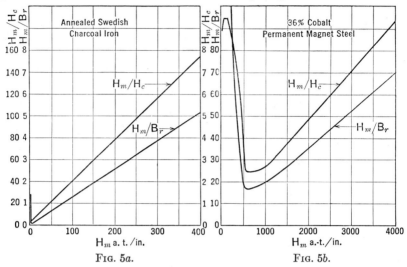

Figs. 5a and 5b. H_m/H_c, and H_m/B_r plotted as a function of the maximum magnetizing intensity.

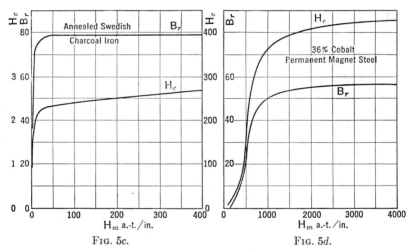

Figs. 5c and 5d. Residual flux density and coercive intensity plotted as a function of the maximum magnetizing intensity.

Figs. 13a and 13b curves are given for B_r and H_c as a function of H_m for various kinds of steel. Table II also gives values of B_r and H_c for various steels.

25. Energy Changes Occurring during Magnetic Cycles

In Fig. 6 is shown a normal hysteresis loop for a sample of commercial mild cold-rolled steel, obtained after the iron was put in the cyclic state. Consider the energy changes as the loop is traversed. Starting at point 6 and going toward 2 it is seen that the flux is rising in the iron and, hence, inducing a voltage acting against the current through the exciting coil. This causes energy to be abstracted from the electric circuit, the value of which is

$$W_{6-2} = \int_{-B_r}^{B_m} H\, dB \quad \text{joules per cubic inch}$$

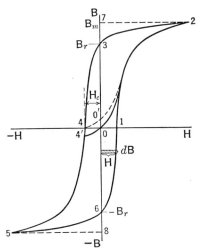

Fig. 6. Energy changes occurring during a normal hysteresis cycle.

and is evidently equal to the area 6–1–2–7–6 of the figure. During the next step of the cycle 2–3, energy is returned to the electric circuit from the iron, as the induced voltage due to the falling flux linkage is in the same direction as the falling exciting current. This energy is given by the integral:

$$W_{2-3} = \int_{+B_m}^{B_r} H\, dB \quad \text{joules per cubic inch}$$

and is equal to the area 3–2–7–3 of the figure. The last two steps of the cycle, namely, 3 to 5 and 5 to 6, are identical with the first two steps because of the symmetry of the loop about the origin. Hence the net energy taken from the circuit during the complete magnetic cycle will be equal to $2W_{6-2} - 2W_{2-3}$, which evidently is the total area inside of the loop.

This energy appears as heat in the iron, and because it is abstracted from the electric circuit by the phenomenon of magnetic hysteresis it is called the hysteresis energy loss. This loss, in joules per cubic inch of iron, is therefore equal to the area inside of a continuous hysteresis loop, the coordinates of which are webers per square inch and ampere-turns per inch.

The normal hysteresis loop is of particular interest in apparatus where the flux goes through a cycle having equal positive and negative values, as occurs in most alternating-current apparatus. Then where the flux variations are periodic one can speak of the hysteresis power loss per cubic inch, equal to the product of the area of the loop by the frequency of the supply in cycles per second. This is evidently the average rate at which energy is being dissipated as heat. For such alternating-flux apparatus a magnetic material having small hysteresis loss is desirable from the point of view of increasing the efficiency and decreasing the size. For this purpose silicon is alloyed with steel producing the so-called "silicon steels," which have a very low hysteresis loss.

It is possible to evaluate the hysteresis loss only for a complete magnetic cycle, that is, where the flux is brought back to its original value by purely electrical means. As an illustration of the lack of meaning of assigning definite energy forms to the various areas of the loop consider the following: Does the area 0–1–2–7–0 of Fig. 6 represent the energy stored in the iron due to the flux density B_m?

If the iron had been originally demagnetized the flux density would have risen to B_m along the curve 0–2, and the energy abstracted from the electric circuit would have been equal to the area 0–2–7–0. This area certainly does not equal area 0–1–2–7–0, but nevertheless the energy in the iron at a flux density of B_m is the same no matter whether one arrives along curve 0–2 or 1–2; hence, area 0–1–2–7–0 cannot be equal to the energy stored in the iron by B_m. However, neither does the area 0–2–7–0 represent the energy stored by B_m, because during the change 0–2 some hysteresis loss has occurred, and therefore the area 0–2–7–0 is greater than the available stored energy due to B_m; that is, some of the energy abstracted from the electric circuit, has already been dissipated as heat in the iron. The fact of the matter is that the only way one could determine the energy stored by B_m would be to subtract, from the area 0–2–7–0, the hysteresis loss occurring during 0–2, which would have to be measured by a calorimeter.

Now consider the area 3–4–0–3. Does this represent the energy stored in the iron by the residual flux density B_r? Since this area represents energy abstracted from the electric circuit when the flux density is decreased from B_r to zero, it cannot be energy stored in the iron but must represent the energy input to the iron necessary to demagnetize it. However, at point 4 the iron is not completely demagnetized because it still has stored energy. Thus, if at 4 the exciting magnetic intensity H_c is removed the flux density will rise to point $0'$, returning to the electric circuit the energy equal to the area 4–0–$0'$–4. Besides returning this energy the iron will still have some stored magnetic energy due to the flux density $0'$. To actually demagnetize the iron it is necessary to apply a negative magnetic intensity greater than H_c which will decrease the flux density to 4′, such that when the magnetic intensity is removed the flux density will rise to point 0 along the curve 4′-0.[10]

In other words, at point 3 the iron has stored magnetic energy due to the residual density B_r which could be evaluated by means of a calorimeter as previously mentioned, and in order to remove this stored energy, more energy equal to area 3-4′-0-3 must be put into the iron. Consequently, the stored energy at 3 due to B_r, plus the area 3-4′-0-3, must go into hysteresis loss when the iron is demagnetized. The area 3-4-0-3, therefore, is merely approximately equal to the energy input necessary to demagnetize it from the residual flux density B_r. In no event, though, can hysteresis loss be evaluated electrically until the magnetic cycle has been completed.

26. Normal Hysteresis Loss

In Fig. 12 are shown a series of normal hysteresis loops for various materials. The total symmetrical hysteresis loss (complete loops) in joules per cubic inch per cycle is shown plotted as a function of the maximum loop density B_m, on log-log paper in Fig. 14, by the curves labeled D. These curves have been obtained by measuring the symmetrical loop areas of Fig. 12. In Fig. 17b are shown similar curves of normal hysteresis loss for samples 8 and 9 of silicon steel. The data for these curves were obtained by exciting the sample with alternating current and measuring the total core loss with a wattmeter. Correction was made for the eddy-current loss occurring.

It will be noticed that these curves can, for a portion of their length, be represented by straight lines, the equation of which will be:

$$\log_e W_h = \log K + n \log B_m$$

where W_h is the hysteresis energy loss in joules per cubic inch per cycle, $\log K$ is the intercept on the axis of ordinates for $B_m = 1$, and n is the slope of the straight line. Or, taking the antilog of both sides of the equation, we have

$$W_h = K B_m^n \text{ joules per cubic inch per cycle} \tag{7}$$

This is Steinmetz's equation for hysteresis loss: in general it can only be applied over a limited range of flux density. In Table I the values for

---

[10] It must not be thought, however, that the magnetic state 0 so obtained will be identical with what would have been obtained had the iron been demagnetized by successive reversals. Actually the slope of the curve 4′-0 at the origin will be greater than the slope of the normal magnetization curve at this point. It is apparent, then, that, even though the iron is demagnetized as far as any of the ordinary tests would show, it has not been returned to its virgin state; that is, it retains a magnetic history.

K and n, evaluated from Figs. 14 and 17, are given for limited ranges of flux density. These constants can be evaluated only from experimental data for the particular iron in question, as they depend not only on the composition of the iron, but also on the heat treatment, mechanical working, etc. For this reason the values given in the table should not be considered as necessarily representative.

TABLE I

Sample	K	n	Range of B_m
1	0.010×10^{-3}	2.00	48 to 80
2	0.0068 "	1.63	48 to 80
3	0.0063 "	1.62	32 to 80
4	0.047 "	1.50	32 to 100
8	0.0047 "	1.57	20 to 80
9	0.0019 "	1.70	25 to 75

27. Total Iron Loss Due to an Alternating Magnetic Field

In the presence of a purely alternating magnetic field there will be, besides the normal hysteresis loss discussed in the last article, a power loss due to the presence of circulating currents in the cross section of the iron flux path. These currents, known as eddy currents, are produced by the voltages induced in the perimeters of the cross sections of the iron path by the alternating magnetic field passing through the cross sections. Thus, in Fig. 7a is shown the cross section of an iron lamination having a thickness t inches and a width large compared to the thickness. The magnetic flux passing normal to the cross section is represented by the dots. The path for eddy currents is normal to the flux lines and parallel to the center line of the lamination as shown by the dashed arrows. The eddy-current density at any distance x from the center of the lamination is equal to the voltage induced in one turn by the flux in area a–b–c–d divided by the resistivity of the metal, or

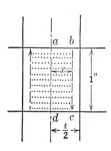

Fig. 7a. Eddy-current loss in an iron lamination.

$$i_e = \frac{x}{\rho}\frac{dB_m}{dt} = \frac{4k_f f B_m x}{\rho}$$

The power loss density due to eddy currents will then be:

$$p_e = \rho i_e^2 = \frac{16 k_f^2 f^2 B_m^2 x^2}{\rho}$$

and the total power loss in the section of lamination will be:

$$P_e = \frac{16 k_f^2 f^2 B_m^2}{\rho} \int_{x=0}^{x=t/2} x^2 dx = \frac{16 k_f^2 f^2 B_m^2 t^3}{3 \times 8 \times \rho}$$

and the loss due to eddy currents per unit volume of material is then

$$p_e = \frac{4}{3\rho} k_f^2 f^2 B_m^2 t^2 \quad \text{watts per cubic inch} \tag{8}$$

where B_m is the maximum value ($\Delta B_m/2$) of the cyclic loop density in webers per square inch, t the thickness of the laminations in inches, and ρ the resistivity of the laminations in ohms in an inch cube.

It is thus seen that for any given lamination the eddy-current loss will vary as the square of: the maximum cyclic flux density, the frequency, or the form factor of the induced wave of voltage producing the eddy currents. The constant $4/3\rho$ is usually determined by measuring the eddy-current loss experimentally, as it is found in actual practice that the eddy-current loss depends not only upon the resistivity of the sheet but also on the grain size, increasing as the grain size increases.[11]

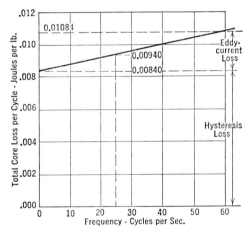

Fig. 7b. Separation of total core loss into eddy-current and hysteresis components.

In Fig. 16a are shown curves of total core loss at two frequencies as a function of B_m for low-, medium-, and high-silicon steel. Figure 16b gives the same data in terms of frequency for different values of B_m.

The eddy-current loss of a sample of iron at any value of B_m can be easily determined from readings of the total core loss for two different

[11] See "Properties and Testing of Magnetic Materials," by Thomas Spooner, McGraw-Hill Book Co., First Edition, 1927, p. 25, also Chapter VII.

frequencies by plotting the energy loss per cycle against the frequency. As the hysteresis loss per cycle does not depend on frequency, and the eddy-current loss per cycle varies directly as the frequency, the plot will be a straight line, the intercept of which on the axis of ordinates will be the hysteresis loss per cycle. If this loss is subtracted from the total loss per cycle the difference will be the eddy-current loss per cycle. Figure 7b shows such a plot for sample 9a of 29 gauge silicon steel, the data being taken from Fig. 16a or 16b. From Fig. 7b it can be calculated that of the total core loss of 0.65 watt per pound for sample 9a at $B_m = 64.5$ kmax. per sq. in. and a frequency of 60 cycles per second; 0.1466 watt, equal to 0.00244×60, is the eddy-current loss; and 0.564 watt, equal to 0.00840×60, is hysteresis loss.

The data of Figs. 16a and 16b, while given for 29 gauge sheet only, may be extended to cover the iron loss for other thicknesses by correcting the eddy-current component of the loss for the change in thickness.

28. Unsymmetrical Hysteresis Loss in Direct-Current Electromagnets, etc.

In a direct-current electromagnet the normal magnetic cycle of the iron is quite complicated. For all practical purposes, however, this cycle is approximated quite closely by a loop having the normal magnetization curve for the rising branch and the demagnetization curve of a normal hysteresis loop for the descending branch. Such a loop is shown in Fig. 6 by the rising line 4'–0–2 and the falling line 2–3–4'. Each time an electromagnet goes through a complete cycle of operation, energy equal to the area of this loop will be dissipated as heat per unit volume of iron.

Consider the simple ring magnet illustrated in Fig. 8a. Let the air gap at the beginning of the working stroke be g_1 and at the end of the working stroke be g_2; assume that the flux density throughout the iron is constant. When the coil circuit is closed (air gap held constant at g_1) the flux in the iron will build up along the curve 0–1, Fig. 8c, to ϕ_1, causing the flux density in the iron to reach the value B_1 of Fig. 8b. If the air gap is now allowed to decrease from g_1 to g_2, the flux in the iron will increase from ϕ_1 to ϕ_2, causing the flux density to rise to B_2. Magnetically the same result would have been obtained by decreasing the air gap first and then applying the magnetomotive force, in which case the curve 0–2 of Fig. 8c would have resulted. This curve is obtained by adding together the rising magnetization curve of the entire piece of iron, computed from curve 0–1–2 of Fig. 8b, and the magnetization curve for the air gap shown by line 0–7 of Fig. 8c. Likewise the demagnetization curve 2–3–4–5 for the entire magnet (iron and air gap) can be determined in exactly the same manner. When the coil circuit is broken the flux will decrease along this demagnetization curve to point 4, corresponding to the flux density B_4. The cycle of the magnet is completed by increasing the air gap from g_2 to g_1. This will require mechanical energy. If the permeance of the air gap at g_2 is large compared to that at g_1 the flux can, for all practical purposes, be considered to fall to zero along curve 4–0, Fig. 8c.

It is now possible to evaluate the effects of hysteresis from a practical point of view. First, if the cycle of the magnet were ideal there would be no losses of energy of any kind and the total area of the rectangle 0–1'–2–8–0 would be available as mechanical work.[12] Of this total area certain portions are unavailable; area 4–3–2–7–8–4 is returned to the

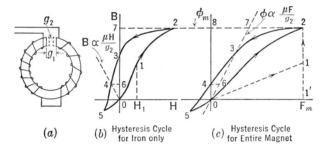

FIG. 8. Unsymmetrical hysteresis cycle of direct-current electromagnets.

electric circuit in the form of a spark when the coil circuit is interrupted, and area 0–2–3–4–6–0 is lost owing to hysteresis. Area 3–2–7–3 represents the energy returned to the electric circuit by the iron and corresponds to area 3–2–7–3 of Fig. 8b. Area 0–2–3–6–0 represents that portion of the hysteresis loss which is supplied directly by the electric circuit and corresponds to area 0–1–2–3–6–0 of Fig. 8b. Area 0–6–3–7–8–4–0 represents the energy stored in the air gap (length g_2) just before the coil circuit is opened. After the coil circuit has been opened the energy of the air gap is represented by area 0–6–4–0. The difference between these two energies is divided into two portions: part 4–3–7–8–4, which is returned to the electric circuit; and part 4–6–3–4, which is consumed in demagnetizing the iron from the residual flux density B_3 (at zero air-gap length) to the actual residual flux density B_4, and corresponds to area 4–6–3–4 of Fig. 8b. When the armature is moved mechanically from g_2 to g_1 the flux will fall from

---

[12] This is explained in detail in Chapter III.

4 to 0 if the permeance in position g_1 can be considered small compared to that in position g_2.

The energy changes involved during this latter process can be evaluated if the change from 4 to 0 is made electrically. This can be done by increasing the coil current in the negative direction (air gap = g_2) until point 5 is reached and opening the coil circuit. This will reduce the flux to zero, and the mechanical motion from g_2 to g_1 can now be carried out without expending any energy. In this cycle of operations the energy left in the air gap (area 0–6–4–0) will be dissipated, and, in addition, energy represented by area 0–5–4–0 will be taken from the electric circuit. These two energies evidently correspond to area 0–6–4–5–0 of Fig. 8b and constitute the energy necessary to demagnetize the iron from B_4 to zero.

Summing up: In a complete cycle of the magnet the hysteresis loss will be equal to area 0–2–3–4–5–0 of Fig. 8c, or the loss per cubic inch of iron will be the area of the loop of Fig. 8b. Of this total loss, part 0–6–3–4–0 is supplied by the stored energy of the air gap, and part 5–0–4–5 by mechanical means. Of the total work available from the ideal magnetic cycle, namely area $\phi_m F_m$, the portion 0–2–3–4–0 may be considered as lost as the result of hysteresis and portion 4–3–2–7–8–4 lost by its return to the electric circuit.

To make possible the predetermination of hysteresis cycles for electromagnets, the loops of Fig. 12 have been plotted from actual data taken on the samples indicated. These loops may be used directly to construct the complete hysteresis cycle as shown in Fig. 8c. Loops for values of B_m other than those given on Fig. 12 may be approximated by finding the values H_c, B_r, and H_m for the particular desired value of B_m from the data of Figs. 11 and 13 and then drawing a curve through these points of the same shape as the adjacent loops of Fig. 12.

To facilitate evaluating the energy losses corresponding to the various loop areas the data given in Fig. 14 have been determined from the loops of Fig. 12. Curve A gives the energy loss of iron corresponding to the complete magnetic cycle, area 0–2–3–5–0 of Figs. 8b or 8c. This curve can be extrapolated to higher values of B_m if desired. Curve C gives the energy required to demagnetize the iron from the residual density B_r, area 0–3–5–0 of Figs. 8b or 8c. This energy approaches a definite maximum because the demagnetization curve quickly approaches a definite limit as saturation is reached. This is shown by the relatively small area inclosed between the dashed magnetization curves (labeled $B_m = B_s$) and those corresponding to $B_m = 100$ of Fig. 12. Curve B gives the energy returned to the electric circuit by the iron and corresponds to area 3–2–7–3 of Figs. 8b or 8c.

This curve cannot be safely extrapolated. Values for B_m greater than 100 can be approximated by adding to the value at $B_m = 100$, a value approximated from the area back of the normal magnetization curve for the increment over $B_m = 100$. Partial areas like 4–6–3–4 or 0–6–3–4–0 can be approximated by taking a part of the value as given by Curve C, which part can be estimated from the residual density B_4 and the shape of the demagnetization curve as given on the loops of Fig. 12. Area 0–2–3–0 is obtained by subtracting the value given on Curve C from that given by Curve A.

29. Unsymmetrical Hysteresis Loops in Alternating-Current Apparatus

Unsymmetrical hysteresis cycles are also of interest in apparatus which carries an alternating flux superposed upon a constant flux. This occurs in transformers or chokes carrying direct current, such as audio-frequency chokes, audio-frequency transformers, and transformers supplying half-wave rectifiers. In these cases the hysteresis loss for a given alternating flux density depends upon the displacement produced by the polarizing magnetic intensity, and may sometimes be as great as several times the normal hysteresis loss produced when the polarizing magnetic intensity is zero. The ratio of the loss for the displaced loop to that of the normal loop is called the displacement factor. A series of such loops having the same B amplitude but different displacements are shown in Fig. 9a,[13] while in Fig. 9b [13] a series of such loops having the same B displacement for a varying B amplitude are shown.

In Fig. 17a the displacement factors for various values of polarizing magnetic intensities H_0 are shown plotted as a function of the maximum cyclic flux density $(\Delta B)/2$ (see Fig. 3) of the displaced hysteresis loops for samples 8 and 9 of silicon steel. In Fig. 17b the normal hysteresis loss for the two samples is shown plotted. The data of Fig. 17 were obtained by making alternating-current measurements on the iron samples in the following manner: The iron samples, in the form of rings with the laminations well insulated with varnish, were provided with three insulated windings. One winding was connected to a 60-cycle alternating-current source to provide an alternating magnetic flux. A wattmeter was used to measure the power input to this winding. Another winding was connected to a high-resistance direct-current voltmeter through a rectifying commutator in order to measure the true value of the flux variation ΔB produced by the alternating current. The third winding was connected

---

[13] Figures 9a and 9b are taken from data from "Properties and Testing of Magnetic Materials," by Thomas Spooner, McGraw-Hill Book Co.

in series with a choke coil to a direct-current supply in order to produce the polarizing magnetic intensity H_0. The wattmeter reading was corrected for all extraneous power loss, that is, copper loss in the exciting winding, loss in the direct-current voltmeter, alternating-current power loss in the direct-current magnetizing circuit, and eddy-current loss in the iron. The eddy-current loss was obtained by measuring the core loss at two different frequencies with zero polarizing magnetic flux density and extrapolating in the usual manner. Flux variations were maintained sinusoidal by keeping the resistance in the exciting winding low and using a variable-voltage source of good wave form.

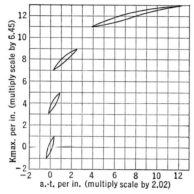 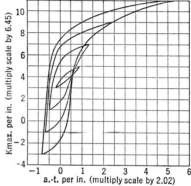

Fig. 9a. Unsymmetrical hysteresis loops having the same B amplitude but different displacements.

Fig. 9b. Unsymmetrical hysteresis loops having different B amplitudes but the same displacement.

Reprinted from "Properties and Testing of Magnetic Materials" by Thomas Spooner, by courtesy of the McGraw-Hill Book Co.

The hysteresis loss for any value of H_0 and $\Delta B/2$ is obtained by multiplying the displacement factor obtained from Fig. 17a by the corresponding normal hysteresis loss from Fig. 17b. It will be noticed that the displacement factor increases as H_0 increases for any constant value of $\Delta B/2$, and decreases as $\Delta B/2$ increases for any constant value of H_0. In other words, the displacement factor increases as the center of the displaced loop is further removed from the origin. This, however, is not true over the entire range of flux densities; it has been found by Edgar [14] that when the alternating flux density $\Delta B/2$ exceeds a certain critical value the displacement factor becomes negative, and that below this density the hysteresis loss for constant values of $\Delta B/2$ first rises as H_0 or B_p increases as shown in Fig. 17, then approaches a maximum, and finally, at high values of minor loop tip densities, $(B_p + \Delta B/2)$, decreases. The data of Fig. 17 do not show the latter two changes for the reason that they have not been carried to high enough values of $(B_p + \Delta B/2)$.

The data of Fig. 17 do not give any information regarding the value of B_p, the polarizing flux density, due to any value of H_0. For any given value of H_0 the polarizing flux density B_p will vary, depending upon $(\Delta B/2)$, the maximum value of the superposed magnetic flux density. In Figs. 18a and b, values of B_p plotted as a function of H_0 for constant values of $\Delta B/2$ are shown. These data were obtained while taking those of Fig. 17 by reversing H_0 with the alternating flux density $\Delta B/2$ superposed and reading ΔB_p on a ballistic galvanometer. By using these curves in conjunction with Fig. 17 the displacement factor in terms of $\Delta B/2$ and B_p may be obtained.[14] B_p may also be obtained by the method outlined at the end of Art. 22.

In Figs. 18c and 18d are given curves of alternating flux density $\Delta B/2$ plotted as a function of the r.m.s. alternating magnetic intensity H_a for various constant values of H_0. These data were obtained while taking those of Fig. 17. It is useful for calculating the required alternating-current excitation for apparatus having a superposed direct excitation, and can be used where the alternating flux wave form is sinusoidal.

Where specific data like those of Figs. 18c and 18d are not available the alternating-current excitation in the presence of direct excitation may be calculated from the data of Fig. 19 as follows: For any given value of $\Delta B/2$ and H_0 the incremental permeability μ_Δ may be found from Fig. 19. The crest-to-crest value of ΔH will then be given by the equation

$$\Delta H = \frac{\Delta B}{\mu_\Delta} \qquad (9)$$

If the minor loop is such that the wave of alternating magnetic intensity can be assumed fairly sinusoidal the r.m.s. exciting magnetic intensity will be

$$(H)_{\text{r.m.s.}} = \frac{\Delta H}{2\sqrt{2}} = \frac{\Delta B}{2\sqrt{2}\mu_\Delta} \quad \text{ampere-turns per inch} \qquad (10)$$

---

[14] A good many data on this subject are given in this way. See Chapter VI, "Properties and Testing of Magnetic Materials," Thomas Spooner, McGraw-Hill Book Co.; "Magnetic Properties of Sheet Steel under Superposed Alternating Field and Unsymmetrical Hysteresis Losses," Yasujiro Niwa and Yoshihiro Asami, Dept. of Communications, Tokyo, Japan, *Researches of the Electrotechnical Laboratory*, 124, June, 1923; "Loss Characteristics of Silicon Steel at 60 Cycles with D-C Excitation," R. F. Edgar, *Trans. A.I.E.E.*, Vol. 52, p. 721, September, 1933.

where ΔB is the total alternating flux pulsation in kilomaxwells per square inch and μ_Δ is the absolute incremental permeability in kilomaxwells per ampere-turn in an inch cube, equal to the values given in Fig. 19 multiplied by 0.00319.

30. Effect of Grain Direction and Machining Strains on the Magnetic Properties of Steels

The permeability and iron loss vary considerably with the grain direction. Spooner states that, for electrical sheet, a flux direction perpendicular to the grain direction will cause the permeability throughout its entire range to be about 75 per cent of that obtained with the flux parallel to the grain, while the hysteresis loss is increased about 14 per cent. Likewise the coercive intensity is about 25 per cent greater. It is generally desirable to use the iron parallel to the grain direction, especially when it is desired to build a sensitive relay where high permeability and low coercive intensity are advantageous. In tractive magnets which usually operate at high flux densities it is not so important.

Recently silicon steels of exceptionally high permeability and low hysteresis loss at both low and high flux densities have been produced by a special method of cold rolling and annealing. When the method is properly carried out a fine-grained material having magnetic and electrical properties approaching those of a single crystal are obtained. Losses as low as 0.46 watt per pound at 60 cycles per second and $B_m = 64.5$ kmax. per sq. in., and maximum permeabilities as high as 22,000 measured in the direction of rolling, have been reported.[15]

Machining a piece of iron is decidedly detrimental to its magnetic properties. It seems that the metal for a considerable depth behind the machined surface is strained, causing it to have a low permeability and a high coercive intensity. Cold working such as hammering or rolling has a similar effect. These undesirable effects can be completely removed by annealing the iron. This is accomplished by heating the iron to a maximum temperature of about 760° C. and then allowing it to cool *very slowly*. The iron should not be allowed to come in contact with oxygen during this process as the resulting oxidation is detrimental to the magnetic properties. The maximum annealing temperature varies slightly with different kinds of steel.

---

[15] "New Development in Electrical Strip Steels Characterized by Fine Grain Structure Approaching the Properties of a Single Crystal," by Norman P. Goss, *Trans. Am. Soc. Metals*, Vol. 23, p. 511, 1935.

The importance of machining strains depends upon the size of the piece of iron. If the dimensions are small, then a considerable percentage of the total volume will be affected and annealing is very necessary, whereas with a large piece it is relatively unimportant. The effect of machining strains can be seen by comparing curves 1 and 2 of Fig. 11a, both of which are for Swedish charcoal iron. Ring sample 1 was machined from a solid bar and has relatively small dimensions but was thoroughly annealed. Ring sample 2 was machined out of $\frac{1}{8}$-inch dead soft plate as received, and was given no heat treatment after machining. Its dimensions are considerably larger than those of sample 1. For low flux densities the magnetization curve of sample 2 is much lower than that of sample 1; at the higher densities the difference is not so marked. Likewise by referring to Figs. 12 and 13 it will be seen that the coercive intensity of sample 2 is much greater than that of sample 1.

31. Magnetic Materials for Electromagnets

American Ingot Iron. This is the purest form of iron commercially refined in open-hearth furnaces. The total impurities do not exceed 0.16 per cent, the carbon content being only of the order of 0.01 per cent. It has high electrical conductivity, high permeability, and low coercive intensity. It can be obtained in the form of bars and plates (hot-rolled), cold-rolled strip, and wire. Magnetic data for this iron are given in Figs. 11a, 12, 13, 14, and Table II. To bring out the best magnetic properties it must be very carefully annealed after machining. A maximum temperature of 1400° F. or 760° C. followed by slow cooling is recommended. Generally speaking, for direct-current electromagnets this material is the best obtainable. Unfortunately this iron, except for sheet and strip stock, can be purchased only in large quantities, usually rolled special to order. In sheet form this iron can be obtained hydrogen-annealed, a form having much higher permeability.

Cold-Rolled Steel. From a practical point of view, when building tractive magnets and other devices that operate at high flux densities, a good-quality mild cold-rolled steel will be found satisfactory. In strip stock, that is, strips having a thickness between 0.010 and $\frac{1}{4}$-inch maximum, the usual steel is a low-carbon steel with a bright finish, designated as S.A.E. 10–10 having between 0.05 and 0.15 per cent carbon. It is furnished in four degrees of hardness: dead soft, $\frac{1}{4}$ hard, $\frac{1}{2}$ hard, and full hard. The dead soft steel will take a 180° bend in either grain direction, the $\frac{1}{4}$ hard will take a 180° bend across the grain only, the $\frac{1}{2}$ hard will take a 90° bend across the grain only, and the full hard cannot be bent in either grain direction without cracking. In bar stock, that is, bars and rods over $\frac{1}{4}$ inch thick, the usual steel is one having slightly more carbon, between 0.15 and 0.25 per cent, designated as S.A.E. 10–20. This steel is not annealed or heat-treated after the final finish. Its hardness corresponds to about that of the $\frac{1}{2}$ hard strip. Data for both of these materials are shown in Figs. 11a, 12, 13, 14, and Table II. It will be noticed that the S.A.E. 10–20 steel has a very low permeability at the low flux densities, without annealing; it can be used without annealing for tractive magnets working at high flux densities. The advantage of using this material is the ease with which it can be obtained in a variety of sizes and shapes, the ease with which it can be machined, and its fine finish. For the most efficient results, the final machined magnet should be annealed.

Where a free machining steel is required, as, for instance, in parts to be made by a screw machine, S.A.E. 1112 may be used. Magnetically this steel is only slightly inferior to S.A.E. 10–10.

Swedish Charcoal Iron. The best grade of Swedish charcoal iron is practically identical in its magnetic properties to pure American ingot iron. For that reason only one set of curves and data is given for these two irons in Figs. 11a, 12, 13, 14, and Table II. These curves have been obtained from a sample of Swedish charcoal iron and check very well for other published curves for American ingot iron. These irons cannot generally be purchased from open stocks but must be specially ordered.

Cast Steel. Where the shell or some other part of a large magnet is of intricate shape, it probably is of advantage to make it of cast steel. Magnetically this material is very superior to cast iron, having a saturation density of 135.5 kmax. per sq. in., which is only slightly less than that of pure iron (see Table II). Owing to the variation in composition and heat treatment, cast steels differ greatly in their magnetic characteristics. A representative magnetization curve for this material is shown in Fig. 11a.

Cast Iron. Cast iron as a magnetic material is rather inferior. Its use can be justified only because of cheapness and ease of casting and machining. Its saturation density is about 90 kmax. per sq. in. This material like cast steel varies greatly, depending upon its composition. A representative magnetization curve is shown in Fig. 11a.

Malleable Cast Iron. Recently there has been developed a highly magnetic form of cast iron which is sold under various trade names.[16]

---

[16] The Newark Malleable Iron Works manufactures this iron under the trade name of Magtiz. Other manufacturers are: Eastern Malleable Iron Co., Delaware; National Malleable and Steel Castings Co., Cleveland. For additional magnetic data see "Symposium on Malleable Iron Castings," June 26, 1931, A.S.T.M. and American Foundrymen's Association.

This iron is made from a white-iron base, the exact composition and process of manufacture being a secret. It can be cast in intricate shapes and requires careful annealing after casting to develop its best magnetic properties. The mechanical properties and machinability of this iron are about the same as for malleable cast iron. Its cost is about twice that of ordinary gray-iron castings. Magnetically, up to densities of 80 kmax. per sq. in., it is superior to ordinary $\frac{1}{2}$ hard machine steel, having a higher permeability and a lower coercive intensity; and up to 60 kmax. per sq. in. it is comparable to cast steel. This cast iron should be very useful for the yokes and pole cores of magnets where an intricate shape is desirable or economical. For detailed data on an annealed casting of this iron see Figs. 11a, 13a, 13b, and Table II.

Electrical Sheet Steel. This material is used in the construction of all electrical machinery in which the flux is rapidly changing or alternating. It is made of high-quality open-hearth steel with varying percentages of silicon. Manufacturers usually grade these sheets on the basis of their core loss (combined hysteresis and eddy-current losses), there being usually about six grades. Those grades having the lowest core loss have the highest silicon content. The silicon content ranges from about $4\frac{1}{4}$ per cent in the highest-grade transformer steel to about 0.5 per cent in so-called "armature" grade which is used for the armatures of small direct-current machines. The term "silicon steel" is usually applied to only those steels having in excess of about 1 per cent silicon. Our interest in these steels is merely confined to their use in alternating-current electromagnets, choke coils, and transformers carrying direct current. For choke coils and alternating-current electromagnets a medium silicon (approximately 2.5 per cent) is suitable; for transformers carrying direct current a high silicon content is desirable.

These sheets are manufactured in thicknesses ranging from 29 to 22 U. S. gauge, except the high-silicon sheet which is not made in the heavier gauges, but is often made in lighter gauges, 32, 36, and 43 for special radio applications. Data for low, medium, and high silicon steels are given in Figs. 15, 16, 17, 18, 19, and Table II.

Electrical Bar Steel. Bar and strip stock of the same manufacture and composition as electrical sheet steel is available in several percentages of silicon. Its low coercive intensity makes it desirable in a relay requiring small residual force, and its high resistivity is useful where it is necessary to mitigate the effect of eddy currents as in high-speed direct-current magnets or in alternating-current magnets. In the latter application it is practical to make the plungers of small alternating-current magnets of solid silicon steel, while those of the larger magnets can also be made of the solid bar stock provided that radial slots are milled in to break up the eddy-current paths.

Iron-Nickel Alloys. There are two rather important iron-nickel alloys, one having approximately 50 per cent nickel, known as "Nicaloi" (General Electric Co.) or "Hipernik" (Westinghouse Electric and Manufacturing Co.), and the other having 78.5 per cent nickel known as "Permalloy" (Western Electric Co.); all these nickel-iron alloys require a very careful annealing process after machine working to develop their best magnetic properties.

Permalloy. This is a very remarkable alloy distinguished particularly for its high initial and maximum permeabilities. The initial permeability is about 9,000, and the maximum permeability is as high as 100,000 at 32.2 kmax. per sq. in. Another property which makes it useful is its extremely low coercive intensity and hysteresis loss. It should therefore be particularly useful for sensitive relays which are to have very low residual forces. At present it is used extensively in the telephone industry to load cables. Its disadvantages are that it requires a very careful heat treatment and is very susceptible to mechanical strains; likewise it is difficult to obtain. Data for this material are given in Fig. 11b.

Fifty Per Cent Nickel-Iron. This alloy, though having lower initial and maximum permeabilities than those of Permalloy, about 5,000 and 32,000, respectively, is in many ways more of a practical commercial material for general use. It does not require such an exacting heat treatment, nor is it so affected by mechanical strains; also its saturation density is higher, being about 100 kmax. per sq. in. Likewise the material can be more readily purchased. It is available in all the usual rolled forms, including sheets, plates, bars, rods, and strips. It, like Permalloy, has a very low coercive intensity and small hysteresis loss. The main use of this metal at present is for the cores of high-quality audio-frequency transformers and chokes, where its high incremental permeability is essential. It is also useful for the cores of particularly sensitive relays where low coercive intensity and high permeability are essential. Data for an alloy of this type (47 per cent Ni) manufactured by the Allegheny Steel Co., under the name of "Allegheny Electric Metal," are given in Figs. 11b, 12g, and Table II.

Iron-Cobalt Alloys. These alloys are unusual because their normal permeability remains high up to high values of flux density and the incremental permeability is much higher than that of other materials in the presence of strong polarizing flux densities, 75 kmax. per sq. in. and up. The saturation density of these alloys is about 12 per cent higher than that of iron. This makes them particularly useful for electromagnets where the space is definitely limited or where high force densities or flux densities are desired. The high incremental permeability at high polarizations is useful in polarized devices like telephone receivers, where it enhances the sensitivity.

There are two iron-cobalt alloys that are useful: an alloy containing 34.5 per cent cobalt corresponding to the compound Fe_2Co, and another containing 50 per cent cobalt. This material cannot readily be cast directly into its final shape as the molten metal is very viscous and has a tendency to form blow holes. It is generally cast in the form of a billet, then hammer-forged into a bar, and then rolled if necessary. Mechanical working is quite essential to develop structure. The process of heat treatment is very important if the alloy is to develop its best magnetic properties. At the present time, owing to the cost of cobalt and also to the difficulty of manufacture, ferrocobalt is relatively expensive.

Data for two commercial samples (12a and 12b) containing 34.5 per cent cobalt are given in Figs. 11b and 12h and Table II.

Data for a sample of Permendur containing 50 per cent cobalt and 2.0 per cent vanadium are given in Fig. 11b and Table II, Chapter II, and Figs. 1 and 2, Chapter I.

Iron-Nickel-Cobalt Alloys. There has been developed recently a group of iron-nickel-cobalt alloys, known as "Perminvar," [17] having with certain heat treatments a very constant permeability and unusually small hysteresis losses at low flux densities. Data for this material are given in Table II.

Iron-Nickel-Chromium-Silicon Alloys. These alloys are characterized by having their critical temperatures [18] depressed into the low-temperature range. They are useful in the construction of temperature-controlled apparatus such as thermal relays or contactors, reactors whose reactance varies with temperature, and special devices like transformers whose output can be made to vary with temperature. An alloy of 45 per cent Ni, 5 per cent Cr, and 50 per cent iron has a Curie point of 325° C.; one of 45 per cent Ni, 15 per cent Cr, and 40 per cent iron has a Curie point of 59° C.[19]

---

[17] "Magnetic Properties of Perminvar," G. W. Elmen, *Bell Sys. Tech. J.*, January, 1929.

[18] The critical temperature or Curie point is the temperature at which a ferromagnetic material becomes non-magnetic. The critical temperatures for nickel, iron, and cobalt, are 352, 780, and 1120° C., respectively. The permeability of these materials remains substantially constant with temperature until about 50° C. below the Curie point and then rapidly decreases to unity at the Curie point.

[19] For a discussion of these materials see "Temperature-Sensitive Magnetic Alloys and Their Uses," by L. R. Jackson and H. W. Russell, *Instruments*, Vol. 11, November, 1938.

Non-Magnetic Steels. In many instances it is desirable to use a steel which is non-magnetic. Stainless steel, S.A.E. 18–8, fulfills this qualification and may be used in place of the non-ferrous alloys.

32. Magnetic Materials for Permanent Magnets

The prime requisites for a permanent magnet steel are high coercive intensity, high residual flux density, and magnetic permanency. Their relative importance depends upon the application, in some cases it being economical to use a magnetically inferior material such as cast iron while in others the use of expensive cobalt steel, or Alnico is justified.

A permanent magnet is useful only because it can produce magnetic flux in an air gap outside of the magnet. The usefulness of a magnet is measured by the quantity of flux it can produce in the gap and the magnetomotive force it can maintain across the gap. One-half the product of these two quantities is the energy stored in the gap. The maximum possible energy of the gap per cubic inch of iron is, therefore, a logical way of evaluating the magnetic efficiency of a permanent magnet steel.

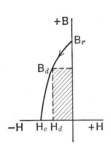

FIG. 10. Air-gap energy available from a permanent magnet.

Figure 10 shows the portion of a hysteresis loop between the residual flux density B_r and the coercive intensity H_c. This section of the loop is called the demagnetization curve and is useful in the discussion of permanent magnets. The residual flux density B_r can exist only in a closed iron sample such as a ring, the total coercive intensity H_c being required to overcome the reluctivity of the iron. If an air gap is introduced into the magnetic circuit part of the available magnetomotive force is required to send the flux across the gap, thereby reducing the magnetomotive force available to overcome the iron reluctance. Thus in Fig. 10 the introduction of a gap will reduce the flux density from B_r to B_d, thereby reducing the reluctivity drop in the iron from H_c to $H_c - H_d$ and making available in the air gap a magnetomotive force equal to $H_d \times$ length of iron. The shaded rectangle having an area equal to $B_d H_d$ will therefore be equal to twice the energy of the gap per unit volume of iron. Obviously, then, the most efficient point of operation of the magnet steel will be where the area $B_d H_d$ is a maximum. For this reason the criterion for the comparison of various magnet steels has become the largeness of the product $B_d H_d$. The method of designing a permanent magnet to operate at this point will be discussed in Chapter IV.

Cast Iron. This material can be used for permanent magnets when properly heat-treated. Its advantages are its cheapness and ease of machining. It has a coercive intensity of about 100 ampere-turns per inch, which is almost as large as that of carbon steel, but it has a much lower residual flux density.

Carbon Steel. Carbon steel for permanent magnets contains about 0.7 to 1 per cent carbon. This steel when properly heat-treated will have a coercive intensity between 100 and 120 ampere-turns per inch and a residual flux density between about 60 and 50 kmax. per sq. in. Commercially this steel has been almost entirely replaced by alloy steels having better properties, particularly less aging. See Fig. 20 and Table III for data.

Chrome Steel. This steel usually contains from 2 to 3 per cent chromium with about 1 per cent carbon. It is an oil-hardening steel with a fairly simple heat treatment. Data for a 2 per cent chrome steel are given in Fig. 20 and also in Table II. This steel is about as stable as tungsten steel but is less expensive and hence has replaced tungsten in many applications. It is reported by Gumlich[20] that an alloy having 6.24 per cent chromium and 1.14 per cent carbon is decidedly superior. See Table III for data. This steel is not generally available commercially.

Tungsten Steel. This steel contains about $5\frac{1}{2}$ per cent of tungsten, from 0.6 to 0.8 per cent of carbon, and sometimes from $\frac{1}{2}$ to 1 per cent of chromium. It can be made either water or oil hardening, and requires some care in heat treatment owing to the possibility of distorting and cracking it. Whereas the available energy of tungsten steel is only a little greater than that of carbon steel its chief advantage is that it is more stable; that is, it is not so subject to loss of magnetic energy due to mechanical shock or heating. See Fig. 20 and Table III for data on this steel.

Cobaltchrome Steels. This name is given to a series of low- and medium-cobalt steels ranging from about 9 to 20 per cent cobalt. They also contain about 9 per cent chromium and from 0.8 to 1 per cent carbon. They are air-hardening steels and contain a small amount of tungsten and molybdenum to assist the air-hardening properties. Their heat treatment is quite complex. Figure 20 gives data for a 15 per cent cobaltchrome steel. The maximum available energy of this steel is not quite so great as that of cobalt steel, but it is considerably less expensive.

Cobalt Steel. This steel contains about 35 per cent cobalt and corresponds to the alloy Fe_2Co previously described as having remarkable

[20] "Chromium-Carbon Steels for Permanent Magnets," E. Gumlich, *Elek. u. Mash.*, Vol. 39, p. 569, Nov. 20; and p. 586, Nov. 27, 1921.

magnetic properties. Besides cobalt it contains tungsten, about 4 per cent; chromium, 2 per cent; and 0.8 per cent carbon. It is an oil-hardening steel and gives little trouble from distortion or cracking. It can either be cast into its final shape or rolled into bar stock. Data for this steel are given in Figs. 5b, 5d, and 20 and Table III.

The cobalt steels are expensive because of the high cost of cobalt and their use generally can only be justified where it is essential to decrease the size or weight of the magnet. Magnetically, they are very stable.

Dispersion-Hardened Alloys.[21] This is a relatively new group of permanent-magnet steels developed in Japan and the United States. There are two widely used types, the aluminum-nickel-cobalt-iron alloy developed by the General Electric Co. and sold under the name of Alnico; and the aluminum-nickel-iron alloy sold under the names Alnic (General Electric Co.) and Nipermag (Cinaudagraph Corp.).

Alnico. This alloy is made in two varieties, 5 and 12 per cent cobalt, and requires very careful heat treatment. The material is fabricated either by casting or sintering. When sintered it can be molded to its final dimensions, and when cast it must be finished by grinding. Any necessary holes must be cored into the casting; soft steel inserts may be cast in for fastening or for any other purpose. It has a comparatively high coefficient of thermal expansion, and due care must be taken in designing the magnet and the mold to make adequate allowances for shrinkage in the casting. It is relatively weak and brittle as compared with other magnet alloys.

This alloy is remarkable because of its high coercive intensity, which is about 870 ampere-turns per inch for the 5 per cent cobalt variety. This is almost twice that of 36 per cent cobalt steel. Its residual flux density is only about 80 per cent that of cobalt steel. Because of the higher maximum available energy of Alnico, a magnet of this material will be of smaller volume than one of other magnet materials for a given amount of energy in the air gap. This reduction is so marked that Alnico is now being employed commercially for many applications formerly served by electromagnets. It is very stable as regards decrease in magnetization due to vibration, superposed alternating fields, or high temperatures. Because of its high coercive intensity it is difficult to magnetize, requiring a magnetizing force of at least 4,000 ampere-turns per inch actually effective in the material. Figures 20, 12i, 12j, and Table III give data on both 5 and 12 per cent Alnico.

The 5 and 12 per cent Alnico's are commercially designated as Alnico's I and II, respectively. The 12 per cent cobalt variety has only

---

[21] See Art. 19, Chapter I.

slightly higher residual flux density but about 25 per cent higher coercive intensity.

Table III also gives data for Alnico III and Alnico IV.

Permanent magnets need not necessarily be made from bar stock. Evershed [22] points out that tungsten and cobalt steels can be cast to form magnets slightly superior magnetically to those made from rolled stock. These steels are very difficult to machine, and casting gives an easy means of economically forming intricate shapes. Alnico, as previously mentioned, can only be cast.

Nipermag. This alloy which contains no cobalt has a higher coercive intensity than Alnico, about 1400 ampere-turns per inch, but a considerably lower residual flux density. Data for this material are given in Fig. 20 and Table III.

33. Magnetic Data for Iron and Iron Alloys [23]

In Figs. 11a and 11b are given magnetization curves for the ordinary magnetic materials used in electromagnets. The samples from which these data were taken are described below:

(1) A carefully annealed ring sample of Swedish charcoal iron. Grain direction one-half longitudinal; one-half transverse. This sample, by comparison, is magnetically identical with the best quality American ingot iron.

(2) A ring sample of Swedish charcoal iron cut from $\frac{1}{8}$-in. annealed sheet. Machined with very light cuts and not annealed after machining. Grain direction one-half longitudinal; one-half transverse.

(3) A ring sample of high-quality, bright-finish, dead-soft, mild cold-rolled $\frac{1}{8}$-in. steel sheet of analysis S.A.E. 10–10. Machined with very light cuts and not annealed after machining. Grain direction one-half longitudinal; one-half transverse.

(4) A ring sample of high-quality, $\frac{1}{2}$-hard, bright-finish, mild cold-rolled $\frac{1}{4}$-in. steel plate of analysis S.A.E. 10–20. Machined with very light cuts and not annealed after machining. Grain direction one-half longitudinal; one-half transverse.

---

[22] "Permanent Magnets in Theory and Practice," S. Evershed, *Jour. I.E.E.*, August, 1925. This paper is particularly recommended to anyone wishing a comprehensive discussion of the properties of permanent-magnet steels.

[23] It must be remembered that such data will vary depending upon the particular sample tested. No such data in this chapter, other than that of Figs. 15 and 16, and where specifically mentioned, are to be considered necessarily as representative: they are merely data taken with care on samples of standard commercial materials made by reputable manufacturers.

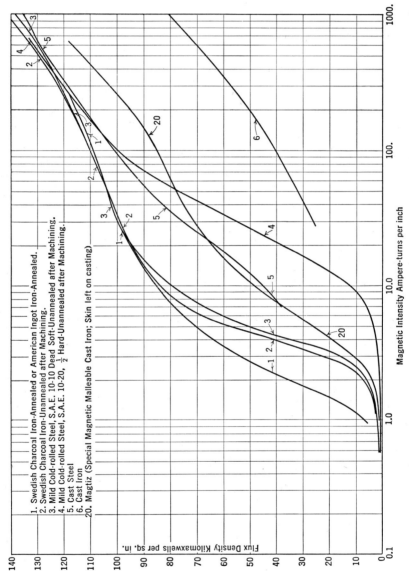

Fig. 11a. Normal magnetization curves for the more common "irons" commercially available.

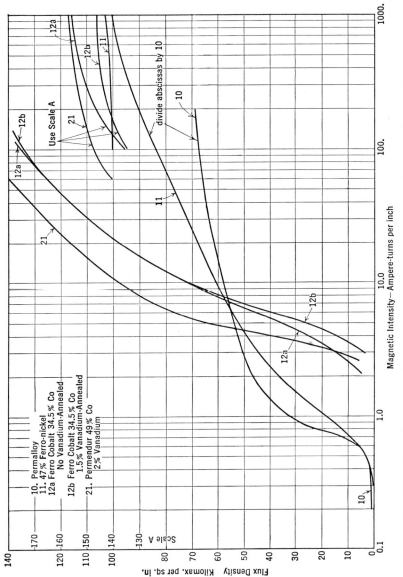

FIG. 11b. Normal magnetization curves for the more common nickel- and cobalt-iron alloys. Data for Permendur have been kindly supplied by the Bell Telephone Laboratories. Data for Permendur have been taken from the *Bell System Technical Journal*, Vol. XVIII, No. 3.

(5) A representative curve for a mild cast steel, averaged from several sources.

(6) A representative curve for cast iron.

(20) A carefully heat-treated ring sample of "Magtiz" (skin left on the casting).

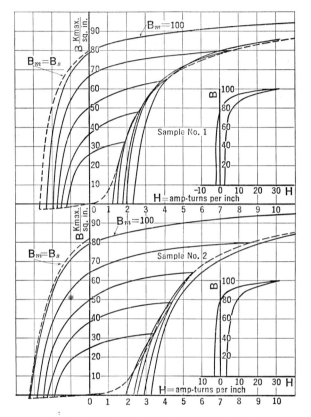

Figs. 12a and b. Normal hysteresis loops for:

Sample 1. Annealed Swedish charcoal iron.
Sample 2. Swedish charcoal iron unannealed after machining.

(10) A curve for Permalloy, kindly supplied by the Bell Telephone Laboratories.

(11) A carefully annealed ring sample of Allegheny Electric Metal (47 per cent ferronickel). Grain direction one-half longitudinal; one-half transverse.

(12a) A ring sample of carefully annealed ferrocobalt containing 34.5 per cent cobalt and no vanadium.

(12b) A ring sample of carefully annealed ferrocobalt containing 34.5 per cent cobalt and 1.5 per cent vanadium.

(21) A curve for Permendur (49 per cent cobalt, 2 per cent vanadium) taken from an article by V. E. Legg, *Bell Sys. Tech. J.*, July, 1939.

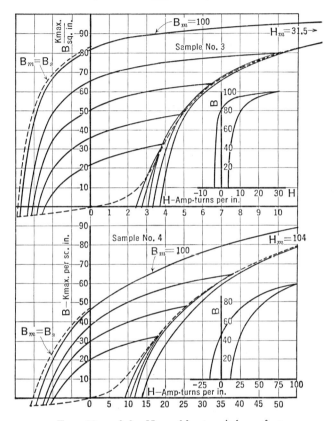

FIGS. 12c and d. Normal hysteresis loops for:

Sample 3. Mild cold-rolled steel, S.A.E. 10–10, dead soft—unannealed after machining.

Sample 4. Mild cold-rolled steel, S.A.E. 10–20, ½ hard—unannealed after machining.

In Fig. 12 are shown a series of hysteresis loops for samples 1, 2, 3, 4, 8,[24] 9,[24] 11, 12b, and Alnico I and II designated as samples 22a and 22b, respectively.

[24] All data on samples 7a, 8a, and 9a given in Figs. 15 and 16 have been taken from booklets issued by Allegheny Steel Company. These samples are their Armature, Super-Dynamo, and Transformer A power grades, respectively. The values

In Figs. 13a and 13b are shown, for samples 1, 2, 3, 4, 11, 20, and samples of 36 per cent cobalt and 0.85 per cent carbon permanent-magnet steels,[25] curves of the residual flux density B_r and the coercive

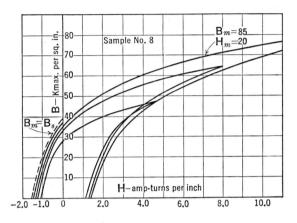

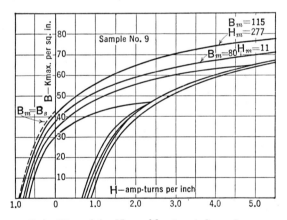

FIGS. 12e and f. Normal hysteresis loops for:

Sample 8. Medium-silicon steel.
Sample 9. High-silicon steel.

above 400 ampere-turns per inch have been extrapolated. Note should be taken of the difference between samples 7, 8, and 9 and 7a, 8a, and 9a. Samples 7, 8, and 9 are ring samples of low-, medium-, and high-silicon steel and are the Armco Electric grade of American Rolling Mills Co., and the Super-Dynamo, and Transformer A grades of the Allegheny Steel Co., respectively. All data on these samples have been taken by the author. The normal permeability curves for these samples, taken by a ballistic galvanometer, may be seen in Figs. 19a, 19b, and 19c, respectively.

[25] *Bur. Standards Sci. Paper* 383, 1920.

intensity H_c, plotted against the maximum magnetic intensity H_m of the normal hysteresis loop.

In Fig. 14 are shown four series of curves: Curves A give the hysteresis energy loss per cubic inch of iron for a complete unsymmetrical magnetic cycle (magnetic flux density increased from zero at zero

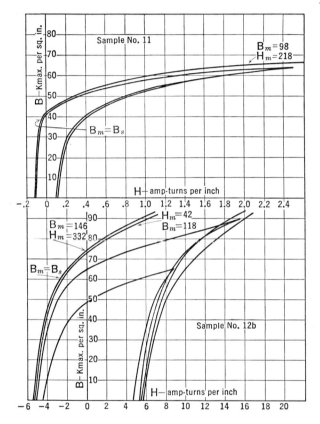

FIGS. 12g and h. Normal hysteresis loops for:

Sample 11. 47% ferronickel.
Sample 12b. Ferrocobalt—34.5% Co, 1.5% V.

magnetic intensity to a maximum and then returned to zero at zero magnetic intensity (area 0–2–3–4′–0 of Fig. 6); Curves B give the energy returned to the electric circuit per cubic inch of iron when the flux density is decreased from its maximum to the residual value (area 3–2–7–3 of Fig. 6); Curves C give the energy required to demagnetize the iron from the residual flux density to zero flux density (area 0–3–4′–0 of Fig. 6); Curves D give the hysteresis energy loss for a complete symmetrical magnetic cycle ($2 \times$ area 0–1–2–3–4–0 of Fig. 6). These data are given for each of the samples 1, 2, 3, and 4, and are computed from Figs. 11a and 12.

In Fig. 15 are given the magnetization curves for three grades of electrical sheet: 7a, low silicon content (0.5 per cent); 8a, medium

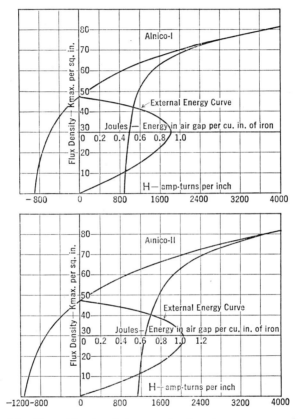

FIGS. 12i and j. Normal hysteresis loops and external energy curves for Alnico I and II. (By courtesy of the General Electric Co.)

silicon content (2.5 per cent); 9a, high silicon content (4.25 per cent). These are representative curves taken from manufacturers' data.[24]

In Fig. 16a are shown curves [24] at 60 and 25 cycles per second, of total core loss (hysteresis and eddy current) for samples 7a, 8b, and 9a plotted against the maximum cyclic flux density. These curves are for annealed sheets of 29 gauge which have no burrs and are well insulated. *Note*: Core losses will vary greatly depending upon whether the laminations

Art. 33] MAGNETIC DATA FOR IRON AND IRON ALLOYS

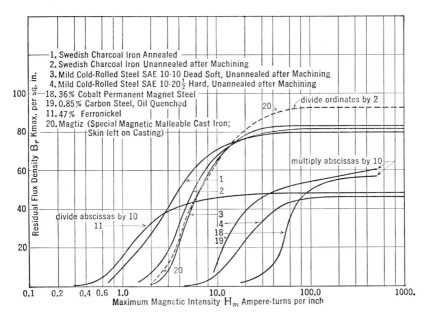

Fig. 13a. Curves of residual flux density as a function of maximum magnetizing intensity for various magnetic materials.

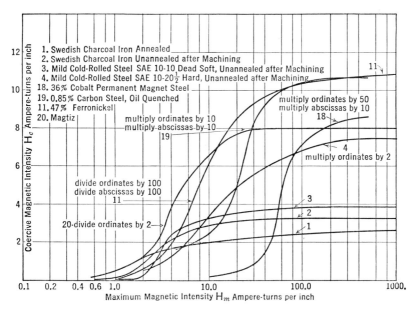

Fig. 13b. Curves of coercive intensity as a function of maximum magnetizing intensity for various magnetic materials.

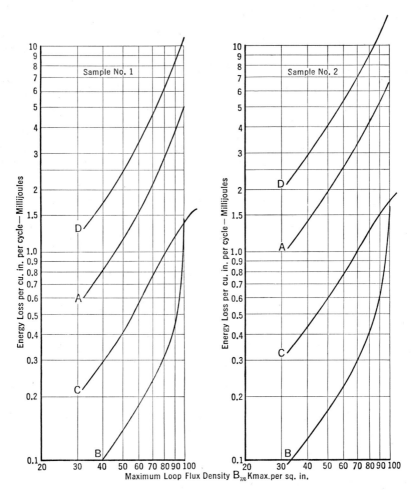

Figs. 14a and b. Hysteresis energy losses as a function of the maximum loop flux density.

Sample 1. Annealed Swedish charcoal iron.
Sample 2. Swedish charcoal iron unannealed after machining.

A. Energy loss, complete unsymmetrical loop.
B. Energy returned to electric circuit.
C. Energy required to demagnetize from B_r to 0.
D. Energy loss, complete symmetrical loop.

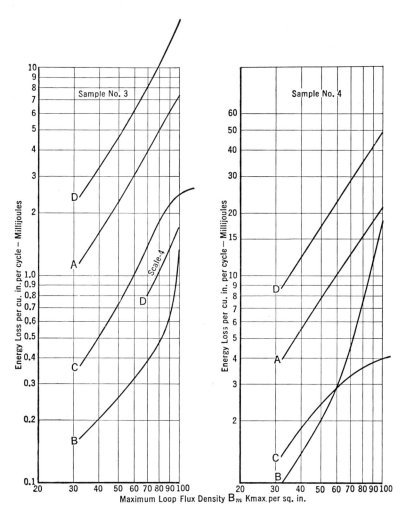

FIGS. 14c and d. Hysteresis energy losses as a function of the maximum loop flux density.

Sample 3. Mild cold-rolled steel, S.A.E. 10-10, dead soft—unannealed after machining.
Sample 4. Mild cold-rolled steel, S.A.E. 10-20, ½ hard—unannealed after machining.

    A. Energy loss, complete unsymmetrical loop.
    B. Energy returned to electric circuit.
    C. Energy required to demagnetize from B_r to 0.
    D. Energy loss, complete symmetrical loop.

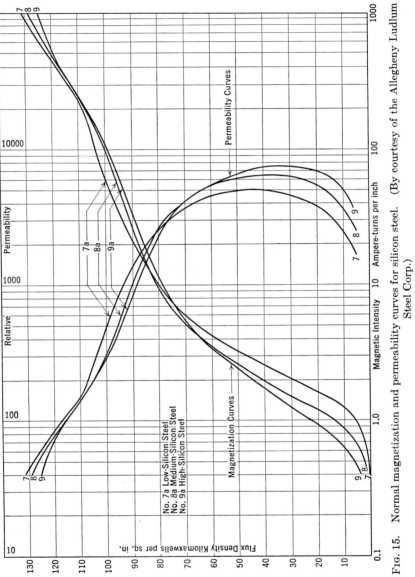

Fig. 15. Normal magnetization and permeability curves for silicon steel. (By courtesy of the Allegheny Ludlum Steel Corp.)

are annealed after punching, whether they are short-circuited by burrs, whether they contain mechanical stresses introduced by clamping, etc. Figure 16b shows similar data plotted against frequency for constant values of maximum cyclic flux density.

In Fig. 17 are shown curves of displacement factors for unsymmetrical

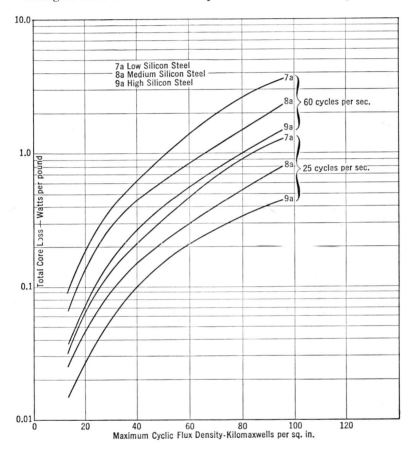

Fig. 16a. Total core loss for silicon steel.
(Courtesy of Allegheny Ludlum Steel Corp.)

hysteresis cycles for samples 8 and 9. The data for these curves were obtained from carefully insulated laminated ring samples made of the standard commercial sheet as designated by samples 8 and 9.[24] (See Art. 29 for details as to method of obtaining data.) Figure 17b gives the normal hysteresis loss for these samples as a function of the maximum cyclic flux density.

In Fig. 18 are given some apparent magnetization curves for ring samples 8 and 9 in the presence of a superposed alternating magnetic flux (see Art. 29 for details as to method of obtaining data).

In Fig. 19 are given curves of incremental permeability for ring samples 7, 8, and 9. (See Art. 22 for details as to method of obtaining data.)

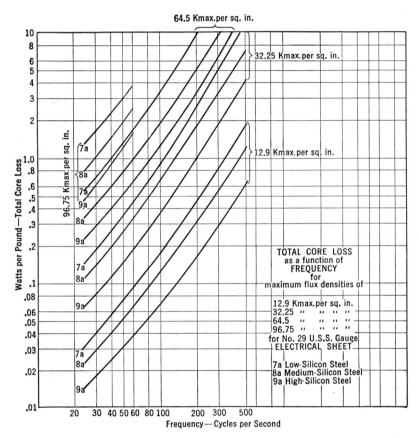

Fig. 16b. Total core loss as a function of frequency for silicon steel. (Courtesy of Allegheny Ludlum Steel Corp.)

The dashed curves shown are normal permeability curves taken by means of a ballistic galvanometer and are given so that a check upon the incremental permeability curve for $H_0 = 0$ may be made.

In Fig. 20 are shown demagnetization curves, and useful air-gap energy curves, for various permanent magnet steels.

Curve 14 is for a 0.7 per cent carbon steel.

ART. 33] MAGNETIC DATA FOR IRON AND IRON ALLOYS

Curve 15 is for a 2 per cent chrome steel.[26]
Curve 16 is for a $5\frac{1}{2}$ per cent tungsten steel.[26]
Curve 17 is for a 15 per cent cobaltchrome steel.[26]

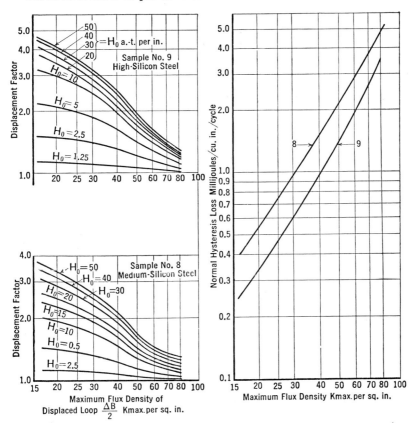

FIG. 17a. Displacement factors for silicon steel.

FIG. 17b. Normal hysteresis loss for silicon steel.

Curve 18 is for a 36 per cent cobalt steel.[26]
Curve 22a is for Alnico I.[27]
Curve 22b is for Alnico II.[27]
Curve 23 is for Nipermag.[28]

---

[26] These data are taken from a paper by E. A. Watson, "The Economic Aspect of the Utilization of Permanent Magnets in Electrical Apparatus," *Jour. I.E.E.*, August, 1925.

[27] These data and those of Figs. 12i and 12j and Fig. 21 for Alnico I and II and also those of Table III for the Alnico's have been supplied by the courtesy of the General Electric Co.

[28] These data have been supplied by the courtesy of the Cinaudagraph Corp.

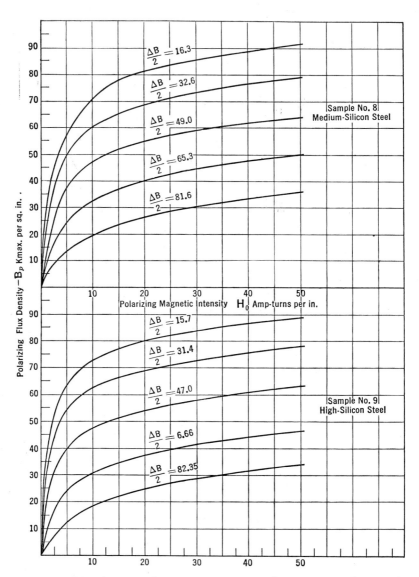

Figs. 18a and b. Apparent direct-current magnetization curves for silicon steel in the presence of a superposed alternating field having a maximum value of $\Delta B/2$ and a sinusoidal wave form.

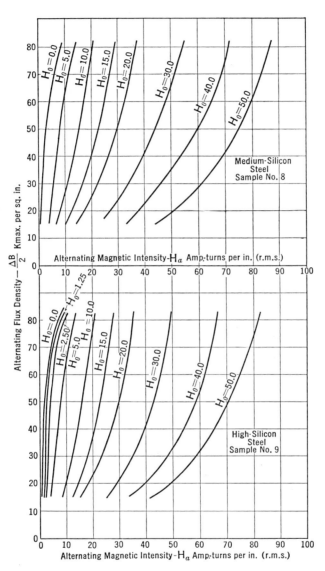

FIGS. 18c and d. Apparent alternating-current magnetization curves for silicon steel in the presence of a superposed direct-current polarizing magnetic intensity of H_0 (alternating-flux wave form sinusoidal).

In Fig. 21 are given normal magnetization and permeability curves for the permanent-magnet steels of samples 22a and 22b, and for samples of 5 per cent tungsten,[29] $3\frac{1}{2}$ per cent chromium steel,[29] and 35 per cent

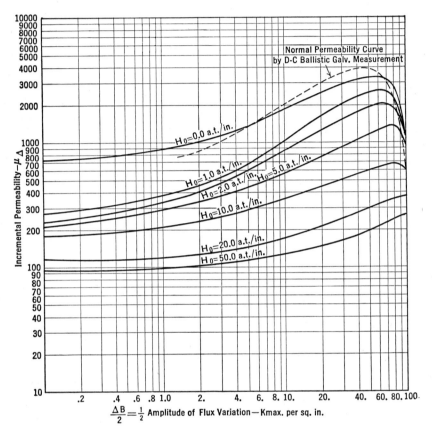

Fig. 19a. Incremental permeability curves for low-silicon steel (sample 7). (ΔB sinusoidal.)

cobalt steel.[30] These data are useful in calculating the incremental permeability of these steels by Spooner's method. (See Art. 22.)

In Table II a résumé [31] of the more important magnetic properties of laboratory and commercial soft magnetic materials is given. The source

[29] By courtesy of the Crucible Steel Co. of America.
[30] From the *Bell Lab. Rec.*, Vol. 13, No. 2.
[31] A very comprehensive résumé of magnetic alloys containing nickel, entitled "Iron-Nickel Alloys for Magnetic Purposes," has been published by the International Nickel Co., New York, N. Y., Development and Research Division.

of the data is given in the column marked "Authority"; where a sample number is listed in this column it indicates that the data were taken on that particular sample by the author.

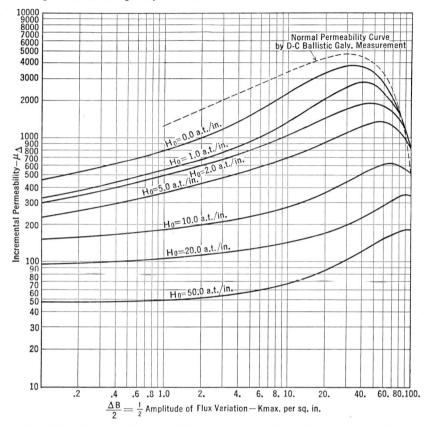

FIG. 19b. Incremental permeability curves for medium-silicon steel (sample 8). (ΔB sinusoidal.)

In Table III a résumé [31] of the more important magnetic properties of laboratory and commercial hard magnetic materials is given. These data have been computed from the curves given for the various materials.

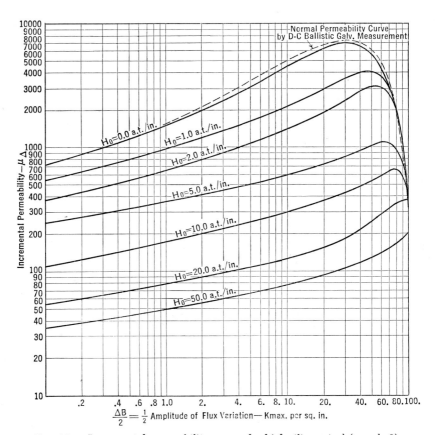

Fig. 19c. Incremental permeability curves for high-silicon steel (sample 9). (ΔB sinusoidal.)

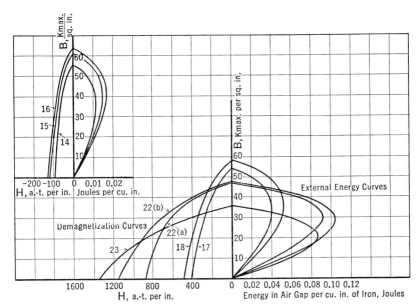

FIG. 20. Demagnetization and external energy curves for permanent-magnet steels.

    14. 0.7% carbon steel.
    15. 2.0% chrome steel (*Jour. I.E.E.*, August, 1925).
    16. 5.5% tungsten steel (*Jour. I.E.E.*, August, 1925).
    17. 15% cobalt chrome steel (*Jour. I.E.E.*, August, 1925).
    18. 36% cobalt steel (*Jour. I.E.E.*, August, 1925).
    22a. Alnico I (General Electric Co.).
    22b. Alnico II (General Electric Co.).
    23. Nipermag (Cinandagraph Corp.).

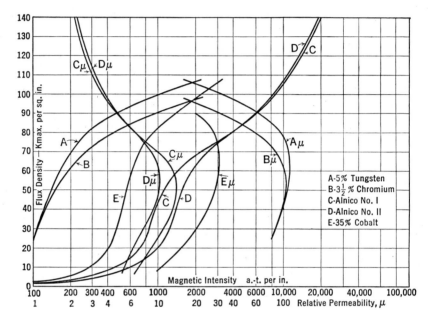

Fig. 21. Normal magnetization and permeability curves for permanent-magnet steels.

Data for curves A and B by courtesy of the Crucible Steel Co. of America.
Data for curves C and D by courtesy of the General Electric Co.
Data for curve E by courtesy of *Bell Laboratories Record* (Vol. XIII, No. 2).

TABLE II
Soft Magnetic Materials

Material	Saturation Density $B_s = B - \mu H$	Residual Flux Density		Coercive Intensity	Normal Hysteresis Loss		Max. Permeability μ_m	Flux Density for Max. Permeability $B_{\mu m}$	Initial Permeability μ_i	Resistivity ρ	Specific Gravity δ	Density	Authority
		B_r	B_m	H_c	W_c	for							
	kmax. per sq. in.	kmax. per sq. in.	← for →	amp.-turns per in.	milli-joules per cu. in. per cycle	B_m for ↓	Relative	kmax. per sq. in.	Relative	microhm inches		lb. per cu. in.	
Irons													
Hydrogenized iron	139	87.8	B_s	0.1	0.492	B_s	275,000		25,000	3.93	7.88	0.284	V. E. Legg, B.S.T.J., Vol. 18, No. 3
Ingot or Swed. char. iron annealed	139	51	64.5	1.77	3.9	64.5	5,200	42	250	4.2	7.85	0.283	Sample 1
C. R. steel S.A.E. 10-10 dead soft		51	64.5	3.05	6.8	64.5	3,200	55					Sample 3
C. R. steel S.A.E. 10-20 ½ hard		37	64.5	11.6	25.0	64.5	590	45		6.0			Sample 4
Cast steel	135						1,500	45					
Cast iron annealed	108	34	108	9.0	18	64.5	520	13 to 32		15.3			
Magtiz Magnetic Mall. C. I. skin on	129	41	64.5	3.8	9.8	64.5	1,680	30		12.6			Sample 20
Silicon steels													
Low-silicon, approx. ½% Si	136	49	64.5	2.0	5.2	64.5	5,000	44	690	7	7.70	0.278	Sample 7
Medium-silicon, approx. 2½% Si	133	33	64.5	1.1	3.2	64.5	6,450	37	400	16	7.60	0.274	Sample 8
High-silicon, approx. 4¼% Si	127	35	64.5	0.7	2.2	64.5	7,500	34	550	23	7.50	0.272	Sample 9
Nickel alloys													
Nickel	39.4	23	B_s	6.8	4.92	B_s	600		110	3.15	8.85	0.318	V. E. Legg
Ferronickel, 47% Ni	100	47	64.5	0.1	0.36	64.5	63,000	25	8,000	18.0	8.25	0.298	Sample 11
Permalloy, 78% Ni	69	38.7	B_s	0.1	0.33	B_s	105,000	32	9,000	6.3	8.60	0.31	V. E. Legg
Cobalt alloys													
Cobalt	116	32	B_s	2.0	3.28	B_s	240		70	3.54	8.90	0.321	V. E. Legg
Ferrocobalt, 34.5% Co. 0.0V	158	33	64.5	2.5	5.55	64.5	2,450	55		3.95	8.22	0.298	Sample 12a
34.5% Co. 1.5V	150	48	64.5	4.5	9.48	64.5	2,310	60		13.90	8.22		Sample 12b
Permendur, 49 Co, 49 Fe, 2V	155	90.4	B_s	4.0	9.84	B_s	4,500		800	10.2	8.32	0.296	V. E. Legg
Nickel-cobalt alloys													
Perminvar, 45 Ni, 25 Co, 30 Fe, baked at 425° C	100.0	21.3	B_s	2.8	6.45	B_s	1,800		365	7.5	8.6		V. E. Legg
Perminvar, 7.5 Mo, 45 Ni, 25 Co	66.4	27.7	B_s	1.2	4.26	B_s	3,800		550	31.5	8.66		V. E. Legg

TABLE III
Permanent-Magnet Steels

Material	Materials Alloyed with Iron, Per Cent	Reversible Permeability μ_r, Relative	Residual Flux Density B_r, kmax. per sq. in.	B_m	Coercive Intensity H_c, amp.-turns per in.	Maximum Stored Energy of Air Gap $(BdHd)_m$, joules per cu. in.	Flux Density for Max. Energy $-B_d$, kmax. per sq. in.	Resistivity ρ, microhm, inches	Specific Gravity δ	Density, lb. per cu. in.
Carbon steel, oil-quenched	....	......	58	141	100	0.011	38	....	7.82	0.282
Chromium steel	2 Cr, 1C	......	62	B_s	118	0.014	41	....	......	......
Chromium steel	6 Cr, 1C	......	77	B_s	132	0.022	45	....	8.17	0.295
Tungsten steel	5.5 W, 0.7 C	30	64	B_s	130	0.017	41	11.8	......	......
Cobalt-chrome steel	15 Co, 11 Cr, 1C	......	54	B_s	405	0.040	33	....	8.27	0.298
Cobalt steel	36 Co, 5W, 2 Cr, 0.8 C	9	58	142	480	0.052	35	....	6.9	0.249
Alnico I	12 Al, 20 Ni, 5 Co	4	47	81.5	890	0.091	30	31.5	7.1	0.256
Alnico II	10Al, 6 Cu, 17 Ni, 12.5 Co	4	47	81.5	1,140	0.104	29	24.4	6.9	0.250
Alnico III	12 Al, 25 Ni	......	44	......	960	0.085	29	....	7.0	0.254
Alnico IV	12 Al, 28 Ni, 5 Co	......	34	......	1,470	0.085	19	....	7.0	0.254
Nipermag	12 Al, 32 Ni, + Ti	......	36	......	1,350	0.087	20	....	7.0	0.254
Platinum-cobalt alloy	77 Pt, 23 Co, 0 Fe	1.1	29	......	5,250	0.244	16	19.7	14.6	0.529

PROBLEMS

1. A sample of soft steel is to be tested for its magnetic qualities up to a density of 120 kmax. per sq. in. It is to be in the form of a ring having a mean diameter of 8 in. and a radial thickness of $\frac{3}{4}$ in. The weight of the sample is to be 4 lb. The ballistic galvanometer to be used has a calibration constant of 50 kmax. turns change per cm. swing, the useful length of the galvanometer scale being 20 cm. If the maximum permissible exciting current is 15 amperes, compute the necessary primary turns. Also compute the best number of secondary turns. Should the secondary be wound next to the iron or over the primary?

2. It is proposed to use a 1 kv-a., 110/440-volt transformer as a choke coil in an experimental set-up. The data for the transformer are as follows:

Primary turns................. 242
Secondary turns............... 968
Core material................. high-silicon steel
Core dimensions, net cross section........ 3.0 sq. in.
  length of magnetic circuit............ 23.5 in.

Compute (neglect effect of joints in core) the effective inductance to alternating currents if the primary and secondary windings are connected in series and carry a direct current of 388 milliamperes, for a maximum value of alternating flux density of (a) 1 kmax. per sq. in., (b) 60 kmax. per sq. in. (c) If an alternating current wave of 1.0 ampere from the positive crest to negative crest is superposed on the direct current what will be the inductance of the transformer? Use the data of Fig. 19c only.

3. Assuming in part (c) of Problem 2 that the alternating exciting current is sinusoidal, calculate the pulsation in flux density (ΔB) by means of the data of Fig. 18d. How does your result check with the value of ΔB obtained in Problem 2? Explain the discrepancy between these two results.

4. Calculate the polarizing flux density which will be produced in a piece of high-silicon steel by a polarizing magnetic intensity of 20 ampere-turns per in., in the presence of an alternating flux density ($\Delta B/2$) of 47.0 kmax. per sq. in. Use the data of Fig. 19c. Compare your answer with the experimental results shown in Fig. 18b.

5. Calculate by Spooner's method the incremental permeability of 36 per cent cobalt magnet steel if it is subjected to a flux pulsation of 1 kmax. per sq. in. about a mean flux density of 35 kmax. per sq. in. A magnetization curve for 36 per cent cobalt steel may be found in Fig. 21.

6. Calculate by Spooner's method the incremental permeability of medium-silicon steel subjected to a polarizing magnetic intensity of 10 ampere-turns per in., and carrying a sinusoidal alternating flux having a maximum density $\Delta B/2$ of 20 kmax. per sq. in. Compare your answer with the experimental results shown in Fig. 19b.

7. Determine the equation for the ferric reluctivity of high-silicon steel (sample 9a) from the data of Fig. 15. What is the saturation density? Check your answer with the value given in Table II. Explain why this is lower than that of pure iron.

8. The ring of Problem 1 is made of S.A.E. 10–20 steel $\frac{1}{2}$ hard. If the primary winding has 600 turns and carries 1 ampere, calculate the residual flux of the ring if the primary current is interrupted.

9. Determine from the hysteresis loops of Fig. 12d by the method of Art. 24 the saturation values of B_r and H_c for S.A.E. 10–20, $\frac{1}{2}$ hard, cold-rolled steel. Compare your answers with the data of Fig. 13. Explain why the result for $(B_r)_s$ does not check.

10. To about what minimum magnetic intensity should the following steels be subjected in order to develop their full possibilities as permanent magnet materials:

carbon steel; 36 per cent cobalt steel; Alnico? What can you say in general regarding this quantity with respect to the maximum coercive intensity the steel develops?

11. For Fig. 6 describe exactly the significance of the area 0–2–3–4′–0.

12. Using equation 7 and the data of Table I compute the hysteresis energy loss per cycle for medium-silicon steel at a flux density of 64.5 kmax. per sq. in. Check your answer with that given in Fig. 17b. What will the hysteresis power loss be at 60 cycles per second in watts per pound?

13. By measuring the area of one-half the hysteresis loop of Fig. 12a, at $B_m = 64.5$ kmax. per sq. in., check one point of Curve D, Fig. 14a, and the corresponding value given in Table II.

14. On the basis of the data of Fig. 16b, calculate the total core loss of 26 gauge (0.0187 in.) medium-silicon steel sheet at a maximum flux density of 64.5 kmax. per sq. in. *Note:* Data supplied by the Allegheny Steel Co. give 1.08 watts per pound at 60 cycles as the commercial maximum core loss guarantee value by the Epstein test.[32]

15. On the basis of the data of Figs. 16a or 16b, calculate the percentages of the hysteresis and eddy-current losses of the total core loss for 29 gauge medium-silicon steel at B_m equal to 64.5 kmax. per sq. in.

16. The core of a 6-volt spark coil is made of a bundle of annealed Swedish charcoal iron wires having a net cross section of 0.25 sq. in. and a length of 4 in. The primary winding consists of 200 turns of wire having a resistance of 2 ohms. If at the instant of breaking the primary one can assume that the current has reached its final value, that all the flux of the core passes through the entire length of the core, and that 20 per cent of the primary magnetomotive force is consumed by the iron core, calculate: (1) the total energy (exclusive of RI^2 loss) taken from the electric circuit while the current builds up; (2) the total hysteresis loss during the complete magnetic cycle as a percentage of (1), and the portions of this loss supplied by the energy absorbed by the iron and by the energy of the air gap, respectively.

17. Determine the hysteresis power loss for the core of a half-wave rectifier, data for which are given below:

Length of magnetic circuit..............	18 in.
Area of magnetic circuit (net)............	2 sq. in.
Material of magnetic circuit............	medium-silicon steel, 29 gauge
Turns on primary winding..............	480
Primary voltage (60 cycles).............	115 volts, r.m.s.
Turns on secondary (rectifier circuit).....	200
Average rectified current in secondary....	4.5 amperes

Assume no air gap in the iron core, and the supply voltage sinusoidal.

18. What will the total core loss be for the core of Problem 17?

19. Calculate the exciting current for the core of Problem 17 using the data of Fig. 18. Check this by the method of Art. 29 which assumes sinusoidal wave form.

20. It is desired to make a permanent magnet having an air gap of 2 sq. in. area and a length along the flux lines of $\frac{1}{8}$ in. The flux density in the air gap is to be 20 kmax. per sq. in. Calculate the minimum amount of (1) 36 per cent cobalt steel, (2) tungsten steel, (3) Alnico I, that can be used. Neglect all fringing and leakage flux. Make a sketch showing how you would arrange the magnetic circuit so that this minimum usage of materials would be practicable.

[32] Epstein Test Specification A-34-28 of the American Society for Testing Materials.

CHAPTER III

THE THEORY OF OPERATION OF ELECTROMAGNETS AND THE FACTORS ENTERING INTO THEIR EFFICIENT DESIGN

34. The Flux-Current Loop Applied to an Electromagnet

The physical action involved in the operation of any electromagnetic device is the conversion of electric energy into work by the motion of a rotor, armature, or plunger in such a way as to change its flux linkage, and thereby induce a voltage in a current-carrying coil. The energy so converted can be represented mathematically as

$$W = \int I \, d(N\phi) \quad \text{joules}$$

or graphically as the area of a flux-current loop. This energy will include, besides that converted directly from the electrical to the mechanical form by the motion, that made available by any change in stored energy of the magnet by the motion. In a tractive electromagnet the flux is ordinarily produced by the current which furnishes the energy (Fig. 1a), and the flux-current loop has the form shown in Fig. 1b by the full-line loop $O1$–$A1$–$B4$–$C4$–$O1$. Inasmuch as loops of this type form the basis for the force analysis for all kinds of electromagnets we will examine it in great detail.

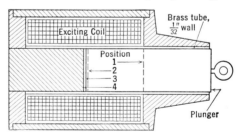

FIG. 1a. Flat-faced iron-clad plunger magnet. See Fig. 15, Chapter VIII, and Fig. 3, Chapter IX, for actual dimensions.

Assume initially that the plunger is at position 1 of Fig. 1a. Now, if the plunger is held stationary and the circuit through the coil is closed, the flux linkage will build up as shown by the line $O1$–$A1$ of Fig. 1b as the current gradually increases to its final value E/R, where R is the resistance of the coil. The designation $A1$ for a point on the loop indicates flux linkage A existing with plunger at position 1. During this time energy is abstracted from the electric circuit and stored in the magnetic field by virtue of the circuit current flowing against the voltage induced in the coil while the flux linkage is rising. The energy so abstracted during this interval is

$$W_{01-A1} = \int_0^A I\, d(N\phi) \quad \text{joules}$$

and is represented by the area $01-A1-A-01$ behind the curve $01-A1$. The next step is to allow the plunger to move very slowly[1] to position 4. During this motion additional energy is abstracted from the electric circuit equal to the area of the rectangle $A-A1-B4-B-A$, as given by

$$W_{A1-B4} = \int_A^B I\, d(N\phi) \quad \text{joules}$$

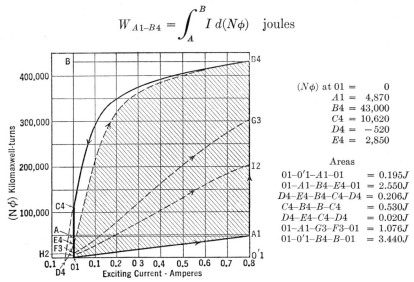

Fig. 1b. Flux linkage-current loop for the flat-faced iron-clad plunger magnet of Fig. 1a.

This area includes, besides the mechanical work done, the change in stored energy of the magnet due to the motion. As this energy change is unknown it is impossible at this stage to evaluate the work done. The cycle is completed electrically by holding the plunger stationary at position 4 and opening the electric circuit, causing the flux linkage to decrease to $C4$ following the curve $B4-C4$.[2] During this interval

[1] For simplicity only: When the plunger moves with appreciable rapidity there will be induced in the coil sufficient voltage to cause the current to decrease appreciably during the motion, making line $A1-B4$ bend in toward the $(N\phi)$ axis.

[2] The value of flux linkage at $C4$ is called the residual flux linkage; the flux which produces it, the residual flux; the force produced, the residual force.

stored energy equal to the area $C4$–$B4$–B–$C4$, as given by the integral

$$W_{B4-C4} = \int_B^C I\, d(N\phi) \quad \text{joules}$$

is returned to the circuit.[3] This, however, does not complete the cycle magnetically because the flux linkage has not been reduced to its original value at 01. This can be accomplished by returning the plunger to its original position mechanically. If this is done the flux linkage will fall from $C4$ to 01 as shown. Even when the plunger has been returned to state 01 the flux linkage has not been reduced to zero, owing to the coercive magnetomotive force of the iron which is producing a very small but appreciable flux through the permeance of the air gap. The state 01 is merely taken as a convenient datum point. Thus, if the plunger is moved from position 1 to 4 with zero magnetomotive force applied, the flux linkage will increase from 01 to $E4$, this increase being

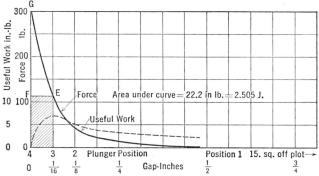

Fig. 1c. Force-stroke curve for the flat-faced iron-clad plunger magnet of Fig. 1a.

caused solely by the increase in permeance of the magnetic circuit on which the coercive magnetomotive force acts.

Some work will be required to move the plunger from $C4$ to 01 as it must be returned against the magnetic pull created by the residual flux. One method of evaluating this work is to measure it by completing electrically the flux-current loop from $C4$ to 01. This is done by sending just enough current (I_d) through the coil in the negative direction to cause the flux linkage to drop along the dashed curve from $C4$ to $D4$ (the plunger is held stationary at position 4). Point $D4$, which is determined experimentally, is so chosen that, when the negative exciting current is removed (flux linkage rises along the dashed curve to $E4$) and the plunger is pulled out to position 1, the same change in flux

[3] This energy is generally not usefully returned to the circuit; that is, it is dissipated as heat usually in the form of a spark at the switch contacts.

linkage will occur but in the opposite direction (change $E4$ to 01), as if the plunger had been pushed in from position 1 to 4 after having been mechanically changed from state $C4$ to 01 (change 01 to $E4$). Under these circumstances the area 01–$C4$–$D4$–01 represents approximately the energy mechanically supplied to move the plunger from $C4$ to 01.[4] The area of the loop $D4$–$E4$–$B4$–$C4$–$D4$ is, except for the extremely small hysteresis losses occurring during the permanent-magnet cycle between $E4$ and 01, the total hysteresis loss occurring during a complete cycle of the electromagnet.

The net energy abstracted from the electric circuit during the completed electric cycle is

$$W_{01-A1-B4-C4-01} = W_{01-A1} + W_{A1-B4} - W_{B4-C4} \quad \text{joules}$$

This area must represent the entire mechanical work done during the useful motion from 1 to 4 as no other forms of energy are involved, and it includes besides the work actually available (that is, the total area under the force-distance curve shown in Fig. 1c) the energy dissipated as heat in the entire volume of iron of the magnet due to magnetic hysteresis during the incompleted magnetic cycle 01–$A1$–$B4$–$C4$.[5]

The loss due to magnetic hysteresis may be isolated by putting the magnet through the same magnetic cycle without producing any available mechanical work. This is done by moving the plunger to position 4 and then closing the circuit, causing the flux to build up along the curve $E4$–$B4$. The rest of the cycle will be the same as before. The area $E4$–$B4$–$C4$–$E4$ will then represent the net energy abstracted from the electric circuit, and must be the hysteresis loss occurring during the first

---

[4] The analysis of the work necessary to move the plunger from $C4$ to 01 is very complicated as it involves the unstable coercive magnetomotive force of the iron which produces the residual flux linkage $C4$ and the stable coercive magnetomotive force of the iron which produces the residual flux linkage $E4$. This subject will be discussed more fully when dealing with sensitive relays; it is not particularly important when dealing with tractive magnets.

[5] Energy loss due to magnetic hysteresis, where the change in flux linkage is caused by the motion of an armature or plunger, always manifests itself as a mechanical drag on the moving part: thus the torque delivered at the shaft of a motor, including friction and windage torques, is less than the electromagnetic torque developed on the armature due to the countertorque necessary to create the hysteresis and eddy-current loss in the armature. Where the flux is not increased by motion of the core, as in a transformer, the hysteresis loss, while supplied electrically, may be considered as a direct conversion from electrical to mechanical work and hence to heat. This is explained on the basis that the hysteresis loss is the actual mechanical work necessary to overcome the internal friction of the molecules of the iron as they are oriented by the magnetic field.

three steps of the cycle (01–A1; A1–B4; B4–C4), if we exclude the small hysteresis loss during the change (01–E4).[6]

Therefore, subtracting the hysteresis loss from the area 01–A1–B4–C4–01 the net mechanical work done will be the shaded area 01–A1–B4–E4–01, which is equal to the area under the force-distance curve of Fig. 1c.

35. Magnetic Efficacy

If a magnet could be built which had zero flux in the open gap position (position 1) with the magnetomotive force applied, zero stored energy (area C4–B4–B–C4 equal to zero) at the closed gap position (position 4), and zero hysteresis loss (area E4–B4–C4–E4 equal to zero), then the rectangular area 01–0'1–B4–B–01 would be available for a given maximum flux and exciting magnetomotive force. The ratio of the shaded area to the area of the rectangle described above is a measure of how effectively the potential work ability of the magnet is realized, and can be properly called the *magnetic efficacy* of the design. Using the symbol η for the magnetic efficacy we may write the following expression for the available mechanical work:

$$W = \eta I(N\phi) \quad \text{joules} \tag{1}$$

where η in a well-designed magnet may vary between 0.4 and 0.7. As the flux ϕ of the magnet is proportional to the area of the core cross section, and as the ampere-turns (NI) are approximately proportional to the core length, it follows that the available mechanical work $\eta I(N\phi)$ is directly proportional to the bulk or weight of the magnet. The value of η depends primarily on the size of the three loss areas relative to that of the rectangle and can be increased only by decreasing the loss areas.

Loss area C4–B4–B–C4, which represents the stored energy of the magnetic circuit which is returned to the electric circuit, is made up of the stored energy of the iron and that of the air gaps which are left in the circuit in the closed-gap position. The former energy depends on the shape of the normal hysteresis loop for the iron, and the latter energy on the area and length of the air gap (or gaps) which is left in the circuit.[7]

The energy stored in the iron will be very nearly equal to the area behind the falling magnetization curve from the maximum flux density

---

[6] The hysteresis loss occurring between any two states of magnetization depends on the change in flux between the two states and not on the manner in which the change is made, provided always that the change is in one direction.

[7] In the magnet of Fig. 1a the only air-gap volume left in the closed-gap position is the fixed cylindrical volume of brass between the plunger and the upper cylindrical piece of iron.

occurring at $B4$, to the residual flux density at $C4$ of a normal hysteresis loop,[8] multiplied by the volume of iron.[9] This energy can be reduced only by changing the properties of iron used. In general the softest grade of iron or steel, annealed after the machining, will give the best results.

The energy stored in the air gaps is given by the following equation:

$$W = \frac{1}{2}\frac{\phi^2}{P} = \frac{\phi^2 l}{2\mu S} \quad \text{joules} \qquad (2)$$

where ϕ is the total flux through the gap having an area of S square inches, and the length (measured along the flux lines) of l inches.

The magnitude of this item is within the control of the designer. From the data given on Fig. 1b, the loss due to the final stored energy of both the iron and the air gap is, for the magnet of Fig. 1, $0.53 \div 3.44 \times 100 = 15.4$ per cent.

Loss area $E4$–$B4$–$C4$–$E4$, which represents the hysteresis loss during the incomplete magnetic cycle 01–$A1$–$B4$–$C4$, is entirely a property of the iron,[10] and for any particular iron can be made a minimum by proper annealing. This loss, for the magnet of Fig. 1a, is equal to $0.186 \div 3.44 \times 100 = 5.4$ per cent.

Loss area 01–$0'1$–$A1$–01, which represents the stored energy of the magnet with the plunger in its initial position at the beginning of its stroke, is very interesting, as it is almost entirely under the control of the designer; its smallness in some measure represents the skill of the designer. It depends primarily on the position of the plunger and on the leakage flux at this position. For a given cross section of iron an increase in the leakage coefficient (due to poor proportion or arrangement) increases the initial flux linkage A without increasing the final flux linkage B. This obviously decreases the magnetic efficacy, and therefore decreases the useful work of the magnet.

Before considering what determines the initial position of the plunger let us see how the force distance curve of Fig. 1c can be derived from the flux-current loop of Fig. 1b. Now the total area under the force-

---

[8] This area per cubic inch of iron is plotted in Fig. 14, Chapter II, as a function of the maximum cyclic flux density for various steels.

[9] If different parts of the iron have different maximum flux densities, these volumes must be handled separately.

[10] For all practical purposes this area can be considered equal to that between the normal magnetization curve and the falling branch of the hysteresis loop between the maximum flux density and the actual residual flux density of the magnet. This area per cubic inch of iron can be evaluated from the data of Fig. 14, Chapter II, by the method outlined in Art. 28.

distance curve is equal to the area 01–$A1$–$B4$–$E4$–01, as explained before; and the average force of the plunger during its motion to 4 will be equal to this shaded area divided by the total displacement 1–4. The average force over a smaller displacement can be found in a similar manner; thus the average force during the motion of the plunger from 2 to 3 is equal to the net area of the flux-current loop available for useful mechanical work during the displacement 2–3 (area $H2$–$I2$–$G3$–$F3$–$H2$) divided by the displacement 2–3. This average force is evidently the mean height of the force-distance curve between ordinates 2 and 3. By the same reasoning, the force at any plunger position may be obtained by allowing the plunger displacement from that position to approach an infinitesimal: thus the force at any position is

$$F = \frac{dW}{ds} \quad \text{joules per inch}$$

where dW is the infinitesimal energy made available for useful mechanical work during the infinitesimal motion ds.

Now consider the numerical values given on the force-distance curve. Suppose that the plunger is to pull a load of 112 lb. This means that the initial position of the plunger must be at 3, and of the amount of work available, the area under the force-distance curve between 1 and 3, or the area 01–$A1$–$G3$–$F3$–01 of the flux-current loop must be wasted. This loss, which is 1.076/2.505 of the total area under the force-distance curve, is unavoidable if this particular magnet is used to pull a 112-lb. load.

One of the problems in the design of an efficient electromagnet is, therefore, to reduce to a minimum the potential mechanical work wasted by the initial plunger position necessary to produce the required starting force. This loss depends primarily on the quotient of the square root of the assigned force and the stroke; and the type of working gap[11] (shape of working gap, and the manner in which the working-gap surfaces approach each other). For the problem considered above, with the plunger starting its stroke at position 3, this loss as a percentage of the maximum potentially available work ($NI\phi$) is equal to (1.076 + 0.195) ÷ 3.44 × 100 = 37.0 per cent and is by far the greatest of the three loss areas.

The magnetic efficacy of the magnet of Fig. 1a, if it starts its stroke at position 3, will then be: 100 − 15.4 − 5.4 − 37.0 = 100 − 57.8 = 42.2 per cent.

---

[11] The proper method of shaping the pole faces in relation to the force and stroke in order to minimize this loss is discussed in detail in Chapter IX.

36. Mechanical Efficacy: Slow- and Rapid-Acting Magnets

The actual useful work that would be done by the above magnet with the given load is equal to 112 lb. $\times \frac{1}{16}$ in. or 7 in-lb., which is equal to the shaded area of the rectangle 4–3–E–F–4 of Fig. 1c. The total mechanical work done on the load, however, will be equal to the area 4–3–E–G–4. The difference between these two areas, namely, F–E–G–F, represents the amount of energy available for accelerating the load during the motion; it is finally dissipated as heat when the plunger strikes the stop.[12] This energy so dissipated cannot necessarily be considered as wasted, because it determines the speed with which the useful work will be accomplished. Thus the ratio of the useful work done (4–3–E–F–4) to the total mechanical work done (4–3–E–G–4) is a measure of the effectiveness with which the available work is employed, and it can be properly called the *mechanical efficacy*. Magnets which complete their useful work with great rapidity (rapid-acting magnets) must of necessity have a low mechanical efficacy, as compared to magnets which carry out their useful work slowly (slow-acting magnets). When magnets act very rapidly eddy currents induced in the solid iron parts will often modify the action, tending to make operation more slow.[13] Also, a relatively low mechanical efficacy tends to make a magnet more reliable; that is, if the plunger or armature encounters any unusual opposing force during its stroke, such as an abnormal increase in friction, it will have sufficient reserve pull to overcome the obstruction. The magnet of Fig. 1a when pulling a 112-lb. load from position 3 has from Fig. 1c a mechanical efficacy of $7.00 \div 12.7 \times 100 = 55.0$ per cent.

Another problem in the efficient design of a magnet is, then, to produce no more mechanical energy for acceleration or reliability than is necessary. The problem both of this article and of the last is handled by changing the shape of the force-distance curve to give the maximum possible magnetic and mechanical efficacy consistent with the required force, stroke, and rapidity of action.

---

[12] When a magnet operates on a constant-voltage circuit, as is usual, rapid acceleration of its plunger or armature causes the area F–E–G–F to decrease, owing to the decrease in current produced by the velocity of the plunger (see footnote 2). In magnets designed for rapid action this tendency for the current to decrease is reduced by increasing the resistance of the winding and raising the supply voltage to compensate. This increases the RI^2 loss.

[13] The effect of the eddy currents in the iron parts of a magnet is simply to delay the building up of the magnetic flux. They do not alter the ultimate value of flux if the magnet does not complete its cycle too rapidly.

37. Time-Delay Action

In the design of magnets and especially relays it is sometimes desirable to cause the response or action (that portion of the motion of the armature or plunger which does the useful work) to occur an appreciable interval of time after the circuit through the magnet is made or broken. This is called *time-delay action,* and it should not be confused with slow-acting, which means that the actual motion is carried out slowly, whereas in a delay-action device there may be no motion for, say, 1 second after the current is turned on, after which time the motion may be very rapid.

This type of action has many practical applications: the automatic door closers of a passenger elevator (lift) are so arranged that the doors close before the elevator starts, when the elevator starting button is closed; in certain devices, too rapid action of relays causes mechanical oscillations—thus, in an automatic battery charger, if the voltage relay across the battery on the instant of breaking contact, due to low voltage, connects the charging circuit across the battery, instability will result, as the rise in voltage due to the charging current will immediately close the voltage relay contact again. This is because sufficient time has not elapsed after the opening of the voltage relay contact to allow its armature to drop completely to its open position. In the selection of certain type of pulse signals it is often desirable to select between pulses of different time durations, which can be easily done by means of a time-delay action relay requiring a predetermined duration of excitation before operation.

Short time delay, of the order of the electrical time constant of the magnet, can often be produced by a special design of the magnet itself.

It is possible, with certain types of auxiliary circuit equipment (entirely non-mechanical), to produce, economically, accurate relay time delays of the order of 0.01 to 30 seconds, followed by practically instantaneous relay operation. Mechanical means may also be used for long time delay (1 second and up); where accuracy is desired some kind of clock mechanism may be employed; where accuracy is not paramount and where the time delay is not too long a dashpot or similar device may be built into the magnet.

38. Heating

Most magnets, other than relay magnets, operating on power circuits are designed with heating as a limitation.[14] In general, if a magnet is

[14] The copper loss in a magnet is identical with that of a motor, except that generally it represents practically the entire power input; that is, the tractive electromagnet is mostly used to produce a force or at the best goes through its magnetic cycle

designed efficiently (magnetically and mechanically) the bulk and weight of the magnet for a given useful work will decrease as the allowable temperature rise is increased. Tractive magnets are usually designed with temperature rise as a limitation, as this gives the magnet which occupies the least space and has the least weight and hence the least cost for a given job. Therefore, when designing tractive magnets the permissible temperature rise becomes a factor as important as the force and stroke, and is accordingly brought into the design formulas at the start.

Relay magnets, on the other hand, are usually designed for maximum efficiency (maximum effort for minimum power expenditure) consistent with the speed of operation desired, and are characterized by having relatively light moving parts. The light weight of the moving parts is necessary to prevent sluggish action which might otherwise result because of the small forces usually employed. Small power expenditure is necessary as these magnets often operate from the end of long lines where the power is limited, or they operate in battery circuits where electrical energy is expensive. For particularly sensitive, and very rapid action, polarized relays employing a permanent magnet as part of the magnetic circuit are often used. These also have the advantage of reversing the direction of the armature motion if the current is reversed.

39. Résumé—Factors Entering into Electromagnet Design

The following tabulation gives the more essential factors in the electrical design of tractive and relay magnets.

Tractive Magnets

1. *Force-Stroke Characteristic.* A definite minimum force through a given stroke.
2. *Heating.* A maximum temperature rise consistent with a reasonable length of life of the coil insulation, etc.
3. *Magnetic Efficacy.* Design of working-gap surfaces and manner of approach of gap surfaces such as to give required force-stroke characteristic with maximum magnetic efficacy.
4. *Mechanical Efficacy.* Shape of force-distance characteristic such as to produce sufficient mechanical energy in excess of force-stroke requirement, as may be necessary for the speed of action desired.

so slowly that the energy consumed by its final stored energy (which is usually not usefully returned to the electric circuit), hysteresis loss, and useful work are inconsequentially small compared to the copper loss. In a motor, however, the current-carrying armature coils link and unlink with the flux so rapidly in a continuous process that the electrical energy converted to work is large compared to the copper loss. Of course, electromagnets are chiefly used to do intermittent work over periods of time short compared to the time during which they are excited, and hence should not be compared to motors, which do continuous work, on an over-all efficiency basis.

Relay Magnets

1. *Force.* Force throughout the stroke such as to be sufficiently in excess of that required to overcome return-spring pull and friction, as is necessary to produce the desired speed of action and contact pressure.
2. *Stroke.* This is determined entirely by the clearance necessary between the relay contacts, which depends on the kind of circuit connected across these contacts.
3. *Power Consumption.* Though this depends somewhat on the force (spring and friction) throughout the stroke, it is in a large measure determined by the weight of the moving armature relative to the speed of action desired.
4. *Residual Force.* This generally undesirable force is determined by the residual flux density in the closed-gap position, which depends on the magnetic properties of the iron (coercive intensity in particular) and the permeances of the air gaps in the closed-gap position.

The various items of the above tabulation will be discussed in later chapters.

PROBLEMS

1. A horseshoe-type magnet made of annealed Swedish charcoal iron (sample 1) has an armature, pole cores, and yoke which may all be considered to have a mean cross section of 0.6 sq. in. and a total length of 11.5 in. It is excited by coils developing a total of 2,300 ampere-turns. Assume that the working gap is sufficiently long to reduce the residual flux in the open-gap position to zero and that at the end of the working stroke the effective gap length is zero. Compute: (1) the total hysteresis loss occurring from the time the coil circuit is closed until the end of the stroke after the coil circuit is opened; (2) the work necessary to return the armature to the open-gap position after the coil circuit is broken; (3) the energy returned to the electric circuit in the form of a spark when the coil circuit is broken. (4) Express items (1), (2), and (3) as percentages of the work ideally available from the magnet.

2. Repeat Problem 1 using S.A.E. 10–20 half-hard steel (sample 4) instead of Swedish charcoal iron. Compare the magnets made of the two materials and state whether much is to be gained by using the low hysteresis loss iron for magnets working at high flux densities.

3. Considering the magnet of Fig. 15, Chapter VIII, to start its stroke for maximum useful work at $x = 1.34$ with a constant load of 9 lb. and finish at $x = 0$, compute for the experimental force-distance curve of Fig. 16b its magnetic efficacy. Use the data of Fig. 16a to determine the ideal work. Compare this with the magnetic efficacy for maximum useful work obtained for the same magnet with a flat-faced plunger as illustrated in Fig. 1a of this chapter. Explain why the stepped cylindrical plunger, even though it starts its stroke with a lower flux linkage than the magnet of Fig. 1a, has almost the same magnetic efficacy. Compare the mechanical efficacy of the two magnets for the strokes as above, and explain why that of the stepped cylindrical plunger magnet is so high. Which magnet will be faster acting?

CHAPTER IV

CALCULATION OF MAGNETIC CIRCUITS CONTAINING IRON AND AIR GAPS OF KNOWN PERMEANCE

40. General

Magnetic circuit calculations can be divided into two classes: those in which the flux of the circuit is specified and the necessary magnetomotive force is required; and those in which the magnetomotive force is given and the flux which will be produced is required. In ordinary electric-circuit calculations in which resistances are the only circuit elements, one would make no distinction between problems in which the current was given and the necessary voltage required, and those in which the voltage was given and the current required. The reason for this is that the relationship between current and voltage in the conducting circuit is linear and can be expressed by a simple equation that can be solved with equal ease for either current or voltage. In the magnetic circuit containing iron, however, the relationship between flux and magnetomotive force is by no means linear, nor can it be expressed by any reasonably simple mathematical equation. For this reason magnetic circuits must very often be solved graphically, or, if the actual solution is not graphical, it is at least carried out by reference to the graphical relation between flux density and magnetomotive force as given on a magnetization curve. It should be remembered that there is no essential difference between graphical and analytical methods in this type of calculation; the analytical method determines the intersection of two functions by a purely symbolic method, while the graphical determines the same intersection by the crossing of the graphical plots of the two functions. In this type of problem the analytical method is by far the simpler if the function is easily expressible as an equation, but unfortunately this is not true for magnetization curves.

To illustrate the above discussion and also the method of carrying out the solution of magnetic circuits, the illustrative problems of the following articles are given. These problems are arranged in the order of increasing complexity.

41. Magnetic Circuits Containing Only Iron of Uniform Cross Section

In Fig. 1 is illustrated the ordinary ring sample of iron used for magnetic testing. This is probably the simplest magnetic circuit that can be constructed: it is uniformly wound with the exciting winding and hence

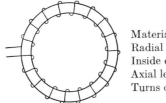

DATA

Material annealed ingot iron
Radial thickness.... $\frac{1}{2}$ in.
Inside diameter.... $11\frac{1}{2}$ in.
Axial length....... $\frac{1}{2}$ in.
Turns on winding... 1,000

FIG. 1. Magnetic circuit containing only iron of uniform cross section.

has no magnetic leakage; if the radius is at least about six times the radial thickness it can be considered as having a uniform flux distribution over its cross section;[1] it has absolutely no air gap; the material can be considered homogeneous throughout.

Problem 1. Find the current required to produce a flux of 25 kmax. in the magnetic circuit of Fig. 1.

Solution. As the ring is homogeneous and of constant cross section, the flux density throughout the magnetic circuit is equal to

$$\frac{\phi}{S} = \frac{25}{\frac{1}{4}} = 100 \text{ kmax. per sq. in.}$$

Referring now to the magnetization curve for annealed ingot iron (curve 1, Fig. 11a, Chapter II), we see that, for $B = 100$ kmax. per sq. in., a magnetic intensity of 30 ampere-turns per inch is necessary. The total magnetomotive force required for the entire magnetic circuit will then be:

$$F = Hl = 30 \times \pi \times 12 = 1{,}131 \text{ ampere-turns}$$

The exciting current will then be:

$$I = \frac{F}{N} = \frac{1{,}131}{1{,}000} = 1.131 \text{ amperes}$$

[1] Although the ring is of uniform cross section the flux density will not be exactly uniform over the entire cross section. This is because the magnetic intensity will be greater at the inner radius than at the outer radius. However, if the radial thickness is small compared to the diameter the effect of the non-uniform distribution can be neglected.

This represents the simplest type of magnetic-circuit calculation; direct reference can be made to the magnetization curve. The converse, though generally not so straightforward, is for this particular case just as simple; thus:

Problem 2. Find the flux that will be produced in the magnetic circuit of Fig. 1 by a current of 1 ampere.

Solution. The total magnetomotive force effective in the circuit will be:

$$F = NI = 1{,}000 \times 1 = 1{,}000 \text{ ampere-turns}$$

Since the ring is homogeneous and of constant cross section, the magnetic intensity throughout the ring will be:

$$H = \frac{F}{l} = \frac{1{,}000}{\pi \times 12} = 26.5 \text{ ampere-turns per inch}$$

Referring now to the magnetization curve for annealed ingot iron, we see that B, corresponding to $H = 26.5$, is 98.0 kmax. per sq. in., and the total flux will be

$$\phi = BS = 98.0 \times \tfrac{1}{4} = 24.5 \text{ kmax.}$$

42. Series Magnetic Circuit Containing Only Iron of Different Cross Sections

In Fig. 2 is illustrated the magnetic circuit of a bipolar magnet with the armature in the closed-gap position. If the armature is not too highly saturated the leakage in this circuit can be considered zero (for the closed-gap position only); likewise, if the magnet is well made we may neglect the small air gaps that must exist between the pole faces and armature, pole cores and yoke, etc.[2] This circuit, practically, is equiv-

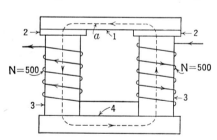

Fig. 2. Series magnetic circuit containing only iron of different cross sections.

[2] Ewing, in his treatise on magnetism, "Magnetic Induction in Iron and Other Metals," gives data showing that a perfectly faced joint without compression is equivalent to an air film of about 0.0014-in. thickness regardless of the flux density, but at a compressive stress of about 3,000 lb. per sq. in. the thickness of the equivalent air gap becomes zero. He also states that for comparatively rough joints, that is, bars which were simply cut in the lathe without having the ends afterwards scraped to the form of true planes, the equivalent air gap is about 0.0018 in. and is reduced only slightly by compression.

Art. 42]   SERIES MAGNETIC CIRCUIT CONTAINING IRON

alent to that of Fig. 1 except that the cross section is variable. Data for Fig. 2 are as follows:

Name of Part	Part No.	Length, in.	Cross Section Dimensions	Material
Armature	1	2*	$1'' \times \frac{1}{4}''$	Mild
Pole face	2	$\frac{1}{8}$ each	$1'' \times 1''$	cold-rolled
Pole core	3	2 "	$\frac{3}{4}''$ dia.	steel
Yoke	4	2*	$1\frac{1}{4}'' \times \frac{3}{8}''$	S.A.E. 10–10

* This is the distance between pole centers which can be considered the effective magnetic length of these parts. If the saturation of the armature is very high, that part in contact with the pole faces should be treated separately.

Problem 3. For the circuit of Fig. 2 calculate the exciting current that will be necessary to produce a flux of 27.5 kmax.; also find the magnetomotive force existing between the pole faces.

Solution. The mean path of the flux is shown by the dashed line (*a*), which links with both coils. This is obviously a magnetic circuit in which the various parts are in series; that is, all the flux of one part flows through the entire length of that part and then enters in its entirety the next part and so through the entire circuit; it is referred to as a series magnetic circuit. The simplest way to handle this problem is to tabulate the flux densities and corresponding magnetic intensities for the various parts of the circuit. The magnetomotive force for each part can then be easily computed by multiplying each length by the corresponding magnetic intensity. The summation of the magnetomotive forces for each part will give the total magnetomotive force from which the exciting current can be computed. The tabulation and computation are shown below and need no further explanation. Refer to magnetization curve 3, Fig. 11*a*, Chapter II.

Part No.	S sq. in.	l in.	ϕ kmax.	B kmax. per sq. in.	H ampere-turns per inch	F ampere-turns
1	0.25	2.0	27.5	110.0	105.0	210
2	1.00	0.25	27.5	27.5	3.7	1
3	0.442	4.0	27.5	62.2	6.2	25
4	0.469	2.0	27.5	59.4	5.9	12
						Total $F = 248$

$$\text{Exciting current } I = \frac{F}{N} = \frac{248}{1{,}000} = 0.248 \text{ ampere}$$

The magnetomotive force existing between the pole faces will evidently be the reluctance drop in the armature over this distance or 210 ampere-turns.

Problem 4. For the circuit of Fig. 2 compute the flux that will be produced by an exciting current of 0.6 ampere.

Solution. This is the inverse of Problem 3 and is decidedly more difficult because the calculation is no longer direct. It is solved by plotting the saturation curve of the magnet (total ϕ against either F or I) and finding the intersection between this function and a straight line corresponding to the constant F or I given. Of course, it is necessary to plot only that portion of the saturation curve in the region of the given F or I. The solution of Problem 3 gives us one point on the saturation curve. By referring to the magnetization curve of Chapter II we can readily see that the flux density of the armature cannot exceed 120 kmax. per sq. in. because the magnetic intensity corresponding to this value of B is 280 ampere-turns per inch, which would require a total magnetomotive force of 560 ampere-turns for the armature alone, while the given magnetomotive force is only 600 ampere-turns. Therefore it will be sufficient if we plot the saturation curve between the values of $B = 110$ to 120 for the armature, necessitating the calculation of only two additional points, say at 115 and 120 kmax. per sq. in. The computations are shown in the following tabulation. The units will be omitted in the tabular forms as it will be understood that the units used throughout the text apply.

Part No.	ϕ	B	H	F	ϕ	B	H	F
1	28.75	115	175	350	30	120	280	560
2	28.75	28.75	3.8	1	30	30.0	3.8	1
3	28.75	65.0	6.7	27	30	67.9	7.1	30
4	28.75	61.3	6.2	12	30	64.0	6.5	13
			Total F =	390			Total F =	604

The final graphical solution is shown in Fig. 3, giving the answer $\phi = 29.96$ kmax. Where the magnetization curve shows large curvature, more than three points should be plotted to insure drawing the proper-shaped curve.

It will be noted from the examination of Fig. 3 that because of the saturation of the armature the magnetization curve is quite flat, and hence one could with accuracy extrapolate between points at $F = 390$ and $F = 604$, thus:

$$\phi = 28.75 + 1.25 \times \frac{600 - 390}{604 - 390} = 29.96 \text{ kmax.}$$

If in a problem of this type the cross section of a part of the circuit is continuously variable, it becomes necessary to break up that part into a

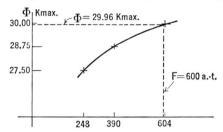

Fig. 3. Magnetization curve for the graphical solution of Problem 4.

sufficient number of short lengths, so that each may be considered with reasonable accuracy to have a constant cross section.

43. Symmetrical Parallel Magnetic Circuit Containing Only Iron of Different Cross Sections

This type of circuit is illustrated in Fig. 4 and, as labeled, represents the magnetic circuit of an alternating-current magnet in the closed-gap position. It could equally well represent the magnetic circuit of

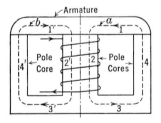

Fig. 4. Symmetrical parallel magnetic circuit containing only iron of different cross sections.

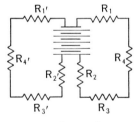

Fig. 5. Electrical equivalent circuit for magnetic circuit of Fig. 4.

a shell-type audio-frequency or power transformer, or the direct-current circuit of a variable reactance for alternating currents. This type of circuit is usually symmetrical about the center line through the center pole core as shown, and the area of the center pole core, except for the reactor application, is usually at least twice that of the outer cores. This circuit, like that of Fig. 2, can be considered to have no air gaps or leakage flux. The path and direction of the flux are shown by the paths (a) and (b), both of which have the total magnetomotive force of the coil effective on them. The two magnetic circuits represented by the flux lines (a) and (b) are magnetically independent of each other and can be computed separately in exactly the same manner as Problems 3 and 4. Of course, if the two paths are identical as illustrated, only one-half the area of the center core belongs to each path, and it is necessary only to compute the flux for one path and merely double it to get the total flux of the center core.

In Fig. 5 is shown an ordinary electric circuit equivalent to that of Fig. 4. Here the battery, equivalent to the coil of Fig. 4, is shown supplying two electric circuits in parallel, consisting of resistances labeled to correspond to the reluctances of the various parts of the circuit of Fig. 4. Note that the resistance corresponding to the reluctance of the center pole core is drawn as two separate resistors, R_2 and R_2', each in series with its own circuit. These must each have a resistance corresponding to twice the reluctance of the entire center pole core, while the other resistances have values corresponding directly to the reluctances they represent. If a circuit of this type is not symmetrical about the center of the exciting coil, the solution is more difficult. This type of problem is the subject of the next article.

44. Unsymmetrical Parallel Magnetic Circuit Containing Only Iron of Different Cross Sections

This type of circuit, illustrated in Fig. 6, differs from that of Fig. 4 only in that the two outer pole cores are of different areas. It is sometimes used where it is desired to keep the flux of one branch, the smaller one, relatively constant while that of the other branch varies considerably. We will again assume that there are no air gaps or leakage fluxes in the circuit. The material is medium-silicon steel laminations; the core has an effective stacking height of 1 in.

Problem 5. Determine for the circuit of Fig. 6 the saturation curve (flux of the center core as a function of the exciting magnetomotive force) between the values of $B = 42.5$ to $B = 57.5$ for the center core 1, and the distribution of flux between the two outer legs.

Solution. Since this problem is more difficult than those preceding, it is best to draw an electrical equivalent circuit. This is shown in Fig. 7; it differs from that of Fig. 5 only in that the center core is not split between the two paths. This reluctance cannot be divided to make the simpler circuit of Fig. 5 because no definite portion of it can be assigned to either path owing to the dissymmetry of the circuit. The method of solution is based upon the following facts: the flux of the center leg is

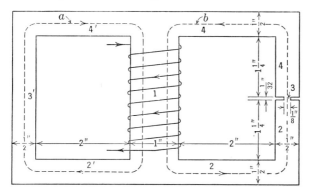

FIG. 6. Unsymmetrical parallel magnetic circuit containing only iron of different cross sections.

equal to the sum of the fluxes of the outer legs ($\phi_a + \phi_b = \phi_0$), and the magnetomotive force across the outer legs must at all times be equal to $F_a = F_b$. The range of variation of ϕ_0 is known, but the distribution of ϕ_0 between paths (a) and (b) is indeterminate, and hence the range of variation of either ϕ_a or ϕ_b is unknown. If the range of variation of, say, ϕ_a were known, the magnetization curve for that branch could be plotted between its known flux limits, thereby determining the range of variation of F_a. If this is known, the magnetization curve for branch (b) could be plotted between the same limits of magnetomotive force. The sum of these two curves would give ϕ_0, and the addition of the reluctance drop F_1 in the center leg for any value of ϕ_0, to the corresponding F_a or F_b, would give the total exciting magnetomotive force for the particular value of ϕ_0.

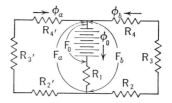

FIG. 7. Electrical equivalent circuit for magnetic circuit of Fig. 6.

The first step in the solution, then, is to estimate the range of variation of either ϕ_a or ϕ_b. At high flux densities the flux will divide, very roughly, in proportion to the smallest areas of the paths; thus, ϕ_a might be as high as 0.8 ϕ_0, while at lower densities ϕ_a would be a smaller part of ϕ_0. We can estimate ϕ_a to vary between, say, 0.5 ϕ_0 minimum to

0.75 ϕ_0 maximum. As will be seen later, this estimate, whether right or wrong, will not influence the accuracy of the final result; a poor guess will merely increase the amount of computation slightly.

These limits being taken, the saturation curve must be computed between the limits of $0.5 \times 42.5 = 21.25$ and $0.75 \times 57.5 = 43.0$. This computation is shown in tabular form in Table I. The dimensions taken for the various parts of the circuit are also shown in this table; as parts $2'$, $3'$, and $4'$ have the same areas and carry the same flux they can be handled as a unit. Magnetization curve $8a$, Fig. 15, Chapter II, was used.

Table I shows that the magnetomotive force across circuit (a) will vary between 16.8 and 129 as the flux varies between 20 and 42.5. The next step in the computation is, therefore, to compute the saturation curve of circuit (b) between the same limits of magnetomotive force. This computation is shown in Table II. It will be noticed that the computation is not carried out as far as the upper limit of 129 ampere-turns, because a comparison of the tables indicates that the distribution at high flux densities is different from that estimated—circuit (a) does not carry as much as 0.75 ϕ_0 maximum. The magnetic data at high flux densities were obtained from Fig. 4, Chapter II, which was obtained by extrapolating the data of curve $8a$, Fig. 15.

TABLE I

Tabular Computation: Magnetization Curve
Branch (a)—$(2', 3',$ and $4')$; $l = 8.6$; $S = 0.5$

Part No.	ϕ	B	H	F
$2' + 3' + 4'$	20	40	1.95	16.8
$2' + 3' + 4'$	25	50	2.6	22.4
$2' + 3' + 4'$	30	60	3.55	30.5
$2' + 3' + 4'$	35	70	5.1	44
$2' + 3' + 4'$	37.5	75	6.6	57
$2' + 3' + 4'$	40.0	80	9.4	81
$2' + 3' + 4'$	42.5	85	15.0	129

In Fig. 8 these two magnetization curves and their sum ϕ_0 are shown plotted against F_a or F_b as abscissas. Table III gives the distribution of the flux ϕ_0 between the two outer legs and the magnetomotive force of the outer branches for equal increments in ϕ_0 between the assigned limits of 42.5 and 57.5. These data (columns (1) to (4)) are read directly from Fig. 8. The magnetomotive force of the center leg for the various values of ϕ_0 is shown in column (5). It is computed directly from the dimensions

TABLE II

Tabular Computation: Magnetization Curve
Branch $(b) — (2 — 3 — 4)$. Parts 2 and 4: $l = 8.37$; $S = 0.5$
Part 3: $l = 0.03125$; $S = 0.125$

Part No.	ϕ	B	H	F	Total F
2 + 4	12.5	25	1.33	11.1	14.3
3	12.5	100	100	3.2	
2 + 4	13.75	27.5	1.4	11.7	
3	13.75	110	245	7.5	19.2
2 + 4	15.0	30	1.5	12.0	
3	15.0	120	490	15.0	27.0
2 + 4	15.75	31.5	1.55	13	
3	15.75	126	730	23	36
2 + 4	16.1	32.2	1.6	13	
3	16.1	128.8	1000	31	44
2 + 4	17.0	34	1.65	14	
3	17.0	135.7	2000	61	75
2 + 4	17.5	35	1.70	14	
3	17.5	140.2	3000	92	106

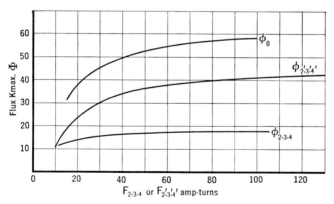

Fig. 8. Magnetization curves for branches (a) and (b) of the magnetic circuit of Fig. 6.

of this path ($l = 3.5$; $S = 1.0$), and magnetization curve 8a. In column (6) the total exciting magnetomotive of the coil is tabulated.

TABLE III
Tabular Solution of Problem 5

(1) ϕ_0 (1)	(2) ϕ_b (2–3–4)	(3) ϕ_a (2'–3'–4')	(4) F_a or F_b (2–3–4) or (2'–3'–4')	(5) F_0 (1)	(6) F_0 Exciting Coil Item (4) + (5)
42.5	14.7	27.8	26.3	7.4	33.7
45.0	15.3	29.7	30.0	7.9	37.9
47.5	15.7	31.8	35.0	8.5	43.5
50.0	16.0	34.0	41.5	9.2	50.7
52.5	16.3	36.2	50.0	9.9	59.9
55.0	16.7	38.3	65.0	10.6	66.6
57.5	17.2	40.3	90.0	11.4	101.4

The problem is completed by merely plotting from Table III the flux of the center leg ϕ_0, and its components ϕ_a and ϕ_b, against the total exciting magnetomotive force F_0. This is shown in Fig. 9. It must be

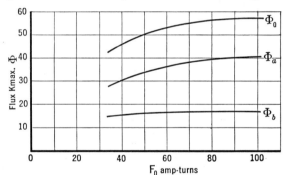

Fig. 9. Magnetization curve for flux of center leg and its components, for the magnetic circuit of Fig. 6.

remembered that if other current-carrying coils, such as transformer secondaries, surround the outer legs the solution will be quite different.

45. Simple Series Magnetic Circuit of Constant Cross Section Containing an Air Gap of Known Permeance

A simple form of this circuit is illustrated in Fig. 10, which is identical to that of Fig. 1, except for the radial air gap. The material of the ring is annealed ingot iron, and the permeance of the air gap, including all fringing flux, etc., is 0.01 kmax. per ampere-turn; all other data are given in the figure.

Problem 6. If the flux of the magnetic circuit of Fig. 10 is to be 20 kmax., compute the required exciting current.

Solution. The magnetomotive force equation for this circuit is:

$$NI_{coil} = F_i + F_a$$

where the subscripts i and a designate the iron and air gap, respectively. Or, rewriting,

$$NI_{coil} = H_i l_i + \frac{\phi}{P_a}$$

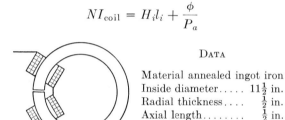

DATA

Material annealed ingot iron
Inside diameter..... $11\frac{1}{2}$ in.
Radial thickness.... $\frac{1}{2}$ in.
Axial length........ $\frac{1}{2}$ in.
Turns per coil...... 500

Fig. 10. Simple series magnetic circuit of constant cross section containing an air gap.

This equation may be solved directly as the iron is of constant cross section; hence, B_i is equal to $20 \times 4 = 80$; H_i, equal to 7.6, may be read directly from magnetization curve 1, Fig. 11a, Chapter II. Then, neglecting the air-gap length when computing the length of the iron path:

$$NI_{coil} = 7.6 \times 12\pi + 20 \div 0.01 = 286 + 2{,}000 = 2{,}286 \text{ ampere-turns}$$

$$I_{coil} = \frac{2{,}286}{1{,}000} = 2.286 \text{ amperes}$$

Problem 7. If the exciting current of the magnetic circuit of Fig. 10 is 1.5 amperes, compute the flux of the ring.

Solution. This problem is not a direct one like Problem 6 because H_i is an unknown analytical function (known graphically, however) of the unknown flux ϕ. Resort must be made to either a "cut-and-try" or a graphical solution. Because of the simplicity of the circuit the cut-and-try method is easier here, but when the circuit is more complicated, or where the air-gap permeance is to take a series of values, the graphical method is more satisfactory. We shall solve it by both methods for the purpose of illustration.

Cut-and-Try Method. As the magnetomotive force of the air gap is large compared to that of the iron, a rough estimate of the total flux of the core may be obtained by neglecting the reluctance of the iron; thus,

$$\phi = FP_a = 1.5 \times 1{,}000 \times 0.01 = 15 \text{ kmax.}$$

With this value of ϕ, B_i is equal to 60 and H_i is equal to 3.5 from curve 1. Then,

$$NI_{coil} = 3.5 \times 12\pi + \frac{15}{0.01} = 132 + 1{,}500 = 1{,}632 \text{ ampere-turns}$$

This value is, of course, too high by F_i, but it indicates that a choice of $\phi = 14$ kmax. will be about correct. Let us try this value; $\phi = 14$, then $B_i = 56$ and $H_i = 3.25$, and

$$NI_{coil} = 3.25 \times 12\pi + \frac{14}{0.01} = 122 + 1{,}400 = 1{,}522 \text{ ampere-turns.}$$

We now have two points on the magnetization curve: $\phi = 15$, $F = 1{,}632$; and $\phi = 14$, $F = 1{,}522$. By extrapolating we can with fair accuracy find ϕ at $F = 1{,}500$. Thus

$$\phi = 14 - 1 \times \frac{22}{1{,}632 - 1{,}522} = 14 - \frac{22}{110} = 14 - 0.2 = 13.8$$

Let us check this value: $\phi = 13.8$, then $B_i = 55.2$ and $H_i = 3.2$, and

$$NI_{coil} = 120 + 1{,}380 = 1{,}500 \text{ ampere-turns}$$

A check to this precision is usually unwarranted because the magnetization curves used are generally average curves, and the leakage permeance of the air gap at the best cannot be computed closer than 2 or 3 per cent—in usual cases nearer 5 per cent.

Graphical Method. This method consists in plotting the magnetization curve of the iron part of the circuit in the usual way, and then drawing the magnetization curve of the air gap either through the point of required flux on the iron magnetization curve if flux is given, or through the point of given magnetomotive force if this quantity is given. In Fig. 11 is shown the magnetization curve (*a*) of the iron of the ring alone. To solve Problem 6 it is necessary merely to draw (with negative slope) the magnetization curve of the air gap (*b*) through the point on curve (*a*) where $\phi = 20$ as shown. The intersection of curve (*b*) with the F axis gives the required number of ampere-turns, which is 2,286 as before. To solve Problem 7 it is necessary merely to draw curve (b) through the F axis at the point of given magnetomotive force, as shown by curve (b').

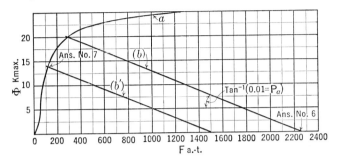

FIG. 11. Graphical construction for the solution of Problems 6 and 7.

The intersection of (b') with (a) gives the answer $\phi = 13.8$ as before. This method is particularly applicable where the air-gap permeance takes on a series of values, or where it is desired to know on what part of the saturation curve of the iron the circuit is operating.

46. Leakage Flux between Pole Cores Having Exciting Coils

Figure 12a represents a bipolar magnet which has an exciting coil of $N/2$ turns per inch distributed uniformly along each pole core. Let P'_L represent the direct leakage permeance between pole cores per inch of axial length; assume that the pole cores extend far beyond the pole faces shown, so that the distortion of the leakage field produced by the pole faces may be neglected; also, that the iron has high permeability. Then, the magnetomotive force across the yoke end of the leakage path will be zero, while at the pole face end it will be NIh, the variation between these two values being uniform, as illustrated in Fig. 12b. The total leakage flux ϕ_L up to any point x measured from the yoke will be:

$$\phi_x = \int_0^x F_x P'_L dx = P'_L N I \int_0^x x dx = \frac{P'_L N I x^2}{2}$$

When $x = h$,

$$\phi_x = \phi_L = \frac{P'_L N I h^2}{2} = \frac{P'_L h}{2} \times NIh \tag{1}$$

Or, in words, the total leakage flux is equal to one-half the total leakage permeance between the pole cores multiplied by the total exciting magnetomotive force of both poles. When the iron becomes very saturated, the expression of equation 1 ceases to apply unless corrected for the reluctance drop in the iron. The variation in ϕ_L with x is shown in Fig. 12c.

Figure 12d shows the total flux ϕ of the pole core as a function of h.

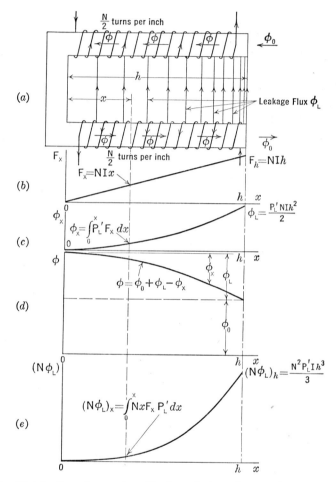

Fig. 12. Distribution of magnetomotive force, leakage flux, and leakage flux linkage produced by a distributed winding.

The flux ϕ_0 may be taken to represent the main flux of the pole cores which issues from the pole faces. When ϕ_L is small compared to the constant flux ϕ_0, the permeability of the pole core may be considered substantially constant over its entire length. In this event the reluctance drop in the pole core can be estimated with fair accuracy by assuming it to carry a uniform flux of $\phi_0 + \frac{2}{3}\phi_L$ throughout its entire length. The proof of this is as follows:

$$(F_i)_x = \int_0^x H_i dx = \int_0^x \frac{\phi}{\mu_i S} dx = \frac{1}{\mu_i S} \int_0^x \left(\phi_0 + \phi_L - \frac{P'_L N I x^2}{2}\right) dx$$

integrating and substituting in the limits, we get:

$$(F_i)_x = \frac{1}{\mu_i S}\left(\phi_0 x + \phi_L x - \frac{P'_L N I x^3}{6}\right)$$

When $x = h$,

$$(F_i)_h = \frac{(\phi_0 + \frac{2}{3}\phi_L)}{\mu_i S} h \qquad (2)$$

The total flux linkage produced by the leakage flux ϕ_L can be evaluated as illustrated in Fig. 12e as follows:

$$(N\phi_L)_h = \int_0^h Nx \cdot F_x P'_L dx = I N^2 P'_L \int_0^h x^2 dx$$

$$= \frac{N^2 P'_L I h^3}{3} \text{ weber-turns} = \frac{2}{3}\Phi_L N h \qquad (3)$$

47. Series Magnetic Circuit Containing Series Air Gaps and a Parallel Leakage Path

A typical circuit of this type is that of the lifting magnet illustrated in Fig. 13. In fact, any electromagnet which is to do work must have

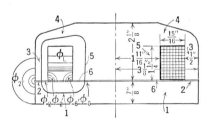

Fig. 13. Section of a circular lifting magnet.

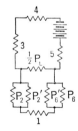

Fig. 14. Equivalent circuit for magnet of Fig. 13.

this type of flux path. Another illustration in a previous chapter of this type of circuit is the flat-faced plunger magnet of Fig. 1a, Chapter III.

The dimensions and data for the magnet of Fig. 13 are listed below:

Name of Part	Part No.	Material	Effective Length, in.	Area Cross Section, sq. in.
Armature........	1	Cast steel	2.6	11.8
Pole Core.......	3	" "	1.25	9.0
Yoke...........	4	" "	2.5	9.0
Pole Core.......	5	" "	1.25	9.0

Useful permeance P_2 Gap 2 = 0.89 kmax. per amp-turn
" " P_6 " 6 = 0.89 " " "
Fringing " P_2' " 2 = 0.20 " " "
" " P_6' " 6 = 0.068 " " "
Leakage " P_L between pole cores = 0.046 kmax. per amp-turn
Exciting winding 180 turns No. 14 wire.

The significance of dividing the working air-gap permeances into two parts, useful and fringing permeances, is that practically all the work of the magnet is done by the useful flux which passes through the useful permeance. The fringing flux contributes practically no work; in fact, it sometimes slightly decreases the work done. However, it is very necessary to evaluate correctly and to take into account all the fringing and other leakage fluxes, because they exist in the same iron circuit as the useful flux and hence determine its saturation.

Problem 8. Compute, for the magnet of Fig. 13, the leakage coefficient [3] and the exciting current required to produce a useful flux of 805 kmax.

Solution. The first step is to draw the equivalent circuit of Fig. 14. In accordance with the work of the last article the distributed leakage permeance P_L is replaced by one of half the value concentrated between the pole faces. The permeances of the air gaps 2 and 6 are shown split into their useful and fringing parts to facilitate the calculation of the leakage coefficient.

The leakage coefficient will be equal to the flux through part 4 divided by that through P_2 and P_6. This will equal:

---

[3] The leakage coefficient of a magnet is defined as the ratio of the useful flux (flux in working air gap) to the maximum flux linking with the exciting coil.

$$\frac{\dfrac{1}{[1/(P_2 + P_2')] + [1/(P_6 + P_6')]} + \dfrac{P_L}{2}}{\dfrac{1}{1/P_6 + 1/P_2}} = \frac{0.510 + 0.023}{0.445} = 1.20$$

if we consider the permeance of part 1 to be infinite, this is a fair approximation because the permeance of part 1 will be very much greater than that of parts 2 and 6 in series.

The calculation of the exciting current required to produce a useful flux of 805 kmax. is carried out below in tabular form. Air gaps 2 and 6 are handled as one: this is permissible if both carry the same flux. The flux through these air gaps, again considering the permeance of part 1 to be infinite, will be

$$805 \times \frac{0.510}{0.445} = 923$$

and the magnetomotive force across them is computed from the relation $\phi = PF$, where P is the joint permeance of the gaps in series. Thus,

$$F_{2+6} = \frac{923}{0.510} = 1{,}810 \text{ ampere-turns}$$

The flux in the armature (part 1) is the same as that in the air gaps. The pole cores 3 and 5 may be handled as one because they have the same flux and the same area. The flux carried by the pole cores, following out the ideas developed in the last article, can be considered equal to $923 + \frac{2}{3}\phi_L$. ϕ_L can be estimated by either one of the following methods: as the leakage coefficient of the entire magnet is 1.20 the total flux through the yoke will be $805 \times 1.20 = 966$. Hence, $\phi_L = 966 - 923 = 43$. Or, referring to the last article, ϕ_L is equal to $\frac{1}{2}P_L$ times the magnetomotive force across the pole faces. This magnetomotive force can be determined by evaluating the first two items in the tabular calculation. From the table, the total magnetomotive force across the pole faces is 1,897. Hence,

$$\phi_L = 0.023 \times 1{,}897 = 44$$

The flux of the pole cores to be entered in the table will therefore be

$$923 + \tfrac{2}{3} \times 44 = 952$$

The remaining computation can be followed directly from the table:

Part No.	Branch	Length	Area	Flux	B	H	F
1	Armature	2.6	11.8	923	78.2	33.5	87
2 + 6	Air gaps	.....	.....	923	.....	.....	1,810
3 + 5	Pole cores	2.5	9.0	952	106	140	350
4	Yoke	2.5	9.0	966	107.5	155	388

Total $F = 2,635$

The exciting current will equal

$$2,635 \div 180 = 14.6 \text{ amperes }[4]$$

Problem 9. (a) Compute for the magnet of Fig. 13 the useful flux that will be produced by an exciting current of 11.1 amperes; (b) repeat (a) neglecting all leakage and fringing fluxes; (c) considering the force to vary as the square of the flux density produced by the useful flux in the working gaps, compute for this exciting current the percentage error that would result by neglecting the fringing and leakage fluxes when computing the force.

Solution. As this problem presents no new difficulties its solution will merely be indicated. Part (a) may be solved by the methods outlined in Problem 7. The magnetic circuit calculations will be the same as for Problem 8. Part (b) is identical with Problem 7, except that the iron part of the circuit is not of constant cross section. Part (c) needs no special comment.

48. Calculation of Magnetic Circuits Involving Incremental Permeability

Typical apparatus, the magnetic circuits of which fall into this class are filter chokes, audio-frequency transformers used for coupling vacuum tubes, half-wave rectifiers, reactors which are made variable by means of a superposed constant flux, etc.

---

[4] The data for this magnet are taken from one designed by L. A. Hazeltine and built in the Electrical Engineering Laboratory at Stevens. Experiments with this magnet at $\frac{1}{32}$-in. gap (the gap permeances as given on page 100 were computed for $\frac{1}{32}$-in. gap; for the method of computing see Art. 54, Sec. 3) show that the exciting current for 805 kmax. useful flux is about 15 amperes. This magnet, with no gap, will exert a force of more than 1 ton when supplied by an ordinary dry cell. See Problem 18, page 133, of *Electrical Engineering* by L. A. Hazeltine, Macmillan Co.

Figure 15 shows the magnetic and electric circuits for an ordinary filter choke. The purpose of the air gap is to control the direct-current magnetizing intensity in the iron, so that the over-all incremental permeance of the core may be a maximum.

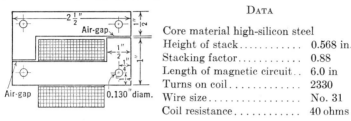

DATA

Core material high-silicon steel
Height of stack............ 0.568 in.
Stacking factor............ 0.88
Length of magnetic circuit.. 6.0 in
Turns on coil.............. 2330
Wire size.................. No. 31
Coil resistance............ 40 ohms

FIG. 15. Section of a filter choke.

Problem 10. For the choke of Fig. 15 compute the air gap necessary to produce maximum inductance if the direct current is to be 100 milliamperes, and the maximum total flux variation is assumed to be 1 kmax. per sq. in.[5] Neglect fringing in computing the air-gap permeance.

Solution. The inductance of a coil is

$$L = \frac{N\phi}{I} \text{ henries}$$

where ϕ is the flux in webers produced by the current I amperes; by substituting FP for ϕ, we get

$$L = N^2 P \text{ henries} \qquad (4)$$

where P is the permeance of the core in webers per ampere-turn. When speaking of choke coils, the inductance referred to is that effective to alternating currents. Hence ϕ and I will be alternating quantities, and P will be the effective incremental permeance. This permeance for the ordinary type of choke can be expressed as

$$P = 3.19 \times 10^{-8} \frac{\mu_d S}{l} \text{ weber per ampere-turn} \qquad (5)$$

where μ_d is the effective incremental permeability of the iron and series air gap combined. By substituting equation 5 into equation 4 we get

$$L = 3.19 \times 10^{-8} \frac{\mu_d N^2 S}{l} \text{ henries} \qquad (6)$$

---

[5] This is purposely taken small, because the incremental permeability increases with increasing flux variation; hence the answer will correspond to the least inductance which will occur when the alternating voltage across the choke is low.

As μ_d is the only variable, L will be a maximum when μ_d is a maximum. μ_d depends upon the length of the air gap and the amplitude B_m of the alternating flux. As the latter is stated to be constant it is only necessary to find the air gap corresponding to maximum μ_d. This is done as follows: As the direct-current magnetomotive force is constant, it is possible to compute the polarizing magnetic intensity H_0 as a function of the gap length l_a, and hence the incremental permeability of the iron μ_i as a function of gap length. To do this the magnetization curve [6] of the

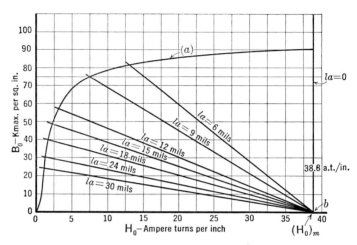

FIG. 16a. Magnetization curves for determining the polarizing magnetic intensity in the iron for the choke of Fig. 15 for different air-gap lengths.

core iron, curve (a), Fig. 16a, is plotted up to the maximum possible polarizing magnetic intensity $(H_0)_m$, equal to

$$\frac{F}{l_i} = \frac{0.1 \times 2{,}330}{6} = 38.8 \text{ ampere-turns per inch}$$

$H_0 = (H_0)_m$ occurs when the air-gap length is zero, and the corresponding $(B_0)_m$ can be determined by the intersection of curve (a) and the magnetization curve of the zero-length air gap, designated as $(l_a = 0)$ and plotted from point (b) as an origin. Values of H_0 at other gap lengths are obtained by plotting the magnetization curves for these

---

[6] In general this magnetization curve should be the anhysteretic magnetization curve of the iron core (Fig. 18b, Chapter II), but as the superposed alternating flux density is so small in this problem the normal magnetization curve for sample 9a of Fig. 15, Chapter II, has been used.

air gaps and determining H_0 from their intersection with curve (a). These air-gap magnetization curves are plotted by considering that for each inch length of iron there is an air gap of l_a/l_i inches length, and that the total magnetomotive force across both of these is constant at $(H_0)_m$. Thus if the total air-gap length is 30 mils, l_a/l_i is 5 mils, and the magnetization curve for this gap length is constructed by drawing from point (b) a straight line having a negative slope of $B_a/H_a = \mu_a \times l_i/l_a = 0.00319 \times 200 = 0.638$ kmax. per ampere-turn, as shown in Fig. 16a. This process is repeated for several lengths of air gap.

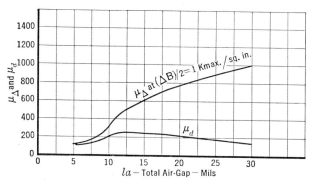

Fig. 16b. Incremental permeability of the iron (μ_Δ), and the effective permeability of the entire magnetic circuit (μ_d), for the choke of Fig. 15 plotted as a function of air-gap length.

It is now possible, by reference to Fig. 19c of Chapter II, to plot for $B_m = 1$ kmax. per sq. in., μ_Δ as a function of l_a. This is done in Fig. 16b.

The permeability μ_d can now be computed as a function of the gap length by the relation

$$\mu_d = \frac{1}{\dfrac{1}{\mu_\Delta} + \dfrac{l_a}{l_i}} \qquad (7)$$

where all the μ's are relative to a vacuum, and is plotted in Fig. 16b.

From this plot it can be seen that μ_d has a maximum of 250 when $l_a = 12$ mils total, or 6 mils for each gap. The inductance of the choke, with 100 milliamperes direct current and an alternating flux density of 1 kmax. per sq. in., will then be, by equation 6:

$$L = \frac{3.19 \times 10^{-8} \times 250 \times 2{,}330^2 \times \tfrac{1}{4}}{6.0} = 1.80 \text{ henries}$$

49. Calculation of Magnetic Circuits Containing Permanent Magnets

There are two distinctly different types of magnet which make up this kind of circuit: the real permanent magnet made of a hardened-steel alloy, and the temporarily permanent magnet made of soft steel.

The real permanent magnet is reversible; that is, it may be partially demagnetized by introducing an air gap, but upon the removal of the air gap it will regain its former strength. In some types of apparatus such a variation in the flux of a permanent magnet occurs periodically; for instance, the rotation of a shuttle-type armature in a magneto produces such a cyclic variation.

The temporarily permanent magnet is not reversible; that is, partial demagnetization due to the introduction of an air gap is, to a great extent, permanent demagnetization. Thus the electromagnet of Fig. 13 has a residual force of the order of 600 lb. in the closed-gap position. If, however, the gap is momentarily opened and then closed again the residual force will disappear almost entirely.

Reversible and irreversible as applied to a permanent magnet are relative terms; a real permanent magnet is not reversible if the demagnetization is carried too far; a temporarily permanent magnet may be reversible if the demagnetization is extremely small. The real point of the matter is that the ability of a permanent magnet to resist demagnetization depends upon the work necessary to demagnetize the iron completely, which, as explained in Art. 25, is equal to the area (4′–0–3–4′ of Fig. 6, Chapter II) under the demagnetization curve. This area in turn is proportional to the area of the symmetrical hysteresis loop. Consequently the value of a steel as a permanent magnet can be estimated from its hysteresis loss as well as from the maximum value of $B_d H_d$ under the demagnetization curve.[7]

Problem 11. Compute the residual flux density in the working gap of the magnet of Fig. 17 after the application of a maximum magnetizing current of 1.5 amperes at zero gap length. Assume that any possible fringing or leakage flux is negligible.[8]

Solution. It is necessary first to determine the flux produced by the maximum magnetomotive force of 1,500 ampere-turns, so that the maximum magnetic intensity of the yoke and armature can be found. This can be determined by the "cut-and-try" method, giving $\phi = 16.4$

---

[7] See Art. 32, and Table III, Chapter II.

[8] This assumption is well justified because at zero gap length the fringing flux will disappear and the leakage flux between poles will be small, owing to the high reluctance drop in the pole cores and yoke.

kmax. and the corresponding H_m's as 12 and 708 ampere-turns per inch, respectively.

From Figs. 13a and 13b, Chapter II, at these values of H_m it is possible to determine the B_r's and H_c's that would exist in the yoke and armature

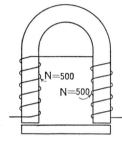

DATA

Material S.A.E. 10–10 dead soft
Yoke (including pole cores).. $\frac{1}{2}$ in. dia.
Effective length............ 7.2 in.
Armature................. $\frac{1}{4}$ in. $\times \frac{1}{2}$ in.
Effective length............ 2 in.

FIG. 17. Soft-steel horseshoe magnet.

if they were individually closed on themselves. These values are listed below along with the corresponding values of total residual flux ϕ_r and total coercive magnetomotive force F_c of the parts.

	H_m	B_r	H_c	ϕ_r	F_c
Yoke.............	12	70	3.4	13.7	24.5
Armature.........	708	84	3.9	10.5	7.8

The next step is to plot the demagnetization curves for both the yoke and armature. Only two points, ϕ_r and F_c, have been determined on these curves. Intermediate points may be determined with sufficient accuracy by interpolation from the series of hysteresis loops for this material given in Fig. 12c or Fig. 2, Chapter II. That for the armature must be extended into the region of positive magnetomotive force for a reason that will be apparent shortly. These curves are shown plotted in Fig. 18.

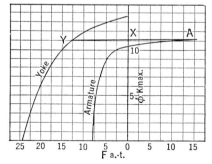

FIG. 18. Demagnetization curves for the yoke and armature of the magnet of Fig. 17.

As the flux of the armature and yoke are equal at all times, it can be seen from Fig. 18 that, as the flux of the magnet falls from its maximum value of 16.4 kmax., that of the yoke will fall below its tabulated residual value 13.7 before that of the armature falls below its tabulated residual value 10.5. Consequently, in this flux range the yoke will tend to magnetize the armature, and equilibrium will be reached when the magnetomotive force made available by demagnetizing the yoke to some value below 13.7 is just sufficient to magnetize the armature to the same value, which will be above 10.5. This equilibrium flux is shown by the line Y–X–A, so drawn that the magnetomotive force YX equals the magnetomotive force XA.

The total residual flux of the magnet as read from the plot is OX, equal to 11.2 kmax. Therefore the residual flux density in the working gap will be [9]

$$B_r = \frac{\phi_r}{S_y} = \frac{11.2}{\pi \times (\tfrac{1}{4})^2} = 57 \text{ kmax. per sq. in.}$$

Problem 12. Repeat Problem 11, assuming that an air gap having a permeance [10] of 0.45 kmax. per ampere-turn is inserted between each pole and the armature.

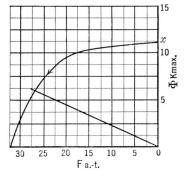

Fig. 19. Demagnetization curve of the entire magnet of Fig. 17.

Solution. The demagnetization curves of the yoke and armature of the magnet of Problem 11 can be added together, as shown in Fig. 19, to form the demagnetization curve of the magnet as a whole. The intersection of this curve with the Φ axis gives the residual flux of the magnet at zero air-gap length, which is the same as that obtained by the method used in Problem 11. However, the form of the demagnetization curve of Fig. 19 makes it simple to take into account the effect of the added air gap. This is done by drawing from the origin with negative slope the magnetization curve of the air gaps in series, that is, a straight line having a slope of -0.225 kmax. per ampere-turn. The intersection of this straight line with the demagne-

---

[9] As the force in this type of magnet is given by the equation $F = B^2/72$ lb. per sq. in., where B is the flux density in working gap in kilomaxwells per square inch, the residual force of this magnet will be equal to

$$\frac{2 \times 57^2 \times \pi \times (\tfrac{1}{4})^2}{72} = 17.7 \text{ lb.}$$

[10] This permeance corresponds to the minimum air gap (0.0014 in.) that can be expected with carefully faced surfaces (see footnote 2).

tization curve will give the total residual flux of the magnet with the air gap. This, as read from Fig. 19, is 6 kmax. Hence the flux density in the air gap will be

$$\frac{6.0}{\pi(\frac{1}{4})^2} = 30.5 \text{ kmax. per sq. in.}$$

The residual force corresponding to this flux density will be $17.7 \times (30.5/57.0)^2 = 5$ lb.

When a magnetic circuit containing a real permanent magnet is subjected to a powerful magnetizing force the magnetic state of the circuit after the removal of the magnetizing force can be calculated by the same method used in the last two problems provided that the permeance of the magnetic circuit is not increased after the removal of the magnetizing force. If the permeance does increase or go through a cyclic variation a subsequent calculation involving the incremental permeability of the permanent magnet steel must be made. The following problems will illustrate the method of solving magnetic circuits involving permanent magnets.

Problem 13. Compute, for the magneto illustrated in Fig. 20, the dimensions of the smallest tungsten-steel permanent magnet which will

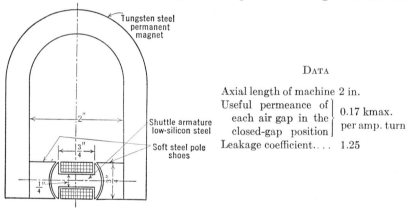

FIG. 20. Magneto.

produce a useful flux in the armature of 50 kmax., assuming that the permanent magnet has just been magnetized in place on the magneto with the armature in the position shown.[11]

[11] Although permanent magnets are usually not magnetized this way, they can be magnetized so that the magnetic cycle will be the same. If the magnet after magnetizing has its magnetic circuit opened (no magnetic conductor between the pole faces) its flux will fall on the normal demagnetization curve and then will rise on a minor loop when the armature assembly is inserted between the poles. Problem 14 is of this nature.

Solution. The total flux in the yoke back of the soft-steel pole shoes,[12] neglecting the direct leakage between the legs of the U, will be 50 × 1.25 = 62.5 kmax. In order to use the magnet steel most economically and hence obtain the smallest magnet, it should work at the flux density B_d corresponding to the maximum energy in the air gap. This density, from Fig. 20, Chapter II, is 41 kmax. per sq. in. for tungsten steel. Hence the cross section of the magnet should have an area of 62.5/41 = 1.52 sq. in. As the axial length of the armature is 2 in. a suitable cross section for the magnet would be 2 by $\frac{3}{4}$ sq. in.

The magnetomotive force F_d required to send the flux through the armature, air gaps, and pole shoes is $F_a + F_i$, which by Art. 32 is equal to H_d times the effective length of the magnet. The flux density in the armature core is equal to $50/\frac{1}{4} \times 2 = 100$ kmax. per sq. in., and the corresponding H_i is 63.0 ampere-turns per inch. F_i for the armature core will then be $63 \times \frac{3}{4} = 47$ ampere-turns. The flux density in the other soft-steel parts is considerably less, and hence their reluctance drop can be neglected in comparison. F_a, equal to ϕ_a/P_a, will be

$$50 \times \frac{2}{0.17} = 590 \text{ ampere-turns}$$

H_d from Fig. 20, Chapter II, is 80 ampere-turns per inch; therefore the effective length (measured to the center of the armature) of the magnet, equal to F_d/H_d, will be

$$\frac{590 + 47}{80} = 8 \text{ in.}$$

The over-all dimensions of the piece of magnet steel will be $\frac{3}{4}$ by 2 by $8\frac{3}{4}$ inches.

Problem 14. Compute for the magneto of Problem 13 the limits of pulsation of the flux density in the permanent magnet due to rotation of the armature if the total permeance between pole shoes is 0.022 kmax. per ampere-turn in the open-gap position and if the direct leakage flux between poles is neglected. Take the incremental permeability of the tungsten steel to be 50 for the conditions of this problem.

Solution. The solution of this problem, unlike that of Problem 13, is indirect. However, it can be readily handled graphically. Thus, in

---

[12] Soft-steel pole shoes are provided because they are easy to shape, and they eliminate the excessive hysteresis loss which would occur if hardened steel were used. This hysteresis loss is caused by the shifting flux produced in the pole faces by the rotation of the armature. The pole faces are sometimes laminated to prevent excessive eddy-current loss due to the shifting flux.

Fig. 21, curve $B_r - H_c$ is the demagnetization curve for the tungsten steel replotted from Fig. 20, Chapter II. Curve 0–1 is the permeance curve in the closed-gap position of the air gap, soft-steel parts, and the leakage path of the magneto per inch cube of effective volume of the permanent-magnet steel. This is computed by multiplying the permeance of these parts by $1.25 l_d/S_d$, where l_d and S_d are the effective length and area of the permanent magnet, respectively, and the factor 1.25 takes the leakage into account. The slope of this curve will therefore be equal to the slope of the magnetization curve of these parts multiplied by $1.25 \times 8 \div 1.5$. The intersection of this curve with the demagnetization curve at point A is, of course, the solution of Problem 13. As the shuttle armature is turned from the position shown in Fig. 20 the permeance of the air gap decreases and the flux falls along the demagnetization curve from A to B, point B being the intersection of the demagnetization curve and curve 0–2, which is the permeance curve of the air gap in the open-gap position per inch cube of the permanent-magnet steel. The slope of this curve will be $0.022 \times 8/1.5 = 0.117$ kmax. per ampere-turn. As the armature is now turned toward the position of maximum permeance the flux will rise along some minor loop such as BEC. As the armature is now rotated back to the position of minimum air-gap permeance the flux will fall on the descending branch of the minor loop CFB, returning to the initial starting point B if the cycle is reversible, as it is here assumed. As the incremental permeability of the permanent-magnet steel is given [13] as 50 the slope of the axis BC of the minor loop may be computed as $50 \times 0.00319 = 0.159$ kmax. per ampere-turn. Laying off the line B–3 with this slope, point C is determined by its intersection with line 0–1.

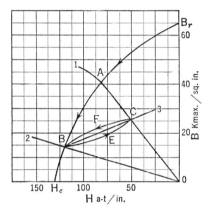

Fig. 21. Graphical construction for the solution of Problem 14.

The limits of pulsation of the flux density of the permanent magnet as the armature is rotated will therefore be

$$\Delta B = B_c - B_B = 25.5 - 14.0 = 11.5 \text{ kmax. per sq. in.}$$

[13] When the incremental permeability is not known it may be found by trial and error, using Spooner's method, provided that the normal permeability of the permanent-magnet steel is known.

PROBLEMS

1. Solve Problem 9 in the text.

2. Verify the statement made in footnote 4 that the magnet of Fig. 13 will exert a force of more than 1 ton on its armature in the closed-gap position when supplied from an ordinary dry cell. Make use of the following data:

Dry cell, generated voltage	= 1.5 volts
Internal resistance	= 0.05 ohm
Resistance of magnet winding	= 0.65 ohm at 20° C.
Force exerted on armature	= $\dfrac{B^2}{72}$ lb. per sq. in., where B is the flux density in the working gap in kilomaxwells per square inch due to the useful flux.

3. Figure 22 is a sketch of a very sensitive low-voltage relay. The current to be detected is passed through the moving coil indicated and, by its reaction on the magnetic field produced by the field coil, causes the coil to move. The force on such a moving coil is given by the following expression:

$$F = 8.85 \times 10^{-5} I B l \text{ lb.}$$

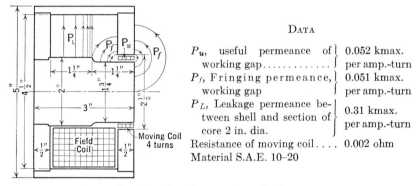

DATA

P_u, useful permeance of working gap	0.052 kmax. per amp.-turn
P_f, Fringing permeance, working gap	0.051 kmax. per amp.-turn
P_L, Leakage permeance between shell and section of core 2 in. dia.	0.31 kmax. per amp.-turn
Resistance of moving coil	0.002 ohm
Material S.A.E. 10–20	

FIG. 22. Sensitive moving-coil relay.

where I is the current of the moving coil in amperes, B is the flux density produced by the field magnet in the region of the coil in kilomaxwells per square inch, and l is the length of the coil in the magnetic field in inches. If the field coil develops 3,000 ampere-turns, compute the force on the coil due to a coil current of 0.25 ampere. Express this force in grams. *Note*: The power input to the moving coil is only 0.000125 watt.

4. Determine the percentage decrease in hysteresis loss for the transformer core of Problem 17 of Chapter II if the transformer has two air gaps inserted in its magnetic circuit, the effective length of each being 0.016 in. corresponding to a permeance of 0.4 kmax. per ampere-turn per gap. *Note*: When plotting the direct-current magnetization curve as in Fig. 16a, the anhysteretic data of Fig. 18a should be used because of the large value of superposed alternating-current flux density.

5. Determine the percentage change in r.m.s. alternating exciting current for the rectifier transformer of Problem 19 of Chapter II, due to the insertion of the air gaps described in Problem 4. Use the data of Fig. 18, and assume that the r.m.s. exciting current for the air gaps and iron core can be added arithmetically. (This is not quite true, as the exciting current for the iron contains harmonics.) Also compute the answer by the method of Art. 29 and explain why this method now gives a better check than in Problem 19, Chapter II.

6. Figure 23 is a sketch of the core of a transformer designed to have a very high leakage reactance. Such transformers are employed practically where the volt-ampere characteristic of the load is negative, making it necessary to limit the current, e.g., neon sign lights. Such a transformer, if properly designed, will regulate for constant current near the point of short circuit, which is the usual operating point. Compute the voltage of the secondary when it is on open circuit. How many per cent greater would it be if the leakage path were absent? Neglect the resistance of the windings and coil leakage fluxes.

7. Compute for the transformer of Fig. 23 the load (secondary) current as a function of the load resistance, between the limits of 0 and 50,000 ohms. *Hint*: This problem may be best handled indirectly. First compute the load voltage as a function

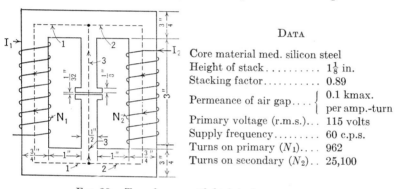

DATA

Core material med. silicon steel
Height of stack.......... $1\frac{1}{8}$ in.
Stacking factor.......... 0.89
Permeance of air gap.... { 0.1 kmax. per amp.-turn
Primary voltage (r.m.s.)... 115 volts
Supply frequency......... 60 c.p.s.
Turns on primary (N_1).... 962
Turns on secondary (N_2).. 25,100

FIG. 23. Transformer with high leakage reactance.

of the load current: that is, assume a value of load current and compute the corresponding value of secondary voltage. This may be done by means of the following relationships which apply to this circuit. The magnetomotive force across path 3 is equal to the reluctance drop of path 2 minus the magnetomotive force of the secondary coil; or $F_3 = F_2 - N_2 I_2$. F_2 may be considered in quadrature with $N_2 I_2$ and may be neglected when $N_2 I_2$ exceeds $2F_2$. F_3 being known, ϕ_3 may be determined, which when substracted vectorially from the constant flux ϕ_1 will give the flux ϕ_2 and hence the load voltage.

8. The variable reactor circuit of Fig. 24 is used as a dimmer to control lights in theaters, etc. Compute the reactance voltage absorbed by this device as the direct current varies from 0 to 2 amperes while the alternating current is held constant at 1 ampere.

9. Discuss the following question: Is there any advantage to be gained by the use of ferronickel (11) in place of the medium silicon steel (8) for the core of a heavy-duty filter choke? Assume both chokes to have the same size core and winding. A heavy-duty filter choke is one that is heavily polarized, e.g., 50 ampere-turns per inch.

10. Redesign the field magnet of the magneto of Problem 13 in the text for 36 per cent cobalt steel. *Note*: This will necessitate a change in the pole shoes if an economical design is to be produced. See the reference of footnote 22 of Chapter II for sketches of the arrangement of economical designs employing this steel.

11. Calculate the magnitude of the flux-density pulsation for the magneto of Problem 10 due to rotation of the armature. Use the same air-gap permeance data as for Problems 13 and 14 of the text. The incremental permeability of the cobalt steel may be found by Spooner's method.

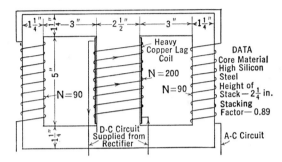

Fig. 24. Variable reactor.

12. Compute the residual flux density in the working gap of the magnet of Fig. 13 for the condition of zero-length air gap if the magnet is made of: (a) annealed Swedish charcoal iron; (b) $\frac{1}{2}$ hard S.A.E. 10–20 steel. Assume the maximum exciting current to be 16 amperes, and neglect leakage and fringing fluxes. Repeat these calculations for an air gap of 5 mils. Using the force equation of Problem 2, compute the residual force for each condition above.

13. The circuit shown in Fig. 25 represents the essential elements of a proposed grid-control rectifier-tube voltage regulator for a 60-cycle alternator. Plot a curve

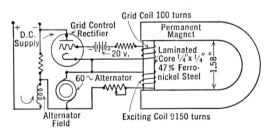

Fig. 25. Voltage regulator for an alternator.

of the instantaneous voltage induced in the grid coil as a function of time for the following two maximum values of alternating current in the exciting coil:

$$(a)\ \frac{220.11 \times 1.58}{9{,}150},\quad (b)\ \frac{220.00 \times 1.58}{9{,}150}$$

amperes, and show that condition (a) will cause the grid-control tube **to trigger** while (b) will not. Assume the following:

1. That the magnet is ideally permanent and impresses a constant magnetomotive force of 220.0×1.58 ampere-turns across the laminated ferronickel core.

2. That the wave form of the current through the exciting coil is sinusoidal and proportional to the voltage of the alternator.

3. That the grid and exciting coils have unity coupling, and are distributed over the full length of the laminated core.

Note: A grid-control rectifier has the property of ionizing and carrying current through its plate circuit when the grid reaches a definite potential with respect to the filament provided that the plate is positive with respect to the filament. This particular grid potential is called the trigger voltage. The tube contemplated for the circuit has a trigger voltage of about -1.5 volt.

CHAPTER V

CALCULATION OF THE PERMEANCE OF FLUX PATHS THROUGH AIR BETWEEN SURFACES OF HIGH-PERMEABILITY MATERIAL

50. General

The precise mathematical calculation of the permeance of flux paths through air, except in a few special cases, is a practical impossibility. This is because the flux does not usually confine itself to any particular path which has a simple mathematical law. For this reason these computations, except in the special cases alluded to, are carried out by making simplifying assumptions regarding the flux paths, or by an entirely graphical method usually referred to as "field mapping." Whichever method is employed, considerable experience is necessary to enable the designer to choose flux paths which are probable. He is assisted in making this choice by the knowledge that flux paths in air between any two surfaces always arrange themselves in such a manner that the maximum possible flux will be produced for a given magnetomotive force; that is, so that the permeance of the air path between the surfaces is a maximum. This fact alone will often enable one to sketch a flux distribution in an air path that is quite probable.

51. Simple Fields Which Can Be Handled Mathematically

1. Parallel Plane Surfaces of Infinite Extent. In this case the flux lines will be perpendicular to the plane surfaces. In Fig. 1 let the two surfaces S represent two opposite areas of S square inches of infinite parallel plane surfaces separated by l inches. The permeance of the flux path between the two surfaces S may then be readily computed, because the path taken by the flux is that of a true mathematical right cylinder; that is, the flux lines everywhere throughout the surface S are parallel to each other and all the equipotential

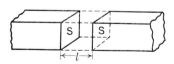

Fig. 1. Parallel plane surfaces.

surfaces are planes parallel to each other and to the surfaces. This permeance will therefore be [1]

$$P = \mu \frac{S}{l} \qquad (1)$$

2. **Non-Parallel Plane Surfaces of Infinite Extent.** Surfaces fulfilling this definition are shown in Fig. 2. Here the two areas S are opposite portions of infinite planes intersecting in the line 0–0 at an angle θ. These areas are bounded by lines parallel to axis 0−0 at radii of r_1 and r_2 and have an axial length of l. With the assumption made regarding the infinite extent of the planes, the flux lines between the areas S will be arcs of circles having their centers along the axis 0−0. The equipotential surfaces will be planes between SS, which contain the axis 0−0. Consequently a cylindrical shell of flux having a radial thickness dr and an axial length l will fulfill the definition of a mathematical right cylinder, and its permeance can be expressed by the equation

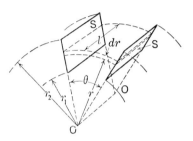

Fig. 2. Non-parallel plane surfaces.

$$dP = \frac{\mu l\, dr}{\theta r}$$

As the permeances of all the cylindrical shells making up the area S will be in parallel with each other, the total permeance between areas SS will be

$$P = \frac{\mu l}{\theta} \int_{r_1}^{r_2} \frac{dr}{r} = \frac{\mu l}{\theta} \log_\epsilon \frac{r_2}{r_1} \qquad (2)$$

3. **Parallel Circular Cylinders of Infinite Extent.** Figure 3 shows a section through two parallel cylinders of infinite extent. The equipotential surfaces are eccentric cylinders having their centers displaced along

---

[1] In this and all other permeance formulas to follow the dimensions may be chosen at will provided that consistency is maintained; thus, if μ is in kilomaxwells per ampere-turn in an inch cube, S should be in square inches and l in inches, making P = kilomaxwells per ampere-turn; if μ is in kilomaxwells per ampere-turn in a centimeter cube, S should be in square centimeters and l in centimeters, making P kilomaxwells per ampere-turn, etc.

the flux line X–X; the flux lines are eccentric circles having their centers displaced along the equipotential line Y–Y. In the general case the cylinders may be of different sizes and may be wholly without or within one another. Any two of the equipotential lines of Fig. 3 may represent the cylinders. Thus if we choose any two equipotential lines on one side of the axis Y–Y we shall have eccentric cylinders one wholly within the other; if we chose equipotential lines on opposite sides of the axis we have cylinders wholly without one another.

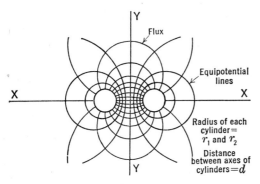

Fig. 3. Field surrounding two parallel circular cylinders.

The general formula [2] for the permeance between the cylinders is

$$P = \frac{2\mu \pi l}{\log_e (u + \sqrt{u^2 - 1})} \qquad (3)$$

where l is the axial length of the cylinders, and u is given by the expression

$$u = \frac{d^2 - r_1^2 - r_2^2}{2 r_1 r_2}$$

r_1 and r_2 are the radii of the two cylinders, and d is the distance between their centers.

Convenient formulas for four special cases which arise in practice are derived below:

Case I. Cylinders of Different Radii, Wholly without One Another. Figure 3a shows the geometrical configuration of the system. The permeance may be determined directly by formula 3.

Case II. Cylinders of Same Radius, Wholly without One Another. Figure 3b shows the geometrical configuration of the system

$$u = \frac{d^2 - 2r^2}{2r^2}$$

[2] This may be derived by the application of potential theory.

Substituting this into equation 3, the following equation will be obtained after the proper simplification:

$$P = \frac{\mu\pi l}{\log_\epsilon\left(\frac{d}{2r} + \sqrt{\left(\frac{d}{2r}\right)^2 - 1}\right)} \tag{3a}$$

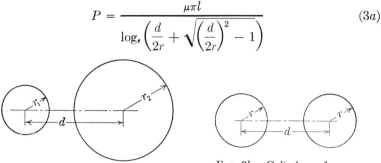

Fig. 3a. Cylinders of different radii, wholly without one another

Fig. 3b. Cylinders of same radius, wholly without one another.

When d is large compared to r, equation 3a may be written in the approximate form:

$$P = \frac{\mu\pi l}{\log_\epsilon \frac{d}{r}} \tag{3b}$$

This approximation is quite accurate when $d > 8r$.

Case III. Cylinders of Different Radii, Wholly within one Another and Eccentric. Figure 3c shows the geometrical configuration of the system. Here

$$u = \frac{r_1^2 + r_2^2 - d^2}{2r_1 r_2}$$

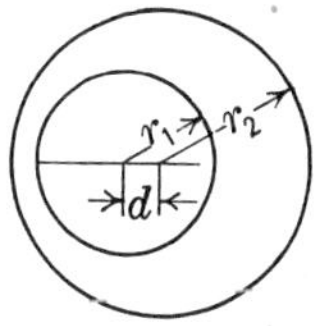

may be substituted directly into equation 3 (see footnote 3).

When d is small compared to r_1 and r_2, the accuracy of the arithmetical work will be greatly increased by substituting $r_2 = r_1 + c$ into the numerator of the expression for u, which will give

Fig. 3c. Cylinders of different radii, wholly within one another and eccentric.

$$u = 1 + \frac{c^2 - d^2}{2r_1 r_2}$$

Letting

$$a = \frac{c^2 - d^2}{2r_1 r_2}$$

[3] The signs of r_1, r_2, and d must be taken so that u is positive.

the permeance will be

$$P = \frac{2\pi\mu l}{\log_\epsilon\left[1 + a\left(1 + \sqrt{\frac{2}{a}+1}\right)\right]} \quad (3c)$$

Case IV. **Cylinders of Different Radii, Wholly within Each Other and Concentric.** Figure $3d$ shows the geometrical configuration of the system. In this case, as $d = 0$ (see footnote 3),

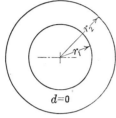

Fig. $3d$. Cylinders of different radii, wholly within each other and concentric.

$$u = \frac{r_1^2 + r_2^2}{2r_1 r_2}$$

Substituting this into (3), we have

$$P = \frac{2\pi\mu l}{\log_\epsilon \frac{r_2}{r_1}} \quad (3d)$$

Another method of deriving this formula will be given in Art. 53, Sec. 6, Path 15.

4. **Parallel Cylinder and Plane of Infinite Extent.** As the equipotential line YY of Fig. 3 is merely one element of the equipotential surface YY, which is at all points equidistant from the axes of the cylinders, it follows that this surface can be considered a plane of infinite extent parallel to a cylinder of infinite extent. Either half of Fig. 3 will therefore represent the configuration of the field. The permeance between this plane and either cylinder will evidently be twice that given by formula $3a$ or $3b$. Using the approximate relation $3b$, the formula for this case will be

$$P = \frac{2\mu\pi l}{\log_\epsilon \frac{d}{r_1}} \quad (4)$$

where d is twice the distance between the center of the cylinder and the plane. This formula will be subject to the same restrictions as $3b$.

5. **Spheres of the Same Radius r Separated by a Distance d Equal to at Least $6r$ between Centers.** Figure 4 shows a section through the centers of two spheres of radius r. The flux lines and equipotential surfaces for this particular case are quite complicated and do not correspond to any of the more familiar curves. This field should be compared very carefully with that of Fig. 3. The bulging of the flux lines of the spherical field of Fig. 4 as compared to those of cylindrical field of Fig. 3 is the

chief difference between fields which extend infinitely in one dimension and those in which this same dimension is produced by revolving a plane. If the distance between centers of the spheres is at least $6r$, the perme-

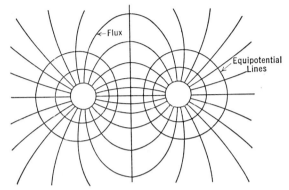

FIG. 4. Field surrounding two spheres.

ance between the spheres is given with sufficient accuracy by the formula:

$$P = \frac{2\pi\mu}{\dfrac{1}{r} - \dfrac{1}{d}} \tag{5}$$

6. Sphere of Radius r and an Infinite Plane Separated by a Distance $d/2$ equal to at Least $3r$ Between Centers. The discussion in this case is identical with that in Section 4. The formula, derived from equation 5, will be:

$$P = \frac{4\pi\mu}{\dfrac{1}{r} - \dfrac{1}{d}} \tag{6}$$

where d is twice the distance between the center of the sphere and the plane.

52. Discussion of Methods of Estimating Permeance of Non-Mathematical Fields

It is probably accurate to state, from a precise point of view, that none of the fields encountered in actual practice satisfy the ideal conditions as to mathematical symmetry necessary for the exact application of the formulas developed in the last article. In fact, many of the usual fields are so complicated that they cannot be readily solved by the method of field plotting. Thus in Figs. 5a and 5b are shown photographs of the leakage field surrounding a bipolar magnet. A casual inspection of this field will show its complexity. The field of Fig. 5a shows the field in a plane through the axes of the two pole cores and actually represents the shape of the flux lines in that plane. This particular plane is the only one that can be drawn containing a flat or two-dimensional field, and hence gives the simplest possible configuration. As we move out of this plane the field becomes three-dimensional; that is, the vector direction of the field at any point must be specified by three coordinates. Figure 5b shows a plane perpendicular to that of Fig. 5a and through the axis

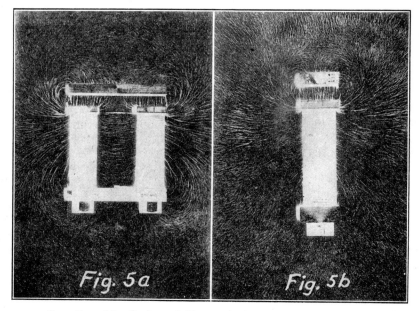

FIGS. 5a and b. Leakage field around a bipolar electromagnet.

of one pole core. The lines of flux shown are merely the projections on the plane of the real lines; that is, the plane does not contain one continuous line of flux because the field at all points in the plane is three-dimensional. In the region of the sharp corners of the square polar enlargements the field is still more complicated. It should be apparent that such a field cannot even be handled by the method of field mapping without overwhelming labor.

From these figures one would get the impression that in order to evaluate the permeance of this field it would be necessary to take into account the leakage flux relatively far away from the magnet. This, however, is not quite true because the field shown is merely qualitative and to no extent indicates the strength at any point. Actually the field at a short distance from the magnet is very weak and need not be taken into account.

In order to get a quantitative idea of the distribution of leakage flux around a magnet reference must be made to Fig. 6. Figure 6a shows the leakage field in a plane containing the axes of the pole cores for the magnet of Fig. 5 with the armature removed. As in Fig. 5a the field for this particular plane is two-dimensional. The line $C\text{-}C$ is the equipotential line midway between the poles. A plane through this line perpendicular to the plane shown will divide the field into two halves exactly alike. The lines A and B represent two other equipotential lines. In order to make the picture quantitative it is necessary to draw flux lines so spaced that the line integral of the flux density along any equipotential line between any two adjacent flux lines is the same. The heavily drawn flux lines, indicated by the numbers 0, 1, 2, 3, etc., satisfy this condition. Consider flux lines 2 and 3. The $\int Bdl$ between these two lines will be the same regardless of whether the integration is carried out along the equipotential line $C\text{-}C$, $A\text{-}A$, or $B\text{-}B$.

The method of determining the distance between these quantitative flux lines is as follows: The flux density in the direction of the flux lines is determined along an equipotential line such as $C\text{-}C$. This is done by reversing the field through a very small exploring coil which is moved from position to position along $C\text{-}C$. The values of flux density so obtained are plotted directly, using the line $C\text{-}C$ as an abcissa, as shown by the curve labeled $B_{C\text{-}C}$. The next step is to draw arbitrarily a flux line to be used as a datum when performing the integration. The flux line so chosen is that designated by numeral 0, drawn midway between the pole faces. Starting from this flux line one must graphically evaluate the area under the curve $B_{C\text{-}C}$ until it reaches a value which has been arbitrarily set as the value of the $\int Bdl$ between adjacent quantitative flux lines. The point on the line $C\text{-}C$ so determined will be one point on the quantitative flux line removed from the datum line 0 by one unit of $\int Bdl$. For convenience we will designate this point as $1C$, indicating the intersection between a flux line removed one unit from the datum and the equipotential line $C\text{-}C$. From this point the integration is started over again, and carried along the same equipotential line until the same arbitrary area is obtained, thus determining the point $2C$, etc. This same process is carried out on the other side of the datum line 0, determining the points $1'C$, $2'C$, etc. In a similar manner the corre-

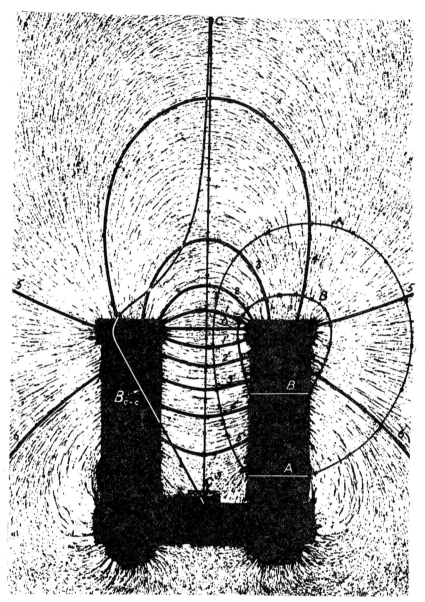

Fig. 6a. Distribution of flux in a plane through the axes of the pole cores of a bipolar magnet. The $\int B dl$ along any equipotential line, as A, between any adjacent heavily drawn flux lines is a constant.

sponding points 1A, 2A; 1B, 2B, etc., are determined for the equipotential lines A and B. By connecting the corresponding points of the

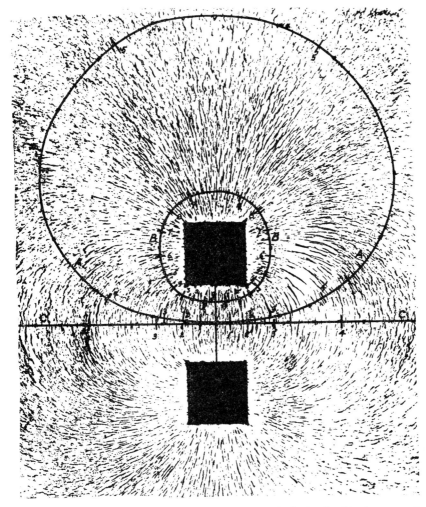

FIG. 6b. Distribution of flux in the plane of the pole faces of a bipolar magnet. The $\int Bdl$ along any equipotential line, as B, between any adjacent heavy marks is a constant. Note: The flux lines shown are merely a projection of the real flux lines, and B in the above integral is the component in the plane of the picture.

three equipotential lines—for example, 1C–1A–1B, with a line having the general shape of the field—the heavily drawn quantitative flux lines are obtained.

Two things are apparent from a careful study of Fig. 6a: (1) that the greater part of the leakage flux is concentrated in the area between the poles and pole faces—thus, by starting with flux line 5′ and including all the flux between it and line 4, over 80 per cent of the leakage flux will be accounted for; (2) that the flux density between the pole cores is a linear function of the distance from the yoke as assumed in Art. 43.

Figure 6b shows a projection of the field on the plane of the pole faces. Because this does not represent a two-dimensional field, it is very difficult to carry out quantitative measurements. However, it is possible to measure the component of the flux density in the plane of the picture. For this particular plane these values will not be much less than those measured parallel to the flux lines because the angle of the field with the plane is small. Some idea of this angle can be obtained from Fig. 6a. The black marks along the equipotential lines are so drawn that the line integral of the component of the flux density in the plane of the picture is a constant between adjacent marks. As in Fig. 6a the heavily drawn flux line midway between the pole faces is used as a datum line. Because the field is three-dimensional, the corresponding marks on the three equipotential lines cannot be joined to form flux lines.

A careful examination of this figure will show that the greater part of the leakage flux is concentrated in the volume immediately surrounding the pole faces. Thus, equipotential line B has 18 units of flux while A has only 10 units of flux, of which 8 units are close to the pole faces. It thus follows that when estimating leakage permeances of magnets it is not in general necessary to include the field at a great distance from the pole faces; it is, however, very important to include the leakage flux in the *entire space* immediately surrounding the pole faces. The exact path assumed for the leakage flux does not seem to be particularly important.

The heuristic method of "estimating the permeances of probable flux paths" is based entirely upon the statements contained in the last paragraph. These statements obviously do not admit of proof. The only attempt that will be made to justify the method, other than by reasoning from Fig. 6, will be to compare for a two-dimensional field the permeance as obtained by the generally accepted method of "field mapping" with that obtained by the method of "estimating the permeances of probable flux paths." In the ultimate analysis, any method of predetermination must be justified by the agreement between the predicted results and the actual results obtained by measurement on the piece of apparatus in question.

As a check between the method of "field mapping" and that of "estimating the permeances of probable flux paths" consider the following problem.

Problem. Calculate the permeance between two infinitely long rectangular plates [4] 1 in. thick, having their 1-in. edges opposite each other and separated at all points by 1 in. Do not consider the permeance of the path farther back than 31/32 in. from the parallel edges.

Solution. Method of "Field Mapping."[5] The method of field mapping consists essentially in sketching the distribution of flux lines and equipotential lines in such a manner that the total volume of the field is broken up into smaller unit volumes each having the same permeance. The permeance of the entire field is then obtained by adding the permeances of the unit volumes in series and parallel until the entire volume of the field is covered. The chief difficulty of the method lies in the fact that it is generally not easy to draw, with the proper distribution, accurate flux and equipotential lines, especially in a three-dimensional field surrounding a pole having sharp corners, like that of the bipolar electromagnet illustrated in Figs. 5 and 6. However, this method in the hands of a patient person having a sound knowledge of the fundamental laws [6] of fields of magnetic flux, dielectric flux, or heat flow will yield very accurate results. The purpose of this article is not to explain the method in any great detail, but to make use of it to substantiate the method of "estimating the permeance of probable flux paths."

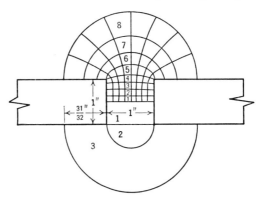

Fig. 7. Fields of illustrative problem. Upper half by "method of field mapping," lower half by "method of estimating the permeance of probable flux paths."

The upper half of Fig. 7 shows the field obtained for the problem by the method of field mapping. This field was drawn by trial and error

---

[4] In problems of this type where the field extends infinitely in one direction it is usual to carry out the computation for unit length along the infinite dimension. This will be understood throughout the solution of this problem.

[5] The method is equally applicable to any lamellar or solenoidal field, such as fields of dielectric flux and heat flow. For an excellent presentation of this method, and also a bibliography of the subject, see "Fundamentals of Electric Design," by A. D. Moore, McGraw-Hill Book Co.

[6] The fundamental law applying to these fields is that the flux lines and equipotential surfaces must form an orthogonal system; that is, they must be mutually perpendicular at all points in space.

until the law of footnote 6 had apparently been satisfied. The areas blocked out by the flux lines and equipotential surfaces are curvilinear squares, and hence for a unit axial length the permeance of the volume determined by each square will be the same. This unit permeance can be determined from the dimensions of the true squares along the center line of the plates; thus, the permeance per square is by permeance formula 1:

$$P_u = \frac{\mu S}{l} = \frac{3.2 \times \frac{1}{8}}{\frac{1}{8}} = 3.2 \text{ maxwells per ampere-turn}$$

As the entire path has 16 of these volumes in parallel and 8 in series, the permeance per inch of length of plate will be:

$$\frac{3.2 \times 16}{8} = 6.4 \text{ maxwells per ampere-turn}$$

Solution: Method of "Estimating the Permeances of Probable Flux Paths." The first step in the application of this method is to break the field up into flux paths which are of simple shape and still probable. The word "probable" as used here means that the assumed path is sufficiently similar to the true path that the permeance as determined from the mean length and area of the former will be somewhat close to the correct value for the latter. The "somewhat close" of the last sentence is purposely vague—some of the assumed paths will have higher estimated permeances than those of the true paths, others will have lower values; but, as the entire path is broken up into several assumed paths, the chances are that the errors in estimating will tend to balance out.

As the upper half of Fig. 7 shows the shape of the true field it is an easy matter to choose volumes of simple shapes which will closely simulate this field. That part of the field represented by flux paths 1, 2, 3, and part of 4 of the upper half of the figure will be replaced in the lower half by the volume 1, which is a right prism having a length along the flux lines of 1 in. and a cross section perpendicular to the flux lines of $\frac{1}{2}$ by 1 in. The permeance of this path will be, by formula 1:

$$P_1 = \frac{\mu S}{l} = \frac{3.2 \times \frac{1}{2}}{1} = 1.6 \text{ maxwells per ampere-turn}$$

Notice that this value is slightly on the high side: the permeance of the first four flux paths of the true field is:

$$\frac{3.2 \times 4}{8} = 1.6$$

while that estimated for volume 1 of the lower half, which does not include quite so much space, is the same.

The bulging part of flux path 4, all of 5, and part of 6 are simulated in the lower half by the semicircular cylindrical volume 2. The permeance of this volume is estimated by reducing it to a right prism having a length equal to the mean length of the flux lines and a cross section equal to the mean area of the flux path. The manner of obtaining the length and area for the prism is explained in Art. 53, Sec. 1, and for this particular prism will be 1.22 in. and 0.322 sq. in., respectively. Thus, the permeance of this volume will be:

$$P_2 = \frac{\mu S}{l} = \frac{3.2 \times 0.322}{1.22} = 0.84 \text{ maxwell per ampere-turn}$$

This value is slightly greater than the permeance of two flux paths in the upper half of the figure, and hence is just about right, as volume 2 is intended to replace about two flux paths.

The remaining part of the field in the upper half is taken care of by volume 3, which has the shape of a half annulus. The permeance of this volume, assuming circular flux lines, can be computed from formula 2, and will be:

$$P_3 = \frac{\mu l}{\pi} \log_\epsilon \frac{r_2}{r_1} = \frac{3.2}{\pi} \log_\epsilon \frac{94}{32} = 1.08 \text{ maxwells per ampere-turn}$$

which corresponds to about $2\frac{1}{2}$ flux paths of the upper half of the figure. As volume 3 is to simulate 8, 7, and part of 6, this value appears to be about correct.

The permeance of the total path will then be:

$$P_0 = 2(P_1 + P_2 + P_3) = (1.6 + 0.84 + 1.08) \times 2$$
$$= 7.04 \text{ maxwells per ampere-turn}$$

If we assume that the permeance obtained by the method of field mapping is correct, then that obtained by the method of estimating the permeances of probable flux paths is $\frac{7.04 - 6.4}{6.4} \times 100 = 10$ per cent high.

This error is only apparent, however. The answer 6.4 is not necessarily correct because the field of the upper half of Fig. 7 is obviously not absolutely perfect. Actually the permeance of this field can be evaluated by mathematical methods employing functions of a complex variable. In applying this method the actual field is broken down into the two simple paths 1 and 3 of the lower half of Fig. 7, and a correction term which takes care of the fact that these simple paths are actually distorted so that flux exists in path 2. When the thickness of the plates is zero the correction term is $0.221 \mu l$ (see equation 7), and when the thickness is large the correction term is $0.241 \mu l$. For ordinary cases $0.241 \mu l$ will apply. Therefore the only error in estimating the permeance for Fig. 7 is the use of $0.26 \mu l = (0.322/1.22)\mu l$ instead of $0.241 \mu l$, a total error of 1 per cent high. Consequently in this case the method of field mapping is 9 per cent low. This shows how difficult it is to draw the field with precision. Like most things of this sort the efficacy of a method can be determined only by experimental check. In the latter part of this chapter calculated and measured values of permeances of various-shaped air gaps are tabulated, enabling the reader to judge for himself. (See Art. 55.)

53. Special Formulas for Use in the Method of "Estimating the Permeances of Probable Flux Paths"

Particularly useful formulas can be derived by considering, in general terms, the calculation of the permeances of flux paths between surfaces which commonly arise in practice. The simplest surfaces to consider are those formed by two identical right prisms which are placed directly opposite each other with their corresponding edges parallel.

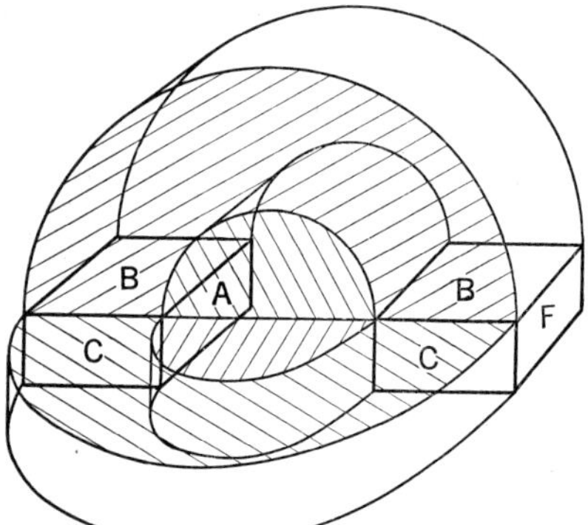

Figure 8 shows two such prisms. The corresponding faces of each prism are designated by the same letter. The back and lower faces which are not visible will be designated by E and D, respectively. Edges will be designated by the letters of the faces which intersect to form them; thus, edge AB is that formed by the intersection of faces A and B. Corners will be designated in a similar manner; thus, corner ABC is that formed by the intersection of faces A, B, and C.

FIG. 8. Diagram showing simple-shaped volumes used to replace field between identical right prisms.

The first step is to decide upon simple-shaped volumes to replace the actual field between the prisms. The method of doing this is to consider separately all the corresponding faces, edges, and corners of the two prisms. The path between faces A will be taken as a right prism. The

ART. 53] SPECIAL FORMULAS FOR ESTIMATING PERMEANCE 131

paths between edges AB, AC, AD, and AE will be taken as semicircular cylindrical volumes. Two of these volumes, for edges AB and AC, are shown. The paths between faces B, C, D, and E will be taken as half annuli. Those between faces B and C are shown. The volumes so far taken fill the entire space between the prisms except for the space between the annuli and semicircular cylinders joining the faces B, C, D, and E. This space can be taken care of by considering the paths between corresponding inner corners, such as ABC, to be quadrants of spheres, and the paths between corresponding edges, such as BC, to be quadrants of spherical shells. The only parts of the prisms now left out of consideration are the faces F, the edges FB, FC, FD, and FE, and the corners FBC, FCD, FDE, and FEB. These parts will be neglected for the reason that the field emanating from them is difficult to simulate by any volume of simple shape, and because the permeance associated with them is usually small compared to that of the rest of the field;[7] refer to Fig. 6.

We shall now derive simplified formulas for the permeances of each of the volumes described above:

1. **Semicircular Cylindrical Volumes (Edges AB, AC, AD, and AE of Fig. 8).** The fringing flux between edges of the magnet of Fig. 13, Chapter IV, follow this path.

The mean length of the flux path for this case can be considered to be equal to the length of a line drawn midway between the diameter and

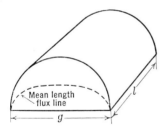 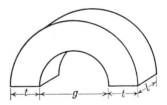

FIG. 9. Semicircular cylinder.   FIG. 10. Half annulus.

the semicircumference as shown in Fig. 9. This by graphical measurement will be found equal to $1.22g$. The mean area of the flux path may now be estimated by dividing the entire volume of the path by this mean

---

[7] This statement must be broadly interpreted; the closer together the prisms are the more accurate it becomes. In the usual calculation of the leakage flux of a magnet such surfaces, though they occur, are of no practical importance as the field is more closely confined. When two prisms, such as those of Fig. 8, are relatively far apart their permeance can be estimated by replacing them by equivalent spheres (spheres having the same surface area) and using equation 5.

length. Thus the mean area will be $\pi g^2/8 \times 1/1.22g = 0.322g$. The permeance can now be calculated by formula 1; it will be:

$$P = \frac{\mu S}{l} = \frac{0.322\mu g l}{1.22g} = 0.26\mu l \tag{7}$$

2. Half Annuli (Faces B, C, D, and E of Fig. 8). The fringing flux between the outer cylindrical faces of the magnet of Fig. 13, Chapter IV, follows this path.

Assume the mean length of the flux path to be $\pi \left(\dfrac{g+t}{2}\right)$ and the average area of the path to be tl, then:

$$P = \frac{\mu S}{l} = \frac{2\mu t l}{\pi(g+t)} = 0.64 \frac{\mu l}{\left(\dfrac{g}{t}+1\right)} \tag{8a}$$

When $g < 3t$, formula 2 should be used, and the permeance will be given by the formula:

$$P = \frac{\mu l}{\pi} \log_e \left(1 + \frac{2t}{g}\right) \tag{8b}$$

3. Spherical Quadrants (Corners ABC, ACD, ADE, and AEB of Fig. 8). The mean length of the flux path can be approximated by considering the mean flux line to be situated 0.65 of the way between the center of the sphere and the circumference. By graphical measurement this length is equal to 1.3g. The volume of the quadrant is $\frac{1}{3}\pi (g/2)^3$, hence the mean area of the flux path will be:

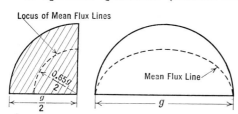

Fig. 11. Spherical quadrant.

$$\frac{\frac{1}{3}\pi \left(\dfrac{g}{2}\right)^3}{1.3g} = 0.1g^2$$

and the permeance by formula 1 will be:

$$P = \frac{\mu S}{l} = \frac{0.1g^2\mu}{1.3g} = 0.077\mu g \tag{9}$$

ART. 53] SPECIAL FORMULAS FOR ESTIMATING PERMEANCE

4. Quadrants of Spherical Shells (Edges BC, CD, DE, and EB of Fig. 8). Figure 12 shows a quadrant of a spherical shell. The mean length of the flux path will be:

$$\frac{\pi}{2}(t+g)$$

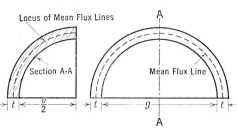

The maximum area of the flux path will be:

FIG. 12. Quadrant of spherical shell.

$$\frac{\pi\left(\frac{g}{2}+t\right)^2}{4} - \frac{\pi g^2}{16} = \frac{\pi}{4}(t^2+tg)$$

The average area of the path may be considered to be $\pi/8\,t(t+g)$, and the permeance by formula 1 will be:

$$P = \frac{\mu S}{l} = \frac{\mu \frac{\pi}{8} t(t+g)}{\frac{\pi}{2}(t+g)} = \frac{\mu t}{4} \tag{10}$$

5. Right Prism and Plane of Infinite Extent Parallel to End of Prism. In Fig. 13, A and B represent two faces of a prism, the end C of which is parallel to a plane D of infinite extent. The simplified flux paths

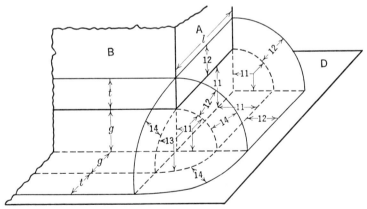

FIG. 13. Right prism and plane of infinite extent parallel to end of prism.

between the lower edge AC, the face A, the corner ABC, and the vertical edge AB to the plane D are shown. These paths are designated by the numbers 11, 12, 13, and 14, respectively.

Path 11. This path is identical with one-half of that of Fig. 9, and hence its permeance will be twice as great:

$$P = 0.52\mu l \qquad (11)$$

Path 12. This path is identical with one-half of that of Fig. 10, and hence its permeance will be twice as great:

$$P = \frac{1.28\mu l}{\frac{2g}{t} + 1} \qquad (12a)[8]$$

When $g < 3t$,

$$P = \frac{2\mu l}{\pi} \log_e\left(1 + \frac{t}{g}\right) \qquad (12b)[8]$$

Path 13. This path is identical with one-half of that of Fig. 11, and hence its permeance will be twice as great:

$$P = 0.308\mu g \qquad (13)[8]$$

Path 14. This path is identical with one-half of that of Fig. 12, and hence its permeance will be twice as great:

$$P = 0.5\mu t \qquad (14)$$

6. Circular Plunger Inserted in Concentric Tube.

The total flux path between the plunger and the surrounding tube can be broken up into five simple paths as shown in Fig. 14. Flux paths of this type occur in many types of plunger magnet; the magnet of Fig. 22, Chapter IV, illustrates a case where paths of the type of 15, 16, and 18 occur.

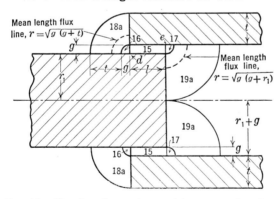

Fig. 14. Circular plunger inserted in concentric tube.

Path 15. If g is small compared to r_1 the permeance of this path can be computed

---

[8] g in these formulas is equal to one-half the g in formulas 8a, 8b, and 9.

Art. 53] SPECIAL FORMULAS FOR ESTIMATING PERMEANCE

directly by formula 1. The mean length of the flux path will be g, and its mean area will be

$$2\pi \left(r_1 + \frac{g}{2}\right) l$$

The permeance will therefore be:

$$P = \frac{\mu S}{l} = \frac{2\pi\mu \left(r_1 + \frac{g}{2}\right) l}{g} \qquad (15a)$$

If g is large compared to r_1 the permeance should be computed in the following manner:

$$dR = \int \frac{dl}{\mu S} = \int_{r=r_1}^{r=r_1+g} \frac{dr}{2\pi\mu l r}$$

$$R = \frac{1}{2\pi\mu l} \log_e \frac{r_1 + g}{r_1}$$

Therefore the permeance will be:

$$P = \frac{2\pi\mu l}{\log_e \frac{r_1 + g}{r_1}} \qquad (15b)$$

Practically, this formula need not be used in place of (15a), unless $g > r_1$.

Path 16. If g is small compared to r_1 the permeance of this path can be computed from formula 11 by letting $l = 2\pi(r_1 + g/2)$. The permeance will therefore be:

$$P = 3.3\mu \left(r_1 + \frac{g}{2}\right) \qquad (16a)$$

When g is large compared to r_1 the mean length of the flux path may be taken as $1.22g$, and the mean area of the path may be taken as:

$$S = \frac{\text{Vol.}}{l} = \frac{2\pi \left(r_1 + \frac{4g}{3\pi}\right) \frac{\pi g^2}{4}}{1.22g}$$

The permeance will therefore be:

$$P = \frac{\mu S}{l} = 3.3\mu(r_1 + 0.425g) \qquad (16b)$$

Path 17. If g is small compared to r_1 the permeance of this path will be the same as that for path 16:

$$P = 3.3\mu\left(r_1 + \frac{g}{2}\right) \qquad (17a)$$

When g is large compared to r_1 the mean length of the flux path may be taken as $1.22g$, and the mean area of the path may be taken as:

$$S = \frac{\text{Vol.}}{l} = \frac{2\pi\left(r_1 + g - \dfrac{4g}{3\pi}\right)\dfrac{\pi g^2}{4}}{1.22g}$$

The permeance will therefore be:

$$P = \frac{\mu S}{l} = 3.3\mu(r_1 + 0.575g) \qquad (17b)$$

Path 18. Assume that the flux lines of path 18 are concentric circles having their center at point d. As the ratio of t to g will usually be large for this type of path, formula 2 should be used as the basis for estimating the permeance; thus

$$P = \frac{2\mu l}{\pi}\log_e \frac{t + g}{g}$$

where l is the effective length of the path measured normal to the flux lines. l can be taken equal to the circumference of a circle the radius of which is equal to r_1 plus the radius of the mean flux line. The radius of the mean flux line will be equal to the geometric mean between g and $t + g$. Thus:

$$l = 2\pi\left(r_1 + \sqrt{g(t + g)}\right)$$

and the permeance will be:

$$P = 4\mu\left(r_1 + \sqrt{g(t + g)}\right)\log_e \frac{t + g}{g} \qquad (18a)$$

Path 19. The permeance of this path is calculated on the same basis as that of path 18. The radius of mean flux line will be $\sqrt{g(r_1 + g)}$, and l will equal

$$l = 2\pi\left(r_1 + g - \sqrt{g(r_1 + g)}\right)$$

and the permeance of the path will be:

$$P = 4\mu \left(r_1 + g - \sqrt{g(r_1 + g)} \right) \log_e \frac{r_1 + g}{g} \tag{19a}$$

7. Circular Plunger Close to End of Concentric Tube. The total flux path between the plunger and the concentric tube can be broken up into three parts as shown in Fig. 15. Flux paths of this type occur in cylindrical-, stepped-cylindrical-, and truncated-conical-faced plunger magnets.

Path 18. This path is made up of concentric circular flux lines having their center at point d, and it is identical with path 18 of Fig. 14 except

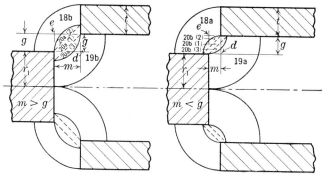

Fig. 15. Circular plunger close to end of concentric tube.

that the inner radius is m instead of g. The permeance of this path, when $m > g$, will be given by the formula

$$P = 4\mu \left(r_1 + \sqrt{m(t + g)} \right) \log_e \frac{t + g}{m} \tag{18b}$$

When $m < g$, use formula 18a.

Path 19. This path is made up of concentric circular flux lines having their center at point e, and it is identical with path 19 of Fig. 15 except that the radius of the shortest line is m instead of g. The permeance of this path, when $m > g$, will be given by the formula:

$$P = 4\mu \left(r_1 + g - \sqrt{m(r_1 + g)} \right) \log_e \frac{r_1 + g}{m} \tag{19b}$$

When $m < g$, use formula 19a.

Path 20a ($m > g$; $m < r_1$). For convenience in deriving a formula this path can be broken into three parts as shown by the dashed lines.

Considering now the center part (1), assume the mean length of the flux path to be m and its mean area to be $(\sqrt{2}/2)(m - g)2\pi(r_1 + g/2)$; then

$$P_{20a}(1) = 4.45\mu \left(\frac{m - g}{m}\right)\left(r_1 + \frac{g}{2}\right) \qquad [20a(1)]$$

assuming g small compared to r_1. The mean area of the outer part can be taken as $(\frac{1}{2})m(1 - \sqrt{2}/2)2\pi(r_1 + m/2)$, and taking its mean length as m, then

$$P_{20a}(2) = 0.92\mu \left(r_1 + \frac{m}{2}\right) \qquad [20a(2)]$$

Likewise the permeance of the inner path will be

$$P_{20a}(3) = 0.92\mu \left(r_1 + g - \frac{m}{2}\right) \qquad [20a(3)]$$

Path 20b. When $m < g$; $m > 0$, the above formulas will be modified as follows:

$$P_{20b}(1) = 4.45\mu \left(\frac{g - m}{g}\right)\left(r_1 + \frac{g}{2}\right) \qquad [20b(1)]$$

$$P_{20b}(2) = P_{20b}(3) = 0.92\mu \left(r_1 + \frac{g}{2}\right) \qquad [20b(2) \text{ and } (3)]$$

8. Cylindrical Plunger and Surrounding Cylindrical Shell with Their Ends in the Same Plane.

The flux path for this case can be broken into two parts as shown in Fig. 16. Figure 22 of Chapter IV illustrates a case where flux paths of this type occur.

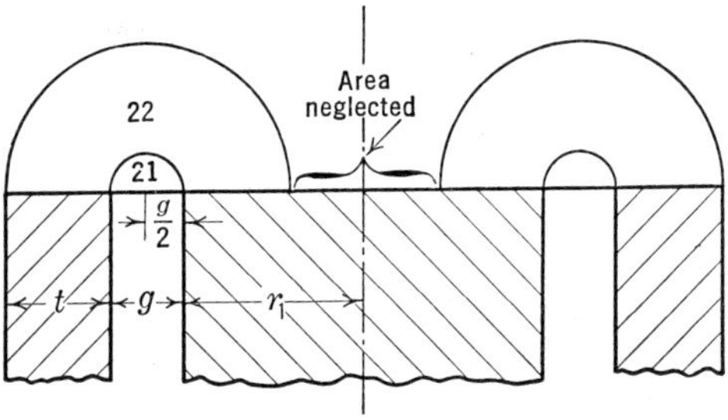

FIG. 16. Cylindrical plunger and concentric shell.

Path 21. The permeance of this path can be computed by formula 7, by letting l equal $2\pi(r_1 + g/2)$. The permeance will then be given by the formula:

$$P = 1.63\mu \left(r_1 + \frac{g}{2}\right) \qquad (21)$$

Path 22. The flux lines in this path are assumed to be concentric circles having their center midway between the plunger and shell.

When t is less than r_1 a small area at the center of the plunger will be neglected, and when t is greater than r_1 a small area at the outside of the shell will be neglected. In either case the omission will produce a negligible error, because the field will be very weak over the areas neglected.

The permeance can be computed by means of formula (8b) by letting

$$l = 2\pi\left(r_1 + \frac{g}{2}\right)$$

The permeance will then be given by the formula:

$$P = 2\mu\left(r_1 + \frac{g}{2}\right)\log_\epsilon\left(1 + \frac{2t}{g}\right) \qquad (22a)$$

when $t < r_1$. When $t > r_1$, the formula must be changed to:

$$P = 2\mu\left(r_1 + \frac{g}{2}\right)\log_\epsilon\left(1 + \frac{2r_1}{g}\right) \qquad (22b)$$

9. Other Special Permeance Formulas. Other special permeance formulas which are of value in evaluating the working air-gap permeances of taper plunger and conical-faced plunger magnets are as follows:

> Coaxial truncated conical surfaces. See equation 11, Chapter VIII.
> Coaxial full conical surfaces. See equation 12, Chapter VIII.
> Coaxial cylindrical and conical surfaces. See equation 13, Chapter VIII.

These formulas are of particular interest when it is desired to plot the working air-gap permeance as a function of stroke in order to evaluate the force-stroke curve, by the method discussed in Art. 77, Chapter VIII.

54. Application of the Method of "Estimating the Permeance of Probable Flux Paths" to the Determination of the Leakage Coefficient of Some Common Types of Electromagnets

1. **The Flat-Faced Cylindrical Plunger Magnet.** Figure 17 is a section through the axis of the magnet showing the various simplified flux paths. These paths are numbered to correspond with the formulas which will be used to evaluate their permeances. The useful flux follows path 1; the fringing flux paths 7 and 8b; the leakage flux path 15a.

The leakage coefficient is obtained by finding the ratio of the permeance of all the air paths to the permeance of the useful path; thus,

$$\nu = \frac{P_1 + P_7 + P_{8b} + P_{15a}}{P_1}$$

Permeance Path 1.

$$P_1 = \frac{\mu S}{l} = \frac{\mu \pi r_1^2}{g}$$

Permeance Path 7.

$$P = 0.26 \mu l$$

where $l = 2\pi(r_1 + g/4)$.
Therefore, $P_7 = 1.63\mu(r_1 + g/4)$.

Permeance Path 8b.

$$P = \frac{\mu l}{\pi} \log_\epsilon \left(1 + \frac{2t}{g}\right)$$

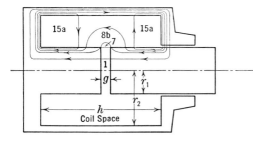

FIG. 17. Section through flat-faced cylindrical plunger magnet showing leakage paths.

t in this formula is determined as follows: The longest flux line in this path must equal twice the radial thickness of the winding space, for if it exceeds this length it will be easier for the flux to go from the plunger to the shell and back to the plunger again, that is, to follow path 15. Thus:

$$\pi\left(t + \frac{g}{2}\right) = 2(r_2 - r_1)$$

and

$$\left(t + \frac{g}{2}\right) = \frac{2(r_2 - r_1)}{\pi}$$

l may be taken equal to $2\pi r_1$. Therefore, $P_{8b} = 2\mu r_1 \log_\epsilon \dfrac{4(r_2 - r_1)}{\pi g}$.

Permeance Path 15a. The total permeance of path 15a for one side only will be:

$$P = \frac{\pi \mu (r_1 + r_2)}{r_2 - r_1} l$$

where l will be $\dfrac{h}{2} - \dfrac{2(r_2 - r_1)}{\pi}$

$$P = \frac{\pi\mu(r_1 + r_2)h}{2(r_2 - r_1)} - 2\mu(r_1 + r_2)$$

The effective value of this permeance with respect to the leakage flux it produces, however, is much smaller because the magnetomotive force across one end of the path is zero, while at the other end only

$$\frac{\dfrac{h}{2} - \dfrac{2(r_2 - r_1)}{\pi}}{h}$$

of the total magnetomotive force of the coil is available. Hence the effectiveness of this path compared to the others will be:

$$\frac{1}{4} - \frac{r_2 - r_1}{\pi h}$$

and the effective permeance of the path will be:

$$P = \left[\frac{\pi\mu(r_1 + r_2)h}{2(r_2 - r_1)} - 2\mu(r_1 + r_2)\right]\left(\frac{1}{4} - \frac{r_2 - r_1}{\pi h}\right)$$

Therefore,

$$P_{15a} = \mu\left[\frac{\pi h}{8} \cdot \frac{r_2 + r_1}{r_2 - r_1} - (r_2 + r_1) + \frac{2(r_2^2 - r_1^2)}{\pi h}\right]$$

The total permeance of all flux paths will be:

$$P_0 = P_1 + P_7 + P_{8b} + P_{15a} = \mu\left[\frac{\pi r_1^2}{g} + 1.63\left(r_1 + \frac{g}{4}\right) + 2r_1 \log_\epsilon \frac{4(r_2 - r_1)}{\pi g} + \frac{\pi h}{8}\frac{(r_2 + r_1)}{(r_2 - r_1)} - (r_2 + r_1) + \frac{2(r_2^2 - r_1^2)}{\pi h}\right]$$

Dividing this by P_1 and simplifying, we will obtain the following leakage coefficient:

$$\nu = 1 + \frac{g}{r_1}\left[0.52 + 0.13\frac{g}{r_1} + \frac{r_2 + r_1}{\pi r_1}\left(\frac{\pi h}{8(r_2 - r_1)} + \frac{2(r_2 - r_1)}{\pi h} - 1\right) + \frac{2}{\pi}\log_\epsilon \frac{4(r_2 - r_1)}{\pi g}\right]$$

By noting that $2/\pi \log_\epsilon 4/\pi = 0.15$, and changing the natural logarithm to the base 10, we will obtain:

$$\nu = 1 + \frac{g}{r_1}\left[0.67 + 0.13\frac{g}{r_1} + \frac{r_2 + r_1}{\pi r_1}\left(\frac{\pi h}{8(r_2 - r_1)} + \frac{2(r_2 - r_1)}{\pi h} - 1\right)\right.$$
$$\left. + 1.465 \log_{10}\frac{r_2 - r_1}{g}\right] \quad (23)$$

This formula does not apply if $g > \dfrac{4(r_2 - r_1)}{\pi}$.

2. Bipolar Magnet with Square Polar Enlargements. Figure 18a shows a bipolar magnet. The various simplified flux paths are numbered to correspond with the formulas which will be used to evaluate their permeances. The useful flux follows path 1; the fringing fluxes paths 7, 8b, 9, 10, 11, and 12b; the leakage flux between poles, path 3b. It is convenient for a magnet of this type to separate the leakage coefficient into two parts: ν_a, the ratio of the flux leaving the polar enlargements and passing through the armature to the useful flux in the working gap; and ν_L, the ratio of the leakage flux between pole cores to the useful flux in the working gap. The leakage coefficient for the entire magnet is the ratio of the flux through the yoke to the useful flux in the working gap; it will be

$$\nu_y = \nu_a + \nu_L$$

Leakage Coefficient ν_a.

$$\nu_a = \frac{P_1 + P_7 + P_{8b} + P_9 + P_{10} + P_{11} + P_{12b}}{P_1}$$

Permeance [9] *of Path 7 (Edges AB, AC, and AD).*

$$P = 0.26\mu l; \; l = 3f, \text{ hence } P_7 = 0.78\mu f$$

Permeance of Path 8b (Faces B, C, and D).

$$P = \frac{\mu l}{\pi}\log_\epsilon\left(1 + \frac{2t}{g}\right); \; l = 3f, \text{ hence } P_{8b} = 0.956\mu f \log_\epsilon\left(1 + \frac{2t}{g}\right)$$

Permeance of Path 9 (Corners ABD, ACD, ABE, and ACE).

$$P = 0.077\mu g$$

---

[9] To avoid unnecessary complication the permeance calculations are carried out for only one-half of the magnetic circuit. This is possible because a center line parallel to the pole core axes will divide the magnet into two exactly similar parts.

This formula does not apply to corners ABE and ACE, which are not corner-to-corner permeances; however, we shall assume that the permeance between a corner like ABE and an edge like AB is equal to twice that of the corresponding corner-to-corner permeance, hence:

$$P_9 = 6 \times 0.077 \mu g = 0.46 \mu g$$

Permeance of Path 10 (Edges BD, CD, BE, and CE). Here again we shall assume that the permeance of an edge like BE to a plane like B is twice that of the corresponding edge-to-edge permeance, thus:

$$P_{10} = 6 \times 0.25 \mu t = 1.50 \mu t$$

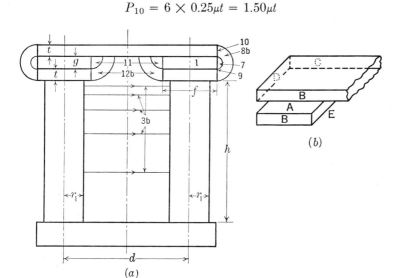

Fig. 18. Section through a horseshoe magnet showing leakage paths.

Permeance of Path 11 (Edge AE to face A).

$$P = 0.52 \mu l; \quad l = f; \quad \text{hence } P_{11} = 0.52 \mu f$$

Permeance of Path 12b (Face E to face A).

$$P = \frac{2\mu l}{\pi} \log_\epsilon \left(1 + \frac{t}{g}\right); \quad l = f; \quad \text{hence } P_{12b} = 0.637 \mu f \log_\epsilon \left(1 + \frac{t}{g}\right)$$

Permeance of Path 1 (Face A). $P = \mu S/l$; $S = f^2$; and $l = g$. Hence, $P_1 = \mu f^2/g$.

The leakage coefficient ν_a can now be obtained by summing these various permeances, dividing by P_1, and then simplifying:

$$\nu_a = 1 + \frac{g}{f^2}\left[1.5t + 0.46g + 1.3f + 1.47f \log_{10}\left(1 + \frac{t}{g}\right)\right.$$

$$\left. + 2.2f \log_{10}\left(1 + \frac{2t}{g}\right)\right] \quad (24)$$

If the thickness of the armature is somewhat greater than that of the polar enlargement, their average thickness should be used for t. This formula should not be used if the distance between the polar enlargements is less than $\pi(g + t)$.

Leakage Coefficient ν_L.[9]

$$\nu_L = \frac{P_4}{P_1}$$

Permeance of Path 4 (Pole Core to Plane, Midway between Poles).

$$P = \frac{2\mu\pi l}{\log_\epsilon d/r_1}$$

l will equal the length of the pole core h, plus the distance between the polar enlargement and the beginning of the coil winding at that end. Let this distance $= h'$. This permeance is only one-half effective, hence:

$$P_4 = \frac{\mu\pi h'}{\log_\epsilon d/r_1}$$

and

$$\nu_L = \frac{\pi g h'}{f^2 \log_\epsilon d/r_1} \quad (25)$$

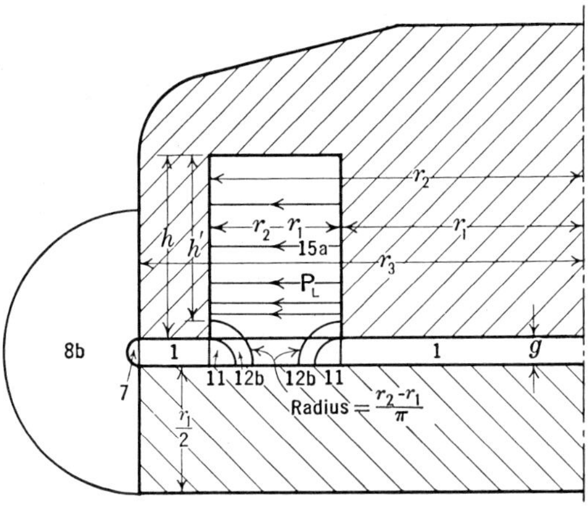

Fig. 19. Half-section through flat-faced cylindrical lifting magnet showing leakage paths.

3. Flat-Faced Lifting Magnet. Figure 19 shows one-half of a section of a flat-faced cylindrical lifting magnet. The iron parts of the magnet are so designed that the cross section of the inner pole core equals that of the outer, and the armature thickness such that cross-sectional area for flux at the radius r_1 is equal to that of the pole cores. The various simplified flux paths are numbered to correspond with the formulas which will be used to evaluate their permeances. The useful flux follows path 1; the fringing fluxes, paths 7, 8b, 11, and 12b; the leakage flux between poles, path 15a. The equivalent circuit for this magnet is shown in Fig. 14, Chapter IV.

Fringing Permeances Outer Pole $(P_f)_o$.

Path 7. $P = 0.26\mu l$, where $l = 2\pi r_3$. Therefore,

$$P_7 = 1.63\mu r_3$$

Path 8b. $P = \mu l/\pi \log_\epsilon (1 + 2t/g)$, where $l = 2\pi r_3$, and $t = r_1/2$. Therefore

$$P_{8b} = 2\mu r_3 \log_\epsilon \left(1 + \frac{r_1}{g}\right)$$

Path 11. $P = 0.52\mu l$, where $l = 2\pi r_2$. Therefore,

$$P_{11} = 3.26\mu r_2$$

Path 12b. $P = 2\mu l/\pi \log_\epsilon (1 + t/g)$, where $l = 2\pi r_2$, and $t = (r_2 - r_1)/\pi - g$. Therefore,

$$P_{12b} = 4\mu r_2 \log_\epsilon \left(\frac{r_2 - r_1}{\pi g}\right)$$

Fringing Permeances Inner Pole $(P_f)_i$.

Path 12b. $P = 2\mu l/\pi \log_\epsilon (1 + t/g)$, where $l = 2\pi r_1$, and $t = (r_2 - r_1)/\pi - g$. Therefore:

$$P_{12b} = 4\mu r_1 \log_\epsilon \left(\frac{r_2 - r_1}{\pi g}\right)$$

Path 11. $P = 0.52\mu l$, where $l = 2\pi r_1$. Therefore,

$$P_{11} = 3.26\mu r_1$$

Useful Permeance, Working Gap.

$$P_1 \text{ (outer)} = P_1 \text{ (inner)} = \frac{\mu S}{l} = \frac{\pi \mu r_1^2}{g}$$

Total Useful Permeance of Magnet (P_u).

$$P_u = (P_1) \text{ outer} + (P_1) \text{ inner, in series} = \frac{\pi \mu r_1^2}{2g}$$

Leakage Permeance between Pole Cores (P_L).

$$P_{15a} = \frac{\pi(r_2 + r_1)h'}{2(r_2 - r_1)} \mu$$

where the factor 2 in the denominator takes care of the fact that this permeance is only one-half effective, and where h', the effective axial length of this leakage path, is equal to

$$h - \left(\frac{r_2 - r_1}{\pi} - g\right)$$

Therefore:

$$P_L = \left[1.57h \frac{(r_2 + r_1)}{(r_2 - r_1)} - \frac{(r_2 + r_1)}{2}\left(1 - \frac{\pi g}{r_2 - r_1}\right)\right] \mu$$

Total Permeance between Pole Cores by Way of Armature (P_a).

$$P_a = \frac{1}{\dfrac{1}{P_1 + (P_f)_i} + \dfrac{1}{P_1 + (P_f)_o}}$$

Leakage Coefficient

$$\nu = \frac{P_a + P_L}{P_u} = \frac{2g}{\mu \pi r_1^2}(P_a + P_L)$$

In a magnet of this type there are certain proportions which will produce an economical design. It is possible by deciding upon such a set of proportions to obtain a leakage formula which will be very simple and highly accurate if the proportions are adhered to. Thus a set of proportions which produce a good design in this type of magnet are as follows:

$$\frac{h}{r_2 - r_1} = 2.5; \quad \frac{r_2}{r_1} = 1.7$$

then

$$\frac{r_3}{r_1} = \sqrt{1^2 + 1.7^2} = 2$$

Using these values and performing the operations indicated we will obtain:

$$\nu = 1 + \frac{g}{r_1}\left(\frac{35.3 + 228g/r_1}{4 + 35.3g/r_1} + 5.1\right) \tag{26}$$

If for the value of g/r_1 in the bracket some mean value is substituted, a very simple formula can be obtained. This procedure will not lead to great error because the fraction in the bracket does not vary fast with g/r_1. Thus letting g/r_1 equal 0.02, which is a representative value for this type of magnet,

$$\nu = 1 + 14 \frac{g}{r_1} \tag{27}$$

The preceding three cases illustrate how the method of "estimating the permeance of probable flux paths" can be applied to derive general leakage formulas for special types of magnets. Such a derivation of a special formula is warranted only when the particular type of magnet has to be designed frequently, or when it is necessary to determine the general effect of proportions on the overall magnetic efficiency.

55. Experimental Check of Leakage Formulas

1. **Flat-Faced Cylindrical Plunger Magnet.** The leakage coefficient for the magnet of Fig. 1a, Chapter III, was measured by finding the ratio of the flux through an exploring coil surrounding the yoke end of the stationary plunger to that through an exploring coil surrounding the air gap. The flux measured by the latter coil was corrected to take into account the slightly larger diameter of the exploring coil as compared to the plunger diameter. The exciting magnetomotive force was low enough to produce no appreciable saturation (0.1 ampere through 2,550 turns). The computed values of the leakage coefficient were obtained by using formula 23.

Length air gap, inches	$\frac{1}{64}$	$\frac{1}{32}$	$\frac{1}{16}$	$\frac{1}{8}$	$\frac{1}{4}$	$\frac{1}{2}$	1
Leakage coefficient, computed	1.09	1.17	1.29	1.49	1.83	2.36	3.20
Leakage coefficient, measured	1.13	1.20	1.31	1.46	1.85	2.41	3.03

2. **Bipolar Magnet with Square Polar Enlargements.** The leakage coefficients for the bipolar magnet of problem 4, at the end of the chapter, were measured by finding the flux through the armature, yoke, and the useful gap. The flux through the armature and yoke was determined by an exploring coil wound around the center of each of these parts. The flux density in the useful gap was measured by means of an exploring coil wound on a Bakelite washer $\frac{15}{16}$ in. in diameter and 0.066 in. thick, and the flux of the useful gap was obtained by multiplying this density by the area of the polar enlargement. The exciting magnetomotive force was low enough to produce no appreciable satura-

tion in the iron parts (0.2 ampere through 2,940 turns on each coil). The computed values of the leakage coefficients were obtained by using formulas 24 and 25.

It will be noticed by comparing the computed coefficients with the measured ones that the computed values of ν_a are high while those for ν_L are low. In order to check the nature of this error the flux actually leaving the polar enlargements was measured by winding a test coil

Length Air Gap, Inches	Leakage Coefficients						
	Measured				Computed		
	ν_{pe}	ν_a	ν_L	ν_y	ν_a	ν_L	ν_y
0.066	1.41	1.30	0.37	1.67	1.28	0.25	1.53
0.129	1.68	1.45	0.70	2.15	1.46	0.49	1.95
0.193	1.91	1.55	1.05	2.60	1.60	0.73	2.33
0.258	2.13	1.62	1.42	3.04	1.76	1.01	2.77

around the pole core directly under the polar enlargements. The tabulated leakage coefficient ν_{pe} was obtained by dividing this flux by that passing through the useful gap. If ν_{pe} is subtracted from ν_y a new value of ν_L will be obtained. This value of ν_L will be a true measure of the leakage flux between the cylindrical pole cores and, as can be seen by performing the subtraction indicated, will be very close to the computed value. The difference between the ν_L based on ν_a and that based on ν_{pe} evidently represents leakage flux passing directly between polar enlargements. The possibility of such a flux, directly between polar enlargements, was not considered in deriving the leakage formula and hence explains why the computed value of ν_a is higher than the measured one.[10]

3. **Flat-Faced Lifting Magnet.** The leakage coefficient for the circular lifting magnet of Fig. 13, Chapter IV, was measured by finding the ratio of the permeance of the entire magnet[11] $(P_a + P_L)$ to that of two useful gaps in series (P_u). The permeance of the entire magnet was determined by measuring the flux through the yoke, and dividing this by the exciting magnetomotive force. This flux was measured by means of an exploring coil wound around the inner pole core between the

---

[10] For further check data on leakage formulas applying to this type of magnet see Problem 4.

[11] See the equation for the leakage coefficient, Art. 54, Sec. 3.

ART. 55] EXPERIMENTAL CHECK OF LEAKAGE FORMULAS

exciting winding and the yoke. The exciting magnetomotive force was low enough (4 amperes through 180 turns) to produce no appreciable saturation in the iron parts. The permeance of the two useful gaps in series was determined by computation. The computed values of the leakage coefficient were obtained by the method outlined in Art. 54. The simplified formula 27 cannot be used because the dimension ratios of the magnet are not the same as those used in deriving (27).

Note: For this particular type of magnet it is impossible to measure the leakage coefficient directly as was done for the other types.

Length air gap, inches	$\frac{1}{64}$	$\frac{1}{32}$	$\frac{1}{16}$	$\frac{1}{8}$
Leakage coefficient, computed	1.13	1.20	1.32	2.04
Leakage coefficient, measured	1.11	1.24	1.41	1.67

PROBLEMS

1. Derive permeance equation 5.
2. Compute, by the method of "field mapping" and also by the method of "estimating the permeance of probable flux paths," the permeance per inch of axial length between the infinitely long parallel plates, a section of which is shown in Fig. 20.

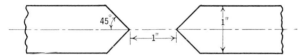

Fig. 20. Section through infinitely long parallel plates for Problem 2.

Consider the flux path to extend back of the upper and lower edges only 1 in. (Refer to footnote 5.)

3. A bipolar magnet has the following dimensions:

   Yoke: $1\frac{1}{2}$ by $2\frac{1}{4}$ in. wide by $10\frac{1}{2}$ in. long.
   Pole cores: 2 in. diameter by 7 in. long.
   Polar enlargements: $2\frac{1}{2}$ in. diameter by $\frac{3}{8}$ in. long.
   Armature: $\frac{7}{8}$ by $2\frac{3}{4}$ in. wide by $10\frac{1}{4}$ in. long.
   Distance between centers of cores: 7 in.
   Length of air gap under armature: $\frac{1}{4}$ in.

Compute the ratio of the flux through the armature to the useful flux in the circular working gap, and also the ratio of the leakage flux between pole cores to the useful flux.

4. The bipolar magnet of Fig. 5 or 18, Chapter V, has the following data:

   Material: Swedish charcoal iron unannealed after machining.
   Yoke: $\frac{1}{2}$ by $1\frac{1}{4}$ in. wide by $3\frac{15}{16}$ in. long.
   Pole cores: $\frac{7}{8}$ in. diameter by 3 in. long.
   Polar enlargements: $1\frac{3}{16}$ in. by $1\frac{3}{16}$ in. by $\frac{1}{4}$ in. long.
   Distance between centers of pole cores: $2\frac{3}{4}$ in.

Compute the flux through (a) the yoke, (b) the section of the pole core next to the pole face, if the magnet is excited by a coil of 2,940 turns on each leg, the coils being connected cumulatively in series and carrying 0.4 ampere. The armature is removed to a great distance.

Note: The pole faces can be treated like the prisms of Fig. 8, neglecting, of course, the faces in contact with the pole cores. The effective length of the pole cores, as regards leakage between pole cores, should then be taken equal to their actual length less the radius of the flux path 7 between the inner lower parallel edges of the pole faces.

(c) Compute the minimum percentage loss of the potential work ability of this magnet due to the initial armature position, if the armature is $\frac{3}{8}$ by $1\frac{3}{16}$ in. in cross section and the material of the magnet is Swedish charcoal iron unannealed after machining.

Hint: The loss in potential work ability of the magnet will be a minimum when the armature is removed to infinity; the maximum flux of the magnet can be computed by assuming a zero-length air gap between the armature and the pole faces.

Ans.: (b) 22.8 kmax. (a) 37.2 kmax., by computation.
(b) 23.3 kmax. (a) 37.7 kmax., by experiment.

5. Compute the permeances P_u, P_f, and P_L for the magnet of Fig. 22, Chapter IV.

6. Compute the leakage coefficient for the magnet of Fig. 1a, Chapter III, when the working air gap has a length of $\frac{1}{8}$ in. See Art. 55, Sec. 1, for the answer.

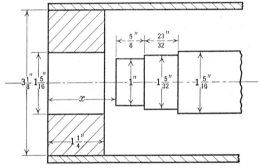

Fig. 21. Working gap of stepped cylindrical plunger magnet.

7. Compute the leakage coefficient for the magnet of Fig. 13, Chapter IV, at an air-gap length of $\frac{1}{16}$ in. See Art. 55, Sec. 3, for the answer.

8. Compute the leakage coefficients ν_a and ν_L for the bipolar magnet described in Problem 4 if the length of the air gap is $\frac{1}{4}$ in. See Art. 55, Sec. 2, for the answer.

9. Plot a curve of the permeance between the cylindrical plunger and plate of Fig. 21, as x varies from $1\frac{1}{2}$ in. to zero. See Art. 77, Chapter VIII, for the answer.

Note: This calculation is very important in the design of cylindrical-faced plunger magnets.

10. Compute the air-path permeances as illustrated in the equivalent circuit of Fig. 14, Chapter IV, for the lifting magnet of Fig. 13. Assume a working gap length of 0.0323 in. Check your answers with those given in connection with the above figures.

CHAPTER VI

COILS[1]

56. General—Types of Coils

Coils are used in practically all electrical machinery, usually forming the exciting source for magnetic fields, and also the current-carrying circuit reacting with magnetic fields.

A coil usually consists of turns of wire wound like a helical thread to form a layer, there being one or more layers to the coil. Insulation, such as paper, is sometimes placed between the layers. The cross section of the coil wall is generally rectangular, and the cross section of wire is almost always round except in coils made of heavy wire where a square, or a rectangular section with rounded corners, is used. In other respects coils differ very considerably. It is convenient to classify them on the basis of the type of construction; thus we have:

Paper-section coils.
Cotton-interwoven coils.
Bobbin-wound coils.
Form-wound coils.
Ribbon-wound coils.
Strap-wound coils.

Paper-Section Coils. The paper-section type of coil is the most widely used of any form of electrical winding for small apparatus, largely because it combines qualities of high insulation resistance and electrical reliability with the lowest possible cost.

These coils are wound on automatic or semi-automatic machines, in "stick" form, several at a time. A supporting tube of paper or special board, called the core tube, is placed on the winding machine mandrel, and the wires for each coil on the stick are shellacked to the core tube. The mandrel is rotated at high speed, and the wire is guided to each coil on the stick by as many sheaves mounted on a traverse rod as there are coils. The length of the traverse is equal to the winding length of one coil. At the end of each layer of wire a sheet of paper is wound into the coil either by hand or by automatic means. In this manner the coils are

---

[1] For a very complete treatment of this subject see "Coils and Magnet Wire" by C. R. Underhill, McGraw-Hill Book Co.

built up of alternate layers of wire and paper until the desired number of turns is reached. The paper serves the triple purpose of making the coils self-supporting, of making the layers even, and of providing insulation between layers. The coils are spaced on the mandrel so that there is a dead paper space between the coils. This unwound paper space, called the paper margin, is necessary so that the individual coils may be cut apart. Figure 1 shows a stick of paper-section coils. The cutting is usually done with a gang of knives so spaced as to cut through the center of each paper margin. After cutting, the beginning and end turns of wire are pulled out so that the leads may be attached. The leads in fine

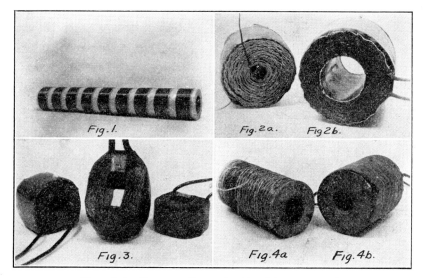

FIGS. 1-4. Paper-section and cotton-interwoven coils.

wire coils are usually of a material different from that of the coil wire, and are firmly anchored to the outside of the coil. The outside of the coil is covered with a paper cover, and the coils are finished by impregnating with a suitable compound, or by dipping in a varnish followed by baking.

Figure 2b shows a medium-sized (this is the coil of the plunger magnet of Fig. 1a, Chapter III) coil, the paper ends of which have been sealed air-tight by a special process. Figure 2a shows a similar coil of heavy wire (No. 17) before the paper ends have been treated. Figure 3 shows an assortment of small varnish-impregnated paper-section coils.

The chief advantages of the paper-section coil are the ease with which it can be produced in quantities, its good appearance due to the smooth and even layers of wire, its high dielectric strength obtained by means of the interlayer paper, and finally its mechanical ruggedness. The disadvantage of the paper-wound coil is its low space factor; however, for fine wire coils (Nos. 44 to 28) the space factor for paper-wound enameled wire coils is better than for single silk-covered wire. The space factor is often decreased by winding an excessive amount of paper between layers; thus it sometimes is the practice to cut the interlayer paper into groups of lengths (for semi-automatic winding) so that the first layer of any group will have too much overlap and the last layer just enough overlap. This unnecessarily increases the average thickness of paper between layers and also increases the length of wire necessary for a given number of turns.

Cotton-Interwoven Coils. Cotton coils are wound on special machines which wind the wire (usually plain enameled) in even layers and at the same time interweave a cross-wound cotton yarn between the wires and wire layers. The cotton yarn not only separates the layers of wire but also forms the end of the coil, there being from $\frac{1}{16}$ to $\frac{3}{16}$ in. of cotton at each end corresponding to the paper margin of paper-section coils. A core tube of paper or fiber combined with the interwoven cotton make the coil entirely self-supporting. These coils have the same general properties as paper-section coils; they have a slightly better space factor, a better appearance, and are more readily impregnated owing to the wick action of the cotton. They are particularly recommended for use where the climatic or atmospheric conditions are severe. The cotton coils for average sizes are more expensive than the paper-wound coils, but are generally less expensive than bobbin-wound coils of equivalent characteristics. Figure 4a shows a cotton-interwoven coil with a few layers removed, and Fig. 4b shows a finished coil.

Bobbin-Wound Coils. This is the oldest type of coil; it consists of wire wound into a specially prepared bobbin. The bobbin may be made of metal or of some insulating material such as Bakelite or fiber. When using a heavy wire with cotton insulation it is not customary to place any insulating material between layers unless the voltage is quite high; however, with the finer wires paper between layers is necessary if the layers are to be smooth and if wires having relatively large voltages between them are to be separated. These coils are generally the most expensive type to build.

Bobbin-wound coils can be divided into three kinds:
 a. Those wound on molded spools.
 b. Those wound on a built-up spool consisting of an iron core (part of the magnetic circuit) with pressed-on fiber end flanges.
 c. Those wound on built-up fiber or metal spools.

The molded spool is used mostly for small, low-voltage, fine-wire coils where the insulated wire can be wound directly on the spool by a hand feed method without any particular regard to the regularity of turns or other insulation than that provided by the spool and the wire insulation. The disadvantage of this type of coil is the haphazard type of winding which tends to stress the insulation and the difficulty of anchoring the leads. The advantage of this mode of construction is the low cost per coil if the quantities involved are large enough to absorb the cost of the molding dies. Figure 5a shows some molded spools, while Fig. 5b shows the same spool random wound with enamel wire.

FIGS. 5–8. Bobbin- and form-wound coils.

The built-up spool having an iron core with pressed-on fiber end flanges is used mostly as a matter of economy or convenience. It gives the maximum possible space for wire, but does not allow the removal of the coil from the pole core. The first layer of wire is insulated from the iron pole core by wrapping the pole core with several layers of paper or other insulating material. If even layers are specified the cost will usually be quite high owing to the large amount of hand work necessary. Where the space is available this type of winding is not employed in the best class of work. Figure 6 shows coils of this type.

The built-up bobbin of metal or fiber is seldom used nowadays, except for special work. Where very few coils are wanted and facilities for manufacturing other types are not available it provides a very convenient method. Experimental coils are often built this way. Neatly made bobbin-wound coils can be finished to give an excellent appearance, and are often preferred in exposed positions or where unusual mechanical protection is desired. Figure 7 shows coils of this type.

Form-Wound Coils. The larger types of coils used on the fields of motors, etc., occur in such a variety of shapes that they cannot be wound economically by any of the methods described, owing to the small quantities involved and the irregular shapes used. This type of coil is therefore wound on an adjustable *wooden* form. After winding the form is removed and the coil is made rigid by impregnating and taping. For small-sized coils the chief recommendation of this method of construction is that the largest number of turns can be put in the space available. However, unless layer-wound the small coils of this type will generally show more short-circuited turns than paper-section or cotton-interwoven coils. Figure 8 shows a form-wound coil for the field of a small motor.

Ribbon-Wound Coils. This type of coil is made of rectangular copper strip wound in the form of a spiral on the flat side. The coil thus has only one turn per layer, the layers being insulated by a strip of asbestos or other insulating material wound along with the copper. This type of coil is used in many kinds of high-current apparatus, such as power transformers, current coils for circuit breakers, and large lifting magnets.

Strap-Wound Coils. This type of coil is made from the same material as the ribbon-wound coil except that the copper strip is wound on edge, forming a single-layer coil. Coils of this type are used for very heavy currents. In open locations the edges of the strap are left uninsulated to facilitate heat radiation.

57. Coil Material

1. **Conductors.** The usual material for the conductors of a coil is copper. In some special cases aluminum is used. Both these materials are soft and ductile and can be drawn into fine sizes of wire. The usual wire cross sections are round, square, or rectangular. Table II gives data on the resistance of the standard sizes of copper and aluminum wire. These resistances are calculated upon the basis of the following standard resistivities:

Copper,   0.6788 microhm-inch at 20° C.
Aluminum, 1.113   "   "   " 20° "

2. **Wire Sizes.** In this country all copper and aluminum wire for electrical use is drawn in accordance with the American wire gauge. This gauge is so designed that the wire diameters increase in geometric progression. In developing any such system the diameters corresponding to two gauge numbers must be selected arbitrarily. It thus happens that No. 0000 A.w.g. is equal to 0.4600 in. in diameter and that No. 36 A.w.g. is equal to 0.0050 in. in diameter. As these two gauge numbers are 39 apart, the ratio of the diameters of successive gauge numbers will be equal to

$$\sqrt[39]{\frac{0.460}{0.005}} = 1.1229$$

As the area varies as the square of the diameter, the ratio of the areas of successive gauge numbers will be

$$1.1229^2 = 1.261$$

From these relations the following simple rules, which greatly facilitate the use of the wire tables, can be deduced:

1. Changing the wire size by three gauge numbers either halves or doubles the cross section.
2. Changing the wire size by six gauge numbers either halves or doubles the diameter.
3. Changing the wire size by ten gauge numbers either increases or decreases the cross section by a factor of ten.
4. Changing the wire size by twenty gauge numbers either increases or decreases the wire diameter by a factor of ten.

Table I gives complete data on the American wire gauge.

3. **Change of Resistance with Temperature.** The temperature coefficients of resistance for standard annealed copper (100 per cent conductivity) and hard-drawn aluminum are 0.00393 and 0.0039 per degree Centigrade at 20° C., respectively. For copper the temperature coefficient is directly proportional to relative conductivity over quite a wide range. This is also true for aluminum, except not over so great a range.

The resistance of a conductor at any temperature may be computed from the resistance at 20° C. by the following relation:

$$R_T = R_{20}[1 + \alpha_{20}(T - 20)] \tag{1}$$

where R_T and R_{20} are the resistances at any temperature T and 20° C., respectively; α_{20} is the temperature coefficient at 20° C.; and T is any temperature in degrees Centigrade. The reciprocal of the temperature coefficient gives the so-called "inferred absolute zero" or temperature

TABLE I

Wire Diameters, Mils

Wire Size, A.w.g.	Area, circular mils	Bare Wire	Enamel	Single Silk Enamel	Single Cotton Enamel	Single Silk	Double Silk	Single Cotton	Double Cotton
44	4.00	2.0	2.4						
43	4.84	2.2	2.6						
42	6.25	2.5	2.9						
41	7.84	2.8	3.2						
40	9.6	3.1	3.6	5.6	7.6	5.1	7.1	7.1	11.0
39	12.2	3.5	4.0	6.0	8.0	5.5	7.5	7.5	11.5
38	16.0	4.0	4.6	6.6	8.6	6.0	8.0	8.0	12.0
37	19.4	4.4	5.1	7.1	9.1	6.5	8.5	8.5	12.5
36	25.0	5.0	5.7	7.7	9.7	7.0	9.0	9.0	13.0
35	31.4	5.6	6.3	8.3	10.8	7.6	9.6	10.1	14.1
34	39.7	6.3	7.1	9.1	11.6	8.3	10.3	10.8	14.8
33	50.4	7.1	7.9	9.9	12.4	9.1	11.1	11.6	15.6
32	62.4	7.9	9.0	11.0	13.5	10.0	12.0	12.5	16.5
31	79.2	8.9	9.9	11.9	14.4	10.9	12.9	13.4	17.4
30	100	10.0	11.2	13.2	15.7	12.0	14.0	14.5	18.5
29	128	11.3	12.5	14.5	17.0	13.3	15.3	15.8	19.8
28	159	12.6	13.8	15.8	18.3	14.6	16.6	17.1	21.1
27	202	14.2	15.4	17.4	19.9	16.2	18.2	18.7	22.7
26	253	15.9	17.1	19.1	21.6	17.9	19.9	20.4	24.4
25	320	17.9	19.4	21.4	23.9	19.9	21.9	22.4	26.4
24	404	20.1	21.6	23.6	26.1	22.1	24.1	24.6	28.6
23	511	22.6	24.1	26.1	28.6	24.6	26.6	27.1	31.1
22	640	25.3	27.1	29.1	31.6	27.3	29.3	29.8	33.8
21	812	28.5	30.3	32.3	35.3	30.5	32.5	33.5	38.0
20	1,024	32.0	33.8	35.8	38.8	34.0	36.0	37.0	41.5
19	1,289	35.9	37.9	39.9	42.9	37.9	39.9	40.9	45.4
18	1,624	40.3	42.3	44.3	47.3	42.3	44.3	45.3	49.8
17	2,052	45.3	47.3	49.3	52.3	47.3	49.3	50.3	54.8
16	2,581	50.8	52.8	54.8	57.8	52.8	54.8	55.8	60.3
15	3,260	57.1	59.6		64.6			62.1	66.6
14	4,109	64.1	66.6		71.6			69.1	73.6
13	5,170	71.9	74.5		79.5			77.0	81.5
12	6,529	80.8	83.3		88.3			85.8	90.3
11	8,226	90.7	93.2		98.2			95.7	100.2
10	10,360	101.8	104.4		110.4			107.9	112.9

TABLE II
Copper and Aluminum Wire Resistances (20° C.)

Wire Size A.w.g.	Area		Copper			Aluminum		
	Circular Mils	Millionths of Square Inches	Ohms per 1,000 in.	Feet per Ohm	Feet per Pound	Ohms per 1,000 in.	Feet per Ohm	Feet per Pound
44	4.000	3.142	222.4	0.3850	82,600.0	....	....	....
43	4.850	3.805	176.6	0.4670	66,400.0	....	....	....
42	6.250	4.91	140.0	0.6050	52,800.0	....	....	....
41	7.845	6.16	111.1	0.7630	42,000.0	....	....	....
40	9.888	7.77	87.42	0.9534	33,410.0	143.3	0.581	110,000.0
39	12.47	9.79	69.30	1.202	26,500 0	113.4	0.733	87,280.0
38	15.72	12.4	54.95	1.516	21,010.0	90.00	0.924	68,960.0
37	19.83	15.6	43.59	1.912	16,660.0	71.50	1.17	55,200.0
36	25.00	19.6	34.56	2.411	13,210.0	56.75	1.47	43,500.0
35	31.52	24.8	27.41	3.040	10,480.0	45.00	1.85	34,450.0
34	39.75	31.2	21.74	3.833	8,310.0	35.67	2.34	27,400.0
33	50.13	39.4	17.24	4.833	6,591.0	28.24	2.95	21,700.0
32	63.21	49.6	13.68	6.095	5,227.0	22.42	3.72	17,225.0
31	79.70	62.6	10.84	7.685	4,145.0	17.75	4.68	13,645.0
30	100.5	78.9	8.60	9.691	3,287.0	14.08	5.91	10,825.0
29	126.7	99.5	6.82	12.22	2,607.0	11.16	7.45	8,583.0
28	159.8	126.0	5.41	15.41	2,067.0	8.835	9.39	6,802.0
27	201.5	158.0	4.290	19.43	1,639.0	7.033	11.8	5,389.0
26	254.1	200.0	3.400	24.50	1,300.0	5.582	14.9	4,269.0
25	320.4	252.0	2.698	30.90	1,031.0	4.425	18.8	3,388.0
24	404.0	317.0	2.139	38.96	817.7	3.508	23.7	2,687.0
23	509.5	400.0	1.696	49.13	648.4	2.783	29.9	2,137.0
22	642.4	505.0	1.345	61.95	514.2	2.209	37.8	1,695.0
21	810.1	636.0	1.066	78.11	407.8	1.750	47.6	1,340.5
20	1,022.0	802.0	0.8458	98.50	323.4	1.392	60.0	1,064.0
19	1,288.0	1,010.0	0.6708	124.2	256.5	1.100	75.7	844.0
18	1,624.0	1,280.0	0.532	156.6	203.4	0.8750	95.5	668.0
17	2,048.0	1,610.0	0.4221	197.5	161.3	0.6922	120.0	528.7
16	2,583.0	2,030.0	0.3346	249.0	127.9	0.5492	152.0	422.1
15	3,257.0	2,560.0	0.2653	314.0	101.4	0.4350	191.0	333.2
14	4,107.0	3,230.0	0.2104	396.0	80.4	0.3450	241.0	264.6
13	5,178.0	4,070.0	0.1670	499.3	63.8	0.2742	304.0	209.7
12	6,530.0	5,130.0	0.1323	629.6	50.6	0.2175	384.0	167.0
11	8,234.0	6,470.0	0.1050	794.0	40.1	0.1725	484.0	132.2
10	10,380.0	8,160.0	0.0832	1,001.0	31.8	0.1366	610.0	104.6

at which the resistance of the conductor would become zero if the linear temperature-resistance relation were to hold for low temperatures. Thus for 100 per cent conductivity copper the inferred absolute zero T_0 will be:

$$\left(-\frac{1}{0.00393} + 20\right) = -254 + 20 = -234° \text{ C}.$$

and the resistance of this conductor material as a function of temperature may be expressed by the following simple relation,

$$\frac{R_T}{R_{20}} = \frac{T + 234}{254} \tag{2}$$

or, in general,

$$\frac{R_T}{R_{20}} = \frac{T + T_0}{T_0 + 20} \tag{3}$$

To facilitate temperature corrections, the factor by which the resistance at 20° C. must be multiplied by to get the resistance at any other temperature is listed in Table III.

4. **Wire Insulation.** *Enamel.* The most common form of insulation for the smaller sizes of magnet wire is enamel. This material is thin, hard, tough, and inert to many chemicals. Its dielectric strength is of the order of 600 volts per mil. It is so elastic that the insulated wire may be stretched and bent without causing the enamel to crack or separate from the wire. Enamel will stand much higher temperatures than cotton, silk, or paper. In the larger sizes of wire the enamel is more subject to abrasion and is frequently covered with silk or cotton which acts as a protective coating and also as an absorbent for the impregnating compounds. Enamel-covered wires can be obtained in two thicknesses of enamel known as single- and double-enameled wire. The single-covered is standard and is generally used. However, copper wire when it is drawn has a tendency to splay, causing fine feathers of copper to stick out from the sides of the wire. Single enamel, though it covers most of these feathers does not cover all, and therefore, in applications where mechanical vibration is excessive and where one or more short-circuited turns might cause a cumulatively destructive effect, it is advantageous to use double-enameled wire. This is particularly true in alternating-current coils where a short circuit between turns is aggravated by the induced voltage in the shorted turns.

Cotton and Silk. Cotton or silk alone are often used for insulation. The covering may consist of one, two, or three layers. Silk is generally used on the smaller wires where it is necessary to reduce the thickness of

TABLE III

FACTORS BY WHICH THE RESISTANCE AT 20° C FOR 100 PER CENT CONDUCTIVITY COPPER MUST BE MULTIPLIED TO GET THE RESISTANCE AT ANY TEMPERATURE

Temp. °C.	Resistance Factor	Temp., °C.	Resistance Factor	Temp., °C.	Resistance Factor	Temp., °C.	Resistance Factor	Temp., °C.	Resistance Factor
1	0.9253	21	1.0039	41	1.0825	61	1.1611	81	1.2398
2	0.9293	22	1.0079	42	1.0865	62	1.1650	82	1.2438
3	0.9332	23	1.0118	43	1.0904	63	1.1690	83	1.2478
4	0.9371	24	1.0157	44	1.0943	64	1.1729	84	1.2517
5	0.9410	25	1.0197	45	1.0983	65	1.1769	85	1.2556
6	0.9450	26	1.0236	46	1.1022	66	1.1808	86	1.2594
7	0.9489	27	1.0275	47	1.1061	67	1.1848	87	1.2633
8	0.9528	28	1.0314	48	1.1100	68	1.1887	88	1.2672
9	0.9568	29	1.0354	49	1.1140	69	1.1926	89	1.2711
10	0.9607	30	1.0393	50	1.1180	70	1.1965	90	1.2751
11	0.9646	31	1.0432	51	1.1218	71	1.2005	91	1.2790
12	0.9686	32	1.0472	52	1.1258	72	1.2045	92	1.2830
13	0.9725	33	1.0511	53	1.1297	73	1.2083	93	1.2869
14	0.9764	34	1.0550	54	1.1336	74	1.2122	94	1.2909
15	0.9803	35	1.0590	55	1.1375	75	1.2162	95	1.2949
16	0.9843	36	1.0629	56	1.1415	76	1.2201	96	1.2988
17	0.9882	37	1.0668	57	1.1455	77	1.2240	97	1.3027
18	0.9921	38	1.0707	58	1.1494	78	1.2280	98	1.3066
19	0.9961	39	1.0747	59	1.1533	79	1.2320	99	1.3105
20	1.0000	40	1.0786	60	1.1572	80	1.2359	100	1.3144

insulation to a minimum. On the larger sizes cotton is more common. Unless the wire is enameled these insulations should be used double. For wires of very large size having fairly high voltages between turns triple cotton is often used. One of the chief advantages of the fabric insulations is the ease with which they can be impregnated.

Asbestos and Chromoxide. Where very high temperatures are encountered asbestos-covered wire (single or double) or chromoxide-covered wire is used. Chromoxide is a relatively recent development in wire insulation, combining the refractory qualities of Class B insulation with the high space factor and flexibility of enameled wire. It consists of a finely divided dielectric heat-resistant inorganic insulation (mainly chromium oxide), combined with an inorganic binder effective at high temperature and an organic water-insoluble insulator added for structural purposes. Mechanically, chromoxide insulation meets all the flexibility requirements of enameled wire. The thickness of the standard coating of chromoxide insulation is the same as that of standard single-enameled wire. It can also be obtained with a thickness corresponding to double-enameled wire. The dielectric strength of this insulation is only about 75 per cent of that of enamel at room temperature; however, it maintains this dielectric strength up to high temperatures, whereas that of enamel decreases. It will stand very high temperatures without damage for short periods of time, and can be used to 125° C. (the limit for Class B insulation) continuously.

Glass Insulation. Although glass has been used for a long time as an insulator it is only recently that it has been made flexible. For insulating purposes the glass is drawn into fibers of the order of 0.0002 in. in diameter, either of staple length or continuous. The staple-length fibers are best for heavy cloths and tapes, while the continuous-fiber yarns are used for making braided sleevings and insulating wires. All sizes of wire are being successfully covered with glass yarn, different thicknesses being used to obtain the same over-all diameters as the standard insulations of silk, cotton, and asbestos. The glass fiber itself can stand very high temperature without damage, but the organic binders which are required to impregnate the insulation are subject to deterioration at the higher temperatures and thus the allowable temperature rise of the insulation as a whole is limited by these organic materials. It is necessary to impregnate the glass fiber used for insulation to improve its dielectric strength and abrasion-resisting qualities. The dielectric strength of a fibrous material which has a large surface area compared to its volume is largely determined by atmospheric conditions and fiber-surface contamination. An impregnating material fills the voids between the fibers and prevents the absorption of dirt and moisture. The usual impregnating material is a heat-resistant electrical varnish. When properly impregnated a glass-insulated wire may be operated continuously at temperatures of 350° F. and retain a good dielectric strength, exceeding 1,000 volts per mil. Mechanically, properly made glass-insulated wires have sufficient resistance to abrasion and enough flexibility for all ordinary winding operations. The non-hygroscopic nature of the glass fiber makes it particularly adaptable to severe moisture or high humidity conditions, while its ability to withstand chemical action makes it highly resistant to attack by acids, oils, and corrosive vapors.[2]

Table I gives the diameter over the insulation for all the common types of wire.

[2] For a complete summary of the properties of glass fabrics for electrical insulation see "Fiber Glass—An Inorganic Insulation," by F. W. Atkinson, *Trans. A.I.E.E.*, 1939.

5. Coil Insulating Materials. Aside from the cotton and silk used on magnet wire practically all the other insulation is of sheet form, usually cotton, linen, silk, glass cloth, or paper treated with varnishes, gums, or other compounds to render them non-hygroscopic and to increase their dielectric strength. They are known as varnished cloth, varnished cambric, empire cloth, varnished silk, impregnated cloth, varnished glass cloth, etc., depending upon the material and the process. Their dielectric strength is of the order of 1,000 volts per mil. Various types of fiber board, hard rubber, Bakelite, Micarta, laminated mica, asbestos, etc., are also used.

Untreated cellulose materials break down at a temperature of 120° C. and should not be subjected to continuous temperatures in excess of 90° C. When treated these materials and also enameled wire may be continuously subjected to temperatures not in excess of 105° C. Micarta, laminated mica, asbestos, and Bakelite will stand higher temperatures.

58. Space Factor—Turn Density—Resistance Density

Space Factor. The space factor of a coil is defined as the ratio of the total cross section of copper in a coil to the gross cross-sectional area of the coil wall.[3] In general, a high space factor is desirable as the size of a coil to satisfy given requirements varies inversely as the space factor. The highest space factor is obtained when the minimum amount of wire insulation is used; thus for a given wire size, enamel wire will have a higher space factor than a fabric insulation. Likewise, square wire will give a higher space factor than the corresponding round wire. However, square wires are not practicable in sizes smaller than No. 10. Paper between layers reduces the space factor; embedding, which is always present with fabric-covered wires without paper between layers, increases the space factor.

Turn Density. The turns per square inch of actual winding space (space occupied by wire, wire insulation, interlayer paper, and voids between wires) is sometimes called the turn density. This factor greatly facilitates the computation of the total turns of a coil. Thus, the total turns are equal to the turn density multiplied by the net winding area of the coil.

Resistance Density. The resistance per cubic inch of actual winding (assuming all wires in space of 1 cu. in. to be laid end to end) is sometimes known as the resistance density of the winding. This factor greatly facilitates the computation of the total resistance of a coil. Thus,

---

[3] This is only one of several ways of defining space factor.

the total resistance is equal to the resistance density multiplied by the net winding volume of the coil.

59. Data for the Economical Manufacture of Low-Voltage, Paper-Section, Enamel Wire Coils

General. This type of coil is generally used for the smaller pieces of electromagnetic apparatus such as magnets, audio-frequency transformers, small power transformers for use in radio equipment, choke coils, and electric clocks. The actual choice of a coil type for any particular application naturally depends upon the judgment and individual preference of the designer.

Core Tubes. Core tubes are generally built up of several layers of paper (usually Kraft) or a special paper board wound directly on the coil mandrel. Fiber tubes are not practical on account of cutting difficulties. The following table gives data for the thickness and construction of core tubes:

Core Dimensions	Tube Construction
Up to $\frac{1}{2}$ in. square or round	2 wraps 0.020 in. special paper board
$\frac{1}{2}$ " $\frac{5}{8}$ " " " "	6 " 0.005 " Kraft paper
$\frac{5}{8}$ " $\frac{3}{4}$ " " " "	7 " 0.005 " " "
$\frac{3}{4}$ " $\frac{7}{8}$ " " " "	8 " 0.005 " " "
$\frac{7}{8}$ " 1 " " " "	9 " 0.005 " " "
1 " $1\frac{1}{4}$ " " " "	10 " 0.005 " " "
Rectangular	2 " 0.024 " special paper board

For any particular application, where the coil will be subjected to abnormally high voltages to ground, special allowances should be made.

Interlayer Material. Glassine paper is usually employed in paper-section coils between wire layers. Dielectric strength of glassine paper is about 300 volts per mil.

Kraft paper is often used as the interlayer material on coils made of the heavier wires, usually below 22 A.w.g. Its dielectric strength is about 100 volts per mil.

Turns per Inch Length of Winding. This in general is not equal to the reciprocal of the wire diameter in inches, for the following reason: As several coils are wound simultaneously on the same mandrel, as many different spools of wire must be used, which wires, though of the same nominal diameter, differ slightly in actual diameter, and if the lead were properly set for one spool it would be incorrect for the others. It is therefore necessary to spread the feed to take care of the largest possible variation in wire diameter.

Paper Margin. The wire in the layers of a paper-section coil cannot be wound to the edge of the coil. There must be a "paper margin," for support, beyond the actual winding. The necessary length of this margin varies principally with the wire size being used. It is also dependent on the shape of the coil cross section, and in some types of coil is dependent upon the thickness of the coil wall.

The data of Table IV supplied by the General Cable Co. give good representative values for "interlayer paper thickness," "paper margins," and "turns per inch." See page 165.

Coil Covers. Kraft papers in different colors (brown and black are most common), red rope paper, red rag paper, and the many shades and designs of bookbinder's cloth are all good materials for coil cover. Allow about 0.005 in. for a Kraft paper cover.

Impregnating Materials. For wax-impregnated paper-section coils, compounds using a paraffin or ceresine base, with proportions of rosin or beeswax added at times, are common. Wax-impregnated coils cannot be used if the coil is to be subjected to heat.

For varnish-impregnated paper-section coils a high-grade varnish which will not attack the enamel of the wire is used. Either air drying or baking varnish is suitable. This is ordinarily obtainable in two colors: black and clear. This type of treatment is recommended for coils which may be subjected to heat or which are electrically self-heated.

Crushing the Paper Margin. It is sometimes the practice when space is at a premium to crush the paper margin of a coil, thus reducing the gross length. Some manufacturers condemn the practice as damaging the coil. The following table gives an approximate idea of the reduction of paper margin obtainable by crushing:

Paper margin before crushing	$\frac{1}{16}$	$\frac{3}{32}$	$\frac{1}{8}$	$\frac{5}{32}$	$\frac{3}{16}$	$\frac{7}{32}$	$\frac{1}{4}$	$\frac{5}{16}$	$\frac{3}{8}$
Paper margin after crushing	$\frac{1}{32}$	$\frac{1}{32}$	$\frac{3}{64}$	$\frac{3}{64}$	$\frac{1}{16}$	$\frac{5}{64}$	$\frac{3}{32}$	$\frac{1}{8}$	$\frac{5}{32}$

Allowance in Coil Thickness. When a coil is designed to fit into a restricted space such as the window of a lamination, its total computed wall thickness must be made less than the available window opening by a definite amount if the coil is to be manufactured economically. This allowance is necessitated by the variations in manufacture, bulging of

TABLE IV

INTERLAYER PAPER, PAPER MARGINS, TURNS PER INCH FOR ENAMELED-WIRE PAPER-SECTION COILS*

Wire Size, A.w.g.	Wire Diameter over Enamel, mils	Interlayer Paper Thickness, mils	Minimum Paper Margins			Maximum Turns per Inch		
			Round Coils, inches	Square or Rectangular Coils < 50% off Square, inches	Rectangular Coils > 50% off Square, inches	Round Coils, inches	Square or Rectangular Coils < 50% off Square, inches	Rectangular Coils > 50% off Square, inches
44	2.4	0.7 G†	$\frac{1}{16}$	$\frac{1}{16}$	$\frac{5}{64}$	388	383	379
43	2.6	"	"	"	"	355	351	347
42	2.9	"	"	"	"	316	312	308
41	3.2	"	"	"	"	278	274	271
40	3.6	"	"	"	$\frac{3}{32}$	254	251	248
39	4.0	"	"	"	"	221	219	216
38	4.6	1.0 G	$\frac{3}{32}$	$\frac{3}{32}$	$\frac{1}{8}$	198	195	193
37	5.1	"	"	"	"	178	176	174
36	5.7	"	"	"	"	158	157	155
35	6.3	"	"	"	"	144	143	141
34	7.1	"	"	"	"	130	128	127
33	7.9	1.3 G	"	"	"	114	113	112
32	9.0	"	"	"	"	103	102	101
31	9.9	1.5 G	$\frac{1}{8}$	$\frac{1}{8}$	$\frac{5}{32}$	93	92	91
30	11.2	"	"	"	"	82	81	80
29	12.5	"	"	"	"	74	73	72
28	13.8	"	"	"	"	66	65	64
27	15.4	2.2 G	"	"	"	59	58	57
26	17.1	"	"	"	"	53	52	52
25	19.4	"	"	"	"	47	46	46
24	21.6	3.5 G	"	"	"	42	42	41
23	24.1	"	"	"	"	38	38	37
22	27.1	"	"	"	"	34	34	33
21	30.3	5.0 K‡	"	"	"	30	30	29
20	33.8	"	"	"	"	27	27	27
19	37.9	7.0 K	"	"	"	24	24	24
18	42.3	"	"	$\frac{5}{32}$	$\frac{3}{16}$	21	21	21
17	47.3	"	"	"	"	19	19	19

* The paper margins given in this table are for single winding coils of low voltage. Where the voltage is quite high or where there are multiple windings such as occur in power transformers these paper margins must be increased.
† Glassine.
‡ Kraft.

the sides of rectangular coils, bulging due to leads, etc. Experience indicates that for economical manufacture the ratio of the calculated coil wall thickness to the available coil space must be equal to or less than the values given in the table on page 166.

Wire Size	Factor	Wire Size	Factor	Wire Size	Factor
40	0.75	34	0.78	28	0.81
39	0.75	33	0.79	27	0.82
38	0.76	32	0.79	26	0.82
37	0.76	31	0.80	25	0.83
36	0.77	30	0.80	24	0.84
35	0.77	29	0.81	23	0.85

Here again the judgment and experience of the designer are important. Coils that are decidedly rectangular should have a more liberal allowance than that shown in the table. If taps are brought out on one side of a square or rectangular coil the factor for that side is considerably reduced. Likewise, two-section coils such as those used in small transformers will have a lower factor.

Specifications. In the production manufacture of paper-wound coils it is not possible to make rigid specification as to turns or resistance of a coil. The turns are usually held to about 3 per cent of the specified value, and the resistance will vary ± 5 per cent on the medium-sized wires to as much as ± 10 per cent on the finer wires. If it is desired to specify the turns or resistance to closer values than these the cost will be increased. However, both turns and resistance cannot be specified for one coil. Thus, if the turns are rigidly specified a large variation in resistance can be expected, with the finer wire sizes, and if the resistance is rigidly specified a large variation in turns can be expected. This is due to the variation in wire diameter normally allowed in drawing, and the reduction in wire diameter produced by the tension when winding a coil. This

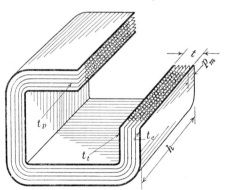

Fig. 9. Cross section of a paper-section coil showing coil symbols.

latter effect can make a very appreciable difference in a fine wire coil.

Space Factor. Figure 9 shows a paper-section coil cut through the coil wall to show the interlayer paper, paper margins, core tube, and coil cover. The following notation will be used when referring to coils:

h, over-all coil length, inches.
t, over-all coil wall thickness, inches.
p_m, length of paper margin, inches.
t_c, cover thickness, inches.
t_t, tube thickness, inches.
t_p, interlayer paper thickness, inches.
d, bare wire diameter, mils.
d', insulated wire diameter, mils.
P_m, perimeter of mean turn, inches.
n_i, turns per inch.

n_s, turns per square inch.
R_i, ohms per inch of wire.
R_c, ohms per cubic inch of winding.
S_w, net winding area, square inches.
V_w, net winding volume, cubic inches.
f_w, winding space factor.
f, space factor of entire coil.
R, coil resistance, ohms.
N, total turns on coil.

The winding space factor f_w (space factor exclusive of the paper margins, core tube, and cover) for a paper-section coil is given by the formula:

$$f_w = \frac{\pi d^2 n_i}{4(d' + t_p)} \qquad (4)$$

The space factor for the entire coil, f, is easily computed as follows:

$$f = \frac{\pi d^2 N}{4ht} = \frac{f_w(h - 2p_m)(t - t_c - t_t)}{ht}$$

or

$$f = f_w \left(1 - \frac{2p_m}{h}\right)\left(1 - \frac{t_c + t_t}{t}\right) \qquad (5)$$

where the quantity in the first parenthesis represents the fraction of the total coil length occupied by the wire, and that in the second parenthesis represents the fraction of the total coil thickness occupied by the wire.[4]

Table V gives the space factor f_w,[5] exclusive of paper margins, core tubes, and cover thickness, for enamel wire paper-section coils computed from the data of Table IV.

[4] It is sometimes customary to serve a paper-section coil with a layer of tape if there is any possibility of mechanical abrasion. An extra allowance of $\frac{1}{32}$ in. all around the coil should be made for a layer of tape.

[5] The values of winding space factor, turn density, and resistance density of Table V should be used for preliminary calculations only. All final calculations should be made by actually computing the number of layers and turns per layer of the coil. This is because a coil as actually wound must have a whole number of layers and turns per layer, whereas Table V assumes fractional layers and turns per layer possible.

Turn Density. The turn density [5] will be equal to the turns per inch multiplied by the layers per inch, thus:

$$n_s = \frac{n_i}{d' + t_p} \text{ turns per square inch} \quad (6)$$

or

$$n_s = \frac{4f_w}{\pi d^2} \text{ turns per square inch} \quad (7)$$

Table V gives the turn densities for paper-section enamel wire coils computed by means of equation 7, using the values of f_w from Table V.

Resistance Density. The resistance density [5] will be equal to the turn density multiplied by the resistance per inch of wire, thus

$$R_c = n_s R_i \text{ ohms per cubic inch} \quad (8)$$

Table V gives the resistance density for paper-section enameled copper wire coils at 20° C. computed by means of equation 8, taking the values of n_s from Table V and the values of R_i from Table II.

60. Data for Form and Bobbin-Wound Coils

Other Coil Types. When designing coils other than paper-section coils, the possibility of standardization becomes small owing to the latitude of choice of wire insulations, types of coil insulation, interlayer paper (if used), coil shape, etc., so that it is not feasible to give detailed information.

Wire Insulation. In order to maintain a reasonably high space factor, the smaller sizes of wire should have correspondingly thinner insulation than the larger sizes. The following table gives in a very general way the kinds of insulation suitable for different wire sizes:

Wire Size A.w.g.	Insulation
Larger than No. 10	Double cotton
No. 10 to 22, inclusive	Single cotton, or enamel and cotton
No. 23 to 28, inclusive	Single silk, or enamel and silk
Smaller than No. 28	Enamel

The combination of enamel and cotton, or enamel and silk, is used for the higher voltages, say over 100 volts. Plain enamel wire may also be used in sizes larger than given above but should have paper between layers for voltages over 100 volts. Such enamel coils are usually not impregnated as are those with silk or cotton insulation.

Coil Insulation. In the bobbin-wound coil it is necessary, if a metal bobbin is used or if the coil is wound directly on a part of the

TABLE V

Winding Space Factors f_w, Turn Densities n_s, and Resistance Densities R_c, at 20° C., for Paper-Section Enameled Copper Wire Coils, Based on the Winding Data of Table IV

Wire Size, A.w.g.	Round Coils			Square or Rectangular Coils < 50% off Square			Rectangular Coils > 50% off Square		
	f_w	n_s	R_c	f_w	n_s	R_c	f_w	n_s	R_c
44	.406	129,200	28,700	.401	127,500	28,380	.398	126,600	28,170
43	.422	110,900	19,600	.417	109,600	19,350	.412	108,250	19,120
42	.444	90,400	12,660	.437	89,000	12,460	.432	88,000	12,320
41	.450	73,100	8,120	.444	72,080	8,014	.440	71,420	7,940
40	.456	58,700	5,134	.452	58,200	5,084	.445	57,300	5,008
39	.462	47,200	3,270	.457	46,690	3,235	.452	46,180	3,200
38	.460	37,100	2,040	.454	36,600	2,012	.450	36,300	1,995
37	.472	30,260	1,320	.467	29,930	1,305	.462	29,610	1,290
36	.477	24,330	841	.475	24,240	838	.468	23,880	825
35	.493	19,880	545	.490	19,760	542	.483	19,480	534
34	.512	16,410	357	.505	16,180	352	.501	16,050	349
33	.502	12,740	220	.497	12,610	217.5	.493	12,510	216
32	.518	10,450	143	.513	10,350	141.5	.508	10,250	140
31	.516	8,240	89.3	.511	8,160	88.4	.506	8,080	87.6
30	.523	6,630	57.0	.518	6,564	56.5	.511	6,478	55.7
29	.542	5,450	37.2	.535	5,378	36.7	.527	5,300	36.1
28	.552	4,380	23.7	.541	4,294	23.2	.532	4,220	22.8
27	.536	3,392	14.6	.527	3,336	14.3	.518	3,279	14.1
26	.548	2,740	9.32	.538	2,690	9.14	.538	2,690	9.14
25	.555	2,202	5.94	.545	2,162	5.83	.545	2,162	5.83
24	.537	1,694	3.62	.537	1,694	3.62	.523	1,650	3.53
23	.557	1,392	2.36	.557	1,392	2.36	.542	1,355	2.30
22	.565	1,118	1.505	.565	1,118	1.505	.549	1,087	1.462
21	.548	862	0.919	.548	862	0.919	.530	833	0.888
20	.564	703	0.595	.564	703	0.595	.564	703	0.595
19	.546	541	0.363	.546	541	0.363	.546	541	0.363
18	.549	429	0.228	.549	429	0.228	.549	429	0.228
17	.566	352	0.148	.566	352	0.148	.566	352	0.148

magnetic circuit, to provide insulation between the bobbin tube and the coil, consisting of two or more layers of oiled linen or mica, depending upon the voltage (5 to 10 mils each). The insulation outside the coil should consist of a layer of oiled linen or mica and a protecting layer of tape (15 mils). The insulation between the coil ends and the bobbin flanges should consist of three or more thicknesses of oiled linen or mica, and (usually) a press-board washer (about $\frac{1}{32}$ in.) to take up the wear in winding. If it is desired to bring the inside coil lead to the outside along the inside of the bobbin flange an allowance must be made for the thickness of an insulated lead.

Form-wound coils after being wound are always served with one or more layers of protective coating. It is quite customary to impregnate such a coil, or dip it in varnish and bake it before applying the outer protective coating. After application of the outer coating which may consist of oiled linen, cotton tape, or friction tape it is customary to dip the coil in varnish and bake it.

Space Factor. It is best to compute the space factor of bobbin- and form-wound coils from the given over-all dimensions, allowances listed under coil insulation, and computed number of layers and turns per layer, making due allowance for clearances that must be provided around the bobbin or coil when it is to be ironclad. When computing the number of layers that can be wound in a given net winding depth the possibility of embedding should be taken into account; from 5 to 10 per cent (depending on the thickness of the insulation relative to the wire diameter) more layers may be wound in a given space due to compression of the insulation and the natural tendency of the wires of a layer to fall into the grooves between the wires of a lower layer. Likewise, when computing the number of turns per layer a deduction of 5 to 10 per cent in the computed turns per layer must be made to allow for the fact that the wires will not lie exactly close to each other as assumed; a further deduction of one turn per layer must be made to allow for the space lost at the ends of a layer.

Table VI[6] gives space factors which are averages taken from the data of a number of manufacturers. These space factors are for solid windings, assuming no embedding and no paper between layers. The values of turn density and resistance density are calculated from the tabulated values of space factor.

61. Coil Calculations (Not Considering Temperature Rise)

Most coil calculations, when heating is neglected, are relatively simple. In the following paragraphs formulas are developed for all the more important coil characteristics. The formulas which are based upon the previously tabulated values of space factor, turn density, and resistance density are necessarily approximate because such formulas assume the winding space to be completely filled with wire. It is obvi-

---

[6] These data are taken from "Coils and Magnet Wire" by C. R. Underhill, McGraw-Hill Book Co.

ous that this would almost always necessitate a fractional number of layers and turns per layer, which is impossible. Formulas of this type should be used for preliminary calculations. For more accurate results the formulas based upon the actual number of turns (computed from number of layers and turns per layer) and the mean length of turn should be used.

It should be remembered that all the coil data given in the tables are for 20° C.; if any other temperature is to be considered a suitable correction factor can be determined from Table III.

Turns. If the net winding cross section S_w of a coil is known the turns of a coil can be computed approximately by multiplying this cross section by the turn density n_s, thus:

$$N = S_w n_s \qquad (9)$$

The more accurate way of determining the turns is to compute the actual number of layers and turns per layer.

Coil Resistance. A simple way of computing this approximately is to multiply the net winding volume of the coil V_w (exclusive of all allowances for insulation, mechanical clearance, bobbins, etc.) by the resistance density R_c, thus:

$$R = V_w R_c \text{ ohms} \qquad (10)$$

or, expressing V_w in terms of the net winding cross section S_w and the perimeter of the mean turn P_m, we can write:

$$R = P_m S_w R_c \text{ ohms} \qquad (11)$$

For more accurate results the resistance should be determined from the following formulas:

$$R = P_m N R_i \text{ ohms} \qquad (12)$$

where N is the actual number of turns on the coil and R_i is the resistance of the wire per inch of length, or:

$$R = \rho \frac{P_m N}{\frac{\pi d^2}{4}} = \frac{4\rho P_m N}{\pi d^2} \text{ ohms} \qquad (13)$$

where ρ is the resistivity of the conductor material in microhm-inches and d is the bare wire diameter in mils.

Ampere-Turns. The ampere-turns of a coil are computed by simply multiplying the coil current by the turns, thus:

$$NI = N\frac{E}{R} \text{ ampere-turns} \qquad (14)$$

172 COILS [Chap. VI

TABLE VI

Wire Size, A.w.g.	Winding Space Factor, f_w, per cent							E.	S.S.	E.S.S.	D.S. or S.C.	E.S.C.	D.C.C.	E.	S.S.	E.S.S.	D.S. or S.C.	E.S.C.	D.C.C.
	E.	S.S.	E.S.S.	D.S. or S.C.	E.S.C.	D.C.C.			Turn Density, n_s						Resistance Density, $R_c - 20°C.$					
40	59	..	..	..	..	..	76,100	..	..	..	..	..	6,660	..	..	..	..	..
39	60	..	..	..	..	..	61,200	..	..	..	..	..	4,250	..	..	..	..	..
38	61	34	29	24	21	..	49,400	27,500	23,500	12,200	10,700	..	2,710	1,510	1,290	422	370	..
37	62	37	32	27	24	..	39,800	23,700	20,500	10,900	8,700	..	1,740	1,030	895	300	266	..
36	63	40	34	30	26	..	32,000	20,400	17,300	9,600	8,310	..	1,110	704	598	209	181	..
35	65	43	37				26,300	17,400	15,000	12,200	8,700		721	477	410			
34	66	45	40				21,100	14,400	12,800	10,900	8,310		460	313	278			
33	67	48	42	32	29		17,000	12,200	10,700	8,120	7,360		293	210	184	140	127	
32	67	50	44	35	31		13,500	10,100	8,880	7,060	6,250		185	138	121	96.5	85.5	
31	67	53	46	37	33		10,700	8,470	7,350	5,920	5,280		116	91.6	79.5	64.0	57.0	
30	69	55	49	40	37	24	8,700	6,930	6,170	5,040	4,660	3,020	75.2	60.0	53.4	43.6	40.3	26.2
29	69	56	51	43	39	27	6,900	5,600	5,100	4,300	3,900	2,700	47.3	38.4	34.9	29.5	26.7	18.5
28	70	59	53	46	42	29	5,580	4,710	4,220	3,670	3,350	2,310	30.2	25.4	22.8	19.8	18.1	12.5
27	70	60	55	48	44	32	4,420	3,790	3,470	3,030	2,780	2,020	19.0	16.3	14.9	13.0	11.9	8.66
26	71	62	57	50	46	35	3,560	3,100	2,860	2,500	2,300	1,750	12.1	10.5	9.70	8.50	7.82	5.95
25	71	64	58	52	48	38	2,820	2,540	2,300	2,060	1,910	1,510	7.66	6.91	6.26	5.61	5.18	4.10
24	71	65	59	54	50	40	2,240	2,050	1,860	1,700	1,570	1,260	4.78	4.38	3.98	3.64	3.37	2.70
23	71	66	60	56	52	43	1,780	1,650	1,500	1,400	1,300	1,080	3.01	2.80	2.48	2.38	2.20	1.82
22	72	67	62	58	54	46	1,430	1,330	1,230	1,150	1,070	910	1.91	1.78	1.65	1.54	1.43	1.22
21	73	68	64	60	57	48	1,150	1,070	1,000	940	895	754	1.23	1.14	1.07	1.01	0.957	0.806
20	73	69	65	62	58	50	912	862	813	775	725	625	0.774	0.731	0.69	0.657	0.615	0.530
19	72	..	..	60	56	48	748	..	..	592	553	479	0.477	..	..	0.397	0.370	0.318
18	73	..	..	63	58	51	572	..	..	493	454	399	0.304	..	..	0.263	0.242	0.213
17	74	..	..	64	60	53	460	..	..	397	372	329	0.194	..	..	0.168	0.157	0.139
16	74	..	..	65	62	55	365	..	..	321	306	271	0.122	..	..	0.107	0.102	0.0907
15	75	..	..	67	64	57	293	..	..	262	250	223	0.078	..	..	0.0697	0.0665	0.0592
14	75	..	..	68	64	59	232	..	..	211	198	183	0.049	..	..	0.0443	0.0417	0.0385
13	75	..	..	69	66	61	184	..	..	170	162	150	0.0308	..	..	0.0283	0.0271	0.0250
12	75	..	..	70	67	62	146	..	..	136	131	121	0.0193	..	..	0.0181	0.0173	0.0160
11	75	..	..	71	68	63	116	..	..	110	105	97.5	0.0121	..	..	0.0115	0.0110	0.0102
10	75	..	..	71	69	65	92.2	..	..	87.5	84.8	80.0	0.00765	..	..	0.00725	0.00704	0.00662

ART. 61]   COIL CALCULATIONS   173

By substituting equation 12 into 14, we get the following formula:

$$NI = \frac{E}{P_m R_i} \text{ ampere-turns} \qquad (15)$$

This equation shows that for a given wire size and fixed length of mean turn the ampere-turns of a coil are constant, regardless of the length of winding or number of turns.

The ampere-turns of a coil may also be expressed in terms of the wire diameter by substituting equation 13 into 14, thus:

$$NI = \frac{\pi d^2 E}{4 \rho P_m} \text{ ampere-turns} \qquad (16)$$

Coil Voltage. Formulas for coil voltage necessary to develop a given number of ampere-turns may be obtained by solving equations 14, 15, and 16.

Wire Diameter. It is often necessary to compute the diameter of the wire with which a coil of given dimensions must be wound in order to have a certain resistance. This may be done approximately as follows:

The wire diameter and number of turns are related to the net winding cross section thus

$$\frac{\pi d^2}{4} N = f_w S_w$$

This equation can be solved for N and the result substituted into equation 13, giving

$$R = \frac{\rho P_m S_w f_w}{\frac{\pi d^2}{4} \cdot \frac{\pi d^2}{4}} = \frac{16 \rho P_m S_w f_w}{\pi^2 d^4}$$

Solving for d and simplifying, this equation can be written:

$$d = 35.65 \sqrt[4]{\frac{\rho P_m S_w f_w}{R}} \qquad (17)$$

where d is in mils and ρ in microhm-inches.

In using this equation it is necessary to estimate the winding space factor f_w. As f_w does not change rapidly as d changes, a suitable value can be obtained from Tables V or VI; if the value of f_w assumed does not correspond closely enough to the value of d computed a new choice for f_w may be made and the value d recomputed. The final choice of wire size should be checked by computing the resistance by formula 12.

The wire size may also be found, if NI is given and P_m is known, by solving equation 15 for R_i and looking up the nearest corresponding wire size in Table II, thus:

$$R_i = \frac{E}{P_m NI} \qquad (17a)$$

62. Coil Problems

Problem 1. A round coil which is to slip over a 1-in. round rod has an outside diameter of $2\frac{1}{2}$ in. and a length of 3 in. If the winding is to be of the paper-section type, compute the number of turns of No. 28 wire which can be wound into the space available.

Solution. The paper margin for No. 28 wire is, from Table IV, $\frac{1}{8}$ in.; hence, the net winding length will be $3 - 2 \times \frac{1}{8} = 2\frac{3}{4}$ in. Referring to the data for core tubes on page 163 it will be seen that 0.045 in. must be allowed for the core tube, and from page 164 that the allowance for the cover will be 0.005 in.; hence, the net winding depth of the coil will be:

$$t - t_t - t_c = \frac{2\frac{1}{2} - 1}{2} - 0.045 - 0.005 = 0.700 \text{ in.}$$

The net winding space will, therefore, be $2\frac{3}{4} \times 0.70 = 1.93$ sq. in. From Table V the turn density for No. 28 enamel wire when wound in a round paper-section coil is 4,380. By means of equation 9 the turns may now be computed:

$$N = S_w n_s = 1.93 \times 4{,}380 = 8{,}450 \text{ turns}$$

Check: the turns per layer will be $(h - 2p_m) \times n_i = 2\frac{3}{4} \times 66 = 181$. ($n_i$ is found by reference to Table IV.) The number of layers will be equal to:

$$\frac{t - t_t - t_c}{d' + t_p} = \frac{0.70 \times 10^3}{13.8 + 1.5} = 45.7$$

which we can call 46. Therefore,

$$N = 46 \times 181 = 8{,}320 \text{ turns}$$

Problem 2. Compute the space factor of the coil of Problem 1.

COIL PROBLEMS

Solution. Using equation 5 (see Table V for f_w):

$$f = f_w\left(1 - \frac{2p_m}{h}\right)\left(1 - \frac{t_c + t_t}{t}\right)$$

$$= 0.552\left(1 - \frac{2 \times \frac{1}{8}}{3}\right)\left(1 - \frac{0.005 + 0.045}{\frac{3}{4}}\right)$$

$$= 0.552 \times 0.9167 \times 0.9334 = 0.472$$

Check: Using the definition of space factor on page 162, we have:

$$f = \frac{\pi d^2 N}{4ht} = \frac{\pi \times 12.6^2 \times 10^{-6} \times 8{,}320}{4 \times 3 \times 0.75} = 0.461$$

Problem 3. Compute the resistance of the coil of Problem 1 at 20° C.

Solution. The perimeter of the mean turn on this coil is:

$$P_m = \pi(1.0 + 2 \times 0.045 + 0.70) = 5.61 \text{ in.}$$

and the resistance by equation 11 will be

$$R = P_m S_w R_c = 5.61 \times 1.93 \times 23.7 = 256 \text{ ohms}$$

where R_c is read from Table V. If equation 12 is used the resistance will be:

$$R = P_m N R_i = 5.61 \times 8{,}320 \times 5.41 \times 10^{-3} = 252 \text{ ohms}$$

where R_i is read from Table II.

Problem 4. Determine the ampere-turns of the coil of Problem 1 if it is supplied from a 50-volt supply.

Solution. Using equation 15, the ampere-turns will be

$$NI = \frac{E}{P_m R_i} = \frac{50}{5.61 \times 5.41 \times 10^{-3}} = 1{,}650$$

If equation 14 is used the answer will be:

$$NI = N\frac{E}{R} = 8{,}320 \times \frac{50}{252} = 1{,}650$$

Problem 5. A paper-section coil is to have a diameter of $\frac{7}{8}$ in. inside of the core tube, an outside diameter of 2 in., and a length of 2 in. With what size of wire should it be wound if it is to develop 1,000 ampere-turns at 20° C. when supplied from a 12-volt source?

Solution. The winding depth will be

$$t - t_t - t_c = \frac{(2 - \frac{7}{8})}{2} - 0.045 - 0.005 = 0.513 \text{ in.}$$

And the perimeter of the mean turn will be

$$\pi(\tfrac{7}{8} + 2 \times 0.045 + 0.513) = 4.64 \text{ in.}$$

Using equation 17a the resistance R_i per inch of wire may be determined, thus:

$$R_i = \frac{E}{P_m N I} = \frac{12}{4.64 \times 1{,}000} = 2.59 \times 10^{-3} \text{ ohm per inch}$$

Referring to Table II it will be seen that the nearest wire size is No. 25, which has a resistance of 2.698×10^{-3} ohm per inch.

Check: The number of turns per layer will be $(h - 2p_m)n_i = 1\tfrac{3}{4} \times 47 = 82$. The number of layers will be:

$$\frac{t - t_t - t_c}{d' + t_p} = \frac{0.513 \times 10^3}{19.4 + 2.2} = 23.8$$

which we can call 24. The turns will therefore be $82 \times 24 = 1{,}970$, and the coil resistance will be (equation 12):

$$R = P_m N R_i = 4.64 \times 1{,}970 \times 2.698 \times 10^{-3} = 24.6 \text{ ohms}$$

Therefore NI by equation 14 will be:

$$NI = N\frac{E}{R} = 1{,}970 \times \frac{12}{24.6} = 962 \text{ ampere-turns}$$

This value is lower than the required 1,000 because No. 25 wire has slightly more resistance per inch than was desired. The next larger size of wire will produce considerably more than 1,000 ampere-turns.

When a desired wire size is between two gauge numbers the coil is sometimes wound with both sizes of wire, the length of each being chosen so as to get the desired resistance. This is usually done only when the larger sizes of wire are being used, where an error of one-half a gauge number will materially change the resistance.

Problem 6. Determine the wire size for a coil having the dimensions of that of Problem 5 if it is to have a resistance of 100 ohms at 20° C.

Solution. The coil of Problem 5 is wound with No. 25 wire and has 24.6 ohms resistance. If the resistance is to be changed to 100 ohms the resistance density must be four times as great; hence, by reference to

Table V, No. 28 wire would be chosen because R_c for No. 28 is 23.7 while R_c for No. 25 is 5.94. This result can be determined approximately without reference to the tables because it is known that for an increase of three gauge numbers the cross section of a wire is reduced to one-half and hence the resistance density is quadrupled.

However, if no previous knowledge of this particular coil was available, the solution could not be made in this manner. The solution could then be worked out by means of equation 17:

$$d = 35.65 \sqrt[4]{\frac{\rho P_m S_w f_w}{R}}$$

where f_w would have to be estimated. By reference to Table V, a probable value of f_w can be selected. Let us assume that we have picked the value 0.516 corresponding to No. 31 wire, then

$$d = 35.65 \sqrt[4]{\frac{0.6788 \times 4.64 \times 1.75 \times 0.513 \times 0.516}{100}} = 35.65 \sqrt[4]{0.0146}$$
$$= 35.65 \times 0.348 = 12.4 \text{ mil}_s$$

By reference to Table I it will be seen that this corresponds almost exactly to No. 28 bare wire, despite the fact that estimated space factor does not correspond to that for No. 28 wire.

Check: Assume No. 28 enamel wire. The turns per layer will equal

$$(h - 2p_m)n_i = 1\tfrac{3}{4} \times 66 = 115$$

The number of layers will be equal to:

$$\frac{t - t_t - t_c}{d' + t_p} = \frac{0.513 \times 10^3}{13.8 + 1.5} = 33.6, \text{ say } 36$$

Therefore N will equal $115 \times 36 = 4{,}140$ turns. The resistance by equation 12 will be:

$$R = P_m N R_i = 4.64 \times 4{,}140 \times 5.41 \times 10^{-3} = 104 \text{ ohms}$$

CHAPTER VII

HEATING OF MAGNET COILS

63. Temperature Rise of a Magnet Coil under Ideal Conditions

The entire energy input to a magnet coil, as soon as the motion of the armature has ceased, is dissipated as heat. Thermally the magnet coil has two properties: the ability to store energy due to an increase in temperature, and the ability to lose its stored heat energy from its external surfaces due to a temperature difference between these surfaces and the surrounding air. These two properties are known as the thermal capacity and heat-dissipation capacity, respectively, and will be designated by C and K.

If θ equals the difference between the average coil temperature and the surrounding air, the instantaneous rate at which heat energy is absorbed by the thermal capacity of the coil will be $C(d\theta/dt)$, and the instantaneous rate at which heat energy is dissipated from the coil surfaces will be $K\theta$. Starting with the coil at room temperature and applying a constant power of P watts we have the following: at the instant the power is first applied θ will be zero and the total power input will be absorbed by the thermal capacity causing the coil temperature to rise at a rate equal to P/C degrees Centigrade per second; after the power has been applied for a long time the coil will have attained a constant temperature, making $d\theta/dt$ equal to zero, and the total power input will be accounted for by the heat-dissipation capacity, making θ equal to P/K degrees Centigrade. Therefore, any time between these two limits the instantaneous temperature difference will be given by the following equation:

$$P = C\frac{d\theta}{dt} + K\theta \qquad (1)$$

where $C(d\theta/dt)$ is the part of the total power P watts absorbed by the thermal capacity, and $K\theta$ is the part dissipated by the coil.

Rearranging, we have

$$\frac{P - K\theta}{d\theta} = \frac{C}{dt}$$

or
$$\frac{K d\theta}{P - K\theta} = \frac{K dt}{C}$$

Integrating both sides, we obtain

$$-\log_\epsilon (P - K\theta) = \frac{K}{C} t + C_1, \text{ a constant of integration}$$

When $t = 0$, $\theta = 0$, and
$$-\log_\epsilon P = C_1$$

Substituting, we get
$$-\log_\epsilon (P - K\theta) + \log_\epsilon P = \frac{K}{C} t$$

or
$$\log_\epsilon \frac{P - K\theta}{P} = -\frac{K}{C} t$$

Taking the antilogarithm of both sides, we have

$$P - K\theta = P \epsilon^{-\frac{K}{C} t}$$

or
$$+ K\theta = P \left(1 - \epsilon^{-\frac{K}{C} t}\right)$$

and
$$\theta = \frac{P}{K} \left(1 - \epsilon^{-\frac{K}{C} t}\right) \qquad (2)$$

When $t = \infty$, $\epsilon^{-\frac{K}{C} t}$ will be zero and the final temperature rise will be P/K, which checks with the result previously deduced. Likewise, if equation 2 is differentiated, we obtain

$$\frac{d\theta}{dt} = +\frac{P}{C} \epsilon^{-\frac{K}{C} t} \qquad (3)$$

which on substituting $t = 0$ gives $d\theta/dt = P/C$, which also checks with the previous deduction. Equation 2 is shown plotted below in Fig. 1.

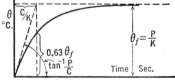

Fig. 1. Temperature rise-time curve of a coil under ideal conditions.

It is a simple exponential curve which approaches its final value P/K as an asymptote. The initial slope of the curve is P/C degrees Centigrade per second. If the temperature were to continue to change at its initial rate the final temperature rise θ_f would be reached in a time equal to $(P/K)/(P/C) = C/K$ seconds. Actually in this time interval the temperature rise will be, by equation 2,

$$\theta = \frac{P}{K}(1 - \epsilon^{-1}) = \frac{P}{K}\left(1 - \frac{1}{2.718}\right) = 0.63\theta_f$$

Or, in words, in a time equal to C/K seconds the temperature rise will be 63 per cent of its final value. The value C/K may be called the thermal time constant of the magnet coil system.

The cooling equation for the magnet coil system may be obtained from equation 1 by making P equal to zero and noting that, when $t = 0$, $\theta = \theta_f$; thus:

$$0 = C\frac{d\theta}{dt} + K\theta$$

or

$$-\frac{K}{C}dt = \frac{d\theta}{\theta}$$

Integrating both sides, we obtain

$$-\frac{K}{C}t = \log_\epsilon \theta + C_1, \text{ a constant of integration}$$

When $t = 0$, $\theta = \theta_f$; therefore

$$C_1 = -\log_\epsilon \theta_f$$

and

$$-\frac{K}{C}t = \log_\epsilon \frac{\theta}{\theta_f}$$

or

$$\frac{\theta}{\theta_f} = \epsilon^{-\frac{K}{C}t}$$

and

$$\theta = \theta_f \epsilon^{-\frac{K}{C}t} \tag{4}$$

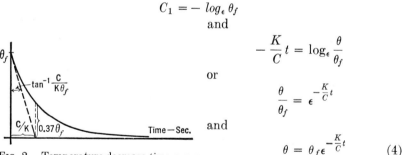

Fig. 2. Temperature decrease-time curve (cooling curve) for a magnet coil under ideal conditions.

where θ_f is the coil temperature at the instant the cooling starts, and C/K will be the time required for the coil system to cool to 37 per cent of its original temperature rise θ_f. Equation 4 is shown plotted in Fig. 2.

If we take the logarithm of both sides of equation 4 we will obtain

$$\log \theta = \log \theta_f - \frac{K}{C}t \tag{5}$$

If this equation is plotted on semi-logarithmic paper it will be a straight line, having a negative slope of K/C and an intercept on the axes of ordinates of θ_f. Equation 2 may also be plotted on semi-logarithmic paper if it is first subtracted from its asymptote $\theta_f = P/K$. Figure 3 shows equation 5 plotted on semi-logarithmic paper.

Practically, plotting on semi-logarithmic paper is advantageous, when analyzing this type of experimental data, to determine how closely the data follow the exponential law and also the best mean value of the exponent which can be used to represent the curve analytically.

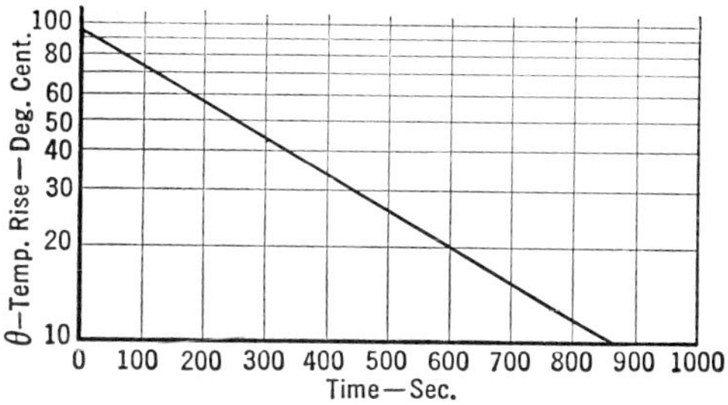

FIG. 3. Cooling curve of Fig. 2 plotted on semi-logarithmic paper (equation 5) for a particular magnet coil.

The foregoing discussion is confined specifically to cases where the coil may be considered as an isothermal body with a constant heat capacity and constant heat-dissipation capacity. Actually this condition never obtains; even though the heat is uniformly evolved throughout the coil structure it is necessary to have a temperature gradient in the coil in order that the heat released on the inside may flow to the outside and be dissipated by the external coil surfaces. The consequence of this is that actual heating curves of a magnet coil deviate more or less from the exponential curve based on the premises of constant power input and on an isothermal coil body.

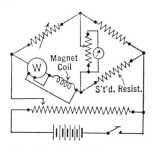

FIG. 4. Circuit for continuously measuring the temperture of a coil winding.

64. Deviation of Actual Magnet Heating and Cooling Curves from the Ideal

Before the results of the previous article can be applied to actual temperature-rise problems we must investigate how closely actual temperature-rise curves of magnets follow the simple exponential law. In order to do this, temperature-rise data at constant power input on several magnets were taken. The temperature rise of the coil was always taken as the difference between the average coil temperature determined by the resistance of the coil and the temperature of the surrounding air. The circuit used in making these measurements is shown in Fig. 4. This is a simple Wheatstone bridge so arranged that it can be operated at about unity ratio with an impressed voltage sufficiently high that the magnet coil will receive the requisite power input. The wattmeter W is arranged to read the power input to the coil, which is held constant by the input potentiometer. During a temperature run the bridge is maintained continuously balanced by adjusting the rheostat arm, thereby enabling the resistance of the magnet coil arm to be read at predetermined time intervals. The magnet coil resistance is obtained by correcting the measured resistance of the magnet coil arm for the resistance of the wattmeter voltage and current coils. The resistance of the magnet coil at some known temperature can be determined by a bridge measurement after the coil is allowed to remain in a constant-temperature room for several hours with no power applied. The temperature rise of the coil may then be determined by making the proper reference to Table III, Chapter VI.

Some of the data of these experiments are shown plotted on semilogarithmic paper in Fig. 5. The curves of Fig. 5a are plotted for various constant power inputs to the paper-section coil for the magnet of Fig. 1a, Chapter III, the coil being suspended in the vertical position in still air. A picture of this coil is shown in Fig. 2b, Chapter VI. If the final values of average coil temperature rise θ_f are plotted against the power input as is done in Fig. 6, curve 1, it will be seen that the relationship between these two variables is not linear and hence K equal to P/θ_f is a function of the final temperature rise, increasing as the final temperature rise increases. This is the natural thing to expect as the heat is dissipated from the coil surfaces by convection currents of air and by radiation. The ability of a surface to lose heat by natural convecting currents increases as the temperature rise increases and the ability to radiate also increases with the temperature; hence the heat dissipated by the coil per unit rise in temperature will increase as the coil temperature increases. This explains why the value of K should increase as the temperature increases.

The curves of Fig. 5c are for the same coil as those of Fig. 5a except that the coil is now encased by the iron magnet shell and plunger as illustrated in Fig. 1a, Chapter III. It will be seen that for the same power inputs the final temperature rise is the same and hence the values of K are the same. Apparently, because the coil has a small amount of clearance between itself and the surrounding iron, the air film so introduced makes the coil act, as far as heat dissipation is concerned, the same as if it were in open air. It will also be noticed that at the start of the heating period the curves have the same slope as those of Fig. 5a. This indicates that the effective thermal capacity is that of the coil only.

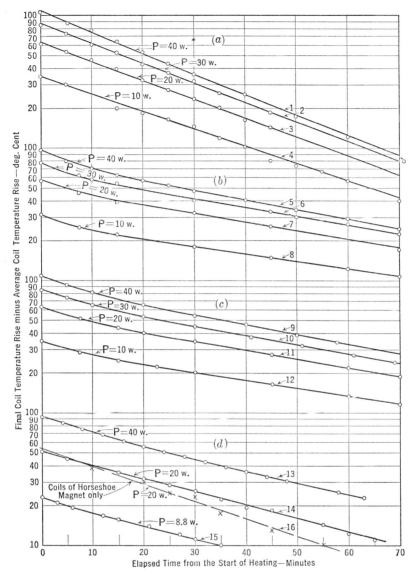

Fig. 5. Temperature-rise-time curves for various magnet coils plotted on semilogarithmic paper by subtracting the actual temperature rise from the final temperature rise, $(\theta_f - \theta) = \epsilon^{-\frac{K}{C}t}$.

(a) Paper-section coil of magnet of Fig. 1a, Chapter III, only (see Fig. 2b, Chapter VI) cylindrical coil surface = 49.5 sq. in.

(b) Ironclad magnet of Fig. 1a, Chapter III. Coil in good thermal contact with iron (high conductivity pottery compound between coil and iron).

(c) Ironclad magnet of Fig. 1a, Chapter III. Coil in poor thermal contact with iron (air film between).

(d) Horseshoe magnet of Problem 4, Chapter V. Coils in poor thermal contact with iron. 2 coils: $\frac{7}{8}''$ I.D.-$2\frac{1}{2}''$ O.D.-$2\frac{15}{16}''$ long, total cylindrical coil surface = 62 sq. in.

As the heating progresses the slope of the curves becomes less, indicating that the thermal capacity of the surrounding iron is becoming effective. The curves of Fig. 5b give the same data for the same magnet coil and surrounding iron shell as that of Fig. 5c, except that the air space between the coil and the surrounding shell has been filled by a high-conductivity potting compound. It will be observed that these curves exhibit the same characteristics as those of Fig. 5c but show lower final temperature rises for the same power inputs.

The curves in Fig. 5d show heating data on a horseshoe type of magnet where the paper-section coils are a loose fit on the steel pole core. It will be noticed that curve 16, for the coils only, shows about the same final temperature rise for the same power input as does curve 14 for the entire magnet, indicating that the heat-dissipation coefficient is only slightly affected by the iron.

Summing up: (1) when considering the heating of a magnet coil only, one may safely assume that the heating follows the simple exponential law of equation 2, and the heat-dissipation constant used in this equation is a function of the final temperature rise; (2) when the magnet coil is surrounded by an iron shell or otherwise in contact with iron parts the final temperature rise will be the same as if the coil is in the open air provided that the coil is in poor thermal contact with the iron parts, as occurs when a paper-section coil is loosely fitted around a pole core or within a shell; (3) when the magnet coil is surrounded by iron parts and is in good thermal contact with these parts, as occurs when a paper-section coil is potted or when a coil is wound on a brass bobbin which is a tight fit over iron parts, the ability to dissipate heat will be greater than in (2) above; (4) when a magnet coil is surrounded by iron parts the effective thermal capacity will be variable, starting at a value equal to the thermal capacity of the coil only and gradually increasing as the iron parts become warmed.[1]

65. Heat-Dissipation Coefficients

In Fig. 6 is plotted from the data of Fig. 5 the final temperature rise θ_f as a function of the power input. These curves are drawn to show that θ_f is not a linear function of the power input. When calculating the heat-dissipating ability of a coil it is convenient to use a

---

[1] A way of handling this effect with fair accuracy would be to consider the coil itself as an isothermal body which loses its heat to another isothermal body (the iron parts) through the medium of a heat conductor between the two bodies (the air film or compound between the coil and shell). The difficulty of this scheme is that the heat conductance of the air film or compound will be somewhat indeterminate owing to the inaccuracy of estimating the clearance between the coil and shell.

factor called the heat-dissipation coefficient, or the watts which can be dissipated per square inch of effective coil surface per degree Centigrade temperature difference between the average coil temperature and the surrounding air. This coefficient, as can be seen from Fig. 6, is a function of the final temperature rise or the power input to the coil. In Fig. 7a this coefficient for varnish-impregnated, black-finish, enamel-wire, paper-section coils of normal proportions is shown plotted as a function of the power input per unit of effective heat-dissipating surface,

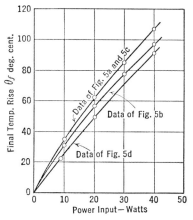

Fig. 6. Final coil temperature-rise data of Fig. 5 plotted as a function of the power input.

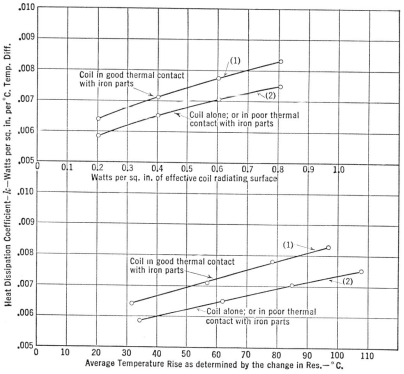

Figs. 7a and b. Heat dissipation coefficient for varnish-impregnated, black-finish, enamel-wire, paper-section coils plotted as a function of: (a) the power input per square inch of effective coil radiating surface; (b) the average temperature rise.

while in Fig. 7b the same coefficient is plotted as a function of the final temperature rise. In general one of these two quantities will be specified in a practical problem and hence a proper value of k may be chosen. When computing the effective dissipating surface of a paper-section coil, only the cylindrical surface areas should be reckoned as the end surfaces are ineffective due to thermal insulating properties of the paper margin. For coils wound on brass bobbins the upper curves of Fig. 7a and b will be found to hold quite well, provided that the coils are impregnated with varnish and the bobbin is in good thermal contact with the iron parts. Under these conditions the entire coil surface should be considered effective.

66. Final Temperature Rise of a Coil in Terms of the Magnetomotive Force and Dimensions of the Coil

When designing an electromagnet it is very convenient if the final temperature rise can be specified in terms of the coil magnetomotive force and dimensions. Using the notation of page 167, we have

$$P = I^2R = \theta_f K = 2\theta_f khP_m$$

or

$$\theta_f = \frac{I^2R}{2khP_m} \tag{6}$$

where $2hP_m$ is the total cylindrical coil surface. By equation 13, Chapter VI, R may be expressed as follows:

$$R = \frac{4\rho P_m N}{\pi d^2}$$

The bare wire diameter d may be expressed in terms of the gross coil length h and over-all coil wall thickness t as follows:

$$d = \sqrt{\frac{4htf}{\pi N}}$$

where f is space factor of the entire coil. Substituting this into the above equation, we have

$$R = \frac{\rho P_m N^2}{htf} \tag{7}$$

Substituting this value for R into equation 6 we get for the final temperature rise of the coil

$$\theta_f = \frac{\rho}{2kft}\left(\frac{NI}{h}\right)^2 \quad \text{degrees Centigrade} \tag{8}$$

In this equation NI/h is the magnetic intensity of the coil winding, ρ is the resistivity of the conductor material in ohm-inches at the final coil temperature, k is the heat-dissipation coefficient at the final coil temperature, and t is the gross coil thickness. If the coil is excited for only a fraction of the time equal to q, equation 8 will be modified to the following form:

$$\theta_f = \frac{q\rho}{2kft}\left(\frac{NI}{h}\right)^2 \qquad (9)$$

Equation 9 can be used only if the period of heating is small (not greater than about one-quarter) compared to the thermal time constant. Thus a magnet having a thermal time constant of 1 hour cannot be considered as having an intermittent excitation if it has a continuous cycle of 2 hours on and 6 hours off.

67. Thermal Capacity

The thermal capacity of a magnet coil can be easily computed from the data of Table I. As an example consider the coil on which the data of Fig. 5a were taken. This coil weighs 3.33 lb. If we consider the entire weight to be made up of copper,[2] the thermal capacity will be $3.33 \times 180 = 600$ joules per degrees Centigrade rise. This computed thermal capacity can be checked by the data of Fig. 5a. Thus the slope of curve 1 is equal to $\log_\epsilon 10.8/(66.5 \times 60) = 0.00060$. This slope, however, is equal to K/C. The heat-dissipation coefficient k for this coil for the final temperature rise of 108° as shown on curve 1 will be found, by referring to the data of Fig. 7b, to be 0.0075 watt per square inch per degree Centigrade temperature difference. Multiplying this by the cylindrical coil area of 49.5 sq. in., K will equal 0.371 watt per degree Centigrade temperature difference. The thermal capacity C will therefore be $0.371 \div 0.0006 = 620$ joules per degree Centigrade rise. Column (2) of Table II gives values of C computed as above for the various curves of Fig. 5a. It will be noted that the values of C computed are substantially constant even though the slopes of the various curves are different. This is because the value of K varies as the temperature rise changes. The average of the four values of C listed in column (2) is 603, which is very close to the computed value of 600.

The effective thermal capacity of a magnet coil and the surrounding iron shell on the basis of using equation 2, however, is more difficult to

---

[2] For relatively large coils very little error is introduced by considering the entire coil weight to consist of copper, as the weight of the paper and other insulation is relatively small; for small coils, however, the thermal capacity of the paper and insulation should be calculated separately.

estimate. If an ironclad magnet is designed with heating as a limitation the effective thermal capacity of the iron will depend on whether the operation of the magnet is intermittent or continuous. If the magnet operation is intermittent the rate of heating will be high and there will be insufficient time for the thermal capacity of the iron parts to become as effective as if the heating were continuous. Thus referring to Fig. 5, it will be seen that for the first 10 minutes of heating the slopes of curves 9 and 5 are practically the same as the slope of curve 1. As the heating

TABLE I

Material	Thermal Capacity, joules per lb. per deg. Cent.	Density, lb. per cu. in.	Material	Thermal Capacity, joules per lb. per deg. Cent.	Density, lb. per cu. in.
Aluminum	433	0.093	Paper and cotton insulation	700	0.035
Copper	180	0.32	Impregnating compound	1000	0.035
Steel	225	0.28	Paraffin	1125	0.029
Brass	200	0.307	Phenolic material	....	0.0491

progresses, however, the thermal capacity of the iron becomes effective. This can be seen by noting that the curves of Figs. 5b and 5c become straight at about $t = 20$ minutes, the slopes of the straight part being much less than that of curves of Fig. 5a. Using curve 5 between the time of 20 and 70 minutes the effective thermal capacity will be found to be 1,470 joules per degree Centigrade rise. Actually the iron of this magnet weighs 7.5 lb., which corresponds to a thermal capacity of $7.5 \times 225 = 1,690$ joules per degree Centigrade rise. Subtracting from 1,470 the thermal capacity of the coil [3] 550, we will obtain 920 joules per degree Centigrade rise as the effective thermal capacity of the iron. Thus the iron is $(920 \div 1,690) \times 100 = 55$ per cent effective. The reason that the iron is not 100 per cent effective is that its average temperature is considerably less than that of the coil and the thermal capacities here computed are based on the coil temperature. This difference in temperature is necessary in order that the heat will flow from the coil through the space between the coil and the iron shell to the shell. As

[3] The coil actually used in the magnet is slightly smaller than that of Fig. 5a.

the heat conductivity of the space between the coil wall and the shell is decreased the effective thermal capacity of the iron will decrease. Thus, if we use the data of curve 9, Fig. 5c, for the case where the coil is in poor thermal contact with the iron, the thermal capacity for the coil and iron will be found to be 1,320 between $t = 20$ and 70 minutes, making the effective thermal capacity of the iron $1,320 - 550 = 770$, and the effectiveness of the iron $770 \div 1,690 = 46$ per cent. In Table II, columns (3), (4), and (5), are listed the effective thermal capacities of the entire magnet and the iron of the magnet computed as above for the various curves of Figs. 5b and c. It will be noted that, when the coil is in good thermal contact with the iron parts, these parts are about 55 per cent effective, while if the coil is in poor thermal contact with the iron parts they are only about 45 per cent effective. These results are for the particular case where the iron completely surrounds the coil. It appears, however, that even in the horseshoe magnet, where the coil is not completely surrounded by the iron, the iron has about the same effectiveness.

The effective thermal capacity of the iron just discussed is not immediately effective, as was pointed out before. A study of Figs. 5b and c shows that after a time equal to about one-third of the thermal time constant (20 minutes) it becomes effective. However, the degree of its effectiveness during the elapse of this interval is more difficult to determine. In Figs. 5b and c the iron is about 25 per cent effective (about one-half of its final effectiveness) during this time interval, while in Fig. 5d the iron is practically immediately effective at its final value. This difference in the effectiveness of the iron during the initial time interval is apparently dependent on the ratio of total iron thermal capacity to the total area of the coil dissipating surfaces. Thus for Figs. 5b and c there is 34 joules per degree Centigrade thermal capacity per square inch of coil surface, while for Fig. 5d there is only 8.75.

Résumé. When the thermal contact between the coil and iron is good the thermal capacity of iron may be taken as 55 per cent effective after a time of about one-third of the thermal time constant. If the thermal contact is poor it should be taken as 45 per cent effective. During the first part of the heating cycle for a duration of about one-third the thermal time constant the thermal capacity of iron may be taken at one-half its final effective value if the total thermal capacity of the iron per unit of coil heat-dissipating surface is about 30 joules per degree Centigrade per square inch; if it is as low as 8 joules per degree Centigrade per square inch, the iron thermal capacity may be taken equal to its final effective value. For other values of this constant, proportional values of iron effectiveness can be chosen.

TABLE II

Thermal Capacity of Magnet of Fig. 1a, Chapter III

(1) Power Input, watts	(2) Coil only	(3) Coil and Surrounding Iron		(4) Iron (Effective)		(5) Per cent Effectiveness of Iron	
	From data of Fig. 5a	From data of		From data of		From data of	
		Fig. 5b	Fig. 5c	Fig. 5b	Fig. 5c	Fig. 5b	Fig. 5c
40	620	1,470	1,320	920	770	55	46
30	623	1,500	1,315	950	765	56	45
20	593	1,480	1,300	930	750	55	44
10	575	1,490	1,290	940	740	56	44

Thermal Capacity of Horseshoe Magnet of Problem 4, Chapter V

	From data of Fig. 5d, curve 16	From data of Fig. 5d, curve 14		
20	755	1,015	260	48

68. Calculation of Time-Temperature Rise Curve for an Ironclad Magnet

As an example of the method of calculating the temperature rise as a function of time let it be desired to calculate for the ironclad magnet,[4] data for which are given below, the time required to reach a temperature rise (average for the coil) of 70° C. if the power input is maintained constant at 148 watts.

Given:
  Magnet type, ironclad solenoid and plunger.
  Force, 3.5 lb.
  Stroke, 10 in.
  Weight of copper coil, 2.14 lb.
  Weight of iron, including plunger, 4.24 lb.
  Weight of brass bobbin tube, 0.33 lb.
  Total coil surface, 82 sq. in.
  Thermal contact between iron parts and shell may be considered good.

[4] This magnet is illustrated in Fig. 9, Chapter IX.

Solution. Referring to Fig. 7b, the heat-dissipation coefficient at a 70° C. final temperature rise may be taken equal to 0.00755 from curve 1 for good thermal contact. The heat-dissipation constant K will therefore be $0.00755 \times 82 = 0.618$ watt dissipation per degree Centigrade temperature difference between the average coil temperature and the surrounding air. The thermal capacity of the copper coil will be $2.14 \times 180 = 385$, while that of the iron and brass will be $0.55 \times 4.25 \times 225 + 0.55 \times 0.33 \times 200 = 525 + 36 = 561$, assuming the iron and brass to be 55 per cent effective for the condition of good thermal contact with

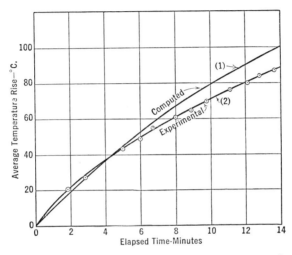

Fig. 8. Computed and experimental temperature-rise-time curves for the magnet of Art. 68.

the coil. The thermal time constant of the magnet will be equal to $C/K = (385 + 561)/0.618 = 1{,}530$ sec. or 25.5 minutes. The total iron thermal capacity per unit of coil dissipating surface will be $(525/0.55)/82 = 11.7$. If the suggestion of the last article for the first 8.5 minutes of heating (about one-third of the thermal time constant) is followed, the iron thermal capacity can be taken as $100 - 50 \times (12 - 8)/(30 - 8) = 91$ per cent of its final effective value, or $0.91 \times 561 = 511$ joules per degree Centigrade rise. Substituting into equation 2, the heating equation for the first 8.5 minutes will be

$$\theta = \frac{P}{K}\left(1 - \epsilon^{-\frac{K}{C}t}\right) = \frac{148}{0.618}\left(1 - \epsilon^{-\frac{0.618}{385+511}t}\right)$$

or

$$\theta = 240(1 - \epsilon^{-0.000690t}), \quad [0 < t < 510]$$

After the elapse of the 8.5-minute interval the heating equation will be

$$\theta = \frac{148}{0.618}\left(1 - \epsilon^{-\frac{0.618}{385+561}t}\right)$$

or

$$\theta = 240(1 - \epsilon^{-0.000653\,t}), \ [510 < t < \infty\,]$$

Substituting into these equations various values of t the temperature rise curve 1 of Fig. 8 will be obtained. Curve 2 of Fig. 8 was obtained by experiment on the actual magnet. The actual time required to produce an average coil temperature rise of 70° C. is 9.75 minutes; the computed time is 8.45 minutes. This difference, though it may appear considerable, is not so bad considering that the solenoid and plunger magnet is of a very different shape from the magnets of Fig. 5. It is very long and narrow with an extremely thin coil wall. Because of the thin coil wall, the surface coil temperature is closer to the average coil temperature than was assumed in Fig. 7, with the result that the heat-dissipation coefficient is greater than that given in Fig. 7. A more accurate solution to a problem of this type can be obtained if a thermal equivalent circuit based on the suggestion of footnote 1, page 184, is used.

PROBLEMS

1. The suggested equivalent circuit of footnote 1, page 184, may be represented by the circuit of Fig. 9, where the condenser C_1 represents the thermal capacity of the coil, the condenser C_3 the thermal capacity of the iron, G_2 the thermal conductivity between the coil and the iron, and G_4 the heat-dissipating capacity of the outer iron shell to the surrounding air. The constant power input to the coil is simulated by the constant current I flowing into the circuit. Derive an expression for the voltage across C_1 as a function of time for the condition of a constant current of I amperes suddenly impressed on the circuit, and then rewrite your answer in terms of thermal quantities. Discuss what information, in addition to that already contained in Chapter VII, is necessary to use this circuit for temperature-rise problems.

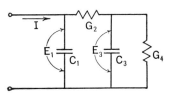

Fig. 9. Electrical equivalent circuit for computing the temperature-rise-time curve for a magnet coil surrounded by iron parts.

Ans.

$$\theta_1 = P\frac{K_2 + K_4}{K_2 K_4} + \theta_{11}\epsilon^{a_1 t} + \theta_{12}\epsilon^{a_2 t}$$

$$\theta_3 = \frac{P}{K_4} + \theta_{31}\epsilon^{a_1 t} + \theta_{32}\epsilon^{a_2 t}$$

where

$$a = \frac{-\left(C_1 + C_3 + C_1\dfrac{K_4}{K_2}\right) \pm \sqrt{\left(C_1 + C_3 + C_1\dfrac{K_4}{K_2}\right)^2 - 4\dfrac{C_1 C_3 K_4}{K_2}}}{\dfrac{2C_1 C_3}{K_2}}$$

$$\theta_{11} = \frac{\left(\dfrac{1}{C_1} + a_2\dfrac{K_2 + K_4}{K_2 K_4}\right) P}{a_1 - a_2}$$

$$\theta_{12} = \frac{\left(\dfrac{1}{C_1} + a_1\dfrac{K_2 + K_4}{K_2 K_4}\right) P}{a_2 - a_1}$$

$$\theta_{31} = \frac{a_2}{K_4(a_1 - a_2)} P$$

$$\theta_{32} = \frac{a_1}{K_4(a_2 - a_1)} P$$

2. The coil of the lifting magnet of Fig. 13, Chapter IV, is wound on a brass bobbin which is tightly fitted over the iron pole core. The length of the brass bobbin is $1\frac{3}{16}$ in. If a constant power input of 40 watts to the coil is maintained, compute the final temperature rise. *Ans.* By experiment, with a constant power input of 40 watts, the final temperature rise is 83° C.

3. If the field coil of the relay of Fig. 22, Chapter IV, is a paper-section coil wound for 120 volts, determine the final temperature rise, if the coil is excited continuously as described. *Note:* It will be necessary to estimate the space factor of the coil, which may be done by the proper reference to Chapter VI.

4. For how long may the magnet of Fig. 15, Chapter VIII, be excited with 100 watts if the average temperature rise of the coil is not to exceed 70° C. The coil is of the paper-section type, which is in poor thermal contact with the iron parts.

5. Calculate, by using the answer to Problem 1, the temperature-rise-time curves for the coil and iron parts of the magnet of Fig. 10, assuming the coil terminal voltage to be held constant at 13.0 volts, and the room temperature constant at 20° C.

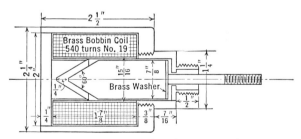

FIG. 10. Magnet designed for short time of excitation.

DATA

Coil resistance at 20° C.................... 1.66 ohm
Total weight of all iron parts............... 1.62 lb.
Total weight of coil and bobbin............. 0.82 lb.

The bobbin is fabricated from a brass tube with Bakelite end flanges. The thermal contact between the coil and iron parts is poor.

Discussion. In order to solve the problem it will be necessary to determine the values of K_2 and K_4 of the thermal equivalent circuit. The value of K_4 (heat-dissipation capacity of the outer iron surfaces to the surrounding air) cannot be determined from the data of the chapter. Reference should be made to "Fundamentals of Electrical Design," by A. D. Moore, McGraw-Hill Book Co.

The heat conductivity K_2 between the coil surface and the surrounding iron surfaces can be estimated if it is possible to predetermine the final temperature rise of the coil and iron parts. This may be done on the basis of the work of Art. 64, namely, that the final temperature rise of a coil in poor thermal contact with the iron is the same as it would be in the open air. By this assumption it will be possible to estimate the final coil temperature rise. This being known, the final iron temperature rise may be estimated on the basis of the statement in Art. 67, that the thermal capacity of the iron in poor thermal contact with the coil is only 46 per cent effective. This is equivalent to saying that the final iron temperature rise is 46 per cent that of the coil. The constant K_2 will then be $P_f/(\theta_c - \theta_i)$, where P_f is the final power transferred between the coil and iron surfaces, and θ_c and θ_i are the final temperature rises of the coil and iron, respectively. This equation can be solved by trial and error as follows: Assume θ_c to be 215° C.; then

$$R = 1.66 \times \frac{469.5}{254.5} = 3.06 \text{ ohms, and } P_f = \frac{13^2}{3.06} = 55.1 \text{ watts}$$

The total surface of the coil is approximately 22 sq. in., therefore the final power dissipation per square inch of coil surface is $55.1 \div 22 = 2.5$ watts. Referring now to Fig. 7a, the data of curve 2 may be extrapolated by assuming the curve to follow

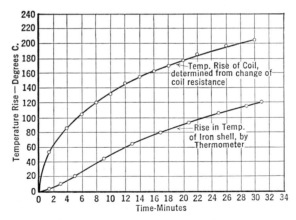

Fig. 11. Experimental temperature-rise-time curves for the coil and shell of the magnet of Fig. 10.

the last two points, or $k = 0.007 + (0.0005 \div 0.2) \times (2.5 - 0.6) = 0.01175$. K will then be $22 \times 0.01175 = 0.258$, and the final temperature rise of the coil with reference to the room temperature will be $55.1 \div 0.258 = 214°$ C. As this checks with the original assumption it will be taken as correct. The final temperature rise of the iron parts will be $214 \times 0.46 = 99°$ C. The heat conductivity K_2 between the coil and iron surfaces will therefore be $P_f/\theta_f = 55.1/(214 - 99) = 0.48$ watt per degree Centigrade temperature difference.

As the power input varies very considerably with time it will be necessary to perform the evaluation of the equations of Problem 1 in steps. For each step a constant power input should be assumed; the length of each step should be so chosen that the constant power assumed does not differ too much from the extreme values of the actual power.

The actual experimental temperature-rise-time curves for the coil and iron parts of this magnet for a constant impressed voltage of 13.0 volts are shown in Fig. 11.

CHAPTER VIII

MAGNETIC FORCES

69. General

In Art. 34, Chapter III, it was shown that the mechanical energy made available by the motion of the armature or plunger of an electromagnet can always be represented by an area on a suitably drawn flux linkage current loop, and that, if the displacement of the armature is made sufficiently small, the force at any particular position of the armature can be obtained as the quotient of the area by the displacement. This method of determining the force is not only theoretically correct, but also practically exact, as it takes into account all changes in stored energy, all leakage and fringing fluxes, and all losses due to hysteresis. Unfortunately, the evaluation must generally be carried out graphically. With certain simplifying assumptions, however, it is possible to derive analytical expressions covering particular special cases. It is the purpose of this chapter to derive these special formulas and to set forth, exactly, their limitations.

70. Derivation of the General Magnetic Force Formula

Consider a magnetic system, of a permeability greater than that of air, having two surfaces approaching each other as illustrated in Fig. 1a. Assume that all the flux ϕ passing between the surfaces links with all the turns of the exciting winding, and that the only change in the magnetic circuit with motion is in the gap between the surfaces under consideration.[1] Suppose the magnetomotive force[2] to be increased from $-F_d$ to the value F_1, thereby causing the flux to increase from zero to the value ϕ_1, as shown by the line 0–1 of Fig. 1b. During this process energy will be abstracted from the electric circuit and partially stored as magnetic energy and the rest dissipated as heat due to magnetic hysteresis. Now let the surfaces approach each other by an infinitesimal dis-

---

[1] This restriction should be carefully noted: it removes from consideration for the time being magnets in which the motion causes the length of the iron part of the magnetic circuit to change, e.g., plunger magnets in general.

[2] The magnetomotive force $-F_d$ is that required to reduce the residual flux to zero.

tance ds inches while the flux is held constant at ϕ_1, as shown by the line 1–2. During this motion mechanical work will be done entirely at the expense of the stored energy in the gap, and the magnetomotive force across the gap will be decreased by dF_a. The same final state 2 (same final stored energy and same hysteresis loss) would have been reached if the motion had taken place first at point 0 and then the flux had been allowed to build up to the value ϕ_1 as shown by the line 0–2. This process, however, would have produced no mechanical work. There-

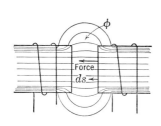

FIG. 1a. General case of two high-permeability surfaces approaching each other.

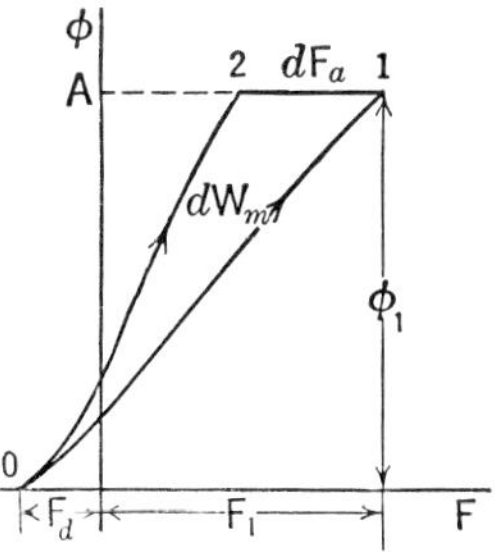

FIG. 1b. Flux-magnetomotive force loop for deriving the force between the surfaces of Fig. 1a based on constant flux during motion.

fore, the area 0–1–2–0, which represents the difference in the energy abstracted from the electric circuit by process 0–1–2 and 0–2, must be equal to the mechanical work dW_m joules performed during the motion 1–2. The difference in magnetomotive force for any value of flux in steps 0–1 and 0–2 is entirely the decrease in the magnetomotive force $(-dF_a)$ required by the gap in the second position due to its increased permeance, and hence will be proportional to ϕ. Therefore, dW_m will be equal to $-\frac{1}{2}\phi_1 dF_a$, and

$$\text{Force} = \frac{dW_m}{ds} = -\frac{1}{2}\phi_1 \frac{dF_a}{ds} \quad \text{joules per inch} \qquad (1)$$

For the air gap, $\phi_a = F_a P_a$, and, since $\phi_1 = \phi_a$ is constant,

$$\frac{dF_a}{ds} = -\frac{\phi_a}{P_a^2}\frac{dP_a}{ds}$$

Substituting this into equation 1, we get

$$\text{Force} = \frac{1}{2}\frac{\phi_a^2}{P_a^2}\frac{dP_a}{ds} = \frac{1}{2}F_a^2 \frac{dP_a}{ds} \quad \text{joules per inch} \qquad (2a)$$

As the joule per inch is equal to 8.86 lb., equation 2a can be written

$$\text{Force} = 4.43 \, F_a^2 \frac{dP_a}{ds} \quad \text{lb.} \tag{2b}$$

In this equation, according to the limitations under which it was derived, F_a is the magnetomotive force across the air gap [3] in ampere-turns, and dP_a/ds is the rate of change of the total air-gap permeance (air gap includes all paths for fringing and other fluxes which link with all the turns of the exciting coils) in webers per ampere-turn per inch. The force will, of course, be in the direction taken for ds, in the term dP_a/ds.

If the system is capable of rotation instead of translation, equation 2a is more useful if expressed in terms of torque, thus

$$\text{Torque} = \frac{1}{2} F_a^2 \frac{dP_a}{d\theta} \quad \text{joules per radian} \tag{3a}$$

where $dP_a/d\theta$ is the angular rate of change of the total air-gap permeance in webers per ampere-turn per radian; or, converting from joules per radian to pound-inches, we have

$$\text{Torque} = 4.43 \, F_a^2 \frac{dP_a}{d\theta} \quad \text{lb-in.} \tag{3b}$$

These equations can be easily rewritten in terms of the magnetomotive force and permeance of the entire magnet system, as follows: For the entire magnet $\phi_m = (F_d + F_1)P_m$, where P_m is equal to the permeance of the entire magnetic circuit, that is, both iron and air gap; and ϕ_m, which is equal to ϕ_a, is the flux of the magnetic circuit. As ϕ_m is constant

$$\frac{d(F_d + F_1)}{ds} = -\frac{\phi_m}{P_m^2} \frac{dP_m}{ds}$$

And as the magnetomotive force across the iron must be constant,

$$\frac{d(F_d + F_1)}{ds} = \frac{dF_a}{ds}$$

Substituting these into equation 1, we have [4]

$$\text{Force} = \frac{1}{2} \frac{\phi_m^2}{P_m^2} \frac{dP_m}{ds} = \frac{1}{2} (F_d + F_1)^2 \frac{dP_m}{ds} \quad \text{joules per inch} \tag{4}$$

[3] The total magnetomotive force across the air gap regardless of source, that is, whether it is derived from the exciting current or from the coercive magnetomotive force of the iron.

[4] In general this formula is more difficult to apply than formula 2 because the calculation of dP_m/ds is more difficult and uncertain than the evaluation of F_a and dP_a/ds.

where dP_m/ds is taken for constant ϕ. Likewise, formula 3a can be written

$$\text{Torque} = \frac{1}{2}(F_d + F_1)^2 \frac{dP_m}{d\theta} \quad \text{joules per radian} \tag{5}$$

If the magnetic system consists of a permanent magnet, F_1 will be zero and F_d will be the magnetomotive force necessary to demagnetize the permanent magnet. If the magnetic system consists of soft steel which requires negligible magnetomotive force to remove its residual flux, F_d will be zero and F_1 will equal the magnetomotive force of the exciting coils.

If in the above derivation it had been assumed that the magnetomotive force of the exciting coils was constant during the infinitesimal motion, the flux-magnetomotive force loop would have been changed to that shown in Fig. 2. Here, during the motion from 1 to 2, energy equal to the area A–1–2–B–A is abstracted from the electric circuit. This energy equal to $F_1 d\phi$ is converted into stored magnetic energy, hysteresis loss, and mechanical work. The same final magnetic state would have been reached, and the same hysteresis loss would have occurred, had the motion taken place first at point 0 and then the magnetomotive force had been allowed to build up from $-F_d$ to F_1, line 0–2; but no mechanical work would have been done. Therefore, the sectioned area 0–1–2–0 must be equal to the mechanical work dW_m performed during the infinitesimal motion.

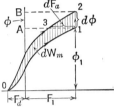

Fig. 2. Flux-magnetomotive force loop for deriving the force between the surfaces of Fig. 1a based on constant magnetomotive force during motion.

If we neglect the second-order area 3–1–2–3, dW_m will equal $-\frac{1}{2}\phi dF_a$ and we will arrive at force formulas already derived on the basis of constant flux during motion.

If we include the second-order area 3–1–2–3, and assume the iron unsaturated,[5] dW_m will equal $\frac{1}{2}(F_d + F_1)d\phi$, where F_d is the magneto-

---

[5] When the iron is unsaturated, the vertical displacement $d\phi$ between curves 0–1 and 0–2 will be proportional to F; hence, $dW_m = \frac{1}{2}Fd\phi$. Thus, under these conditions, the energy abstracted from the electric circuit due to motion (with the magnetomotive force held constant) will be divided into two equal parts, one half increasing the stored energy of the magnet, the other half going into mechanical work. If the magnet is highly saturated this division of energy will no longer hold, but the mechanical force developed will be given by the same formula. The derivation for this case must be carried out by treating the second-order area 3–1–2–3 separately and noting that on evaluation it disappears.

motive force necessary to remove the residual flux and

$$\text{Force} = \frac{dW_m}{ds} = \frac{1}{2}(F_d + F_1)\frac{d\phi}{ds} \quad \text{joules per inch} \qquad (6)$$

For the entire magnetic system $\phi = (F_d + F_1)P_m$, and $d\phi/ds = (F_d + F_1)(dP_m/ds)$, it being remembered that $(F_d + F_1)$ is constant during the motion. Substituting these relationships into equation 6, we get

$$\text{Force} = \frac{1}{2}(F_d + F_1)^2\frac{dP_m}{ds} \quad \text{joules per inch}$$

which formula is identical with (4).

It follows, therefore, in the case of attraction between magnetized surfaces that the instantaneous magnetic force will always be the same, regardless of whether the motion is carried out with the flux constant or the magnetomotive force constant.[6]

71. Loss of Force Occurring When the Magnetic Circuit Changes in Places Other Than in the Air Gap under Consideration

Wherever motion in the air gap under consideration introduces iron into the magnetic circuit which was previously unmagnetized, a loss in force will occur. As an illustration, consider the cylindrical plunger magnet shown in section in Fig. 3.

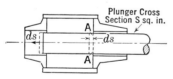

Fig. 3. Diagram for deriving the loss in force that occurs when the magnetic circuit changes in places other than the air gap under consideration.

When the plunger is moved an infinitesimal distance ds (flux constant), it has the effect of introducing into the magnetic circuit a volume of iron Sds, shown at section A–A, which must be magnetized from zero to the maximum value of flux ϕ_m. This will cause the decrement in magnetomotive force $(-dF_a)$, Fig. 1a, to be less by the amount $dF_p = H_p ds$ required to magnetize the length ds of the plunger. Hence the area of the loop dW_m will be less by the amount

---

[6] Whereas, for the same initial values, the instantaneous forces are equal, it does not follow that the total work done during a finite displacement will be the same under the two conditions of motion. Motion carried out under the condition of constant magnetomotive force will always yield more work than when carried out under the condition of constant flux, for the reason that the instantaneous force increases with the finite motion.

$$(dW_m)_p = \int_0^{\phi_m} dF_p d\phi = \int_0^{\phi_m} (H_p ds) d\phi$$

$$= S ds \int_0^{B_m} H_p dB \quad \text{joules}$$

where B_m is the maximum plunger flux density occurring at section A–A. The loss in force will be

$$\text{Loss in force} = \frac{(dW_m)_p}{ds} = S \int_0^{B_m} H_p dB \quad \text{joules per inch} \qquad (7a)$$

or converting from joules per inch to pounds, and from webers per square inch to kilomaxwells per square inch, we have

$$\text{Loss in force} = 8.86 \times 10^{-5} S \int_0^{B_m} H_p dB \quad \text{lb.} \qquad (7b)$$

This loss is quite negligible in flat-faced plunger magnets in which the forces are large, but is appreciable in plunger magnets having low working forces, such as solenoid and plunger types. To evaluate formula 7b the area behind the rising magnetization curve for the particular steel of the plunger must be determined graphically, or from the data of Fig. 14, Chapter II.

72. Formulas for Magnetic Force in Special Cases

In many magnetic systems where the geometry of the air-gap field is relatively simple, or where certain simplifying assumptions can safely be made, it is possible to derive simple analytical expressions for rate of change of permeance of the magnetic system with motion, thus enabling the general magnetic force formula $\frac{1}{2}F_a^2(dP_a/ds)$ to be put in a form which is convenient for analytical use. Where this is not possible the general formula will have to be evaluated graphically. In the following sections special forms of this force formula, to cover the more usual simple field configurations, will be developed.

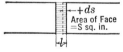

Fig. 4. Special case of force between parallel plane surfaces approaching each other.

1. Parallel Plane Surfaces; Attractive Force Due to Flux Passing Normally between Surfaces; Fringing Flux Neglected. Figure 4 shows the cross section of such a magnetic system comprising two plungers, the opposite faces of which are parallel and are free to move toward each other in a direction normal to the faces. Using the notation shown in the figure, the permeance of the air gap by equation 1, Chapter V, is

$$P_a = \frac{\mu S}{l}$$

Differentiating this expression with respect to the motion ds, we have

$$\frac{dP_a}{ds} = -\frac{\mu S}{l^2}\frac{dl}{ds}$$

A positive increment in l (increase in gap length) represents a negative increment in displacement, in the sense that a positive displacement produces an output of mechanical work, hence $dl = -\,ds$. Substituting the above relations in the general magnetic force formula 2a, we have

$$\text{Force} = \frac{1}{2}F_a^2\frac{dP_a}{ds} = \frac{1}{2}F_a^2\frac{\mu S}{l^2}$$

The flux density in the air gap will be

$$B = \mu H = \frac{\mu F_a}{l}$$

Substituting this into the above force expression, we obtain

$$\text{Force} = \frac{B^2 S}{2\mu} \quad \text{joules per inch} \tag{8a}$$

where B is the flux density in the air gap in webers per square inch, S is the area of the air gap in square inches, and μ is the permeability of air in webers per ampere-turn in an inch cube. By changing B from webers per square inch to kilomaxwells per square inch, and making the proper numerical substitutions for the constants, we obtain the following more useful form:

$$\text{Force} = \frac{B^2 S}{72.0} \quad \text{lb.} \tag{8b}$$

If the approaching surfaces are circular, as in the flat-faced plunger magnet of Fig. 1a, Chapter III, equation 8b can be changed to the more convenient form

$$\text{Force} = \frac{B^2 \pi r_1^2}{72.0} = \frac{B^2 r_1^2}{22.9} \quad \text{lb.} \tag{8c}$$

where r_1 is the radius of the circular surface in inches.

2. **Coaxial Cylindrical Surfaces; Axial Force Due to Radial Flux between Cylinders; Fringing Flux Neglected.** Figure 5 shows a cross section of such a magnetic system comprising a cylindrical plunger and a surrounding concentric cylindrical shell, the plunger being free to move along its axis. Using the notation shown in the figure, the permeance of the air gap (neglecting fringing [7] and assuming g small compared to r_1) by equation 15a, Chapter V, is

$$P_a = \frac{2\pi\mu \left(r_1 + \frac{g}{2}\right) l}{g}$$

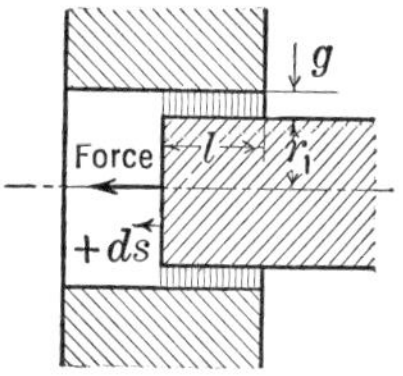

Differentiating this expression with respect to the motion ds, we have

$$\frac{dP_a}{ds} = \frac{2\pi\mu \left(r_1 + \frac{g}{2}\right)}{g} \frac{dl}{ds}$$

FIG. 5. Special case of axial force produced by coaxial cylindrical surfaces.

Noting that $dl = ds$, we will obtain on substituting in the general magnetic force equation 2a

$$\text{Force} = \frac{1}{2} F_a^2 \frac{dP_a}{ds} = \frac{\pi\mu \left(r_1 + \frac{g}{2}\right) F_a^2}{g} \quad \text{joules per inch} \quad (9a)$$

This formula can be expressed in terms of the air-gap flux density, but no particular advantage is obtained. Substituting for the constants in the above equation, the following more useful form is obtained:

$$\text{Force} = \frac{\left(r_1 + \frac{g}{2}\right) F_a^2}{1{,}125{,}000 g} \quad \text{lb.} \quad (9b)$$

where the lineal dimensions can be in any units as any conversion factors cancel, and F_a is the magnetomotive force across the air gap in ampere-turns. This formula will not apply before the plunger enters the surrounding cylinder or after it is relatively close to the open end, because under either of these conditions the rate of change of air-gap permeance is different from that calculated on the basis of the simple radial field assumed above.

---

[7] As long as the plunger has entered the surrounding shell and is relatively far from the open end of the shell the fringing permeances are constant and hence do not affect the force.

Formula 9b can be expressed in terms of the plunger flux density as follows: the flux in the plunger, neglecting fringing and leakage,[8] is

$$\phi = F_a P_a = \frac{2\pi\mu\left(r_1 + \dfrac{g}{2}\right) l F_a}{g}$$

and the plunger flux density is

$$B_p = \frac{\phi}{\pi r_1^2} = \frac{2\mu\left(r_1 + \dfrac{g}{2}\right) l F_a}{r_1^2 g}$$

Substituting this into equation 9b we will have after substituting for the numerical value of the constants

$$\text{Force} = \frac{r_1^4 g B_p^2}{45.8(r_1 + g/2)l^2} \text{ lb.} \tag{9c}$$

where B_p is the flux density in the plunger in kilomaxwells per square inch, neglecting fringing and leakage fluxes. Because the force in this type of magnet is relatively small the correction given by equation 7 should be applied when B_p is high.

3. **Coaxial Cylindrical Surfaces; Torque about Axis of Cylinders Due to Radial Flux between Cylinders; Fringing Flux Neglected.** Figure 6 shows a cross section of such a magnetic system having an axial length l, the inner cylindrical member being free to rotate about its axis. Using the notation shown in the figure, the permeance of both air gaps, in series (neglecting fringing),[7] will be given by equation 15a, Chapter V, if we replace the 2π radians for the entire circumference by θ radians; thus,

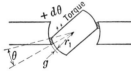

FIG. 6. Special case of torque produced by coaxial cylindrical surfaces.

$$P_a = \frac{\mu\left(r_1 + \dfrac{g}{2}\right) l\theta}{2g}$$

---

[7] See footnote on page 203.

[8] Radial leakage flux between the plunger and outer shell in this type of magnet will create an additional force which can be conveniently calculated by equation 20a. This force will usually be quite an appreciable part of the total force before the plunger enters the surrounding cylinder, that is, when the force due to the main cylindrical gap is small.

Differentiating this expression with respect to the rotation $d\theta$, we have

$$\frac{dP_a}{d\theta} = \frac{\mu\left(r_1 + \dfrac{g}{2}\right)l}{2g}$$

If this expression is substituted in the general magnetic torque equation ($3a$), we will obtain

$$\text{Torque} = \frac{1}{2}F_a^2\frac{dP_a}{d\theta} = \frac{\mu\left(r_1 + \dfrac{g}{2}\right)l}{4g}F_a^2 \quad \text{joules per radian} \quad (10a)$$

Substituting for the constants in the above equation, the following more useful form for the total torque is obtained:

$$\text{Torque} = \frac{\left(r_1 + \dfrac{g}{2}\right)l}{14{,}150{,}000\,g}F_a^2 \quad \text{lb-in.} \quad (10b)$$

where the lineal dimensions are in inches, and F_a is the magnetomotive force across both gaps together in ampere-turns. This formula is subject to the same restrictions as formula $9a$.

4. Coaxial Truncated Conical Surfaces; Axial Force Due to Radial Flux between Cones; Fringing Flux Neglected. Figure 7 shows the cross section of such a magnetic system comprising a truncated conical plunger and a surrounding conical shell. Using the notation shown in the figure, the air-gap permeance (neglecting fringing)[7] will be, by equation 1, Chapter V,

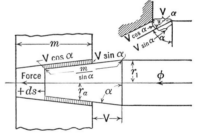

FIG. 7. Special case of axial force produced by coaxial truncated conical surfaces.

$$P_a = \frac{\mu S}{l} = \frac{2\pi\mu r_a}{V\cos\alpha}\left(\frac{m}{\sin\alpha} - V\sin\alpha\right) \quad (11)$$

where r_a is the mean radius of the truncated section inserted in the shell.

Differentiating this expression with respect to the motion ds, we have

$$\frac{dP_a}{ds} = -\frac{2\pi\mu m r_a}{V^2\sin\alpha\cos\alpha}\frac{dV}{ds}$$

[7] See footnote on page 203.

Noting that $dV = -\,ds$, and substituting into the general magnetic force formula 2a, we have

$$\text{Force} = \frac{1}{2}F_a^2\frac{dP_a}{ds} = \frac{\pi\mu m r_a F_a^2}{V^2 \sin\alpha\cos\alpha} \quad \text{joules per inch} \quad (11a)$$

Substituting for the constants in the above equation, the following more useful form is obtained:

$$\text{Force} = \frac{m r_a F_a^2}{1{,}125{,}000 V^2 \sin\alpha\cos\alpha} \quad \text{lb.} \quad (11b)$$

where all the lineal dimensions can be in any units, as any conversion factors cancel, and F_a is the magnetomotive force across the air gap in ampere-turns. This force can also be expressed in terms of the flux density of the plunger, as follows: the total flux of the plunger (neglecting the fringing and leakage [8] fluxes) will be

$$\phi = F_a P_a = \frac{2\pi\mu r_a F_a}{V\cos\alpha}\left(\frac{m}{\sin\alpha} - V\sin\alpha\right)$$

and the flux density in the full part of the plunger will be

$$B_p = \frac{\phi}{S} = \frac{\phi}{\pi r_1^2} = \frac{2\mu r_a F_a}{V r_1^2 \cos\alpha}\left(\frac{m}{\sin\alpha} - V\sin\alpha\right)$$

Substituting into equation 11b, we have

$$\text{Force} = \frac{B_p^2 r_1^4 m \cos\alpha}{45.8 r_a V^2 \sin\alpha\left(\dfrac{m}{V\sin\alpha} - \sin\alpha\right)^2} \quad \text{lb.} \quad (11c)$$

where B_p is in kilomaxwells per square inch.

When the plunger is at the end of its stroke ($V = 0$), this equation will evaluate to

$$\text{Force} = \frac{B_p^2 r_1^4 \sin\alpha\cos\alpha}{45.8 m r_a} \quad \text{lb.} \quad (11d)$$

When α is large (near 90°) the force in this type of magnet will be relatively small, and the correction given by equation 7 should be applied when B_p is large.

5. Coaxial Full Conical Surfaces; Axial Force Due to Radial Flux between Cones. Figure 8 shows the cross section of such a system com-

---

[8] See footnote on page 204.

SPECIAL MAGNETIC FORCE FORMULAS

prising a full conical plunger and a surrounding conical shell. This is, of course, a special case of the truncated conical plunger, and the force will be given by equation 11c. Using the notation of Fig. 8, the force will be

$$\text{Force} = \frac{B_p^2 r_1^4 \tan\alpha \cos\alpha}{45.8 r_a V^2 \sin\alpha \left(\dfrac{r_1 \tan\alpha}{V \sin\alpha} - \sin\alpha\right)^2} \text{ lb.}$$

where B_p is the flux density in the plunger due to the normal flux

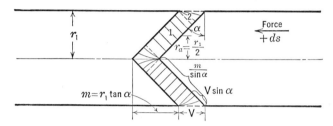

FIG. 8. Special case of axial force produced by coaxial full conical surfaces.

between cones only, that is, the flux in the working gap 1 of Fig. 8. Assuming that $r_a = r_1/2$, canceling, and simplifying, we have

$$\text{Force} = \frac{B_p^2 r_1^2 \cos^2\alpha}{22.9\left(1 - \dfrac{V}{r_1}\sin\alpha\cos\alpha\right)^2} \text{ lb.}$$

This formula may be simplified by substituting for B_p in terms of the total flux between the conical surfaces. This flux will include that in the fringing path 2 besides that in the main path 1, and will be proportional to the sum of the permeances of paths 1 and 2.

$$P_1 = \frac{\pi\mu r_1}{V \cos\alpha}\left(\frac{m}{\sin\alpha} - V\sin\alpha\right) = \frac{\pi\mu r_1}{V \cos\alpha}\left(\frac{r_1}{\cos\alpha} - V\sin\alpha\right)$$

and

$$P_2 = \frac{2\pi\mu(r_1 - \tfrac{1}{2}V\sin\alpha\cos\alpha)\tfrac{1}{2}V\sin\alpha}{V\cos\alpha}$$

$$P_1 + P_2 = \frac{\pi\mu r_1}{V\cos\alpha}\left[\left(\frac{r_1}{\cos\alpha} - V\sin\alpha\right) + \left(1 - \frac{1}{2}\frac{V}{r_1}\sin\alpha\cos\alpha\right)V\sin\alpha\right]$$

$$P_1 + P_2 = \frac{\pi\mu r_1}{V\cos\alpha}\left[\frac{r_1}{\cos\alpha} - \frac{1}{2}\frac{V^2}{r_1}\sin^2\alpha\cos\alpha\right] \quad (12)$$

Letting B be the flux density in the plunger due to both the normal flux between cones and the fringing flux, we have

$$\frac{B_p}{B} = \frac{P_1}{P_1 + P_2} = \frac{\dfrac{r_1}{\cos \alpha} - V \sin \alpha}{\dfrac{r_1}{\cos \alpha} - \dfrac{1}{2}\dfrac{V^2}{r_1}\sin^2 \alpha \cos \alpha}$$

Performing the long division indicated we will obtain the quotient,

$$\frac{B_p}{B} = 1 - \frac{V}{r_1}\cos \alpha \sin \alpha + \frac{1}{2}\frac{V^2}{r_1^2}\sin^2 \alpha \cos^2 \alpha + \cdots$$

All terms beyond the second are negligibly small; hence, substituting

$$B_p = B\left(1 - \frac{V}{r_1}\cos \alpha \sin \alpha\right)$$

in the force formula, we have

$$\text{Force} = \frac{B^2 r_1^2 \cos^2 \alpha}{22.9} \text{ lb.} \qquad (12a)$$

It is interesting to note that when α is made equal to zero the magnet becomes a flat-faced cylindrical plunger magnet and formula $12a$ becomes the same as formula $8c$.

5a. Coaxial Full Conical Surfaces; Total Axial Force Due to Radial and Fringing Flux between Cones. The permeance of path 2 of the flux of Fig. 8 approaches infinity as V becomes zero, and hence produces a positive force. A force formula taking this into account may be derived by substituting directly into the general magnetic force formula $2b$,

$$\text{Force} = 4.43 F_a^2 \frac{dP_a}{ds} \text{ lb.}$$

dP_a/ds may be found from eq. (12) as follows:

$$\frac{dP_a}{ds} = \frac{d(P_1 + P_2)}{ds} = -\frac{d(P_1 + P_2)}{dV} = \mu\pi\left[\left(\frac{r_1}{V \cos \alpha}\right)^2 + \frac{1}{2}\sin^2 \alpha\right]$$

Substituting into equation $2b$, we have

$$\text{Force} = 4.43\, \mu\pi F_a^2 \left[\left(\frac{r_1}{V \cos^{\frac{3}{2}} \alpha}\right)^2 + \frac{1}{2}\sin^2 \alpha\right] \text{ lb.} \qquad (12b)$$

This formula is slightly more optimistic than formula 12a, because it takes into account the force produced by the fringing flux. This correction is the second term $\frac{1}{2}\sin^2\alpha$. In general, unless α is large, formula 12a is to be preferred as being more simple.

6. **Coaxial Cylindrical and Conical Surfaces; Axial Force Due to Radial Flux between Inner Cone and Outer Cylinder; Fringing Flux Neglected.** Figure 9 shows a cross section of such a magnetic system comprising a truncated conical or tapered plunger entering a surrounding cylindrical shell. The permeance between the taper plunger and the surrounding cylindrical shell may be computed approximately as follows: Assume the radial flux lines to follow circular paths having their centers at the intersection of the corresponding element of the cylinder and cone on the line A–A. Consider the annular flux path having an axial length of dx shown in cross section; the permeance of this path will be, by equation 1, Chapter V,

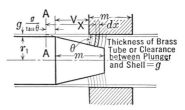

Fig. 9. Special case of axial force produced by coaxial cylindrical and conical surfaces.

$$dP_a = \frac{\mu dS}{l} = \frac{2\pi\mu\left(r_1 + g - \frac{\theta X}{2}\right) dx}{\theta X}$$

assuming that the angle θ is small. The permeance of the entire path will be (neglecting fringing)

$$P_a = \int_{x=V+\frac{g}{\tan\theta}}^{x=m+\frac{g}{\tan\theta}} \frac{2\pi\mu\left[(r_1 + g) - \frac{\theta X}{2}\right] dx}{\theta X}$$

$$P_a = \pi\mu \int \left[\frac{2(r_1 + g)dx}{\theta X} - dx\right] = \pi\mu \int \left[\frac{2(r_1 + g)}{\theta} \cdot \frac{\theta dx}{\theta X} - dx\right]$$

Performing the integration and substituting in the limits, we have

$$P_a = \pi\mu \left[\frac{2(r_1 + g)}{\theta} \log_\epsilon \frac{\theta\left(m + \frac{g}{\tan\theta}\right)}{\theta\left(V + \frac{g}{\tan\theta}\right)} + V + \frac{g}{\tan\theta} - m - \frac{g}{\tan\theta}\right]$$

If θ is small, as it generally is, we can assume $\tan\theta = \theta$, and

$$P_a = \pi\mu\left[\frac{2(r_1 + g)}{\theta}\log_\epsilon\left(\frac{\theta m + g}{\theta V + g}\right) + V - m\right] \quad (13)$$

Differentiating this expression with respect to the motion ds, we obtain

$$\frac{dP_a}{ds} = \frac{-dP_a}{dV} = 2\pi\mu\left[\frac{r_1 + g}{\theta V + g} - \frac{1}{2}\right]$$

which on substituting in the general magnetic force formula 2a, will give

$$\text{Force} = \frac{1}{2}F_a^2\frac{dP_a}{ds} = \pi\mu\left[\frac{r_1 + g}{\theta V + g} - \frac{1}{2}\right]F_a^2 \text{ joules per inch} \quad (13a)$$

Changing from joules per inch to pounds, and substituting for the constants, we obtain

$$\text{Force} = \frac{\left[\dfrac{r_1 + g}{\theta V + g} - \dfrac{1}{2}\right]F_a^2}{1{,}125{,}000} \text{ lb.} \quad (13b)$$

Equation 13b can be expressed in terms of the plunger flux density (neglecting fringing and leakage [8] fluxes) as follows:

$$B_p = \frac{\phi}{\pi r_1^2} = \frac{F_a P_a}{\pi r_1^2} = \frac{\mu}{r_1^2}\left[\frac{2(r_1 + g)}{\theta}\log_\epsilon\left(\frac{\theta m + g}{\theta V + g}\right) + V - m\right]F_a$$

which on substituting into equation 13b will give

$$\text{Force} = \frac{r_1^4\left[\dfrac{r_1 + g}{\theta V + g} - \dfrac{1}{2}\right]B_p^2}{11.45\left[\dfrac{2(r_1 + g)}{\theta}\log_\epsilon\left(\dfrac{\theta m + g}{\theta V + g}\right) + V - m\right]^2} \text{ lb.} \quad (13c)$$

where B_p is in kilomaxwells per square inch.

As the force in this type of magnet is generally small, the correction given by equation 7a or 7b should be applied when B_p is large.

7. Wedge-Shaped Gap with Similarly Wedge-Shaped Plug or Armature; Axial Force Due to Flux Crossing Series Gaps; Fringing

---

[8] See footnote on page 204.

Flux Neglected. Figure 10a shows such a magnetic system. This system is almost identical with that of Fig. 7, except that it comprises two similar gaps which are magnetically in series while that of Fig. 7 can be considered as made up of gaps in parallel only. By the notation of Fig. 10a, the permeance, neglecting fringing, of both gaps in series will be

$$P_a = \frac{\mu\left(\dfrac{m}{\sin\alpha} - V\sin\alpha\right)l}{2V\cos\alpha}$$

Differentiating this expression with respect to the motion ds, we have

$$\frac{dP_a}{ds} = -\frac{\mu l m}{2V^2 \sin\alpha \cos\alpha}\frac{dV}{ds}$$

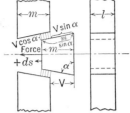

Fig. 10a. Special case of axial force produced by wedge-shaped gap and similarly wedge-shaped armature or plug.

Noting that $dV = -ds$, and substituting into the general magnetic force formula 2a, we have

$$\text{Force} = \frac{1}{2}F_a^2\frac{dP_a}{ds} = \frac{\mu l m F_a^2}{4V^2 \sin\alpha \cos\alpha} \quad \text{joules per inch} \quad (14a)$$

Substituting for the constants in the above equation, the following more useful form is obtained:

$$\text{Force} = \frac{lmF_a^2}{14{,}150{,}000\, V^2 \sin\alpha \cos\alpha} \quad \text{lb.} \quad (14b)$$

where the lineal dimensions can be in any units, as they cancel, and F_a is the magnetomotive force across both air gaps in ampere-turns.

Formula 14a may be expressed in terms of the flux density (neglecting fringing) in the magnet core, as follows:

$$B = \frac{\phi}{S} = \frac{F_a P_a}{ml} = \frac{\mu\left(\dfrac{m}{\sin\alpha} - V\sin\alpha\right)F_a}{2mV\cos\alpha}$$

which on substituting into equation 14a gives

$$\text{Force} = \frac{lm^3 \cos\alpha\, B^2}{\mu \sin\alpha \left(\dfrac{m}{\sin\alpha} - V\sin\alpha\right)^2} \quad \text{joules per inch} \quad (14c)$$

Substituting for the constants and converting from joule per inch to pounds, we have

$$\text{Force} = \frac{lm^3 \cos \alpha\, B^2}{36.0 \sin \alpha \left(\dfrac{m}{\sin \alpha} - V \sin \alpha\right)^2} \text{ lb.} \quad (14d)$$

where B is flux density in the section ml in kilomaxwells per square inch.

When $V = 0$, equation 14d reduces to

$$\text{Force} = \frac{lm \sin \alpha \cos \alpha\, B^2}{36} \text{ lb.} \quad (14e)$$

When α is large and B high, the correction as given by equation 7a or 7b should be applied.

If the plug or armature in Fig. 10a is rectangular, that is, α equals 90°, equation 14a will reduce to the following:

$$\text{Force} = \frac{lF_a^2}{14{,}150{,}000 g} \text{ lb.} \quad (15)$$

where $g = V \cos \alpha$ is the radial length of the air gap. This formula is a modification of formula 9b and is subject to the same restrictions.

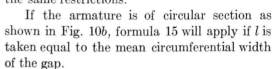

Fig. 10b. Special case of axial force produced by cylindrical-shaped gap and cylindrical armature or plug.

If the armature is of circular section as shown in Fig. 10b, formula 15 will apply if l is taken equal to the mean circumferential width of the gap.

8. **Non-Coaxial Cylindrical Surfaces; Radial Force Due Radial Flux between Cylinders; Fringing Fluxes Neglected.** Such a system is produced in the fixed cylindrical gap of any sliding plunger magnet, because of the necessary tolerance in fit between the plunger and the tube it slides in. Figure 11a shows

Fig. 11a.   Fig. 11b.

Special case of radial force produced by non-coaxial cylindrical surfaces.

a section through the axis of two eccentric plungers, and Fig. 11b a radial section. The radii of the inner and outer plunger are designated by r_1 and r_2, respectively, and the eccentricity by d.

The permeance between the inner and outer surfaces will be, by equation 3c, of Art. 51,

$$P_a = \frac{2\pi\mu l}{\log_\epsilon\left[1 + a\left(1 + \sqrt{\frac{2}{a}+1}\right)\right]}$$

where $a = (c^2 - d^2)/2r_1 r_2$, and $c = r_2 - r_1$.

The force $= \frac{1}{2}F_a^2(dP_a/ds)$ will be, noting that $d(d) = +ds$:

$$\text{Radial force} = \frac{1}{2}F_a^2 \frac{2\pi\mu l}{2\log_\epsilon\left[1 + a\left(1 + \sqrt{\frac{2}{a}+1}\right)\right]a\sqrt{\frac{2}{a}+1}}\left(\frac{d}{r_1 r_2}\right) \quad (16a)$$

If the flux ϕ_a of the radial gap is known, the force equation can be simplified by substituting ϕ_a/P_a for F_a, thus:

$$\text{Radial force} = \frac{1}{2} \times \frac{\phi_a^2}{2\pi\mu l a \sqrt{\frac{2}{a}+1}}\left(\frac{d}{r_1 r_2}\right) \text{ joules per inch} \quad (16b)$$

or

$$= 4.43 \frac{\phi_a^2 \times 10^{-5}}{2\pi\mu l a \sqrt{\frac{2}{a}+1}}\left(\frac{d}{r_1 r_2}\right) \text{ lb.} \quad (16c)$$

where ϕ_a is in kilomaxwells, μ is in kilomaxwells per ampere-turn in an inch cube, and the other dimensions are in inches.

73. Effect of Air-Gap Fringing Fluxes on Force

If a fringing flux is to produce force, the permeance of its path must change with motion; that is, $dP_f/ds \neq 0$. Thus in the cylindrical plunger magnet of Fig. 5, as long as the end of the plunger has entered the surrounding shell and is not too close to the open end, the fringing path will remain unaltered with motion and no force will be developed by the fringing flux. This, however, will not be the case when the plunger is just entering the shell, as the fringing flux is then being established. Likewise, at the end of the stroke the path of the fringing flux is changing. Where the fringing flux is shifting from one type of path to another its rate of change can often be best determined graphically from a curve of fringing permeance. Where only the dimensions of the path are changing the rate of change can be determined by differentiating the fringing permeance formulas of Chapter V.

Consider the flat-faced plunger magnet illustrated in Fig. 17, Chapter V. The total permeance of all flux paths is given by the equation of page 141 as follows:

$$P_0 = P_1 + P_7 + P_{8b} + P_{15a}$$

Of these separate permeances only P_1, P_7, P_{8b} change with the length of the working gap. The total magnetic pull on the plunger will be given by equation 2a

$$\text{Force} = \frac{1}{2} F_a^2 \frac{dP_a}{ds}$$

Since $ds = -dg$, we have

$$= -\frac{1}{2} \mu F_a^2 \frac{d}{dg} \left\{ \frac{\pi r_1^2}{g} + 1.63 \left(r_1 + \frac{g}{4} \right) + 2r_1 \log_\epsilon \frac{4(r_2 - r_1)}{\pi g} \right\}$$

$$= \frac{1}{2} \mu F_a^2 \left\{ \frac{\pi r_1^2}{g^2} - \frac{1.63}{4} + \frac{2r_1}{g} \right\}$$

But, $F_a = Hg = Bg/\mu$; hence

$$\text{Force} = \frac{B^2 \pi r_1^2}{2\mu} \left\{ 1 - \frac{0.41 g^2}{\pi r_1^2} + \frac{2g}{\pi r_1} \right\}$$

The relative increase in force over that obtained when fringing fluxes are neglected is given by the terms

$$\frac{g}{\pi r_1} \left(2 - \frac{0.41g}{r_1} \right)$$

As g/r_1 is small in well-proportioned designs, this correction is usually neglected to be on the safe side.

In a similar manner the effect of fringing and leakage fluxes can be found for other types of magnet. In general, where the main gap flux is large compared to the fringing flux, it is safe to neglect the effect of the fringing flux.

74. Magnetic Force Formulas in Terms of Inductance

The general magnetic force formula 4 of Art. 70 was derived on the premise that all the flux of the magnet linked with all the turns of the exciting winding and took into account the variable permeance of the iron. This was possible because the flux was assumed to remain constant during the infinitesimal motion and consequently no change in magnetomotive force occurred in that portion of the system having variable reluctance. If one considers a magnetic system in which the flux is proportional to the current it is possible to derive a force formula, in terms of the rate of change of inductance of the system, which will take into account all leakage fluxes. This would occur approximately, in a system like that of Fig. 1, if the iron parts were magnetically soft and unsaturated so that most of the reluctance of the magnetic circuit would be in the air gap and hence not dependent on flux. Then a cycle may be passed through by taking the following steps: (1) the current I is increased from zero to its actual value I_m amperes; (2) one element is moved an infinitesimal distance ds inches relative to the other element (e.g., the faces of the magnetic cores of Fig. 1 would be moved toward each other) with or without a change in current; (3) the current is decreased to zero; and (4) the return motion is made.

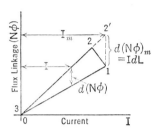

Fig. 12. Flux-linkage current loop for deriving the force in the general case where flux is proportional to current.

The flux-current loop will then have the form shown in Fig. 12, in which the sides 0–1 and 2–0 are separated by a vertical distance $d(N\phi)$ weber-turns representing the increment in flux linkage due to motion and proportional to the current I. The mechanical energy derived from the motion 1–2, as represented by the inclosed area (including the second order infinitesimal 1–2–2′), will be

$$dW_m = \tfrac{1}{2} I_m d(N\phi)_m \quad \text{joules} \tag{17a}$$

and the reacting force in the direction of motion ds will be

$$\text{Force} = \frac{dW_m}{ds} = \frac{1}{2} I_m \frac{d(N\phi)_m}{ds} \quad \text{joules per inch} \tag{17b}$$

where $d(N\phi)_m$ is the increment in flux linkage due to motion, corresponding to the fixed current I_m.

Equation 17b cannot be converted to the form of equation 1, for the reason that the leakage flux, which is now taken into account, does not link with all the turns of the exciting winding and hence N cannot be factored out of the expression $(N\phi)$ which is now an entirety. However, in any system where the flux linkage is directly proportional to the current, the linkage is

$$N\phi = LI \quad \text{weber-turns}$$

where L henries is the self-inductance of the circuit and depends only on the configuration of the system. Substituting into equation 17b and remembering that $d(N\phi)_m$ was taken with I_m constant, we have

$$\text{Force} = \frac{1}{2} I_m \frac{d(LI)_m}{ds} = \frac{1}{2} I_m^2 \frac{dL}{ds} \quad \text{joules per inch} \quad (17c)$$

or

$$\text{Force} = 4.43 \, I_m^2 \frac{dL}{ds} \quad \text{lb.} \quad (17d)$$

In a system where rotation is possible the corresponding formula for torque will be

$$\text{Torque} = 4.43 I_m^2 \frac{dL}{d\theta} \quad \text{lb-in.} \quad (17e)$$

These forms are particularly applicable to alternating-current systems, in which case they give average force or torque if the current is expressed in r.m.s. amperes.

75. Force on a Current-Carrying Conductor in an Independent Magnetic Field

Figure 13a shows such a system in which a wire of length l is free to move in the direction shown. In the initial position, with zero current, there will be flux linkage with the current-carrying circuit owing to the independent magnetic field, as shown by point 0 on Fig. 13b. Now

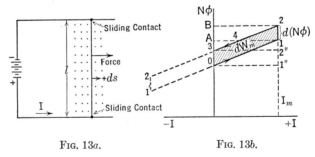

FIG. 13a.   FIG. 13b.

Special case of the force produced on a current-carrying conductor by an independent magnetic field perpendicular to the conductor.

allow the current through the current-carrying conductor to build up from zero to I_m in such a direction as to increase the flux linkage as shown by the line 0–1. During this step energy equal to the area 0–1–A–0 is abstracted from the battery circuit and stored in the magnetic field of the conductor. If the conductor is now moved through a displacement ds, while the current is held constant, the flux linkage will increase from 1 to 2, and energy equal to the area A–1–2–B–A will be abstracted from the electric circuit. If the current is now decreased to zero the flux linkage will fall along the line 2–3, and the stored energy of the circuit equal to the area 3–2–B–3 will be returned to the battery circuit. The cycle is completed by returning the conductor to its original position, thereby causing the flux linkage to decrease to its original value along line 3–0. This step will involve no energy change because the current is zero. The net energy dW_m abstracted from the battery circuit during the complete cycle is area 0–1–2–3–0. This energy, which is equal to that abstracted from the electric circuit during the motion, must be directly converted into mechanical work as no other energy changes are involved.

Therefore the mechanical force developed during the infinitesimal motion ds will be

$$\text{Force} = \frac{dW_m}{ds} = \frac{Id(N\phi)_m}{ds} \quad \text{joules per inch} \tag{18}$$

If the flux linkage had remained constant during the motion, line 1–4, zero energy would have been abstracted from the electric circuit by the motion, and the entire mechanical work dW_m, equal to area 0–1–4–3–0, would have been derived from energy previously stored by the electric circuit. Area 0–1–4–3–0 differs from area 0–1–2–3–0 by area of the second order infinitesimal 4–1–2–4, and hence, whether the current or the flux linkage is held constant during the infinitesimal motion, dW_m will be the same, and force will be always given by equation 18.

If the current is reversed the flux-linkage current loop will lie on the other side of the axis (0–1′–2′–3–0), and the mechanical work dW_m done during the motion will be negative and hence the direction of the force reversed.

Another possibility, which is of considerable interest, is the case where no flux is created by the current of the electric circuit.[9] When this condition obtains, the flux-linkage current loop is shown by the area 0–1″–2″–3–0, which area will equal dW_m, dW_m being obtained by direct conversion of the energy of the electric circuit.

Letting B equal the perpendicular component of the flux density of the independent magnetic field,

$$d(N\phi) = Bl\,ds$$

[9] This condition is partially realized in some machines where compensating windings are employed to neutralize the magnetomotive force of the armature or of the moving conductor.

218 MAGNETIC FORCES [Chap. VIII

and the force on the conductor in the direction of motion will be

$$\text{Force} = \frac{dW_m}{ds} = \frac{I d(N\phi)}{ds} = IBl \quad \text{joules per inch} \tag{19a}$$

where l is the length in inches of the conductor mutually perpendicular to the direction of motion and of the field. Converting from joules per inch to pounds, and from webers per square inch to kilomaxwells per square inch, we have

$$\text{Force} = 8.86 \times 10^{-5} IBl \quad \text{lb.} \tag{19b}$$

76. Force on the Plunger of an Ironclad Solenoid

Figure 14 shows a section through the ordinary type of ironclad solenoid and plunger magnet. In the upper half of the figure the distribution of flux from the plunger passing through the current-carrying

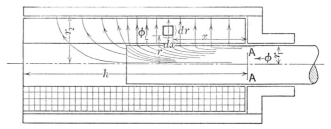

Fig. 14. Force produced on the plunger of an ironclad solenoid.

turns of a coil is shown. It is assumed in the upper half of the figure that the coil space is filled with a solid copper conductor of one turn carrying i amperes per square inch uniformly distributed.

The axial force on an annulus of the copper conductor having a cross-sectional area $drdx$ square inches is due to the reaction between the current of this annulus and the radial component of flux density at the center of its cross section. Special notice should be taken of the fact that this particular flux density is produced only by currents in the section x of the coil. Hence the current through this annulus is independent of the flux density with which it reacts, and the force formula [10] 19a

$$F = IBl$$

[10] The general magnetic force equation derived in Art. 70 cannot be applied directly here because the flux does not link with all the turns of the winding. In cases such as this the more general coordinates of the energy loop must be used, namely, flux linkage and current. This change in coordinates being made, the magnetic cycle used in connection with Fig. 1b will apply, and it will be found that the mechanical energy dW_m derived by the motion with the flux linkage held constant will be

$$dW_m = -\tfrac{1}{2}(N\phi)dI$$

applies. I is the current through the area $drdx$, B is the flux density as defined above, and l is the length of the annulus.

For the annulus in question:

$$I = idxdr$$

$$B = \frac{d\phi_r}{2\pi rdx} \quad \text{where } d\phi_r \text{ is the radial component of leakage flux from the plunger in the axial length } dx.$$

$$l = 2\pi r$$

Force on the annulus $= idxdr \times \dfrac{d\phi_r}{2\pi rdx} \times 2\pi r = idrd\phi_r$:

$$\text{Total force on the copper coil} = i \int_{r_1}^{r_2} dr \int_{x=0}^{x=h} d\phi_r = i(r_2 - r_1) \int_{x=0}^{x=h} d\phi_r$$

But the $\int_{x=0}^{x=h} d\phi_r$ is equal to the total leakage flux of the plunger, assuming the end of the plunger to be far enough away from the open end of the solenoid so that none of the plunger flux passes to the iron shell without passing across the coil. Also $i(r_2 - r_1)$ is the amperes per inch of axial length of coil and is therefore the magnetic intensity H of the solenoid. Hence,

$$\text{Force} = H\phi_L \quad \text{joules per inch} \tag{20a}$$

where ϕ_L is all the flux of the plunger that finally reaches the outer shell radially.

Converting from webers to kilomaxwells and from joules per inch to pounds, this equation can be written

$$\text{Force} = 8.86 \times 10^{-5} H\phi_L \quad \text{lb.} \tag{20b}$$

These equations are obtained in more useful form if they are expressed in terms of the flux of the plunger ϕ instead of the leakage flux ϕ_L. This flux, which is equal to that passing through the plunger at section A–A, will include besides the leakage flux ϕ_L the flux that would exist in the plunger space in the absence of the plunger, namely μHS. Equation 20a can therefore be written

$$\text{Force} = H(\phi - \mu HS) = HS(B_p - \mu H) \quad \text{joules per inch} \tag{20c}$$

where B_p is the maximum plunger flux density occurring at section A–A.

In this magnet, owing to the high plunger flux densities and relatively small forces involved, it is necessary to make a correction for the loss in force due to the increase in length of the magnetized portion of the plunger and shell, which by equation 7a will be

$$\text{Loss of force} = S \int_0^{B_p} H_p dB$$

There will, however, be a corresponding gain in front of the plunger, due to the decrease in length of this space, which will be

$$\text{Gain of force} = S \int_0^B H dB = S \int_0^H H\mu dH = \frac{1}{2} S\mu H^2 \quad \text{joules per inch}$$

Combining these with 20c, the net force is

$$\text{Force} = HS(B_p - \mu H) + \frac{1}{2} S\mu H^2 - S \int_0^{B_p} H_p dB$$

or

$$\text{Force} = S \left[HB_p - \frac{1}{2} \mu H^2 - \int_0^{B_p} H_p dB \right] \quad \text{joules per inch} \quad (20d)$$

When the plunger is inserted far enough in the solenoid so that its flux and force have become constant due to saturation, the value of B_p, in the first term and in the limit of integration of the third term is practically that corresponding to $H = NI/h$ on the magnetization curve for the plunger iron. In this case, all terms are a function of H and this magnetization curve. The last two terms together, for the ordinary soft-steel plunger and values of H suitable for fairly powerful magnets, are found to be about 15 per cent of the first term. Hence an approximate value of force is

$$\text{Force} = 0.85 H B_p S \quad \text{joules per inch} \quad (20e)$$

or

$$\text{Force} = 7.54 \times 10^{-5} H B_p S \quad \text{lb.} \quad (20f)^{[11]}$$

where B_p is the maximum plunger flux density in kmax. per sq. in. In the case of a circular plunger 20f can be conveniently written

$$\text{Force} = 23.6 \times 10^{-5} H B_p r_1^2 \quad \text{lb.} \quad (20g)^{[11]}$$

---

[11] This formula automatically takes into account all leakage flux from the plunger provided that it finally reaches the outside shell by passing through the winding. Likewise equation 20a will take account of the entire flux if ϕ_L includes all the flux passing through the winding and finally reaching the shell radially. When the plunger nears the end of the solenoid the force will be reduced because part of the plunger flux will pass directly out of the hole without passing across the winding.

77. Graphical Evaluation of the General Magnetic Force Formula

In many actual magnetic systems the general magnetic force formula

$$\frac{1}{2} F_a^2 \frac{dP_a}{ds}$$

or its corresponding forms cannot be evaluated analytically, because F_a, the magnetomotive force across the air gap, is dependent on the saturation of the iron; and dP_a/ds varies in such a complicated manner that it can be expressed only graphically. The stepped cylindrical plunger magnet of Fig. 15 is an example of such a system. In this magnet the

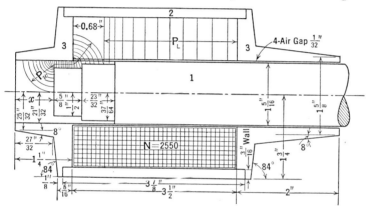

Fig. 15. Stepped-cylindrical plunger magnet.

force is derived not only from the cylindrical working gap but also from the radial leakage flux, and hence it will be necessary to evaluate the force formula $H\phi$ in addition. As the sum of these forces will be relatively small the correction of formula 7b, $S \int_0^{B_p} H_p dB$, must also be applied.

In order to illustrate the method of evaluating these formulas graphically, let it be required to calculate the force-distance curve for the magnet of Fig. 15 from $x = 1\frac{1}{2}$ in. to $x = 0$ in. for an exciting current of 0.8 ampere. This magnet is identical with that illustrated in Fig. 1a, Chapter III, with the exception that the end piece and the plunger have been changed. The material for all the iron parts is S.A.E. 10–20, $\frac{1}{2}$ hard. The mean length and area for each part as determined from Fig. 15 are listed in Table I.

In order to evaluate the force-distance curve it is necessary to obtain F_a, dP_a/dx, ϕ_L, and B_p for several values of x in the interval from $x = 1\frac{1}{2}$ in.

TABLE I

Part No.	Name	Mean Area, sq. in.	Effective Length, in.
1	Plunger............	1.35	$4\frac{1}{2}$
2	Shell..............	1.95	$3\frac{3}{4}$
3	End pieces........	2.50	2
4	Fixed cylindrical gap	Permeance $= \dfrac{0.00319 \times \pi \times 1\frac{11}{32}'' \times 2''}{1/32}$ $= 0.862$, say 0.9 to allow for leakage	

to $x = 0$ in. The permeance P_a of the air gap is made up of two parts: P_w, the permeance of the working gap; and P_L, the permeance of the radial leakage path between the plunger and the shell. The value of P_w is shown plotted in Fig. 16b and is obtained by plotting the solution of Problem 9, Chapter V. As the value of P_L is constant, dP_a/dx will equal dP_w/dx and may be obtained by graphically differentiating the P_w curve. Graphical differentiation at the best is not a very accurate process, but here it is even less satisfactory than usual owing to the difficulty of drawing accurate tangents and to the error in drawing the original curve for P_w through points fairly widely separated. For this reason it is wise to differentiate wherever possible by analytical means. Thus between the limits $x = 0.5$ to $x = 0.2$, and $x = 0.92$ to 1.08, practically the only change in permeance of the working gap is due to the change in axial length of the opposite cylindrical faces. In this range the rate of change of permeance may be determined by differentiating formula 15a of Chapter V. The computed values of P_w and dP_w/dx are listed in columns 2 and 6, respectively, of Table II.

The determination of F_w, the magnetomotive force across the working air gap, is complicated by reason of the fact that it is necessary to deal with the radial leakage flux ϕ_L distributed along the entire length of the plunger in addition to the flux of the working gap ϕ_w. ϕ_L by equation 1, Chapter IV, is equal to $\frac{1}{2}P_L(F_L)_m$, where $(F_L)_m$ is the maximum magnetomotive force across the path P_L, and may be taken equal to $[(3.5 - 0.68)/3.5]F_w = 0.805F_w$, where 0.805 is the part of the total winding length covered by the leakage path P_L. As for the magnetomotive force drop that it produces, ϕ_L can as a convenient approximation (according to Art. 46) be replaced by $\frac{2}{3}\phi_L$ throughout the entire length of the plunger and shell covered by the leakage path P_L. In the fixed cylindrical gap and iron end piece at that end all of ϕ_L is effective, while in the part of the plunger and shell not covered by the leakage path P_L and in the iron end piece at that end ϕ_L is zero effective. As an average it will be assumed that ϕ_L is $\frac{2}{3}$ effective in the entire iron circuit and fixed cylindrical gap. Therefore the value of P_L to be taken in parallel with P_w in computing F_w is equal to $\frac{2}{3} \times \frac{1}{2} \times 0.805 P_L = 0.268 P_L$, as listed in column 3, Table II. P_L is given by formula 15b of Chapter V:

$$P_L = \frac{2\pi \mu l}{\log_\epsilon \frac{r_1 + g}{r_1}} = \frac{2\pi \times 3.19 \times (3.5 - 0.68)}{\log_\epsilon \frac{50}{21}} = 65.1 \quad \text{maxwells per ampere-turn}$$

Column 4 gives the sum of these permeances, or the permeance which may be assumed in series with the entire iron circuit and fixed cylindrical gap in determining F_w.

The next step in determining F_w is to draw the magnetization curve for the entire iron circuit and fixed cylindrical air gap as shown in Fig. 16a. This is computed from the dimensions given in Table I and the magnetization curve for iron sample 4 of Fig. 11a, Chapter II. The value of F_w for any value of effective air-gap permeance ($P_w + 0.268 P_L$) is obtained by finding the intersection between the magnetization curve and a line drawn from the exciting magnetomotive force (2,550 × 0.8 = 2,040 ampere-turns) as an origin with a negative slope of ($P_w + 0.268 P_L$) as shown in Fig. 16a. The magnetomotive force between this intersection point and the origin at 2,040 ampere-turns is F_w. The values of F_w so obtained are listed in column 5. In column 7 the force, due to the rate of change of permeance of the working gap computed as indicated from formula 2b, is given.

The force due to the radial leakage flux ϕ_L is computed directly from formula 20b, the value of H being the magnetic intensity of the coil winding which is equal to the coil magnetomotive force divided by the coil length. ϕ_L is computed directly from formula 1, Chapter IV, as indicated. Values of ϕ_L and the corresponding forces are listed in columns 8 and 9, respectively.

The loss in force due to magnetizing the entering section of the plunger is computed from formula 7b and is listed in column 14. This is computed as shown by evaluating $\int_0^{B_p} H_p dB$, where B_p, the maximum plunger flux density, is listed in column 12 and is computed from columns 8, 10, and 11 as shown. The $\int_0^{B_p} H_p dB$ is evaluated from the data of

Fig. 14d, Chapter II, by adding Curve A to Curve B and subtracting Curve C. For values of B_p above 100 kmax. per sq. in., curve A was extrapolated and Curve B was handled as suggested in Art. 28. The net

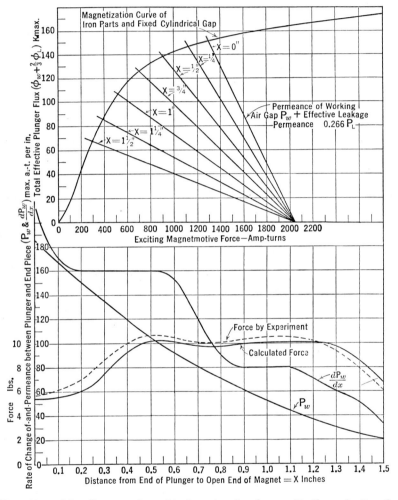

FIGS. 16a and b. Curves and graphical construction for graphically evaluating the force on the stepped-cylindrical plunger magnet of Fig. 15.

pull on the plunger listed in column 15 is computed by adding the forces of columns 7 and 9 and subtracting the force of column 14. This force is shown plotted in Fig. 16b. As a final check on the accuracy and reliability of the calculations the experimental force-distance curve

ART. 77] GRAPHICAL EVALUATION OF MAGNETIC FORCE

TABLE II

1	2	3	4	5	6	7	8
	Permeance of Working Gap	Permeance of Radial Leakage Path Which May Be Considered to Be in Parallel with the Main Working Gap	Total Air-Path Permeance Effective in Determining the M.m.f. across the Working Gap	M.m.f. of Working Gap	Rate of Change of Permeance of Working Air Gap	Force Derived from Working Gap	Radial Leakage Flux ϕ_L
x	P_w	$0.268 P_L$	$P_w + 0.268 P_L$	F_w	dP_w/dx	$4.43 \times 10^{-8} \times F_w^2 \times \dfrac{dP_w}{dx}$	$\tfrac{1}{2} \times 0.805 \times P_L F_w$
in.	max. per a-t.	max. per a-t.	max. per a-t.	a-t.	$\dfrac{\text{max.}}{\text{a-t.}}$ per in.	lb.	max.
$1\tfrac{1}{2}$	20.8	17.5	38.3	1,745	34	4.57	45,700
$1\tfrac{1}{4}$	33.9		51.4	1,650	66	7.96	43,100
1	52.4		69.9	1,525	80	8.25	39,900
$\tfrac{3}{4}$	76.5		94.0	1,345	103	8.26	35,200
$\tfrac{1}{2}$	103.9		121.4	1,140	160	9.20	29,800
$\tfrac{1}{4}$	141.9		159.4	920	160	6.00	24,050
0	185.2		202.7	750	215	5.35	19,600

1	9	10	11	12	13	14	15
	Force Due to ϕ_L $H = \dfrac{2040}{3.5} = 583$	Flux through Working Gap ϕ_w	Maximum Plunger Flux ϕ_p	Maximum Plunger Flux Density B_p	Loss in Work Due to Magnetizing 1 cu. in. of Plunger W_L	Loss in Force Due to W_L	Net Magnetic Force
x	$8.86 \times 10^{-8} H \phi_L$	$P_w F_w \times 10^{-3}$	$\phi_w + 10^{-3}\phi_L$	ϕ_p / S	$\int_0^{B_p} H_p dB$	$8.86 W_L S$	$7 + 9 - 14$
in.	lb.	kmax.	kmax.	kmax. per sq. in.	joules per cu. in.	lb.	lb.
$1\tfrac{1}{2}$	2.36	36.6	82.3	61	0.0105	0.125	6.80
$1\tfrac{1}{4}$	2.22	55.9	99.0	73	0.0154	0.184	10.00
1	2.06	79.9	119.8	89	0.0254	0.304	10.01
$\tfrac{3}{4}$	1.82	103.0	138.2	102	0.0370	0.443	9.64
$\tfrac{1}{2}$	1.54	118.5	148.3	110	0.045	0.538	10.20
$\tfrac{1}{4}$	1.24	130.5	154.5	114	0.058	0.694	6.55
0	1.01	138.9	158.5	117	0.077	0.920	5.44

obtained by test on the actual magnet of Fig. 15 is shown plotted in Fig. 16b. It will be noticed that this curve checks reasonably well as regards shape and magnitude with the computed one.

PROBLEMS

1. Compute the greatest force per square inch of surface that can be obtained by the direct attraction between two magnetized surfaces of (a) soft iron, (b) ferro-cobalt, (c) ferronickel, (d) Permalloy.

2. Considering the rotating magnetic system of Fig. 6, is the torque limited by the saturation of the armature and pole pieces as the force of the magnetic system of Fig. 4 is?

Compute (neglect the effect of fringing and leakage fluxes) for a soft-iron system like that of Fig. 6 the maximum torque that can be produced per inch of axial length of a 12-in. diameter armature for the two following gap lengths: (a) $g = \frac{1}{8}$ in.; (b) $g = \frac{1}{4}$ in.

3. (a) Derive force formula 20a, basing the derivation on the statements made in footnote 10, page 218. *Hint*: The flux linkage ($N\phi$) produced by the entire radial component of leakage flux may be computed by the method of Art. 46 by considering only the permeance between the opposite cylindrical faces of the plunger and shell.

(b) Derive force formula 20a directly from the general magnetic force formula 2.

4. Calculate a force-distance curve for the magnet of Fig. 15 between the limits $x = 1$ in. and $x = \frac{1}{4}$ in. if the stepped cylindrical plunger is replaced by a tapered plunger, the taper (length m of Fig. 9) being $1\frac{1}{4}$ in. long, and the diameter at the end of the taper being $1\frac{1}{32}$ in. Formula 13b may be used in calculating the force due to the taper section. When evaluating F_a of this formula all fringing permeances and the radial leakage permeance of the plunger must be taken into account. The magnetization curve of Fig. 16a will still be applicable, as the plunger change does not change the iron circuit materially.

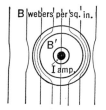

FIG. 17. Current-carrying conductor encased in iron pipe in uniform magnetic field.

5. By what percentage would the force on the stepped cylindrical plunger magnet of Fig. 15 be increased at $x = \frac{1}{4}$ in. if the magnet were made of annealed Swedish charcoal iron instead of S.A.E. 10–20 $\frac{1}{2}$ hard steel?

6. Figure 17 shows an iron pipe in a uniform magnetic field of B webers per square inch. The field inside the pipe, due to the external field, has a density of B' webers per square inch. Compute the force on a conductor at the center of the pipe carrying I amperes. Does the presence of current in the conductor produce a force on the pipe? If so, how much is this force?

7. Two coils carrying currents of I_1 and I_2 amperes are so placed that their mutual inductance L_{12} can be changed by rotating the one coil relative to the other. Derive a formula for the torque between the coils. If the two coils are connected in series so that both carry the same current will the formula you have derived reduce to formula 17? Demonstrate.

8. Demonstrate that formula 11a will reduce to formula 9a if α is made equal to $\pi/2$ radians.

9. Compute the maximum possible radial force in the fixed cylindrical gap of the magnet of Fig. 15 if the plunger slides in $1\frac{3}{8}$ in. O. D. No. 22 B. & S. gauge (nominal wall thickness = 0.0254 in.) seamless brass tube.

10. Same as Problem 9, except that the maximum possible radial force in the working gap (neglect fringing fluxes) at $x = 0$ in. is to be computed.

CHAPTER IX

CHARACTERISTICS OF TRACTIVE MAGNETS: SELECTION OF BEST TYPE FOR A SPECIFIC DUTY

78. General

In general, disregarding any especially desirable mechanical features or requirements regarding speed of action, there is a particular type of magnet which will prove most economical (minimum weight per inch-pound of useful [1] work) for any definite ratio of force to stroke. Throughout this chapter it is assumed that the magnets under discussion are tractive magnets for fairly large values of work which are to be operated continuously or intermittently on direct-current circuits of a given voltage with a given limiting temperature rise. This excludes small magnets where mechanical adaptability and ease of fabrication may be prime requisites; fast-acting magnets, where the entire iron structure must be laminated, and where the power consumption must be increased far beyond the ordinary limitations of temperature rise for continuous duty in order to produce speed of action; sensitive relays, where the ability to operate with reasonable speed on a very small power consumption is the design limitation; and alternating-current magnets which not only must be laminated but also are prohibitively uneconomical in their volt-ampere consumption for the larger sizes.

79. Characteristics of Different Types of Tractive Magnets

1. **The Flat-Faced Armature Type.** In Fig. 1 is shown a cross section of an ordinary circular lifting magnet, or flat-faced armature type, along with its actual experimental force-distance curves.[2] This

---

[1] The useful work of the magnet is defined as the product of the minimum force throughout an assigned stroke by that stroke.

[2] This magnet is the same as that of Fig. 13, Chapter IV. The force-distance curve shown, and all the others shown in Figs. 1 to 9 inclusive, are the actual experimental curves, showing the electromagnetic forces developed exclusive of friction. Friction was eliminated from the spring-balance readings by averaging the reading of the pull taken with the plunger rising and falling.

type of magnet is intended primarily to produce a large force through a relatively short stroke. As generally constructed in the larger sizes it is made of cast steel and is used for lifting scrap iron, sheet iron, iron castings, and also as a holding magnet for magnetic clutches for shaft drives. It is characterized by having a magnetic circuit of extremely short length and great sectional area with two working air gaps. These gaps, magnetically in series, are mechanically in parallel, and hence produce a holding surface of large effective area.

Ideally,[3] the force in this type of magnet varies inversely with the gap length; practically, the variation is slightly different, owing to the effects of magnetic leakage and iron saturation. The deviation between

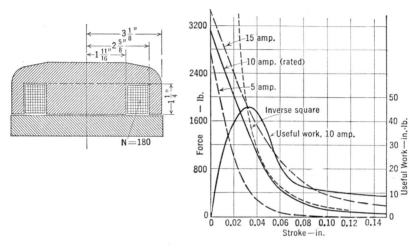

FIG. 1. Flat-faced lifting magnet and its force-stroke curves.

the actual and the ideal force law may be checked as follows: using the curve for rated $I = 10$ amperes, and taking the force of 1,100 lb. at 0.04-in. stroke as a nominal reference point, the dotted curve has been drawn following the inverse square law. The wide discrepancy between this and the experimental curves for the shorter strokes is due primarily to iron saturation which prevents the flux from rising in proportion to the decrease in the gap length and so reduces the force. This effect of saturation to limit the force is somewhat overcome by the change in the leakage field as the gap shortens. Thus at 0.01-in. stroke the flux of this particular magnet is about 17.5 per cent more usefully employed than at 0.04-in. stroke. In other words, if there were no change in the

[3] See equation 8b, Chapter VIII.

leakage field with motion [4] the force at 0.01-in. stroke would be only 1,730 lb. instead of 2,550 lb. It will be noticed that at a lower current when the saturation is less marked ($I = 5$ amperes) the force follows the inverse square law more closely. The other curve shows the useful work plotted as a function of the stroke for the rated current of 10 amperes. This curve ordinarily has no particular significance for a lifting magnet, because the magnet is primarily intended merely to produce a force.

It is interesting to note, however, that the maximum useful work of the magnet, considering it as a tractive magnet, is 46 in-lb. at a stroke of $\frac{1}{32}$ in. with rated current. As the magnet and armature weigh $22\frac{1}{4}$ lb., this corresponds to 0.49 lb. per in-lb. of work. The temperature rise will be 73° C. with this current if the magnet is excited only half the time.

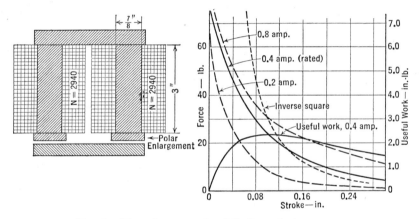

Fig. 2. Horseshoe magnet and its force-stroke curves.

(See data of Table I.) The high weight economy is obtained because of the short time of excitation.

2. The Horseshoe Type. In Fig. 2 is shown a cross section of a horseshoe or bipolar type of magnet, along with its actual experimental force-distance curves.[5] This magnet is also of the flat-faced armature type, though it is not, in general, so economical, nor is it built in such large sizes, as the circular type of Fig. 1. It is often employed in small magnets because of its mechanical adaptability and the ease with which it can be constructed. By means of polar enlargements, so that the

---

[4] See equation 27, Chapter V, for the leakage coefficient for this magnet. Thus at $g = 0.01$ in., $\nu = 1.08$; at $g = 0.04$ in., $\nu = 1.33$. Therefore, other things remaining constant, the force would be $(1.08 \div 1.33)^2 = 0.66$ times as large if the leakage coefficient had not changed.

[5] This magnet is the same as that described in Problem 4, Chapter V.

flux densities in the various points of the magnetic circuit can be adjusted to their optimum values, it can be used, with fair weight economy, over a large range of force-stroke values. It is, therefore, a rather versatile type, and can serve as an alternative for many of the plunger types, when mechanical details, such as a sliding plunger or a fixed cylindrical gap, are undesirable.

Referring to the force-distance curve for the rated current of 0.4 ampere, and using the force of 19 lb. at 0.125-in. stroke as a reference value, the dashed curve, showing how the force should vary if it followed the inverse square law, was constructed. As in Fig. 1, the lack of agreement between these two curves can be explained by the effects of iron saturation and magnetic leakage. The curve labeled "work" shows how the useful work of the magnet varies with the stroke. Thus, a maximum work of 2.4 in-lb. at rated current will be performed at a stroke of 0.1 in.

As this magnet and armature weigh 7.0 lb., the weight economy is 2.9 lb. per in-lb. of useful work. The temperature rise for continuous excitation with rated current is only 28° C. (see Table I). This weight economy is low and can be accounted for by two facts: (1) The temperature rise is low, and if it were increased the coil and pole core length could be shortened resulting not only in a decrease in weight but also in a decrease in magnetic leakage which would increase the force and hence the useful work. (2) The magnet is small, and, in general the weight per unit of work decreases for all apparatus as the size increases. Very frequently, this type, because of its mechanical adaptability and ease of construction, is applied to jobs where the ratio of force to stroke is not large, without taking the precaution of using polar enlargements. In these cases, because it is uneconomically employed, it will have an apparently poor weight economy.

3. **The Flat-Faced Plunger Type.** In Fig. 3 are shown a cross-section view of a cylindrical flat-faced plunger magnet [6] and its actual experimental force-distance curves. This type is adapted to produce, with economy, a relatively smaller force through a longer stroke than the flat-faced armature type. Its magnetic circuit is generally quite short and heavy and has only one working air gap. For the same cross section of iron as the flat-faced armature or horseshoe type, its air-gap permeance will increase only one-half as fast with decrease in gap length and hence will develop only one-half the force. However, as the magnetomotive force of the exciting coil is effective across one working gap, instead of two in series, it can develop this smaller force through twice the stroke.

---

[6] This magnet is the same as that of Fig. 1a, Chapter III.

Ideally, the force should vary inversely as the square of the gap length; practically, as is shown by the dotted inverse square curve, based on a nominal force of 112 lb. at $\frac{1}{16}$-in. stroke at rated $I = 0.8$ ampere, there is a wide divergence from the inverse square law. As before, this may be explained as the effects of iron saturation and magnetic leakage. However, if the curve for the smaller current is checked in this manner, it will be found to follow the ideal much more closely because of the lower saturation.

The work curve shows that this particular magnet will perform at rated current a maximum work of 7 in-lb. at a $\frac{1}{16}$-in. stroke. As the

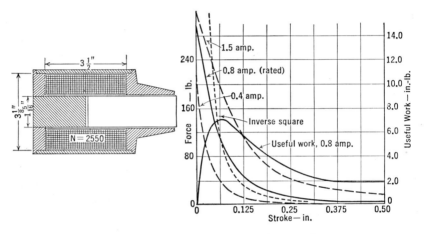

Fig. 3. Flat-faced plunger magnet and its force-stroke curves.

entire magnet weighs 10.5 lb., this corresponds to a weight economy of 1.5 lb. per in-lb. of useful work.[7]

4. Conical-Faced Plunger Magnet. Figure 4 shows a 60° conical-faced plunger magnet[8] with its actual experimental force-distance curves. This type is identical with the flat-faced plunger magnet except that its economical range of usefulness extends over relatively smaller forces and longer strokes. This is due to the conical faces which cause the permeance of the working gap to increase less rapidly with motion, and hence decrease the force.[9] Its force-distance curves can be compared quantitatively to those for the flat-faced plunger magnet of Fig. 3

---

[7] This weight economy is low for this type of magnet. The reason is that the magnet is an experimental one and was designed to take all types of plungers.

[8] This magnet is the same as that of Fig. 3 except that the flat-faced plungers have been replaced by coned ones.

[9] See force-formula derivation of Sec. 5, Art. 72, Chapter VIII.

because they were taken experimentally on the same magnet by using a different set of plungers. Thus, referring to the figures, the flat-faced plunger magnet shows a maximum work of 7 in-lb. at $\frac{1}{16}$-in. stroke while the same magnet with 60° conical-faced plungers has a maximum work of 9.5 in-lb. at a stroke of $\frac{1}{4}$ in. This increase in maximum work available is mostly due to the increase in magnetic and mechanical efficacies.[10] The weight economy of this magnet will therefore be higher; it is equal to 1.05 lb. per in-lb. of work.

Ideally, these force-distance curves also should follow the inverse square law. However, for the reasons explained before, though they follow the ideal in general shape, they are widely divergent from a real inverse square law.

5. Cylindrical-Faced Plunger Magnet. Figure 5 illustrates a cylindrical-faced plunger magnet[11] with its actual experimental force-distance curves. This type is suitable for relatively longer strokes and

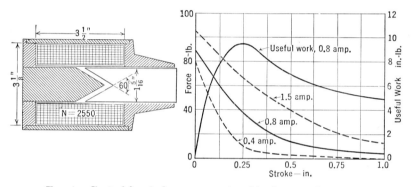

FIG. 4. Conical-faced plunger magnet and its force-stroke curves.

smaller forces than the conical-faced plunger magnet, or where a force-distance characteristic which is high at the beginning of the stroke and tapers off toward the end of the stroke is desired. The shape of the force-distance curve without the stop is explained as follows:[12] from the beginning of the stroke shown at $x = 2$ in., up to $x = 1\frac{1}{4}$ in., the plunger is outside the cylindrical shell and the permeance between the plunger end and the shell increases rapidly as x decreases. This produces a rate of change of permeance which increases as x decreases. When this

---

[10] See Arts. 35 and 36, Chapter III.

[11] This magnet also is the same as used for obtaining the data of Figs. 3 and 4. Actually it was built as illustrated in Fig. 5 with protruding iron shells at each end so as to be universally adaptable to all the types illustrated from Fig. 3 to 8 inclusive. In the illustration of Figs. 3 and 4 it is shown as it would be normally built.

[12] See force formulas 2a or 9a, Chapter VIII.

rate of change of permeance is multiplied by the square of the magnetomotive force between the plunger and shell a rising force-distance curve is produced.[13] From the point where the plunger enters the shell until it commences to protrude from the other end, the distance rate of change of permeance is constant, and if the magnetomotive force between the plunger and shell were constant the force would be constant. This ideal condition is obtained for only a small part of this distance, from $x = 1\frac{1}{4}$ in. to $x = \frac{7}{8}$ in. Beyond this point the iron commences to saturate, and the magnetomotive force across the gap decreases, causing the force to fall off. When the stop is added the initial part of the curve up to $x = \frac{3}{4}$ in. is not materially changed. However, toward the end of the stroke the rate of change of permeance is rapidly increased because

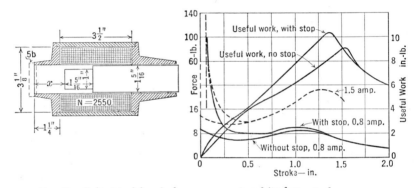

Fig. 5. Cylindrical-faced plunger magnet and its force-stroke curves.

of the direct approach of the flat end surface of the plunger and the stop surface. This produces a force characteristic at the end of the stroke exactly the same as that of the flat-faced plunger magnet. The use of the stop is generally to be recommended in this type of magnet as it produces a large increase in magnetic efficacy and an increase in weight economy, and prevents the force-distance characteristic from falling off at the end of the stroke; sometimes, however, the stop is undesirable because of the residual force which will be produced. This, of course, can be partially eliminated by placing a non-magnetic washer between the plunger and the stop.

[13] This is not the complete story. A large portion (of the order of 25 per cent) of the force between $x = 2$ in. and $x = 1\frac{1}{4}$ in. is derived from the flux of the plunger that leaks radially across the coil to the iron shell surrounding the coil. In this magnet this force is of the order of 2 lb. between $x = 2$ in. and $x = 1\frac{1}{4}$ in. and gradually decreases to about 1 lb. at $x = 0$. See column 9 of Table II of Art. 77, page 225, for actual values of this force in the stepped cylindrical plunger magnet of Fig. 6.

The useful work-stroke curves show maximum useful[14] works of 10.3 and 9.1 in-lb. occurring at strokes of $1\frac{3}{8}$ and $1\frac{1}{2}$ in., respectively, for the magnet with and without the stop, corresponding to weight economies of 1.01 and 1.13 lb. per in-lb. of work.

6. Stepped Cylindrical Faced Plunger Magnet. The falling off of the force with the straight cylindrical plunger at about the middle of the stroke, due to the decrease in gap magnetomotive force as the iron saturates, as occurs in the force-distance curve of Fig. 5, can be minimized by increasing the rate of change of permeance between the cylindrical plunger and the shell as the stroke progresses. One simple way to do this is by stepping the cylindrical plunger as is illustrated in Fig. 6. The force-distance curves of Fig. 6 show exactly the effect

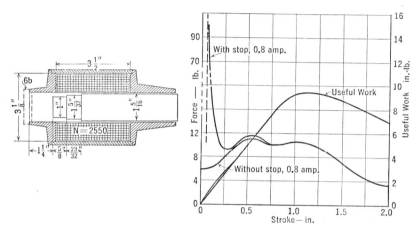

Fig. 6. Stepped-cylindrical-faced plunger magnet and its force-stroke curves.

that is obtained when the 1-in.-diameter cylindrical plunger of Fig. 5 is stepped up to $1\frac{5}{32}$ in. (decreasing the radial gap length to one-half) for the last half of its length. As can be seen at $x = \frac{3}{4}$ in., where the force due to the 1-in.-diameter step is falling off, the curvature changes and the force commences to rise as x gets smaller owing to the $1\frac{5}{32}$-in. step entering the shell. Despite the larger-diameter plunger the force falls off again at $x = \frac{3}{8}$ in., until at about $x = 0$ the force again commences rising because of the step produced by the main body of the plunger coming close to the shell. This last dip could be eliminated by inserting

---

[14] The useful work curve for this magnet and the succeeding ones with similarly shaped force-distance curves is taken as the product of the least force throughout the stroke by the stroke; however, the actual plunger positions for the assigned stroke are taken so as to give the maximum product.

another step at about $x = \frac{1}{4}$ in. When the stop is used the last dip is eliminated and a large sealing-in pull is obtained at the end of the stroke.

The useful work-stroke curves show a maximum work of 9.5 in-lb. at a stroke of $1\frac{1}{8}$ in. for both the magnet with and without the stop, giving a weight economy of 1.1 lb. per in-lb. of work.

7. Taper Plunger Magnet. It is obvious that, as the number of steps of the cylindrical plunger magnet is increased, the force (no stop) may be kept more constant throughout the useful stroke. The logical consequence of increasing the number of steps is to make the plunger tapered, thus getting the effect of an infinite number of steps. If the angle of the taper is chosen correctly, it is possible to get almost constant force. In Fig. 7, the magnets of Figs. 5 and 6 are shown with a tapered

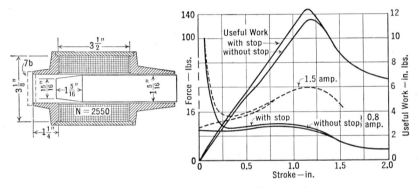

Fig. 7. Taper plunger magnet and its force-stroke curves.

plunger. The force-distance curve for this magnet at rated current without the stop shows practically constant force [15] from $x = 1\frac{1}{8}$ in. to $x = 0$. The addition of the stop as before merely increases the pull at the end of the stroke and produces a high sealing-in pull.

The useful-work-stroke curves show maximum work of 12.4 and 11.5 in-lb. at strokes of $1\frac{1}{8}$ in. and $1\frac{3}{16}$ in., respectively, for the magnet with and without the stop. This corresponds to weight economies of 0.77 and 0.88 lb. per in-lb. of work, respectively.

[15] It is sometimes possible to get an even more flat force-distance curve by combining the cylindrical and taper plungers. Thus if the cylindrical plunger of Fig. 5 is allowed to remain a 1-in. cylinder from the end for a length of $\frac{1}{2}$ in. and then is tapered from 1 in. to $1\frac{5}{16}$ in. in the next $\frac{13}{16}$ in. a force-distance curve which lies between 12 lb. maximum and 10 lb. minimum will be obtained from $x = 1\frac{1}{4}$ in. to $x = 0$. This will increase the useful work to $10 \times 1.25 = 12.5$ in-lb., which is an 11 per cent increase. Usually, however, such special plunger shapes are not warranted.

8. Truncated Conical Plunger Magnet.

If the cylindrical shell of the taper plunger magnet of Fig. 7 is made in the form of a conical shell having the same angle of taper as the plunger, the so-called truncated conical magnet of Fig. 8 is obtained. This particular shape of working gap is theoretically ideal for a long-stroke small force magnet as it makes the stored energy of the working gap zero at the end position of stroke and thereby makes the greatest possible energy available as mechanical work. This can be seen by noting that the experimental force-distance curve of Fig. 8 has more area under it than any of the curves of Figs. 5, 6, and 7 for the same current, and that the magnetic efficacy has a high value of 0.52 (see Table I). Also, the force can be made substantially constant over the greater part of the stroke. Referring to the force-distance curve, the hump at $x = 1$ in. could be flattened out and the

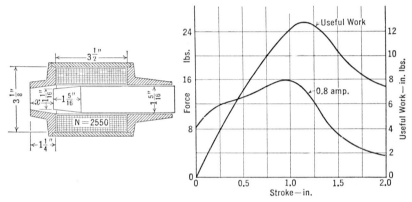

Fig. 8. Truncated conical plunger magnet and its force-stroke curves.

force in the region of $x = \frac{1}{4}$ in. increased by making the angle of the taper greater. As the rate of change of permeance of the working gap [16] varies inversely as the square of the distance x, the decrease in force due to saturation, varying as the square of the gap magnetomotive force, may easily be compensated for. In practice, however, it is extremely difficult to realize an actual available force-distance curve near that of Fig. 8. The reason for this is that if the two conical faces are slightly eccentric the friction toward the end of the stroke due to side pull of the plunger becomes excessively great.[17]

The useful-work-stroke curve shows a maximum work of 12.6 in-lb. at a stroke of $1\frac{3}{16}$ in. This corresponds to a weight economy of 0.83 lb. per in-lb. of work.

[16] See the force formula derivation of Art. 72, page 205.
[17] In testing the magnet of Fig. 8 the net available force as actually measured on a spring balance became zero at $x = \frac{1}{4}$ in.

9. Ironclad Solenoid and Plunger or Leakage Flux Type.

As was mentioned in footnote 13, the cylindrical plunger magnet and all similar long-stroke plunger magnets derive a considerable portion of their force from flux leaking radially from the plunger to the surrounding iron shell. When it is desired to make a magnet with a relatively long stroke and small force, this type of force action must be resorted to for almost the entire length of the stroke. It is generally economical to retain a taper plunger force action for the last portion of the stroke and thereby avoid extending the coil length. In Fig. 9 is illustrated an ironclad solenoid and plunger magnet employing the taper end for the last inch of stroke. The experimental force-distance curve shows that from $x = 3\frac{1}{2}$ in. to $x = 10\frac{1}{2}$ in. the force is practically constant. This is because the force [18] is given by the equation $F \propto H\phi$, where H, the magnetic intensity of the solenoid winding is constant; and ϕ, the flux of the plunger, has

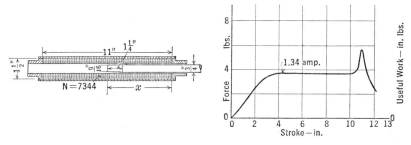

FIG. 9. Ironclad solenoid and plunger magnet and its force-stroke curves.

become constant owing to saturation. In the region between $x = 0$ and $x = 3\frac{1}{2}$ in., the flux leaking from the plunger increases approximately as the square of x, for the reason that the radial leakage permeance between the plunger and the shell is proportional to x, and the effective magnetomotive force also is proportional to x; thus the force increases as the square of x. At the end of the stroke when the taper end becomes effective the variation in force is determined almost entirely by the length and angle of the taper. In this type of magnet there are no magnetic limitations to the length of stroke. If desired the iron shell around the coil may be eliminated and the magnet becomes the ordinary solenoid with a soft-iron plunger. Where the stroke is very short this will result in a large increase in weight and power consumption; where the stroke is very long the difference is less marked. Referring to Table I, it is interesting to note the high values of magnetic and mechanical efficacy obtainable with this type of magnet. The magnetic efficacy

---

[18] See force formula 20a, Art. 76, page 219.

TABLE I

	Magnet type	Magnet Figure No.	Fraction of Time Excited	Power Consumption (Hot) watts	Rated Current amp.	Coil Resistance (Hot) ohms	Final Temp. Rise (Amb. T = 20°C) °C	Weight of Magnet and Armature lb.	Ideal Work Available in-lb.	Total Mechanical Work Available in-lb.	Maximum Useful Work in-lb.	Force Optimum for Maximum Useful Work lb.	Stroke Optimum for Maximum Useful Work in.	Mechanical Work under Force-Distance Curve for Optimum Stroke in-lb.	Magnetic Efficacy for Maximum Useful Work Col. 13 / Col. 8	Mechanical Efficacy for Maximum Useful Work Col. 10 / Col. 13	Weight Economy lb. / in-lb.	Index Number $\frac{\sqrt[3]{W}}{S}$
1	Flat-faced armature	1	$\frac{1}{2}$	70	10	0.70	73	$22\frac{1}{4}$	161	110	46	1,400	0.033	68	.42	.68	0.49	1,133
2	Horseshoe	2	1	10	0.4	62	28	7.0	12.9	6.5	2.4	24	0.1	4.5	.35	.53	2.9	49
3	33° conical-faced plunger	*	$\frac{1}{2}$	152	2.0	38	72	46	602	342	170	97	1.75	242	.40	.70	0.27	5.6
4	Flat-faced plunger	3	1	20.5	0.8	32	65	10.0	31.6	21.7	7.0	112	0.0625	13.1	.41	.54	1.43	169
5	60° conical-faced plunger	4	1					10.0		24.4	9.5	38	0.25	15.8	.50	.60	1.05	25
6	Cylindrical-faced plunger	5a	1					10.3		14.5	9.1	5.8	1.562	11.8	.37	.77	1.13	1.5
7	" " with stop	5b	1					10.4		18.4	10.3	7.5	1.375	16.3	.52	.63	1.01	2.0
8	Stepped cylindrical f.p.	6a	1					10.5		16.3	9.5	8.4	1.125	11.3	.36	.84	1.10	2.6
9	" " with stop	6b	1					10.4		18.8	9.5	8.4	1.125	13.6	.43	.70	1.10	2.6
10	Taper plunger	7a	1					10.5		17.8	11.5	9.7	1.188	13.1	.42	.88	0.90	2.6
11	" " with stop	7b	1					10.4		20.0	12.4	11.0	1.125	16.2	.51	.77	0.84	2.9
12	Truncated conical plunger	8	1					10.5		21.7	12.6	11.6	1.188	16.5	.52	.76	0.83	2.9
13	Tapered-end leakage flux	9	†	149	1.34	83.2	70	7.0	50.2	38.6	31.5	3.5	9.0	33	.66	.95	0.22	0.22

* Not illustrated.   † 70° C. temperature rise after 10 minutes' excitation.

of this type of magnet will increase toward unity (neglecting hysteresis) as the stroke is increased.

80. Shaping of the Force-Distance Characteristic

In order to make an effective comparison of the effects of changing the shape of the working gap, the data of Table I were prepared. All these data were obtained by actual measurement on the magnets listed and illustrated in Figs. 1 to 9 inclusive.[19] The only columns requiring special explanation in this table are 8 and 9. Column 8, the ideal work available, was determined as outlined in Chapter III; it represents the product of the maximum flux leakage obtainable (leaving no gaps in the circuit) by the exciting current. Column 9 is the total area under the force-distance curve as plotted in Figs. 1 to 9.

Examining the data of Table I, it is seen that the magnets have optimum forces and strokes that range from 1,400 lb. through 0.033 in., to 3.5 lb through 9 in. Looking at column 14, it is seen that this range of force and stroke can be obtained at a substantially uniformly high value of magnetic efficacy. If we further examine the limited range including magnets 4 to 12 which cover designs having the same temperature rise conditions, and about the same value of useful work, it is seen that the weight economy is also substantially independent of the force and stroke.

In general, it may be stated that the magnetic efficacy[20] and the weight economy are independent of the force and stroke in well-designed magnets if the temperature-rise conditions and useful work are held constant.

This is accomplished by shaping the force-distance characteristic of the magnet to suit the particular useful force and stroke desired, so that a high magnetic efficacy, with the desired mechanical efficacy, is obtained. The force-distance characteristic can, within limits, be shaped at will by properly combining the three elementary force-distance characteristics: (a) the inverse-square type obtained by the direct approach of the working gap surfaces; (b) the constant-force type, with the force falling off at the end of the stroke because of saturation, produced by increasing the area of the working gap surfaces only;

---

[19] Magnet 3 is not illustrated. This magnet and also magnet 1 were designed by Dr. L. A. Hazeltine at the time he was associated with the Department of Electrical Engineering at Stevens Institute. Magnet 3 is illustrated on page 228 of "Magnets" by Charles R. Underhill.

[20] This excludes the leakage-flux type, the magnetic efficacy of which (neglecting hysteresis) approaches unity as the stroke is increased.

(c) the truly constant-force type using the leakage-flux principle. Magnets 1, 2, and 4 are entirely of type (a); magnets 6 and 8 are entirely of type (b); magnet 13 is entirely of type (c), except at the very end of its stroke; magnets 3, 5, 7, 9, 10, 11, and 12 are combinations of types (a) and (b), and of these magnets 7, 9, 10, 11, and 12 have a small proportion of type (c).

81. Index Number—Choice of Magnet Type for Any Particular Force and Stroke

It is quite apparent from a study of Table I and the discussion of the last article that, if one is to design an efficient magnet (high weight economy), the type of working-gap surface must be correctly chosen for the particular force and stroke.

In general, a long-stroke magnet will be long and a short-stroke magnet short; a large-force magnet will be large in diameter and a small-force magnet small in diameter, if they are to be of economical design. Thus, the square root of the force is proportional to the diameter and the stroke is proportional to the length. The ratio of these two quantities, namely $\sqrt{\mathscr{F}}/s$, will be an indication of the shape of the magnet; it is called the index number.[21] Experience shows that the index number forms a logical basis for determining the proper type of magnet for maximum weight economy. Column 17, Table I, gives the index numbers for the various magnets discussed so far.

In order quantitatively to correlate weight economy with magnet type and index number, the graph of Fig. 10 has been prepared.[22] This graph shows the weight economy for each of seven types of magnet as a function of the index number. The data were obtained by computing a series of designs for each type covering a range of index numbers. The actual designs were carried out in such a manner that each magnet, except those of the horseshoe type, performed approximately the same maximum useful work (10 in-lb.) at its rated stroke with the same final temperature rise (70° C.) for continuous excitation. The actual weight economies shown on the graph, for these types, apply only for magnets rated at about 10 in-lb.

In choosing the magnet types, it was necessary to take only a sufficient number to cover the range of useful index numbers with a high weight economy. Also, some types which would appear from their

[21] This term and concept were introduced by Dr. L. A. Hazeltine, formerly head of the Department of Electrical Engineering at Stevens Institute.

[22] The work of computing the various designs and correlating them was carried out by several graduate students under the supervision of the author.

force-stroke characteristics and their other characteristics as tabulated in Table I to be almost ideal have been omitted because of practical reasons. Thus, the truncated conical plunger magnet is the most efficient listed, but practically, if the plunger is not very carefully centered with reference to the conical end piece, the side pull toward the end of the stroke is prohibitive. The cylindrical and stepped cylindrical plunger types have been omitted because the taper plunger type, by altering the angle and diameter of the tapered part, can be made to give the performance of the other two with a greater degree of flexibility. The horseshoe type of magnet is more involved than the others because

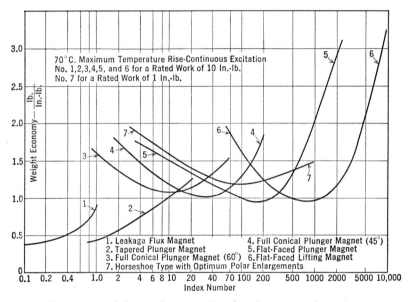

Fig. 10. Comparison of the weight economies of various types of tractive magnets.

one more variable enters into its design. This additional variable is the polar enlargement. In the other types, in order to maintain a high weight economy as the index number is decreased, the shape of the pole faces is changed so as to keep the force as uniform as possible over the longer stroke. In the horseshoe magnet, this same effect is obtained by changing the polar enlargements. Thus a horseshoe magnet with a low index should have big polar enlargements, and one with a high index number should have no polar enlargements. In this manner a fair weight economy can be maintained over a wide range of index numbers. In Fig. 10 are shown the weight economies of horseshoe magnets with optimum polar enlargements, as a function of the index number.

Because this type of magnet is not generally built for as much work as the other types, the data have been computed for a constant useful work of 1 in-lb. instead of 10. The temperature rise for continuous excitation has been kept at 70° C.

Although the seven types chosen for the graph of Fig. 10 will generally be suitable for most designs, it will often be desirable when special force-distance characteristics are to be obtained to combine some of the types or to use a cylindrical type.

Referring now to Fig. 10, the following tabulation of the range of index numbers economically covered by each type may be made:

TABLE II

Best Magnet Type	Index Number Range	As an Alternate Type
Flat-faced armature	Above 350	The horseshoe may be used between index numbers of 1,000 to 10
Flat-faced plunger	350 to 62	
Conical-faced plunger 45°	62 to 14	
Tapered plunger	14 to 0.8	
Leakage flux	Below 0.8	

The 60° conical plunger has been left out of the tabulation because its apparent range of maximum economy is limited. Actually it can be used with good results in the range of index numbers from about 5 to 20.

It should be remembered that the data of Fig. 10 and Table II are to be used only as a guide in selecting the proper type and that practical considerations such as ease of building or the preference of the designer in regard to mechanical details may modify the choice. If speed of action is any consideration, care must be exercised when choosing either the taper plunger or leakage-flux type to see that the mechanical efficacy is not made too high. Magnets designed especially with regard to their speed of action will be considered in detail in Chapter XII.

82. Effect of Temperature Rise and Rated Work on the Weight Economy

Increased allowable temperature rise of a magnet, though it does not decrease the necessary cross section of the iron, will decrease the space necessary for the coil, which will generally allow a reduction in the length of the iron circuit to be made. This results in a reduction in weight and an increase in weight economy. Thus, referring to Table I, the low weight economy of magnet 2 is partly due to its low temperature rise of 28° C., while the high weight economy of magnets 1, 3, and 13 is partly due to their high temperature rise (70° C. with a short period of excitation). In general, unless power consumption is a prime consideration, a magnet should be designed for a temperature rise as high as standard practice will permit for the materials used. Where a magnet is used intermittently, the peak cyclic temperature rise should be chosen as above.

In magnets, as with all other machinery, the weight per unit of work performed decreases as the capacity of the machine increases. Thus, the fact that magnet 3 has four times the weight economy of magnet 5 is due primarily to its large rated work of 170 in-lb. as compared to 9.5 in-lb. for magnet 5. Likewise, the lower weight economy of the horseshoe type as compared to the others of Fig. 10 is partly due to their smaller size. This fact should be kept in mind when using the weight economies given in Fig. 10.

PROBLEMS

1. Figure 11 shows a desired force-stroke characteristic. Sketch a suitable type of magnet, showing in detail the type of pole faces necessary to produce the rise in force at the end. Give approximate dimensions for the pole faces based on magnetic values which are reasonable.

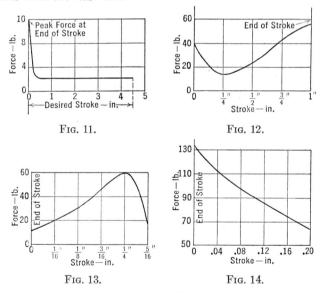

Fig. 11. Fig. 12. Fig. 13. Fig. 14.

2. Figure 12 shows a desired force-stroke characteristic. Sketch a suitable type of magnet, showing in detail the type of pole face necessary. Explain why the type of pole face you have selected should give the desired results.

3. Same as Problem 2, referring to Fig. 13.

4. Same as Problem 2, referring to Fig. 14.

CHAPTER X

DESIGN OF TRACTIVE MAGNETS: DESIGN PROCEDURE: ILLUSTRATIVE DESIGNS

83. General

It is essential for the designing engineer to have some general rules and fixed standards to guide him in his work. In the field of electrical-machine design this resolves itself into a compilation of physical laws governing the operation of the particular machine plus empirical rules derived from experience in designing and building many similar devices. The empirical rules will set economical design limits for flux and current densities and for the proportions of the magnet. These laws and rules plus the natural intuition and feeling of any experienced designer will generally result in a suitable design. The first design so obtained is usually referred to as a preliminary design. This preliminary design must first be modified in order to accommodate stock sizes, and must then be subjected to the inverse process; i.e., upon the basis of the dimensions, winding data, etc., the performance characteristics of the magnet must be computed. These characteristics will include, besides the force-stroke curve and temperature rise, a check to determine how closely to the optimum the materials of the magnet are being worked at the rated stroke and current, and under certain conditions the maximum time of continuous excitation allowable, or the residual force that may occur. If the results of the calculations are satisfactory, the design may be considered finished.

It is the purpose of this chapter to set forth the criterion for optimum economy, the fundamental physical relationships governing the design of tractive magnets, the method whereby these relationships may be put together to obtain a suitable preliminary design, detailed empirical information suitable for choosing design values which would otherwise have to be obtained by laborious cut-and-try methods, and examples of actual design calculations for each magnet type.

The design procedure and details will be worked out for only the six types of magnets covered in Table II of Chapter IX. It will be assumed that a suitable type of magnet has already been selected by the method outlined in Art. 81, and that there are no special requirements regarding speed of action.

84. Fundamental Physical Relationships Governing the Design of Tractive Magnets—Generalized Scheme for Design

1. **Physical Relationships.** For the purpose of illustrating the symbols used in this article and a practical mechanical form of construction, a typical flat-faced cylindrical-plunger magnet is shown in Fig. 1. The cylindrical shape is chosen for the purposes of discussion because it is simpler to formulate than a square or rectangular shape and is the type generally employed unless special considerations, such

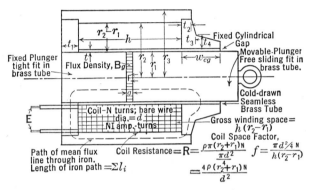

Fig. 1. Cross section through a flat-faced cylindrical plunger magnet, illustrating the symbols of Art. 84.

as quick action, require a laminated type. Mechanically it is very simple to build, as all the parts are turned.

Force Equation. The magnet must exert an assigned minimum force throughout an assigned stroke, this force being expressed by the force equation for each type considered, as discussed in Chapter VIII. This equation is of the form:

$$\text{Force} = \frac{\pi B_g^2 r_1^2}{2\mu} \tag{1}$$

for the attractive type magnets, where B_g is the flux density in the working gap and πr_1^2 is the area of the working gap.

Magnetic Circuit Equation. The exciting magnetomotive force NI, which must provide the flux required for the given force and stroke, is determined by the magnetic circuit equation for each type. This equation is of the following form:

$$NI = \frac{B_g g}{\mu} + \Sigma H_i l_i \tag{2}$$

where $B_g g/\mu$ represents the magnetomotive force necessary to establish the flux at a density of B_g across an air gap of length g, and $\Sigma H_i l_i$ represents the magnetomotive force necessary to establish the flux in the iron parts of the circuit.

Heating Equation. The coil must have such sectional area and heat conducting surface as to keep the temperature rise within an assigned limit, as given by the heating equation 9 derived in Art. 66.

$$\theta_f = \frac{q\rho}{2kf(r_2 - r_1)} \left(\frac{NI}{h}\right)^2 \tag{3}$$

where θ_f is the final temperature rise of the coil, q the fraction of the total time the coil is excited, ρ the volume resistivity of the copper wire, k the heat-dissipation coefficient, f the space factor, $r_2 - r_1$ the gross coil wall thickness, and h the gross coil length.

Voltage Equation. The copper of the coil must be arranged in a suitable number of turns for the voltage assigned, as given by the voltage equation 16 derived in Art. 61.

$$E = IR = \frac{4\rho(r_2 + r_1)NI}{d^2} \tag{4}$$

where E is the assigned coil voltage, $r_2 + r_1$ the mean diameter of a turn on the coil winding, and d the bare wire diameter.

2. Preliminary Design Procedure. In the four formulas listed the following items are usually given in the design data:

$\mathscr{F}$, the required force.
g, the length of the working air gap or stroke of the magnet.
θ_f, the required temperature rise.
q, the fraction of the total time the magnet is excited.
E, the supply voltage.

Of the other design items, some vary between only very narrow limits and, hence, can be assumed for the purposes of the preliminary calculation; the remaining ones must be evaluated from the equations.

The procedure is as follows: The value of the working air-gap flux density B_g in equation 1 will vary between relatively narrow limits depending upon the index number. B_g must be so chosen that the iron in the yoke will be operating at an economical degree of saturation. The saturation in the yoke section relative to that in the working air gap is determined by the leakage coefficient, which is smallest for the large index numbers of any one type of magnet. Thus, B_g can be assumed higher for the high end of the index number range for each type. In general, also, magnets designed for large amounts of work and excited for a small fraction of the time will have higher values of B_g. Data for representative values of B_g for magnets doing medium amounts of work will be given later for each magnet type. This flux density may therefore logically be assumed, and equation 1 solved for r_1.

In equation 2, the first term, $B_g g/\mu$, may now be evaluated. The second term, $\Sigma H_i l_i$, cannot be determined because nothing is known as yet about the dimensions of the iron circuit. However, experience shows that, when the materials of the magnet are used economically, the magnetomotive force required for the iron will be a reasonably definite percentage of that consumed in the working air gap at the beginning of the stroke. This number will vary between 10 to 25 per cent, depending on the magnetic properties of the iron used. If there is a fixed cylindrical gap in the magnet, as illustrated in Fig. 1, another 5 to 10 per cent of the working gap magnetomotive force should be allowed for it. Making these substitutions, equation 2 may be solved for NI as

$$NI = (1.15 \text{ to } 1.35) \frac{B_g g}{\mu}$$

In the heating equation, ρ and k will be the resistivity of the copper wire and the heat-dissipation coefficient, respectively, both at the final temperature corresponding to θ_f. f is the space factor of the coil and, as can be seen from the data of Table V, Chapter VI, does not vary much as the wire size varies between wide limits. With a little experience one can guess the final wire size within a few numbers and hence can estimate the space factor very closely. This will leave two unknowns in the heating equation, r_2 and h. In order to solve for these it will be necessary to introduce another relationship. This is most conveniently done by specifying the shape of the coil. As $r_2 - r_1$ is the radial thickness of the coil and h its length, the shape will be specified by setting a value for $h/(r_2 - r_1)$. It has been found by actual experience that for any type of magnet there is a limited range of this ratio that will produce a high weight economy. Values of this ratio will be given for each magnet type later. Using this, the heating equation may be solved and both r_2 and h determined.

The voltage equation may now be solved for the bare wire diameter d.

3. **Modification of Preliminary Design to Stock Dimensions.** It is almost always necessary to change the bare wire diameter in order to suit the standard sizes of the American wire gauge. The standard

size just larger than the preliminary value should be chosen if it is felt desirable to increase the ampere-turns above the preliminary value or the next smaller size if fewer ampere-turns are desired. It should be remembered, however, that if the iron is relatively unsaturated the force will vary very rapidly with a change in wire diameter. Thus, an increase in one wire size will change the coil resistance by a factor of approximately 1.26^2 or to 159 per cent of the original value; will decrease the coil current by a factor of $1/1.26^2$ or to 63 per cent of the original value; will change the coil magnetomotive force by a factor of $1/1.26^2 \times 1.26$ or to 79 per cent of the original value; and the force by a factor of $1/1.26^2$ or to 63 per cent of the original value; and vice versa for a decrease in wire size. It is sometimes necessary for this reason to split a wire size by winding half the coil with one wire size and the rest of the coil with a wire one size larger or smaller.

The only other dimension that it is generally necessary to modify is the plunger radius r_1 if the plunger is to slide in a brass tube. If because of the choice of wire size the coil magnetomotive force is considerably greater than the amount required in the preliminary design, it is wise to choose r_1 slightly less than the preliminary value to prevent the force from being too much in excess of the desired value.

Cold-drawn seamless brass tubes are made[1] in integral fractions of an inch on the outside diameter, varying by $\frac{1}{16}$'s from $\frac{1}{8}$ up to $1\frac{1}{8}$ in. in diameter, by $\frac{1}{8}$'s up to about $2\frac{3}{4}$ in., and by $\frac{1}{4}$'s up to about $6\frac{1}{2}$ in. Wall thicknesses are made to correspond approximately to a limited range of the B. & S. gauge or A.w.g. for copper wire, and the Stubs iron gauge for steel wire or sheets. Round steel bars are rolled from about $\frac{3}{16}$ to $1\frac{1}{2}$ in. in diameter varying by $\frac{1}{16}$'s, above $1\frac{1}{2}$ in. in diameter varying by $\frac{1}{8}$'s up to $4\frac{3}{4}$ in., and above this by $\frac{1}{4}$'s. The plunger radius r_1 should be chosen, if possible, of such a stock size that it can slide with the proper clearance in a standard brass tube. If this is not practicable, the plunger should be chosen oversized and turned down to give the desired clearance in stock size of brass tube. This clearance should be kept as small as possible to eliminate slide pull at the fixed cylindrical gap. In very special jobs it is sometimes necessary to ream the brass tube, turn the plunger to a specified clearance, and then plate the plunger with a hard chromium finish to prevent the possibility of stickiness between the plunger and tube.

4. Check of Preliminary Design Using Stock Sizes. At this point the magnet has been completely designed and the only work remaining is to check the design. There are several factors which will cause the

---

[1] It is best to refer to manufacturer's catalogues for exact information on materials available.

performance of the "preliminary design," were it built, to deviate from that desired. They are as follows:

1. The (15 to 35 per cent) air-gap magnetomotive force allowed for the iron (and the fixed cylindrical gap, if there is one) may or may not be close, depending on whether the iron parts have been saturated to the degree anticipated. This in turn depends upon the leakage coefficient, which coefficient varies greatly with the ratio $h/(r_2 - r_1)$. In the preliminary design the value of the leakage coefficient used has been absorbed into the value for B_g, initially assumed. It is for this reason that preliminary values of working air-gap flux density are relatively low compared to the saturation density of iron.

2. The changes produced by using stock sizes of wire and steel rods, etc., may be sufficient to throw the design off.

In order to check the design it is necessary to make a sketch of the magnet. To do this it will be necessary to calculate the thicknesses t, t_1, t_2, t_3, t_4, and $(r_3 - r_2)$, of the various parts of the steel shell. This may be done by making the cross-sectional area of the iron path through these points equal to at least that of the plunger with the exception of t_4, where the area may be made less as all the flux does not pass through this section. If the shell is made of a magnetically inferior steel as compared to the plunger, its area should be correspondingly increased to conform to the magnetomotive force allowed for these parts. The width of the fixed cylindrical gap, W_{cg}, may be determined by designing this gap to consume the magnetomotive force assumed for it in solving the magnetic circuit equation 2.

After the sketch has been completed, the actual leakage coefficient and saturation curve of the magnet can be computed. In computing the saturation curve it will be necessary to evaluate $\Sigma H_i l_i$ for various values of plunger flux. The correct value of coil ampere-turns (hot), final temperature rise, force at the beginning of the stroke, may then be computed. The degree to which the optimum has been approached (explained in the next article) in working the magnetic materials may also be computed. Should these values be satisfactory the design, other than for the necessary mechanical details, may be considered finished. If not, changes must be made as indicated by the trial results.

The fewness of the number of the changes necessary to effect a satisfactory design reflects more than anything else the experience and shrewdness of the designer in making the original choice for B_g, $\Sigma H_i l_i$, and f in the preliminary design. There is, however, one consolation, that regardless of how inexperienced the designer is, he can with sufficient patience, by following the rules, ultimately turn out a design just as efficient as the more experienced man.

85. Optimum Conditions for the Economical Use of Magnet Materials

In a tractive magnet the useful work done is considered to be the product of the force at the beginning of the stroke by the stroke.[2]

For a given magnet having a given number of ampere-turns excitation, the useful work can be changed by either changing the stroke or by changing the effective area of the working air gap. An increase of stroke will decrease the initial force, and the product of the two, which is the useful work, may decrease or increase. The area of the working air gap may be changed by the use of polar enlargements, as in a horseshoe magnet, or by changing the angle of the plunger face in the generalized conical plunger magnet. This change in area will change the flux density in the air gap, but because of the change in area of the working gap the resulting force may either increase or decrease. Therefore, there evidently is an optimum stroke for a magnet having a fixed pole face area, or an optimum gap area for a magnet having a fixed stroke, which will give the maximum useful work.

The force of a tractive magnet at the beginning of the stroke can be expressed by equation 8a, Chapter VIII

$$\text{Force} = \frac{B^2 S}{2\mu} \quad \text{joules per inch} \tag{5}$$

where B is the flux density of the useful flux in the working gap at the beginning of the stroke. The useful work of the magnet will equal

$$\text{Force} \times g = \frac{B^2 S g}{2\mu} \quad \text{joules} \tag{6}$$

where g is the gap length or stroke in inches. The expression $B^2 S g / 2\mu$, however, is that for the magnetic energy stored in the working gap. It may, therefore, be concluded that the useful work will be a maximum for the conditions of stroke or gap area that make the magnetic energy of the gap a maximum. Then, assuming the gap area to remain constant, the relation which determines the optimum gap length may be found as follows: The stored magnetic energy of the air gap is $W = B^2 S g / 2\mu = $ maximum, or $B^2 g = $ maximum; differentiating with respect to B and considering both B and g variable, we have

$$2Bg + B^2 \frac{dg}{dB} = 0$$

$$2g + B \frac{dg}{dB} = 0$$

$$\frac{dg}{dB} = -\frac{2g}{B} \tag{7}$$

---

[2] See Art. 36, page 80, and Arts. 78 and 79, pages 228 and 235.

The magnetic circuit equation of the magnet will be $NI = \Sigma H_i l_i + (Bg/\mu)$ = a constant, where H_i is the magnetic intensity in the iron, l_i is the length of the iron parts, and Bg/μ is the ampere-turns across the air gap. Differentiating as before, we have

$$0 = \mu l_i \frac{dH_i}{dB} + g + B \frac{dg}{dB}$$

Substituting from 7,

$$\mu l_i \frac{dH_i}{dB} = -g + 2g$$

$$\frac{dH_i}{dB} = \frac{g}{\mu l_i}$$

$$\frac{dB}{dH_i} \times \frac{S}{l_i} = \frac{\mu S}{g}$$

or

$$\frac{d\phi}{dF_i} = \frac{\mu S}{g} \qquad (8)$$

This equation states that the optimum condition is reached when the differential permeance ($d\phi/dF_i$) of the iron circuit equals the permeance of the working air gap.

If the gap area is allowed to vary while the gap length remains constant, the same optimum will be reached.

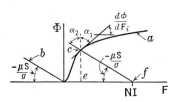

Fig. 2. Graphical construction for determining the optimum operating point.

In Fig. 2 is shown a graphical construction for determining how closely a magnet is working to its optimum. Curve a represents the magnetization curve (useful flux in working air gap against ampere-turns) for all the iron parts of the magnet and any fixed air gaps that might exist. Curve b is the magnetization curve for the working air gap drawn with a negative slope. The slope of this curve is the useful permeance of the air gap. To satisfy equation 8 the magnet must be operated at such a point on saturation curve a that the slope at this point equals the useful permeance of the working gap. If NI equals the magnetomotive force of the exciting coil, the operating point may be determined by finding the intersection of a line drawn from NI having a slope of $-\mu S/g$ with the magnetization curve, as shown at c. Then, if the tangent to the saturation curve at c makes the same angle (α_1)

with the vertical as does the air-gap permeance line (α_2), the magnet is working at its optimum.

The degree to which the optimum is approached by an actual design can be determined by comparing the useful work of the design, area of the triangle *cef*, where *c* is the actual operating point, to the area of the same triangle where *c* is the optimum operating point determined by graphically finding the point on the saturation curve where $\alpha_1 = \alpha_2$.

86. Design of a Flat-Faced Armature Type of Magnet

1. General. The flat-faced armature type of magnet is used primarily as a lifting magnet or holding magnet. Thus it is used for lifting loads of scrap iron, steel plate, steel castings, etc. The magnets used for this purpose are very large, lifting loads of the order of 10 tons and having a diameter of the order of 4 ft. Magnetically, the design of these large magnets is no different from or more difficult than that of the small ones. The problems involved with respect to heat radiation, insulation, and mechanical construction, however, are vastly different and require the application of empirical data gained from experience on actual magnets of this type. The heating data of Chapter VII cannot be considered applicable to such large magnets.[3]

The flat-faced armature type of magnet is also used for magnetic clutches and brakes on both large and small machines. In general it is used for any application requiring a large force to be exerted through a small distance, as is indicated in Table II, Chapter IX, by large index numbers—above 350.

2. Design Data. For the purpose of illustration let it be required to design a magnet having given the following data:

Stroke $= 0.025$ in. $= g$.   Excitation $=$ continuous.
Force $= 400$ lb. $= \mathscr{F}$ Material $=$ Swedish charcoal iron (unannealed
Voltage $= 6$ volts $= E$ after machining) sample 2.
Temperature rise, 70° C.; average temperature rise (by change of resistance) of
 coil above surrounding ambient temperature of 20° C.

The index number of this required design is $\sqrt{\mathscr{F}}/s = \sqrt{400/0.025} = 800$, which indicates a flat-faced armature magnet as the most economical type. Figure 3 is a sketch of the conventional type of small flat-faced armature magnet, showing all the dimensions indicated in the formulas.

---

[3] A comprehensive discussion of the problems involved in the design of large lifting magnets may be found in "Fundamentals of Electrical Design," by A. D. Moore, McGraw-Hill Book Co.

3. Design Equations. The four fundamental design equations of Art. 84, modified to suit the flat-faced armature magnet, are as follows:

Force Equation.

$$\text{Force} = \frac{B^2 r_1^2}{11.45} \text{ lb.} \tag{9}$$

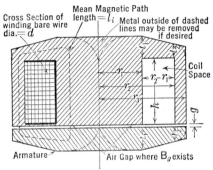

Fig. 3. Cross section through a circular flat-faced armature magnet, illustrating design symbols.

from equation 8c, Chapter VIII; the denominator must be reduced to one-half as the flat-faced armature has two working faces.

Magnetic Circuit Equation. In the flat-faced armature type of magnet, the magnetomotive force used in the iron parts (whose length so far is unknown) may be provisionally estimated as $200 r_1$. This is permissible as in proper designs the required magnetomotive force is found to be nearly independent of the shape of the magnetic circuit but, of course, is proportional to the linear dimensions. Thus, making this substitution in equation 2 for $\Sigma H_i l_i$, and noting that there are two air gaps in series, we have

$$NI = \frac{2Bg}{\mu} + 200 r_1 \tag{10}$$

Heating Equation.

$$\theta_f = \frac{q\rho}{2kf(r_2 - r_1)} \left(\frac{NI}{h}\right)^2 \tag{3}$$

remains unchanged.

Voltage Equation.

$$E = \frac{4\rho(r_2 + r_1)NI}{d^2} \tag{4}$$

remains unchanged.

4. Preliminary Design. Equation 9 may now be solved for r_1 if the value of B_g can be assumed. Referring to Fig. 4, data are given for B_g as a function of the index number, for medium-sized magnets.[4]

---

[4] Large magnets of this type, excited for only a fraction of the time, will use higher values of B_g, ranging up to about 100 kmax. per sq. in. as an upper limit.

Using these data, a preliminary value of 72 kmax. per sq. in. is obtained for B_g corresponding to an index number of 800. Substituting into (9), we have, for r_1,

$$r_1 = \sqrt{\frac{400 \times 11.45}{72^2}} = 0.94 \text{ in.}$$

Substituting this value for r_1 into (10), we have for NI

$$NI = \frac{2 \times 72 \times 0.025}{0.00319} + 200 \times 0.94 = 1{,}317 \text{ ampere-turns}$$

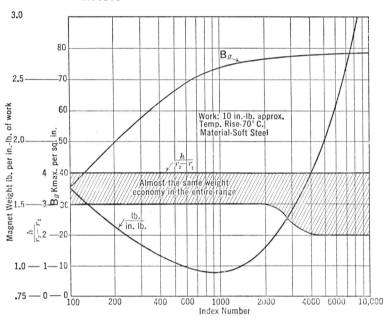

Fig. 4. Design constants for the circular flat-faced armature magnet illustrated in Fig. 3.

Equation 3 may now be solved by making the following substitutions:

$\theta_f = 70°$ C.
$q = 1$.
$\rho = 0.865 \times 10^{-6}$ ohm-inch at 90° C. obtained for the value at 20° C. given in Sec. 1, Art. 57, page 155, and applying the correction factor of Table III, Chapter VI.
$k = 0.0076$ watt per sq. in. per ° C. temperature difference from Fig. 7b, Chapter VII, for a temperature rise of 70° C. and for the coil in good thermal contact with iron parts. In this type of magnet the

coil will generally be wound on a brass bobbin for the smaller sizes, which will fit snugly in the coil space and, hence, be in good thermal contact with the iron parts.

$f = 0.50$, assumed: The space factor of bobbin-wound coils varies from 0.6 for large low-voltage coils, to 0.4 for small high-voltage coils. As this coil will be small and for low voltage a value of 0.50 has been assumed.

$\dfrac{h}{r_2 - r_1} = 4$, for this type of magnet. It will be noticed that this ratio may vary between 4 and 3 without producing an appreciable change in weight economy. Coil shape for large magnets is dictated by the problem of heat radiation, the flat type of coil being preferable where the heat-dissipation problem is severe.

Thus substituting $h/(r_2 - r_1) = 4$ into (3), and solving for h, we have

$$h = \sqrt[3]{\dfrac{2q\rho(NI)^2}{kf\theta_f}}$$

And substituting for the quantities under the radical and solving we get

$$h = \sqrt[3]{\dfrac{2 \times 1 \times 0.865 \times 10^{-6} \times 1{,}317^2}{0.0076 \times 0.50 \times 70}}$$

$$h = 2.24 \text{ in.}$$

and

$$r_2 - r_1 = \dfrac{h}{4} = 0.56 \text{ in.}$$

$$r_2 = 0.56 + 0.94 = 1.50 \text{ in.}$$

The rest of the dimensions of the iron path may now be computed on the basis of maintaining the iron cross section at all points equal to that of the center core, πr_1^2; r_3 and t_2 are best computed after the final value of r_2 has been obtained. To find t_1:

$$2\pi r_1 t_1 = \pi r_1^2$$

$$t_1 = \dfrac{r_1}{2}$$

$$t_1 = 0.47 \text{ in.}$$

Equation 4 may now be solved for d:

$$d = \sqrt{\dfrac{4\rho(r_2 + r_1)NI}{E}}$$

$$= \sqrt{\dfrac{4 \times 0.865 \times 10^{-6} \times 2.44 \times 1{,}317}{6}}$$

$$= 0.043 \text{ in.}$$

5. **Modification to Suit Stock Sizes.** The preliminary design has now been finished, and the next step will be to modify such dimensions as can conveniently be replaced by stock sizes.

Wire Size. From Table I, Chapter VI, the nearest wire diameter to 43.0 mils is 45.3 mils, corresponding to No. 17 A.w.g.

Iron Dimensions. The radius r_1 can be made $\frac{15}{16}$ in., $h = 2\frac{1}{4}$ in., and $t_1 = \frac{15}{32}$ in., giving standard fractional dimensions to these parts where no particular precision is involved. The radii r_2 and r_3 are best left until the coil has been designed, because it is often desirable in such small coils as this one will be to modify r_2 slightly in order to obtain a whole number of layers.

The bobbin in this type of magnet, because it merely supports and protects the winding and does not perform any other mechanical function such as guiding a plunger, may be fabricated from sheet brass, about $\frac{1}{32}$ in. thick. A 2-in. O.D. seamless brass tube could be used but then it would have to have a rather thick wall—slightly less than $\frac{1}{16}$ in. If it were desirable to use the brass tube it would be better to use a 2-in. tube with a wall thickness corresponding to No. 20 A.w.g. and make r_1 larger to fit inside of the tube.

6. **Check of Preliminary Design Using Stock Sizes.**

Coil Design—Space Factor.[5] The gross winding depth $(r_2 - r_1)$ will include besides the thickness of the winding itself: (a) the thickness of the bobbin tube; (b) the insulation between this and the coil, consisting of two or more layers of oiled linen or mica (5 to 10 mils each), according to the voltage; (c) the insulation outside of the coil, consisting of a like thickness of oiled linen or mica and a protecting layer of tape (about 15 mils); (d) an allowance of about 10 mils between the insulated coil and the outside iron for irregularities in winding, etc. Subtracting this total thickness from $(r_2 - r_1)$ and dividing by d_1 (diameter of the insulated wire plus thickness of paper between layers if there is any), the number of layers is determined, assuming no embedding or compression of the insulation; it is found in practice, however, that, owing to these effects, from 5 to 10 per cent more layers may be wound in a given space, depending on the thickness of the insulation relative to the wire diameter. The number of layers should be a whole number and so, if small, may require a slight change in r_2 in order to get whole number of layers. Also it is sometimes desirable to have the number of layers even so that the beginning and ending of the winding will be at the same end of the coil, as this often facilitates the mechanical problem of bringing the leads to the coil through the surrounding iron case.

---

[5] See Art. 60, page 170.

258   DESIGN OF TRACTIVE MAGNETS   [Chap. X

Thus making the following allowances:

(a) 0.03125.
(b) 0.015.
(c) 0.030.
(d) 0.010.

Total allowances = 0.08625; we will therefore have as the net winding depth, using the preliminary value of $r_2 - r_1$, 0.5625 in. $-$ 0.0863 in. = 0.4762 in., which divided by 0.0473, the diameter d_1 of enameled-covered No. 17 wire, gives 10.07 layers. If a 5 per cent allowance is made for embedding, 10.6 layers may be wound; r_2 may therefore conveniently be made $1\frac{17}{32}$ in., allowing 11 full layers to be wound with a 2.5 per cent allowance for embedding.

Besides the insulated wire, the axial length of the coil h includes: (a) the thickness of two flanges of the bobbin (if used), each being chosen as a stock size of sheet brass having approximately the same thickness as the tube; (b) three or more thicknesses of oiled linen or mica and (usually) a press-board washer at each end (about $\frac{1}{32}$ in.) to take the wear in the winding; (c) an allowance of about 10 mils total for imperfect fit of the bobbin in the iron shell, or (in the armature type only) clearance between the outer surface of the bobbin and the armature, $\frac{1}{32}$ in. or more; and (d) the thickness of an insulated lead from the inside of the coil, in case it is desired to bring this out to the outside of the coil (otherwise it may be brought out through a hole in the flange). Subtracting this total thickness from h and dividing by diameter d_1 of the insulated wire, the number of turns per layer is determined, assuming the coil to be closely wound; it is found that from 5 to 10 per cent fewer turns may actually be wound in a given space, and an allowance of 1 turn per layer must also be made for space lost at the ends.

Thus, making the following allowances:

(a) 0.0625.
(b) 0.0925.
(c) 0.0031.
(d) 0.0000.

Total allowances = 0.1581 in.; we will therefore have a net winding length of 2.25 in. $-$ 0.158 in. = 2.092 in., which divided by d_1 = 0.0473 in. will give 44.2 turns per layer; deducting 5 per cent for the spread of the winding and then 1 turn for the loss at the ends, we will have 41.0 turns.

The total turns on the coil will therefore be $11 \times 41 = 451$ turns.

The cross-sectional area of the bare copper wire of the coil will be

ART. 86] FLAT-FACED ARMATURE TYPE OF MAGNET 259

$451 \times 0.00161 = 0.727$ sq. in., where 0.00161 sq. in. is the cross section of one No. 17 wire as determined from Table II, Chapter VI.

The space factor will be

$$f = \frac{0.727}{2.25 \times 0.59375} = 0.545$$

which is slightly higher than the value assumed in the preliminary design of 0.50. This will tend to make the temperature rise lower, but will probably be balanced by the fact that d is larger than the preliminary value.

The coil resistance may now be computed as follows: The mean radius of a turn will be r_1 + the winding thickness allowances (a) and (b) + one-half the net coil winding thickness, $= 0.9375 + 0.03125 + 0.25373 = 1.24$ in. The total length of the wire will therefore be $2\pi \times 1.24 \times 451 = 3{,}515$ in., and the resistance, applying the factor 0.0004221 ohm per inch, at 20° C. for No. 17 wire from Table II, Chapter VI, and the temperature correction factor 1.2751 from Table III, will be $3{,}515 \times 0.0004331 \times 1.2751 = 1.89$ ohms at 90° C.

The coil current will be $6 \div 1.89 = 3.17$ amperes (hot), and the magnetomotive force developed by the coil will be

$$NI = 3.17 \times 451 = 1{,}430 \text{ ampere-turns}$$

Temperature Rise. Substituting the values so far obtained into equation 3 we have

$$\theta_f = \frac{1 \times 0.865 \times 10^{-6}}{2 \times 0.0076 \times 0.545 \times 0.59375} \left(\frac{1{,}430}{2.25}\right)^2$$

$$= 71.0° \text{ C.}$$

This value is close enough to 70° C. to be considered satisfactory. r_3 and t_2 may now be computed:

$$\pi(r_3^2 - r_2^2) = \pi r_1^2$$

$$r_3 = \sqrt{r_1^2 + r_2^2}$$

$$r_3 = \sqrt{0.9375^2 + 1.53125^2} = 1.798 = 1\tfrac{51}{64} \text{ in.}$$

$$2\pi r_2 t_2 = \pi r_1^2$$

$$t_2 = \frac{r_1^2}{2r_2}$$

$$t_2 = \frac{0.9375^2}{2 \times 1.53} = 0.288 \text{ in.}$$

Magnetic Circuit Calculation. At this point a sketch of the magnet should be made so that the mean lengths of the magnetic circuit may be obtained. The sketch is shown in Fig. 5, drawn to scale. The mean

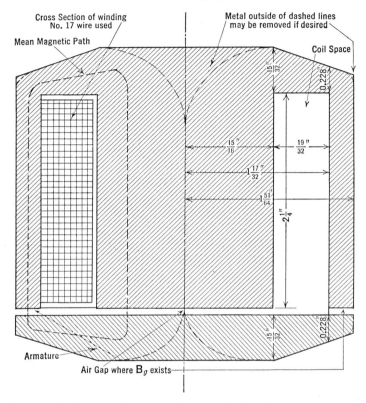

Fig. 5. Final design dimensions of circular flat-faced armature magnet.

lengths of the magnetic circuit as obtained from this sketch by measurement are:

	Length	Area
Armature................	1.20	2.76
Pole core (inner).........	2.25	2.76
Pole core (outer).........	2.25	2.76
Yoke....................	1.20	2.76

FLAT-FACED ARMATURE TYPE OF MAGNET

The mean areas of the yoke and armature are probably greater than those of the inner and outer pole cores, but it is a safe thing to make them the same.

Before the saturation curve of the magnetic circuit can be calculated it is necessary to determine the leakage coefficient so that the distribution of flux in the various parts will be known. Referring now to Sec. 3, Art. 54, page 144, and using the general derivation for the leakage coefficient, we have

$$\nu = \frac{P_a + P_L}{P_u} = \frac{2g}{\pi \mu r_1^2}(P_a + P_L)$$

where P_a, the total effective permeance through the armature between the inner and outer pole core, is

$$P_a = \frac{1}{\dfrac{1}{P_1 + (P_f)_i} + \dfrac{1}{P_1 + (P_f)_o}}$$

The inner pole fringing permeance $(P_f)_i$ is

$$(P_f)_i = P_{11} + P_{12b} = 3.26\mu r_1 + 4\mu r_1 \log_\epsilon \frac{r_2 - r_1}{\pi g}$$

$$= 3.26 \times 3.19 \times 0.9375 + 4 \times 3.19 \times 0.9375 \log_\epsilon \frac{0.594}{\pi \times 0.025}$$

$$= 9.75 + 24.2 = 34 \text{ max. per ampere-turn}$$

The outer pole fringing permeance $(P_f)_o$, is

$$(P_f)_o = P_7 + P_{8b} + P_{11} + P_{12b}$$

$$= 1.63\mu r_3 + 2\mu r_3 \log_\epsilon \left(1 + \frac{r_1}{g}\right) + 3.26\mu r_2 + 4\mu r_2 \log_\epsilon \frac{r_2 - r_1}{\pi g}$$

$$= 1.63 \times 3.19 \times 1.798 + (3.19)(2)1.798 \log_\epsilon 38.5 + 3.26$$

$$\times 3.19 \times 1.53 + 4 \times 3.19 \times 1.53 \times \log_\epsilon 7.57$$

$$= 9.35 + 41.9 + 15.9 + 39.4$$

$$= 107 \text{ max. per ampere-turn}$$

The useful gap permeance (P_1) is

$$(P_1) \text{ outer} = (P_1) \text{ inner} = \frac{\pi \mu r_1^2}{g} = 352 \text{ max. per ampere-turn}$$

P_a will therefore be

$$P_a = \frac{1}{\dfrac{1}{352+34} + \dfrac{1}{352+107}} = 210 \text{ max. per ampere-turn}$$

P_L, the leakage permeance between pole cores, is

$$P_L = \mu\left[1.57h\left(\frac{r_2+r_1}{r_2-r_1}\right) - \frac{r_2+r_1}{2}\left(1-\frac{\pi g}{r_2-r_1}\right)\right]$$

$$= 46.5 \text{ max. per ampere-turn,}$$

and the leakage coefficient will be

$$\nu = \frac{P_a + P_L}{P_u} = \frac{210 + 46.5}{352/2} = 1.46$$

If the simplified equation 27 of Art. 54, which assumes certain proportions in the magnet (see text of Art. 54, Chapter V), had been used the leakage coefficient would have been

$$\nu = 1 + 14\frac{g}{r_1} = 1 + 0.373 = 1.373$$

This is lower than the value above mainly because it is based on a coil proportion of $h/(r_2 - r_1) = 2.5$, while in the actual magnet $h/(r_2 - r_1) = 4.0$. In general where $h/(r_2 - r_1)$ is about 2.5 the simpler formula 27 can be used. The larger ratio $h/(r_2 - r_1)$ the higher will be the leakage coefficient.

The saturation curve for the iron parts (for a constant gap length of $g = 0.025$ in.) may now be computed. This computation will be carried out in the manner of Art. 47, Chapter IV, which should be referred to.

Using ϕ_g, the useful flux of the working air gap, as a reference, the flux of the armature, assuming the permeance of the armature to be infinite,[6] will be

$$\phi_a = \phi_g \times \frac{P_a}{P_u} = \frac{210}{176}\phi_g = 1.192\,\phi_g$$

---

[6] This assumption is generally valid when there is an air gap, and is made so that ϕ_L may be estimated for any value of ϕ_g with reasonable accuracy. When there is no air gap, or when the armature is saturated so that its magnetomotive force is appreciable, ϕ_L must be determined by finding the magnetomotive force effective across P_L: this can be determined by assuming a flux through the armature, and thence determining the magnetomotive force required to force this flux through the armature and total working air-gap permeance P_a. This magnetomotive force will then be that effective across P_L.

The magnetomotive force across the pole faces will then be equal to that across the armature, namely

$$F_a = \frac{\phi_a}{P_a} = \frac{\phi_g}{P_u}$$

The effective flux of the pole cores will be [7]

$$\text{Eff. } (\phi_p) = \phi_a + \tfrac{2}{3}\phi_L$$

where [8]

$$\phi_L = F_a P_L = \phi_g \frac{P_L}{P_u} = \phi_g \frac{46.5}{176} = 0.264 \phi_g$$

Hence

$$\text{Eff. } (\phi_p) = (1.192 + \tfrac{2}{3}\,0.264)\phi_g = 1.37\phi_g$$

The flux of the yoke will be

$$\phi_y = \nu\phi_g = 1.46\phi_g$$

The actual calculation of the saturation curve will be carried out in tabular form; it is started by assuming the value of flux density in the yoke.[9] This is a logical procedure as this is where the magnet will saturate first.

Actual Work and Optimum Work of Design. In Fig. 6 is shown plotted the saturation curve of the iron parts as plotted from the results of Table I. From point f at 1,430 ampere-turns (actual coil NI) on the axis of abscissas there is drawn the working air-gap permeance line b with a negative slope of 176 max. per ampere-turn. The intersection of

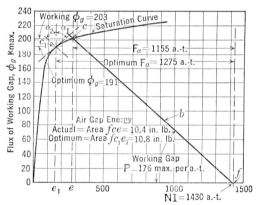

Fig. 6. Determination and check of operating point for the final design of circular flat-faced armature magnet of Fig. 5.

---

[7] See Art. 46, Chapter V.

[8] This is not the same as that of Art. 46, but is the same as that of Art. 54, and represents the leakage permeance between pole cores effective at the ends of the pole cores.

[9] Unless the permeance of the armature cannot be assumed infinite in the first approximation, in which case the procedure of footnote 6 should be followed.

TABLE I

CALCULATION OF SATURATION CURVE OF IRON PARTS OF FLAT-FACED ARMATURE MAGNET (AIR GAP EXCLUDED) FOR A USEFUL AIR-GAP LENGTH OF 0.025 IN.

(See Saturation Curve of Fig. 11a, Chapter II, for Sample 2)

Part	Length, in.	Area, sq. in.	Flux	ϕ kmax.	B kmax. per sq. in.	H a-t. in.	F a-t.
1. Yoke	1.20	2.76	$1.46\phi_g$	331	120	255	306
2. Pole cores	4.50	2.76	$1.37\phi_g$	311	112.7	135	607
3. Useful air gap	$P = \dfrac{352}{2}$ max. per a-t.		ϕ_g	227	82.2	.....	....
4. Armature	1.20	2.76	$1.19\phi_g$	270	97.8	26	31

Total 944

Part	ϕ	B	H	F	ϕ	B	H	F
1	317	115	165	198	304	110	100	120
2	298	108.0	84	378	285	103.2	46	207
3	217	78.8	....	...	208	75.4	....	...
4	259	93.8	19	23	248	89.6	14	17
			Total	599			Total	344
Part 1	276	100	33	40	248	90	14	17
2	259	94.0	19	85	233	84.5	11	50
3	189	68.5	....	...	170	61.6	....	...
4	225	81.5	10	12	203	73.4	7	8
			Total	137			Total	75

line b with the saturation curve at c gives the operating point of the magnet with its rated air gap and current. The optimum operating point for this particular magnet with an excitation of 1,430 ampere-turns (as explained in Art. 85) is shown at c_1.

The energy actually available in the air gap, corresponding to the useful work of the magnet, is equal to the area of triangle fce, which is

$$\tfrac{1}{2} \times 203 \times 10^{-5} \times 1{,}155 \times 8.86 = 10.4 \text{ in-lb.}$$

The energy that would be available if the magnet were operating at its optimum point is

$$\tfrac{1}{2} \times 191 \times 10^{-5} \times 1{,}275 \times 8.86 = 10.8 \text{ in-lb.}$$

The degree to which the actual design approaches the optimum design is

$$\frac{10.4}{10.8} \times 100 = 96.3 \text{ per cent}$$

which is sufficiently close to be satisfactory.

Force. The only factor left to be checked is the force, which according to the stored energy of the air gap should be $10.4 \div 0.025 = 416$ lb. The flux density in the working gap, B_g, will be

$$\frac{\phi_g}{\pi r_1^2} = \frac{203}{2.76} = 73.5 \text{ kmax. per sq. in.}$$

The force by equation 9 will be

$$\text{Force} = \frac{73.5^2 \times 0.9375^2}{11.45} = 415 \text{ lb.}$$

This is satisfactory. Actually a magnet like this if it were required to develop 400 lb. would be designed for more than 400 lb.—say 10 per cent excess as a factor of safety.

Weight. The weight of the copper of the coil will be

$$\frac{3515}{12} \times \frac{1}{161.3} = 1.81 \text{ lb.,}$$

where 161.3 is the number of feet per pound of No. 17 copper wire from Table II, Chapter VI.

The iron volume can be estimated quite closely by multiplying the mean length of the magnetic circuit by the cross section, which cross section is practically constant throughout.

The volume will be

$$2.76 \times 6.70 = 18.5 \text{ cu. in.}$$

and its weight will be
$$18.5 \times 0.28 = 5.18 \text{ lb.}$$
The volume of the brass bobbin will be
$$\pi \times 1\tfrac{29}{32} \times \tfrac{1}{32} \times 2\tfrac{7}{32} + 2\pi(1.53^2 - 1^2)\tfrac{1}{32} = 0.68 \text{ cu. in.}$$
Weight of the brass bobbin will therefore be
$$0.68 \times .32 = 0.22 \text{ lb.}$$
Then allowing $\tfrac{1}{4}$ lb. for insulation, etc., the total weight will be
$$\text{Total weight (magnet and armature)} = 7.5 \text{ lb.}$$
and the weight per unit of work will be
$$\frac{7.5}{10.4} = 0.72 \text{ lb. per in-lb. of work}$$

Final Results. The design is now finished. Figure 5 shows a complete sketch of the magnet with all dimensions. Below are tabulated all the principal data:

Force = 416 lb.   Power (hot) = 19 watts.
Stroke = 0.025 in.   Temperature rise = 71° C.
Voltage = 6 volts.   Weight = 7.5 lb.
Current (hot) = 3.17 amperes.   Useful work = 10.4 in-lb.
            Weight economy = 0.72 lb. per in-lb. of work.

87. Design of a Flat-Faced Plunger Type of Magnet

1. General. The flat-faced plunger type of magnet is used primarily as a part of a machine and as such is not built in the large sizes that the lifting magnets of the last section are. It is essentially a large-force, short-stroke magnet applicable to index numbers from 350 to 62.

2. Design Data. For the purpose of illustration let it be required to design a magnet having given the following data:

Stroke = 0.125 in. = g.   Temperature rise = 70° C., average
Force = 100 lb. = $\mathscr{F}$.   temperature rise (by change in
Voltage = 120 volts = E.   resistance) of coil above surround-
Excitation[10] = 0.1.   ing ambient temperature of 20° C.
Material = Plunger S.A.E. 10–10 sample 3.
Other parts S.A.E. 10–20 sample 4.

[10] Because the period of excitation is so short the complete design must include a calculation of the longest continuous excitation the magnet may receive before the coil will exceed its allowable temperature rise.

The index number of the required design is $\sqrt{\mathscr{F}/s} = \sqrt{100/0.125}$ = 80, which indicates the flat-faced plunger type. In Fig. 7 is shown a sketch of the conventional type of flat-faced plunger magnet, showing all the dimensions indicated in the formulas.

3. **Design Equations.** The four fundamental design equations of Art. 84, modified to suit the flat-faced plunger type of magnet, are as follows:

$$\text{Force} = \frac{B^2 r_1^2}{22.9} \text{ lb.} \tag{11}$$

from equation 8c, Chapter VIII.

Magnetic Circuit Equation. Referring to Fig. 7 it can be seen that the magnetic circuit of this type of magnet includes, besides the working gap and the iron, a fixed cylindrical gap having a length g_c and an axial width of t_5. In the actual calculation this gap must be included in the iron part of the magnet circuit.

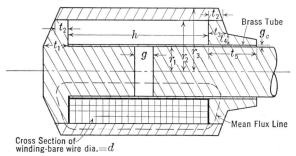

FIG. 7. Cross section through a flat-faced cylindrical plunger magnet, illustrating the design symbols.

The magnetomotive force allotted to this gap is entirely arbitrary and depends merely on what the designer feels can be economically justified. A large percentage will lower the magnetic efficacy of the design but will allow either a cylindrical gap of short axial width t_5, or one of greater radial thickness g_c, to be used. The former will result in a reduction of the overall weight; the latter will tend to reduce the side pull on the plunger due to any eccentricity. If there is any likelihood of the plunger being slightly eccentric in the fixed gap, because of lack of precision in machining or the tolerances allowed, it is wise to increase the radial thickness of the gap. A small percentage of coil ampere-turns across this gap increases the magnetic efficacy, but also increases the weight by reason of the large axial width required unless special means are employed to make the radial length of the gap short.[11]

[11] See the design of the fixed cylindrical gap of the full conical plunger magnet of Art. 88.

Where a magnet is excited for a short part of the time and where it is desirable to reduce the physical size of the magnet a larger percentage should be allotted to this gap. This magnetomotive force will range from 5 to 15 per cent of the coil ampere-turns.

The ampere-turns consumed in the iron parts of the magnetic circuit will vary between 10 and 25 per cent, depending on the kind of iron used and the economy desired. If a relatively hard iron like sample 4 is used 25 per cent should be allotted for this purpose; for the softer irons a lower amount can be used. For large magnets excited for a small portion of the time it is generally economical to use more than on the smaller ones.

Estimating the ampere-turns to be used in the iron, $\Sigma H_i l_i$, provisionally at about 20 per cent of that developed in the coil, and assuming 10 per cent of the magnetomotive force across the fixed cylindrical gap, equation 2 will be

$$NI = \frac{Bg}{\mu} + 0.30NI \tag{12}$$

$$NI = \frac{Bg}{0.70}$$

Heating Equation.

$$\theta_f = \frac{q\rho}{2kf(r_2 - r_1)} \left(\frac{NI}{h}\right)^2 \tag{3}$$

remains unchanged.

Voltage Equation.

$$E = \frac{4\rho(r_2 + r_1)NI}{d^2} \tag{4}$$

remains unchanged.

4. Preliminary Design. Equation 11 may now be solved for r_1 if the value of B, the flux density in the working air gap, can be assumed. Referring to Fig. 8, data are given for B as a function of the index number, for medium-sized magnets.[12] Using these data, a preliminary value of 64 kmax. per sq. in. is obtained for B corresponding to an index number of 80. Substituting into equation 11, we have for r_1

$$r_1 = \sqrt{\frac{22.9 \times 100}{64^2}} = 0.747 \text{ in.}$$

[12] Where a magnet is excited for only a small fraction of the time, and where a high magnetic efficacy is not important but weight economy is, slightly higher flux densities than those of Fig. 8 may be used.

Art. 87] FLAT-FACED PLUNGER TYPE OF MAGNET

Substituting into 12 for NI, we have

$$NI = \frac{64 \times 0.125}{0.70 \times 0.00319} = 3{,}580 \text{ ampere-turns}$$

Equation 3 may now be solved by making the following substitutions:

- $\theta_f = 70°$ C.
- $q = 0.1$.
- $\rho = 0.865 \times 10^{-6}$ ohm-in.
- $k = 0.0076$ watt per sq. in. per °C. temperature difference from Fig. 7b, Chapter VII, for $\theta_f = 70°$ C. and for the coil in good thermal contact with the iron parts. In this type of magnet the coil bobbin can be formed from the brass tube in which the plunger slides and the iron flanges at either end which form the shell. This construction will be assumed in this design. Good thermal contact may therefore be assumed between the coil and the surrounding iron parts.
- $f = 0.45$ assumed. As the voltage is relatively high and the coil small, the wire size will be small. This will necessitate paper between layers, and hence the space factor will be low.
- $\dfrac{h}{r_2 - r_1} = 5.4$ from the data of Fig. 8.

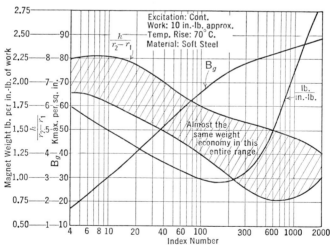

Fig. 8. Design constants for the flat-faced cylindrical plunger magnet.

Thus, substituting $h/(r_2 - r_1) = 5.4$ into equation (3), and solving for h, we have

$$h = \sqrt[3]{\frac{2.7q\rho(NI)^2}{kf\theta_f}}$$

and substituting for the quantities under the radical, and solving we get

$$h = \sqrt[3]{\frac{2.7 \times 0.1 \times 0.865 \times 10^{-6} \times 3{,}580^2}{0.0076 \times 0.45 \times 70}} = \sqrt[3]{14.07}$$

$$= 2.31 \text{ in.}$$

and

$$r_2 - r_1 = \frac{h}{5.4} = 0.429 \text{ in.}$$

$$r_2 = 0.429 + 0.747 = 1.176 \text{ in.}$$

Because the shell and end pieces are made of an inferior magnetic material compared to the plunger they should be operated at a lower maximum flux density, say 0.8 as much. Then, equating the plunger area to 0.8 that of the outer steel shell, r_3 may be found:

$$\pi r_1^2 = 0.8\pi(r_3^2 - r_2^2)$$

$$r_3 = \sqrt{1.25 r_1^2 + r_2^2}$$

$$r_3 = 1.44 \text{ in.}$$

The rest of the iron dimensions and the axial length t_5 of the fixed cylindrical gap can best be computed after the final stock sizes have been chosen for the plunger, shell, and brass tube.

Equation 4 may now be solved for d,

$$d = \sqrt{\frac{4\rho(r_2 + r_1)NI}{E}}$$

$$= \sqrt{\frac{4 \times 0.865 \times 10^{-6} \times 1.923 \times 3{,}580}{120}} = 0.0141 \text{ in.}$$

5. Modification to Suit Stock Sizes. The preliminary design has now been finished, and the next step will be to modify such dimensions as can be conveniently replaced by stock sizes.

Wire Size. From Table I, Chapter VI, 14.2 mils corresponds exactly to No. 27 wire.

Iron Dimensions. Plunger Radius (r_1). This must be chosen in relation to the brass tube. The nearest suitable outside diameter for the brass tube would be 1.5 in. Using a wall thickness of No. 24 A.w.g. (0.020 in.), the inside diameter would be 1.460 in., and the plunger radius (r_1) could be made 0.728 in., leaving a clearance of 0.002-in. all around the plunger.[13] The plunger area will then be 1.66 sq. in.

Steel Shell (r_2 and r_3). The nearest standard size of cold-drawn seamless steel tube to the preliminary dimensions of r_2 and r_3 is an outside diameter of $2\frac{7}{8}$ in. with a wall thickness of $\frac{1}{4}$ in., making the inside diameter $2\frac{3}{8}$ in. Using this,[14] the cross-sectional area of the shell will be

$$\frac{\pi}{4}[(2\tfrac{7}{8})^2 - (2\tfrac{3}{8})^2] = 2.04 \text{ sq. in.}$$

The length of the winding space h can be taken as $2\frac{5}{16}$ in.; t_1 and t_2 can be taken so that their cylindrical areas at the radii r_1 and r_2, respectively, will equal the cross section of the shell:

$$t_1 = \frac{2.04}{2\pi r_1} = 0.446 \text{ in.}$$

$$t_2 = \frac{2.04}{2\pi r_2} = 0.274 \text{ in.}$$

t_4 and t_5 can be computed later when the magnetomotive force of the coil has been determined.

6. Check of Preliminary Design Using Stock Sizes.

Final sizes:
- $r_1 = 0.728$ in.
- $r_2 = 1\frac{3}{16}$ in.
- $r_3 = 1\frac{7}{16}$ in.
- $r_2 - r_1 = 0.459$ in.
- $h = 2\frac{5}{16}$ in.
- $t_1 = 0.446$ in.
- $t_2 = 0.274$ in.

Coil Design—Space Factor. Following the same procedure as in Art. 86, Sec. 6, the gross winding depth ($r_2 - r_1$) will include the following allowances:

(a) 0.022 thickness of brass tube and clearance to plunger
(b) 0.015 insulation between brass tube and coil
(c) 0.025 insulation outside of coil
(d) 0.015 allowance for coil irregularities

0.077 total allowances

---

[13] If it is desired to avoid machining the plunger a No. 22 A.w.g. tube could be used with a $1\frac{7}{16}$-in. diameter plunger. This would give a clearance of 0.005 in. all the way around the plunger, which is somewhat large. In large-force short-stroke magnets more friction due to plunger eccentricity can be tolerated than in those with relatively long strokes and smaller forces, such as occur in conical and taper plunger magnets. In these magnets it is often customary to ream the tube and turn the plunger to a specified tolerance.

[14] Although this size and thickness of tube is catalogued, it is not always stocked. In this event the next thicker tube should be obtained and machined to the thickness desired.

The net winding depth will be 0.459 − 0.077 = 0.382. Allowing 2.2 mils paper between layers and 15.4 mils for the diameter of enameled-covered No. 27 wire, the thickness of a layer will be 0.0154 + 0.0022 = 0.0176 in. The number of layers will equal 0.382 ÷ 0.0176 = 21.7, which can be called 22 if a 1.5 per cent allowance for embedding is made. This being an even number of layers will allow both leads to be brought out at the same end through a single hole or slot.

As the iron end pieces are to form the bobbin flange it will not be necessary to allow for brass bobbin flanges or for imperfect fit of the bobbin, in determining the net winding length of the coil. We shall assume a hole with a rubber bushing at the fixed plunger end to bring the leads out. Then, making an allowance of 20 mils for oiled linen and $\frac{1}{32}$ in. for fiber at each end of the coil, the net winding length will be 2.313 − 0.102 = 2.211 in., and the turns per layer will be

$$\frac{2.211}{0.0154} \times 0.95 - 1 = 135$$

assuming that because of the spread of the winding only 95 per cent as many turns can be wound, and that one turn is lost at the end of each layer.

The total turns will be 22 × 135 = 2,970. As the cross-sectional area of the bare copper wire is 0.000158 sq. in., the space factor will be

$$f = \frac{2{,}970 \times 0.000158}{0.459 \times 2.312} = 0.442$$

which is slightly lower than the assumed value of 0.45.

The coil resistance may now be computed as follows: The mean radius of a turn will be r_1 + the winding thickness allowances $(a) + (b)$ + one-half of the net coil winding thickness = 0.728 + 0.037 + 0.191 = 0.956 in. The total length of wire will be $2\pi \times 0.956 \times 2{,}970$ = 17,850 in., and the resistance at 90° C., applying the factor 0.00429 ohm per inch for No. 27 wire from Table II, Chapter VI, and the temperature correction factor 1.2751 from Table III, Chapter VI, will be

$$17.85 \times 4.29 \times 1.2751 = 97.8 \text{ ohms at } 90° \text{ C.}$$

The coil current will be 120 ÷ 97.8 = 1.227 amperes, and the magnetomotive force developed by the coil will be

$$NI = 2{,}970 \times 1.227 = 3{,}650 \text{ ampere-turns}$$

Temperature Rise. Substituting the values so far obtained into equation 3, we have

$$\theta_f = \frac{0.1 \times 0.865 \times 10^{-6}}{2 \times 0.0076 \times 0.442 \times 0.459} \left(\frac{3,650}{2.3125}\right)^2 = 69.6° \text{ C.}$$

This value is satisfactory.

Fixed Cylindrical Gap. The length of this gap can now be computed so that it will consume 10 per cent of the coil magnetomotive force. The permeance of this gap $P_c = \phi_c/F_c$ cannot be computed until the flux ϕ_c of the gap is known. This flux is equal to the useful flux of the working gap multiplied by the leakage coefficient. The leakage coefficient for this type of magnet is given by equation 23, Chapter V.

$$\nu = 1 + \frac{g}{r_1}\left[0.67 + 0.13\frac{g}{r_1} + \frac{r_2 + r_1}{\pi r_1}\left(\frac{\pi h}{8(r_2 - r_1)} + \frac{2(r_2 - r_1)}{\pi h} - 1\right)\right.$$
$$\left. + 1.465 \log_{10}\frac{r_2 - r_1}{g}\right]$$

which can be applied because

$$g = 0.125 \text{ in.} < \frac{4(r_2 - r_1)}{\pi} = 0.585$$

This gives on substitution

$$\nu = 1.42$$

Applying this, the tentative value of flux through the fixed cylindrical gap is

$$\phi_c = \nu B \pi r_1^2 = 1.42 \times 64 \times \pi \times 0.728^2 = 151.5 \text{ kmax.}$$

where 64 is the tentative value of B in the preliminary design. Therefore

$$P_c = \frac{151.5}{3,650 \times 0.1} = 0.415 \text{ kmax. per ampere-turn}$$

$$P_c = \frac{\mu S}{l} = \frac{\mu\pi(2r_1 + g_c)t_5}{g_c}$$

where g_c will equal the thickness of the brass tube plus the clearance between the plunger and tube = 0.020 in. + 0.002 in. = 0.022 in.

$$t_5 = \frac{P_c g_c}{\mu\pi(2r_1 + g_c)} = \frac{0.415 \times 0.022}{\mu\pi(1.478)} = 0.617 \text{ in., say } \tfrac{5}{8} \text{ in.}$$

Magnetic Circuit Calculation. At this point a sketch of the magnet should be made so that the mean lengths of the magnetic circuit may be obtained. This sketch is shown in Fig. 9, drawn to scale. Because the end pieces have the same area as the shell and may be considered for practical purposes to carry the same flux, these parts can be treated as a unit. The mean lengths and areas are shown in Table II.

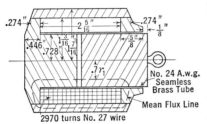

Fig. 9. Cross section through the final design of the flat-faced cylindrical plunger magnet, showing all dimensions.

The saturation curve for the iron parts and fixed air gap (useful flux ϕ_g plotted against magnetomotive force for a constant gap length of $g = \frac{1}{8}$ in.) may now be computed. Using ϕ_g, the useful flux in the working gap as a reference, the flux[15] in the iron parts and fixed cylindrical gap of the magnetic circuit will be equal to $\nu\phi_g$. The actual calculation of the saturation curve will be carried out in tabular form and is started by assuming the value of flux density in the iron plunger.

Actual Work and Optimum Work of Design. In Fig. 10 is shown plotted the saturation curve of the iron parts and fixed air gap as plotted from the results of Table III. From point f at 3,650 ampere-turns (actual coil NI) on the axis of abscissas there is drawn the working air-gap permeance line b with a negative slope of 42.4 max. per ampere-turn. The intersection of line b with the saturation curve at c gives the operating point of the magnet with its rated air gap and current. This also happens to be the optimum operating point.

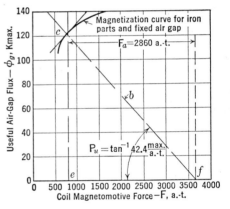

Fig. 10. Magnetization curve and graphical construction for determining the operating point of the magnet of Fig. 9.

---

[15] Actually, of course, the flux in the plunger is variable, being equal to $\nu\phi_g$ only at the ends. At the air gap it is less by the radial leakage flux. However, as the air gap is in the center, the radial leakage flux is much smaller than it would be in the flat-faced armature or horseshoe types, and hence if the saturation in the plunger is not too high the error, which will be on the safe side, will not be appreciable. Should a correction be desired the method of Art. 47, Chapter IV, may be employed.

TABLE II

	Length	Area	Material
Plunger..................	3.08	1.66	S.A.E. 10–10
End pieces and shell...........	3.48	2.04	S.A.E. 10–20

TABLE III

CALCULATION OF SATURATION CURVE OF IRON PARTS AND FIXED AIR GAP OF FLAT-FACED PLUNGER MAGNET (WORKING AIR GAP EXCLUDED) FOR A USEFUL AIR-GAP LENGTH OF 0.125 IN.

(See saturation curve of Fig. 11a, Chapter II, for samples 3 and 4)

Part	Length, in.	Area, sq. in.	Flux, ϕ		B kmax. per sq. in.	H a-t. per in.	F a-t.
			— —	kmax.			
1. Plunger......	3.08	1.66	$1.42\phi_g$	199	120	255	786
2. Ends and shell	3.48	2.04	$1.42\phi_g$	199	97.6	94	327
3. Useful air gap.	$P_u = 0.0424$ kmax. per a-t.		ϕ_g	140.2	84.5	....	....
4. Fixed cylindrical gap	$P_c = 0.415$ kmax. per a-t.		$1.42\phi_g$	199	....	....	480

Total 1,593

Part	ϕ	B	H	F	ϕ	B	H	F
1	191	115	170	524	182.5	110	102	314
2	191	93.5	80	278	182.5	89.5	71	247
3	134.5	......	......	......	128.5	......	......	......
4	191	......	......	460	182.5	......	......	440
			Total	1,262			Total	1,001
1	166	100	31	96	149.5	90	16	49
2	166	81.4	57	198	149.5	73.2	46	160
3	117	......	......	......	105.3	......	......	......
4	166	......	......	400	149.5	......	......	360
			Total	694			Total	569

The energy actually available in the air gap, corresponding to the useful work of the magnet, is equal to the area of triangle fce, which is

$$\tfrac{1}{2} \times 121.5 \times 10^{-5} \times 2{,}860 \times 8.86 = 15.4 \text{ in-lb.}$$

Force. The only other factor left to be checked is the force, which according to the stored energy of the air gap should be

$$\frac{15.4}{0.125} = 123 \text{ lb.}$$

The flux density in the working gap will be

$$\frac{\phi_g}{\pi r_1^2} = \frac{121.5}{1.66} = 73.2 \text{ kmax. per sq. in.}$$

The force by equation 11 will be

$$\text{Force} = \frac{73.2^2 \times 0.728^2}{22.9} = 124 \text{ lb.}$$

This is really larger than is required for a factor of safety, and the magnet should be redesigned using a slightly smaller value of r_1. The excessive force is caused by the coil magnetomotive force being higher than the preliminary value, and by the iron and fixed cylindrical gap using only 22 per cent of the coil ampere-turns instead of the 30 per cent originally assumed. We shall consider it satisfactory for our purpose and proceed with the remainder of the design.

Weight. The weight of the copper of the coil will be

$$\frac{17{,}850}{12} \times \frac{1}{1{,}639} = 0.91 \text{ lb.}$$

where 1,639 is the number of feet of No. 27 wire per pound, from Table II, Chapter VI.

The iron volume can be estimated quite closely by multiplying the mean length of the parts of the magnetic circuit by their areas; referring to Fig. 9:

Plunger $[0.446 + 2\tfrac{5}{16} + \tfrac{5}{8}] \times 1.66 = 3.38 \times 1.66 =$ 5.61
Ends and shell $[0.459 \times 2 + 2\tfrac{5}{16} + 0.274 \times 2] \times 2.04$
                                                   $= 3.78 \times 2.04 =$ 7.71

Total iron volume (cu. in.)   13.32
Weight of iron $= 13.3 \times 0.283 = 3.76$ lb.

The volume of the brass tube will be

$$\pi(2r_1 + g_c)g_c \times (h + t_5) = \pi \times 1.476 \times 0.02 \times 2.94 = 0.273 \text{ cu. in.}$$

and the weight of the tube will be

$$0.273 \times 0.32 = 0.0872 \text{ lb.}$$

Then, allowing $\frac{1}{8}$ lb. for coil paper, insulation, and varnish, we have for the total weight of the magnet:

$$0.91 + 3.76 + 0.09 + 0.12 = 4.88 \text{ lb.}$$

The weight per unit of work will be

$$\frac{4.88}{15.4} = 0.317 \text{ lb. per in-lb. of work}$$

This is much less than the value shown in Fig. 8. The reason is, of course, the short period of excitation.

Maximum Time of Continuous Excitation. There remains only one thing to be determined, and that is how long it can be excited continuously before it reaches its maximum allowable temperature rise of 70° C. This is equivalent to asking how long is the 0.1 time of excitation.

Referring to Art. 67, Chapter VII, it will be seen that, owing to the coil being in good thermal contact with the iron parts, the iron parts may be considered about 55 per cent effective in contributing to the final effective thermal capacity. This thermal capacity will then be:

$$\begin{aligned}
\text{Copper } 180 \times 0.91 &= 165 \text{ joules per } °\text{C. rise} \\
\text{Paper insulation } 700 \times \tfrac{1}{8} &= 88 \\
\text{Iron } 225 \times 0.55 \times 3.76 &= 466 \\
\text{Brass } 200 \times 0.55 \times 0.09 &= 10
\end{aligned}$$

Total final effective thermal capacity = 729

The surface area of the coil effective for dissipating heat will be, according to the first paragraph of Art. 65, Chapter VII, the cylindrical coil surfaces plus the end surfaces:

Cylindrical coil surface $= 2\pi(r_2 + g_c + r_1)h$

$$= 2\pi(1.19 + 0.02 + 0.728)\, 2.31 = 28.1 \text{ sq.in.}$$

End surface $= 2\pi[r_2^2 - (r_1 + g_c)^2] = 5.34$

Total heat-dissipation surface = 33.5 sq. in.

The total iron thermal capacity is $(466 + 10) \times 1 \div 0.55 = 866$. Therefore the ratio of the total iron thermal capacity to the coil heat-dissipating surface is 866 ÷ 33.5 = 26 joules per degree Centigrade rise per square inch. Referring now to the last paragraph of Art. 67, Chapter VII, we may consider that the effective value of the thermal capacity of the iron for an elapsed time equal to no more than $\frac{1}{3}$ of the final effective time constant is $100 - 50[(26-8)/(30-8)] = 59$ per cent of its final effective value, or equal to $476 \times 0.59 = 281$ joules per degree Centigrade rise. Therefore the thermal capacity of the magnet for a short interval of heating will be

Copper and insulation = 253
Iron and brass parts   = 281
Total = 534 joules per degree Centigrade rise

The heat-dissipation capacity of the coil K will be equal to $0.0076 \times 33.5 = 0.255$ watt per degree Centigrade temperature difference between the average coil temperature and the surrounding air. The value of the heat-dissipation coefficient k, chosen in the preliminary design, is 0.0076.

Substituting into equation 2, Chapter VII, the heating equation for the first elapsed time interval equal to $\frac{1}{3}$ of the final thermal time constant is

$$\theta = \frac{P}{K}\left(1 - \epsilon^{-\frac{K}{C}t}\right) = \frac{147.5}{0.255}\left(1 - \epsilon^{-\frac{0.255}{534}t}\right)$$

where $147.5 = 120 \times 1.227$ is the power input to the coil (hot).

$$\theta = 578(1 - \epsilon^{-0.000478 t}).$$

Substituting $\theta = 70°$ C. into this equation and solving, we find $t = 270$ seconds, or 4.5 minutes. If this time is less than $\frac{1}{3}$ of the final thermal time constant, the solution may be considered correct. The final thermal time constant is

$$\frac{C}{K} = \frac{729}{0.255} = 2{,}850 \text{ sec.} = 47.6 \text{ min.}$$

Therefore the maximum period of steady excitation of the magnet from a cold start should not exceed 4.5 minutes, and this period of steady excitation should be repeated not more often than in a $4.5 \times 10 = 45$-minute cycle.

Final Results. The design is now finished. Figure 9 shows a complete sketch of the magnet with all dimensions. Below are tabulated all the principal data:

Force = 123 lb.
Stroke = 0.125 in.
Voltage = 120 volts.
Current (hot) = 1.227 amperes.
Power (hot) = 147.5 watts.
Temperature rise = 69.6° C., after a period of 4.5 minutes continuous excitation from a cold start.
Time of excitation = 0.1.
Weight = 4.88 lb.
Useful work = 15.4 in-lb.
Weight per unit of useful work = 0.317 lb. per in-lb. of work.

88. Design of a Full-Conical Plunger Type of Magnet

1. **General.** The full-conical plunger magnet is used for the same type of work as the flat-faced plunger magnet of Art. 87 but gives, for the same plunger size, less force through a longer stroke. If α is the angle of the cone, as illustrated in Fig. 11, the surface area of the cone will be larger than the plunger cross section by $1/\cos \alpha$. Likewise, the actual length of the flux lines measured normal to the conical surfaces will be equal to the stroke s, multiplied by $\cos \alpha$. Hence, the permeance of the working air gap will be approximately larger by $1/\cos^2 \alpha$ than it would be for a flat-faced plunger magnet of the same diameter and stroke, and likewise the rate of change of permeance of the working gap will be greater by $1/\cos^2 \alpha$. Thus, for the same plunger flux, the force would be in the ratio of $\cos^2 \alpha$ to 1, as compared to the flat-faced plunger. However, if the same magnetomotive force is available for both magnets, it will be possible to pull the coned plunger farther apart, in the ratio of $1/\cos^2 \alpha$ to 1, and still maintain the same flux, because of the increased permeance of its working gap.

If the two magnets are compared on this basis, each will be capable of performing the same useful work, but with force and stroke of the conical magnet smaller and larger in the ratio of $\cos^2 \alpha$, and $1/\cos^2 \alpha$, respectively, as compared to the flat-faced magnet. If the leakage coefficients, at the start of the stroke, are the same, both magnets will have identical flux linkage and the same magnetic energy stored in the air gaps. Thus the magnets can be said to be truly equivalent, and the design of one can be carried out with the equations of the other, provided that the factor $\cos^2 \alpha$ is properly introduced to give the equivalent force and stroke.

The only point left to be considered is whether the leakage coefficient of the conical-faced plunger is equal to that of the equivalent flat-faced one. Referring to Fig. 17, Chapter V, it can be seen that the only alteration in the leakage path will be a slight diminution in the permeance of path $8b$, due to path 7 being longer in the conical-faced plunger than its equivalent flat-faced plunger. Path 7, however, does not change its permeance as its length changes. Therefore, the total net leakage permeance will be slightly less in the conical-faced plunger than in the equivalent flat-faced plunger. However, the permeance of the total working gap of the conical plunger magnet (paths 1 and 2) will be slightly less than that of the equivalent flat-faced plunger by the second term ($\frac{1}{2}v^2/r_1 \sin^2 \alpha \cos \alpha$) of its permeance formula 12, of Chapter VIII. These two effects will tend to counterbalance each other with the result that the leakage coefficient of a conical-faced plunger magnet and its equivalent flat-faced plunger magnet may be considered to be the same.

Cone-faced plunger magnets can be used with economy in the range of index numbers from 62 down to 7. From 62 to 15, α should equal 45°, and from 15 to 7, α should be equal to 60°. Should there be no preference, the range from 12 down can be covered more economically by the tapered plunger magnet of the next article.

2. Design Data. For the purpose of illustration let it be required to design a magnet having given the following data:

Stroke = $\frac{1}{4}$ in. = g.
Force = 40 lb. = $\mathscr{F}$.
Voltage = 12 volts.
Excitation = continuous.
Temperature rise = 70° C., average temperature rise (by change of resistance) of coil above surrounding ambient temperature of 20° C.
Material = iron parts, equivalent of S.A.E. 10–10 sample 3, parts to be machined from hot-rolled stock.
Weight economy. Assume that expense of construction is not to be considered, and that all possible refinements necessary to reduce the weight are to be made.

The index number of the required design is $\sqrt{\mathscr{F}/s} = \sqrt{40/0.25} = 25.2$, which indicates a full conical-faced plunger type of magnet with $\alpha = 45°$. In Fig. 11, there is shown a sketch of a 45° full conical plunger magnet, with special coil arrangements to allow for a very short fixed cylindrical gap, and showing all the dimensions indicated in the formulas. The flat-faced magnet equivalent to the given conical one is shown by the dashed lines, where the plunger separation at the beginning of the stroke, g, is $0.25 \times \cos^2 45° = 0.25 \times 0.5 = 0.125 = \frac{1}{8}$ in. Its force will equal $40/\cos^2 45° = 40/0.5 = 80$ lb.

3. Design Equations. The four fundamental design equations of Art. 84 will be taken directly from Art. 87, Sec. 3, for the equivalent flat-faced plunger magnet; they are as follows:

Force Equation.

$$\text{Force} = \frac{B^2 r_1^2}{22.9} \text{ lb.} \tag{11}$$

Magnetic Circuit Equation. Because of the high magnetic quality of the iron to be used and the special construction of the fixed cylindrica

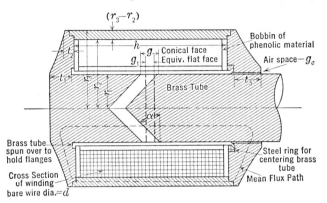

FIG. 11. Cross section through a full-conical cylindrical plunger magnet, illustrating the design symbols.

gap, the ampere-turns to be used in the iron and in the fixed cylindrical air gap can be provisionally estimated at 15 per cent of that developed by the coil. Equation 12 will thus become

$$NI = \frac{Bg}{\mu} + 0.15 NI$$

$$NI = \frac{Bg}{0.85\mu} \tag{12}$$

Heating Equation.

$$\theta_f = \frac{q\rho}{2kf(r_2 - r_1)} \left(\frac{NI}{h}\right)^2 \tag{3}$$

remains unchanged.

Voltage Equation.

$$E = \frac{4\rho(r_2 - r_1) NI}{d^2} \tag{4}$$

remains unchanged.

4. **Preliminary Design.** Equation 11 may now be solved for r_1 if the value of B, the flux density in the plunger due to all the flux in the

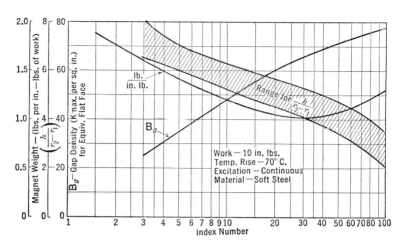

Fig. 12. Design constants for the 45° full-conical cylindrical plunger magnet.

working gap (see Art. 72, Sec. 5), can be assumed. Referring to Fig. 12, data are given for B for 45° full conical plungers, as a function of the index number, for medium-sized magnets. Using these data a pre-

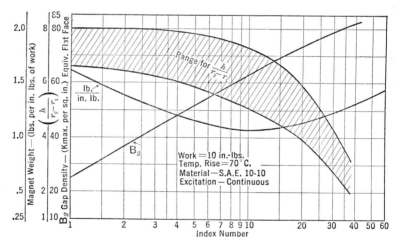

Fig. 13. Design constants for the 60° full-conical cylindrical plunger magnet.

liminary value of 63.5 kmax. per sq. in. is obtained for B, corresponding to an index number of 25.2. Substituting into (11) we have for r_1

$$r_1 = \sqrt{\frac{22.9 \times 80}{63.5^2}} = 0.674 \text{ in.}$$

Substituting into (12) for NI, we have

$$NI = \frac{63.5 \times 0.125}{0.85 \times 0.00319} = 2{,}925 \text{ ampere-turns}$$

Equation 3 may now be solved by making the following substitutions:

$\theta_f = 70°$ C.
$q = 1$.
$\rho = 0.865 \times 10^{-6}$ ohm-in.
$k = 0.0076$ watt per sq. in. per ° C. temperature difference, from Fig. 7b, Chapter VII, for $\theta_f = 70°$ C. and for the coil in good thermal contact with the iron parts. In this particular magnet, as illustrated in Fig. 11, the coil bobbin will be formed by the phenolic flanges held at end as illustrated. This may be considered as good thermal contact.
$f = 0.50$, assumed. As the voltage is low, the wire size will be relatively large, and the space factor high.
$\dfrac{h}{r_2 - r_1} = 4.8$, from the data of Fig. 12.

Thus substituting $h/(r_2 - r_1) = 4.8$ into (3), and solving for h, we have

$$h = \sqrt[3]{\frac{2.4q\rho(NI)^2}{kf\theta_f}}$$

and substituting for the quantities under the radical, and solving, we get

$$h = \sqrt[3]{\frac{2.4 \times 1 \times 0.865 \times 10^{-6} \times 2{,}925^2}{0.0076 \times 0.5 \times 70}} = 4.06 \text{ in.}$$

As the dimensions of the outer shell do not have to be fitted to stock sizes, these dimensions can be left until the final plunger size has been determined. Equation 4 may now be solved for d.

$$d = \sqrt{\frac{4\rho(r_2 + r_1)NI}{E}}$$

$$= \sqrt{\frac{4 \times 0.865 \times 10^{-6} \times 2.19 \times 2{,}925}{12}}$$

$$= 0.043 \text{ in.}$$

5. **Modification to Suit Stock Sizes.** The preliminary design has now been finished and the next step will be to modify such dimensions as can be conveniently replaced by stock sizes.

Wire Size. From Table I, Chapter VI, 43 mils corresponds to a size between No. 17 and No. 18 wire. Choose No. 17 wire, which is the nearest to the desired size. This will produce slightly more magnetomotive force than is desired and may make the temperature rise too high; No. 18 however, will produce too little magnetomotive force and cause the force to fall below the desired value. It may, finally, be necessary to split the wire size.

Iron Dimensions—Plunger Radius. In this particular design, the plunger will be turned to fit the brass tube, and hence the plunger size will be determined by the available sizes of brass tubes.[16] The nearest size tube is $1\frac{3}{8}$-in. outside diameter, and using a thickness of No. 22 A.w.g., the inside diameter will be 1.324 in. If this is reamed out 0.004 in. on the diameter, it will allow a clearance of 0.002 in. over a plunger of 1.324-in. diameter. This final reaming should be done after the brass tube has been spun over to hold the bobbin together. r_1 will therefore be 0.662 in.

Steel Shell (r_2 and r_3). As this is to be turned from an oversize steel tube, there is no restriction on its size. As the material is to be the same as that of the plunger its cross-sectional area should be the same. Using $h/(r_2 - r_1) = 4.8$, and letting $h = 4\frac{1}{16}$ in., r_2 can be made 1.5 in. and r_3 1.64 in. Making the area of end pieces equal to that of the plunger, t_1 will be 0.331 in., t_2 will be 0.146 in. t_4 and t_5 will be computed later when the magnetomotive force of the coil and the leakage coefficient are known.

6. Check of Preliminary Design Using Stock Sizes.

Final Sizes. $r_1 = 0.662$ in.; $r_2 = 1.50$ in.; $r_3 = 1.64$ in.; $h = 4.063$ in.; $t_1 = 0.331$ in.; $t_2 = 0.146$ in.; No. 17 wire.

Coil Design—Space Factor. Following the same procedure as in Arts. 86 and 87, Sec. 6, the gross winding depth ($r_2 - r_1$) will include the following allowances:

(a) 0.021 = thickness of brass tube and clearance to plunger.
(b) 0.062 = phenolic tube between brass tube and coil
(c) 0.030 = insulation outside of coil
(d) <u>0.010</u> = allowance for coil irregularities
0.123 = total allowances

The net winding depth will be $0.838 - 0.123 = 0.715$ in. Allowing 5.0 mils paper between layers and 47.3 mils for the diameter of enameled-covered No. 17 wire, the thickness of a layer will be $0.0473 + 0.005 = 0.0523$ in. The number of layers will be $0.715 \div 0.0523 = 13.8$, which can be called 14 if a 2 per cent allowance for embedding is made. This

---

[16] See Art. 84, Sec. 3.

being an even number of layers, the coil leads may be brought out at the same end through separate holes or a slot.

As the flanges of the bobbin are to be made of phenolic material it will only be necessary to allow for clearance between the flanges and the iron end pieces. Allowing $\frac{1}{16}$ in. for the thickness of each bobbin flange and $\frac{1}{32}$ in. total clearance, the net winding length will be $4.062 - 0.156 = 3.906$ in., and the turns per layer will be $(3.906 \div 0.0473) \times 0.95 - 1 = 78$, assuming that owing to the spread of the winding only 95 per cent as many turns can be wound, and that one turn is lost at the end of each layer.

The total turns will be $14 \times 78 = 1{,}092$. As the cross-sectional area of the bare copper wire is 0.00161 sq. in., the space factor will be

$$f = \frac{1{,}092 \times 0.00161}{0.838 \times 4.0625} = 0.516$$

which is slightly higher than the assumed 0.5 but will tend to compensate for the oversized wire used.

The coil resistance may now be computed as follows: The mean radius of a turn will be r_1 plus the total thickness of the bobbin tube $(a) + (b) + \frac{1}{2}$ the net coil winding thickness $= 0.662 + 0.083 + 0.358 = 1.103$ in. The total length of wire will be $2\pi \times 1.103 \times 1{,}092 = 7{,}450$ in., and its resistance at 90° C., applying the data of Tables II and III, Chapter VI, is

$$0.4221 \times 1.2751 \times 7.45 = 4.01 \text{ ohms}$$

The coil current will be

$$\frac{12}{4.01} = 2.99 \text{ amperes (hot)}$$

and the magnetomotive force will be

$$NI = 2.99 \times 1{,}092 = 3{,}270 \text{ ampere-turns}$$

Temperature Rise. Substituting the values so far obtained into equation 3, we have,

$$\theta_f = \frac{1 \times 0.865 \times 10^{-6}}{2 \times 0.0076 \times 0.516 \times 0.838} \left(\frac{3{,}270}{4.062}\right)^2 = 85.2° \text{ C.}$$

which is too high. This can be made lower by increasing h or by splitting the wire size. Because the coil magnetomotive force is higher than necessary, and because it is desired to keep the weight down, it seems advisable to split the wire size. Let us redesign the coil, using No. 18 for the inner part of the winding and No. 17 for the outer.

Coil Redesign with Two Wire Sizes. By trial it is found that 6 layers of No. 18 having a thickness of 0.284 in. and 8 layers of No. 17 having a thickness of 0.418 may be wound in the space available with 5 mils of paper between layers, with no allowance for embedding. The turns per layer for No. 18 will be 95, and the turns for each section will be, for No. 18, $95 \times 6 = 570$; for No. 17, $78 \times 8 = 624$, making the total turns 1,194. The mean radius of a turn of the No. 18 section will be $0.662 + 0.083 + 0.142 = 0.887$ in., and that of the No. 17 section $0.887 + 0.142 + 0.209 = 1.238$ in. Their mean lengths of turn will be 5.57 in. and 7.76 in., and the lengths of wire 3,180 in. and 4,850 in., respectively. The resistances at 90° C. will be

No. 18—$0.532 \times 1.2751 \times 3.18 = 2.16$ ohms

No. 17—$0.4221 \times 1.2751 \times 4.85 = 2.61$ ohms

The total resistance (hot) will be 4.77 ohms; the current at 12 volts impressed will be 2.51 amperes; and the magnetomotive force, 3,000 ampere-turns. The space factor will be

$$\frac{570 \times 0.00128 + 624 \times 0.00161}{0.838 \times 4.0625} = 0.51$$

and the final temperature rise

$$\theta_f = \frac{0.865 \times 10^{-6}}{2 \times 0.0076 \times 0.507 \times 0.838} \left(\frac{3{,}000}{4.063}\right)^2 = 73° \text{ C.}$$

This value is satisfactory.

Fixed Cylindrical Gap. The length of this gap can now be computed so that it will consume 5 per cent of the coil magnetomotive force. The permeance of the gap $P_c = \phi_c/F_c$ cannot be computed until the flux ϕ_c of the gap is known. This flux is equal to the useful flux of the working gap multiplied by the leakage coefficient. The leakage coefficient for the conical-faced plunger may be evaluated in the manner of Sec. 1, Art. 54, Chapter V, using the gap permeance formulas derived in Sec. 5, Art. 72, Chapter VIII, or by computing it for the equivalent flat-faced plunger magnet. Using the latter scheme, the leakage coefficient for the flat-faced plunger magnet by equation 23, Chapter V, is

$$\nu = 1 + \frac{g}{r_1}\left[0.67 + 0.13\frac{g}{r_1} + \frac{r_2 + r_1}{\pi r_1}\left(\frac{\pi h}{8(r_2 - r_1)} + \frac{2(r_2 - r_1)}{\pi h} - 1\right)\right.$$
$$\left. + 1.465 \log_{10}\frac{r_2 - r_1}{g}\right]$$

which can be applied because the equivalent $g = 0.125$ in. $< 4(r_2 - r_1)/\pi$ = 1.07. This gives on substitution

$$\nu = 1.562$$

Applying this, the tentative value of flux through the fixed cylindrical gap is

$$\phi_c = \nu B \pi r_1^2 = 1.56 \times 63.5 \times \pi \times 0.662^2 = 137 \text{ kmax.}$$

where 63.5 is the tentative value of B chosen in the preliminary design for the equivalent flat-faced plunger magnet. Therefore,

$$P_c = \frac{137}{3{,}000 \times 0.05} = 0.913 \text{ kmax. per ampere-turn}$$

However,

$$P_c = \frac{\mu S}{l} = \frac{\mu \pi (2r_1 + g_c) t_5}{g_c}$$

where g_c is the radial clearance between the plunger and surrounding steel end piece. Because of the method of securing the brass tube which guides the plunger, g_c may be chosen at will. The main consideration in choosing g_c is plunger side pull, which depends on the ratio of g_c to the clearance between the plunger and the brass tube. On this basis let g_c be 0.01-in., which will allow a maximum eccentricity of about 20 per cent. Then

$$t_5 = \frac{P_c g_c}{\mu \pi (2r_1 + g_c)} = \frac{0.913 \times 0.01}{\mu \pi \times 1.334} = 0.685 \text{ in.}$$

Call this $\frac{5}{8}$ in. to allow for the permeance between the steel bobbin ring and the plunger, and the leakage permeances.

Magnetic Circuit Calculation. At this point a sketch of the magnet should be made so that the mean lengths of the magnetic circuit may be obtained. This sketch is shown in Fig. 14, drawn to scale. Because the plunger, shell, and end pieces have the same area they can be treated as a unit. The leakage coefficient will be applied to the entire iron circuit, which is on the safe side as it neglects the distribution of the leakage flux between the plunger and shell.[15] The data on the magnetic circuit are as follows:

	Length	Area	Material
Total iron parts............	10.0 in.	1.37	S.A.E. 10–10
Fixed cylindrical air gap...	Permeance = 0.913		kmax. per a-t.

[15] Refer to footnote 15 of Art. 87.

The saturation curve for the iron parts and fixed air gap (useful flux ϕ_g plotted against magnetomotive force for a constant equivalent flat-faced gap length of $\frac{1}{8}$ in.) may now be computed. Using ϕ_g, the useful

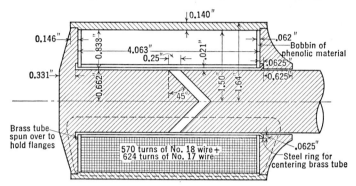

FIG. 14. Cross section through the final design of the 45° full-conical cylindrical plunger magnet, showing all dimensions and method of assembly.

flux in the working air gap, as a reference, the flux in the iron parts and fixed cylindrical gap of the magnetic circuit will be $\nu\phi_g$. The actual calculation of the saturation curve will be carried out in tabular form as

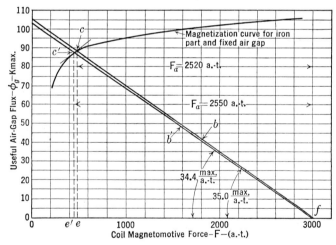

FIG. 15. Magnetization curve and graphical construction for determining the operating point of the magnet of Fig. 14.

shown on page 290 and is started by assuming the value of flux density in the iron.

Actual Work and Optimum Work of Design. On Fig. 15 is shown plotted the saturation curve of the iron parts and fixed air gap as obtained from the results of the tabular calculation. From point f at 3,000 ampere-turns (actual coil NI) on the axis of abscissas there is drawn the equivalent flat-faced working air-gap permeance line b, with a negative slope of 35.0 max. per ampere-turn. The intersection of line b with the saturation curve at c gives the operating point of the magnet with its rated air gap and current. Point c, though not quite the optimum operating point, is so close to it that it may be considered the optimum.

The energy available in the air gap, corresponding to the useful work of the magnet, is equal to the area of the triangle fce, which is

$$\tfrac{1}{2} \times 88 \times 10^{-5} \times 2{,}520 \times 8.86 = 9.90 \text{ in-lb.}$$

Force. The only other factor left to be checked is the force, which according to the stored energy of the air gap should be

$$\frac{9.90}{0.125} = 79.2 \text{ lb.}$$

for the equivalent flat-faced plunger magnet or 39.6 lb. for the actual 45° full conical-faced plunger magnet.

This force may be checked by the formulas of Art. 72, Sec. 5a, without resorting to the artifice of an equivalent flat-faced plunger. Thus the permeance between cones, excluding all fringing fluxes emanating from the outside cylindrical surfaces, will be, by equation 12, of the above reference,

$$P = \frac{\pi \mu r_1}{v \cos \alpha} \left[\frac{r_1}{\cos \alpha} - \frac{1}{2} \frac{v^2}{r_1} \sin^2 \alpha \cos \alpha \right]$$

which on substituting $r_1 = 0.662$ in., $v = 0.25$ in., and $\alpha = 45°$, gives $P = 34.4$ max. per ampere-turn. Drawing line b' with this permeance gives the operating point c', which is almost exactly at the optimum. The available air-gap magnetomotive force F_a will then be 2,550 ampere-turns. Substituting this into the force formula 12b of Art. 72, Sec. 5a,

$$\text{Force} = 4.43 \mu \pi F_a^2 \left[\left(\frac{r_1}{v \cos \alpha} \right)^2 + \frac{1}{2} \sin^2 \alpha \right] \text{ lb.}$$

gives a force of 41.0 lb. This calculation, as mentioned in the above reference, is more optimistic, as it takes into account the force produced by the fringing flux in the conical gap. The permeance calculation by formula 12, above, shows that the equivalent flat-faced plunger magnet almost exactly reproduces the entire permeance within the actual coned gap. We can with safety take the final force to be 40 lb.

TABLE IV

CALCULATION OF SATURATION CURVE OF IRON PARTS AND FIXED CYLINDRICAL GAP, FOR AN EQUIVALENT USEFUL AIR-GAP LENGTH OF 0.125 IN.

(See saturation curve of Fig. 11a, Chapter II, for sample 3)

Part	Length, in.	Area, sq. in.	Flux, ϕ		B kmax. per sq. in.	H a-t. per in.	F a-t.
				kmax.			
1. Iron parts......	10.0	1.37	1.56ϕ_g	164.5	120	270	2,700
2. Fixed cylindrical air gap.......	$P = 0.913$	kmax. per a-t.	1.56ϕ_g	164.5	....	....	180
3. Useful air gap..	$P = 0.035$	kmax. per a-t.	ϕ_g	105.5	....	....	....

Total $F = 2{,}880$

Part	ϕ	B	H	F	ϕ	B	H	F
1	157.5	115	166	1,660	151	110	100	1,000
2	157.5	...	...	172	151	...	...	165
3	100.9	...	...	....	96.5	...	...	....
Total F				= 1,832	Total F			= 1,165
1	137	100	31	310	123	90	16	160
2	137	...	...	150	123	...	...	135
3	87.5	...	...	....	78.5	...	...	....
Total F				= 460	Total F			= 295
1	109.8	80	10	100				
2	109.8	...	...	120				
3	70.0	...	...	....				
Total F				= 220				

Weight. The weight of the copper of the coil will be, for the No. 18 section, $(3{,}180/12)(1/203.4) = 1.30$ lb., and for the No. 17 section, $(4{,}850/12)(1/161.3) = 2.50$ lb., where 203.4 and 161.3 are the number of feet per pound for No. 18 and No. 17 wires, respectively. The total copper weight will therefore be 3.80 lb.

The volume of the brass tube is $\pi \times 1.351 \times 0.0255 \times 4\frac{1}{4} = 0.46$ cu. in. The weight will be $0.46 \times 0.32 = 0.15$ lb.

The volume of the phenolic bobbin will be

$$\text{Flanges, } 2\pi(0.686 + 1.50)(0.814)\tfrac{1}{16} \times 0.05 = 0.034$$

$$\text{Tube, } 2\pi \times 0.717 \times \tfrac{1}{16} \times 3\tfrac{15}{16} \times 0.05 = 0.055$$

$= 0.089$ lb., where 0.05 has been taken as the density of the phenolic material.

The weight of the iron may be calculated by using the mean length and area as for the magnetic calculations.

$$\text{Weight} = 10 \times 1.37 \times 0.283 = 3.88 \text{ lb.}$$

The total weight of the magnet will therefore be $3.80 + 0.15 + 0.09 + 3.88 = 7.92$ lb., which may be called 8.0 lb. to allow for paper, etc. The weight per unit of work will be $8.00 \div 10.0 = 0.8$ lb. per in-lb. of work. This value is less than that shown in Fig. 12 because of the special construction.

Final Results. The design is now finished. Figure 14 shows a complete sketch of the magnet with all dimensions. Below are tabulated all the principal data.

Force = 40 lb.
Stroke = $\frac{1}{4}$ in.
Voltage = 12 volts
Excitation = continuous
Current (hot) = 2.51 amperes
Power (hot) = 30.1 watts
Temperature rise = 73° C.
Weight = 8.0 lb.
Maximum useful work = 10 in-lb.
Weight per unit of maximum work = 0.8 lb. per in-lb. of work

89. Design of a Tapered Plunger Magnet

1. General. The tapered plunger magnet is the proper type of magnet to use when the stroke for a given force exceeds the economical range for the full-conical plunger type but still is not long enough to warrant a leakage-flux type. It is probably the most versatile of all types, as almost any type of force-stroke characteristic can be produced. Thus the force can be made to:

(a) Remain constant throughout the stroke.
(b) Same as (a), except with a sharp rise at the end of the stroke.
(c) Gradually decrease from the beginning to the end of the stroke.
(d) Same as (c), except with a sharp rise at the end of the stroke.
(e) Rise gradually from the beginning to the end of the stroke.
(f) Same as (e), except with sharp rise at the end of the stroke.

(a) and (b) are illustrated in Fig. 7, Chapter IX, by the curves force at 0.8 ampere without and with the stop, respectively; (c) and (d), by the force curves at 1.5 ampere without and with the stop, respectively.

Effects (a), (c), and (e) are controlled entirely by the angle of the taper in relation to the saturation of the plunger. Effect (c) is produced by using a relatively small angle of taper and allowing the plunger to saturate early in the stroke; (e), by using a relatively larger angle of taper and not allowing the plunger to saturate in the beginning of the stroke; and (a), by conditions intermediate between (c) and (e). The sharp rise in force at the end of the stroke is produced by using a stop.

Thus it is possible to produce all the effects obtainable with the cylindrical-faced, stepped cylindrical-faced, and truncated conical plunger magnets. The use of a stop is generally advisable as it assists in holding the force up at the end of the stroke and also produces a high sealing-in force at the end of the stroke. The stop can be flat, as illustrated by the dashed lines of Fig. 7, Chapter IX, or if it is desired to extend its range the stop may be made conical.

The taper plunger type is quite different from those previously considered, in that the optimum useful work conditions derived in Art. 85 do not always apply. This is because the force may decrease toward the end of the stroke and be less than at the beginning; consequently the useful work may not equal the product of the force at the beginning of the stroke by the stroke. This is explained in detail with reference to the actual design in Sec. 6 of this article, "Actual Work and Optimum Work."

2. Design Data. As the constant force characteristic is the one usually desired, let it be required for the purpose of illustration to design a substantially constant force magnet having the following data:

Stroke = $\frac{3}{4}$ in.
Force = 12 lb. = $\mathscr{F}$
Voltage = 12 volts.
Excitation = continuous.
Temperature rise = 70° C. = θ_f, average temperature rise (by change in resistance) of coil above surrounding ambient air temperature of 20° C.
Material = mild cold-rolled steel, S.A.E. 10–10, sample 3.

The index number of this required design will be $\sqrt{\mathscr{F}/s} = \sqrt{12/0.75}$ = 4.62, which, by reference to Fig. 10, Chapter IX, indicates a tapered plunger magnet. Figure 16 is a sketch of this magnet, showing all the dimensions indicated in the formulas.

3. **Design Equations.** This type of magnet cannot be designed in the straightforward manner that the previous ones were, as too many variables are involved. Hence it will be necessary to resort to a cut-and-try method.

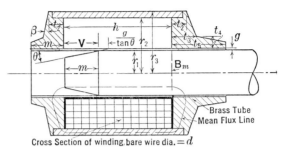

FIG. 16. Cross section through a tapered cylindrical plunger magnet, illustrating the design symbols.

Force Equations. The force due to the radial flux (fringing fluxes excluded) between the tapered plunger and the surrounding steel end piece is

$$\text{Force} = \frac{r_1^4 \left[\frac{r_1 + g}{\theta V + g} - \frac{1}{2} \right] B_p^2}{11.45 \left[\frac{2(r_1 + g)}{\theta} \log_\epsilon \left(\frac{\theta m + g}{\theta V + g} \right) + V - m \right]^2} \quad \text{lb.} \quad (13)$$

from equation 13c, Art. 72, Sec. 6, where B_p is the flux density in the plunger due to this radial flux only. Besides this force, there will be a force due to the distributed leakage flux, which passes across the coil winding, from the plunger to the outer shell. This force is given by equation 20b of Art. 76,

$$\text{Force} = 8.86 \times 10^{-5} H \phi_L \quad \text{lb.} \quad (14)$$

In this type of magnet, owing to the high plunger flux densities and relatively small forces involved, it is necessary to make a correction for the loss in force due to the increase in length of the magnetized portion of the plunger. This correction is given by equation 7b of Art. 71

$$\text{Loss of force} = 8.86 \times 10^{-5} S \int_0^{B_m} H_p dB \quad \text{lb.} \quad (15)$$

Magnetic Circuit Equation. The coil magnetomotive force for this type of magnet can best be estimated by using the work equation 1 of Chapter III.

$$\text{Work} = 8.86 \eta \phi (NI) \quad \text{in.-lb.} \quad (16)$$

where η, the magnetic efficacy, will vary between 0.35 to 0.45 for taper plunger magnets without a stop, and from 0.45 to 0.55 with a stop. The lower values apply to those magnets employing a magnetically inferior steel such as sample 4; the higher values, to the better magnetic steels. Likewise, if the fixed cylindrical gap is assumed to take a large percentage of the magnetomotive force, say 10 to 15 per cent, one should tend to the lower values, while a gap taking only 5 per cent would be in the higher range.

Heating Equation. This remains unchanged.

$$\theta_f = \frac{\rho q}{2kf(r_2 - r_1)} \left(\frac{NI}{h}\right)^2 \tag{3}$$

Voltage Equation. This remains unchanged.

$$E = \frac{4\rho(r_2 + r_1)NI}{d^2} \tag{4}$$

4. Preliminary Design. Equation 13 cannot be solved for r_1 very readily, hence the best procedure is to assume proper values of B_p and θ, and determine by cut and try the proper value of r_1 which will give the required force.

The value of B_p at the end of the stroke will depend upon B_m, the maximum flux density in the plunger, occurring at the beginning of the fixed cylindrical gap as shown in Fig. 16. The plunger should reach saturation at the end of the stroke (H_m about 300 to 400 ampere-turns per inch).

θ, the angle of taper of the plunger, determines how the force will vary throughout the stroke. A value of 0.15 radian is suitable if the force is to remain constant; a smaller value if the force is to decrease toward the end of the stroke, and larger if it is to increase.

Let us substitute the following values into equation 1, to see what force will be produced at the end of the stroke.

$g = 0.025$ in.
$\theta = 0.15$ radian.
$m = 0.75$ in.
$V = 0.0$ in.
$B_m = 120$ kmax. per sq. in.[17]

---

[17] The subsequent work shows that this choice is somewhat low. A value of 125 or possibly 130 would have resulted in slightly higher economy.

$B_p = 84$ kmax. per sq. in.[18] This is obtained by assuming the radial leakage flux between the plunger and shell and the fringing fluxes around the taper gap to be 30 per cent of the total plunger flux, making $B_p = 120 \times 0.7 = 84$.

$r_1 = 0.631$ in. Let us assume a No. 22 A.w.g. brass tube of $1\frac{5}{16}$ in. outside diameter to be used, then if the tube is reamed by 0.003 in. on the diameter, the plunger may be made 1.262 in., making $r_1 = 0.631$.

Making these substitutions, we have

$$\text{Force} \atop (V=0) = \frac{r_1^4 \left[\frac{r_1 + g}{\theta V + g} - \frac{1}{2} \right] B_p^2}{11.45 \left[\frac{2(r_1 + g)}{\theta} \log_\epsilon \left(\frac{\theta m + g}{\theta V + g} \right) + V - m \right]^2} \text{ lb.}$$

$$= \frac{0.631^4 \left[\frac{0.656}{0.025} - \frac{1}{2} \right] 84^2}{11.45 \left[\frac{2 \times 0.656}{0.15} \log_\epsilon \left(\frac{0.1375}{0.025} \right) + 0.0 - 0.75 \right]^2} = 12.5 \text{ lb.}$$

As 12.0 lb. is the desired force the tentative value of 0.631 for r_1 may be considered satisfactory.

Assuming: (1) a value of 0.40 for the magnetic efficacy without a stop, (2) $12 \times \frac{3}{4} = 9$ in-lb. for the work, and (3) $\pi r_1^2 B_m = \pi \times 0.631^2 \times 120 \times 10^{-5} = 150 \times 10^{-5}$ weber for ϕ, equation 16 may be solved for the coil magnetomotive force.

$$NI = \frac{\text{Work}}{8.86 \eta \phi} = \frac{9.0 \times 10^5}{8.86 \times 0.40 \times 150} = 1{,}693 \text{ ampere-turns}$$

Equation 3 may now be solved by making the following assumptions:

$\theta_f = 70°$ C.
$q = 1$.
$\rho = 0.865 \times 10^{-6}$ ohm-in.
$k = 0.0076$ watt per sq. in. per ° C. temperature difference, from Fig. 7b, Chapter VII, for $\theta_f = 70°$ C. and for the coil in good thermal contact with the iron parts. In this particular magnet the bobbin flange will be made up of the brass tube and the iron end pieces.
$f = 0.45$, assumed. As the voltage is relatively low, the wire size will be medium and will necessitate fairly heavy paper between layers.

$\frac{h}{r_2 - r_1}$. A value between 4 and 6 is suitable in this type of magnet, the higher values producing more leakage flux force at the beginning of the stroke.

[18] The proper value to use for B_p is very difficult to estimate, as it depends on the ratio of the permeance of the working gap at the end of the stroke to that of the radial leakage path between the plunger and shell.

Thus substituting $h/(r_2 - r_1) = 5$ into equation 3, and solving for h, we have

$$h = \sqrt[3]{\frac{2.5q\rho(NI)^2}{kf\theta_f}}$$

and substituting for the quantities under the radical, and solving, we get

$$h = \sqrt[3]{\frac{2.5 \times 1 \times 0.865 \times 10^{-6} \times 1{,}693^2}{0.0076 \times 0.45 \times 70}}$$

$$= 2.96 \text{ in.}$$

As $r_2 - r_1 = h/5.0 = 0.592$; $r_2 = 0.592 + 0.631 = 1.223$ in.

Because of the necessity of maintaining a reasonable percentage of the magnetomotive force across the working air gap at the end of the stroke, it is wise not to saturate the shell and end pieces as highly as the plunger. Therefore let us make the area of these parts 25 per cent greater than that of the plunger.

$$1.25\pi r_1^2 = \pi(r_3^2 - r_2^2)$$

$$r_3 = \sqrt{1.25 r_1^2 + r_2^2}$$

$$r_3 = 1.41 \text{ in.}$$

The end pieces can be designed to maintain this area.

Equation 4 may now be solved for d:

$$d = \sqrt{\frac{4\rho(r_2 + r_1)NI}{E}} = \sqrt{\frac{4 \times 0.865 \times 10^{-6} \times 1.854 \times 1693}{12}} = 0.0301 \text{ in.}$$

5. Modification to Suit Stock Sizes. The preliminary design has now been finished and the next step will be to modify such dimensions as can be conveniently replaced by stock sizes.

Wire Size. From Table I, Chapter VI, 30.1 mils corresponds to a size between No. 20 and No. 21 wire. Let us try No. 20, diameter = 32.0 mils, which will produce slightly more magnetomotive force than is desired.

Iron Dimensions. Because of the necessity of careful plunger alignment in the working gap, in order to minimize side pull,[19] the plunger

---

[19] As plunger friction can become an important item in a magnet of this type when the forces are relatively small and the plunger diameters large, it is wise to chromium-plate the plunger in order to reduce the friction and also prevent the stickiness which occurs when steel slides in unlubricated brass. Lubrication is undesirable, as the viscosity of the lubricant changes with temperature and also has a tendency to become gummy.

must be turned to fit the brass tube. Therefore the dimensions already assumed for r_1 and the brass tube may be tried.

The outer shell can be made of a 3-in. seamless tube of $\frac{9}{32}$-in. wall. This will give an inside diameter of 2.44 in., and the outside can be turned to 2.82 in., the required r_3. The end pieces are designed to maintain the shell area throughout.

The length of the winding space h should be changed from the preliminary value of 2.96 to compensate for the larger wire size, if the temperature rise is to be held down to 70° C. In order to do this h must be increased in proportion to NI; as the coil ampere-turns will vary as the square of the wire diameter, h should be made equal to [20]

$$2.96 \times \left(\frac{32.0}{30.1}\right)^2 = 3.35 \text{ in.}$$

6. Check of Preliminary Design Using Final Sizes.

Final Sizes. $r_1 = 0.631$ in.; $r_2 = 1.22$ in.; $r_3 = 1.41$ in.; $h = 3.35$ in.; No. 20 wire; $g = 0.0253$ in.

Coil Design—Space Factor. Following the exact same procedure as in Art. 86, Sec. 5, the gross winding depth $(r_2 - r_1)$ will include the following allowances:

(a) 0.025 in. = thickness of brass tube and clearance to plunger
(b) 0.022 in. = insulation between brass tube and coil
(c) 0.025 in. = insulation outside of coil
(d) 0.010 in. = allowance for coil irregularities

0.082 in. = total allowances

The net winding depth will be $0.589 - 0.082 = 0.507$ in. Allowing 5.0 mils of Kraft paper between layers and 33.8 mils for the diameter of enameled-covered No. 20 wire, the thickness of a layer will be $0.0338 + 0.005 = 0.0388$ in. The number of layers will be $0.507 \div 0.0388 = 13.05$, which can be called 14 if a 7 per cent allowance for embedding is made. An even number of layers is desirable as it brings the leads out at the same end.

As the iron end pieces are to form the bobbin flange it will not be necessary to allow for brass bobbin flanges or for imperfect fit of the bobbin, in determining the net winding length of the coil. We shall assume a hole with a rubber bushing at the fixed plunger end to bring the leads out. Then making an allowance of 20 mils for oiled linen and

---

[20] This increase in length can be avoided if the wire size is split in order that an effective diameter of 30.1 mils is obtained. See Art. 88, Sec. 6, "Coil Redesign with Two Wire Sizes," for the method of computing this type of winding.

$\frac{1}{32}$ in. for fiber at each end of the coil the net winding length will be $3.35 - 0.102 = 3.248$ in., and the turns per layer will be

$$\frac{3.35}{0.0338} \times 0.95 - 1 = 93$$

assuming that because of the spread of the winding only 95 per cent as many turns can be wound, and that one turn is lost at the end of each layer.

The total turns will be $14 \times 93 = 1{,}302$. As the cross-sectional area of the bare copper wire is 0.000802 sq. in., the space factor will be

$$f = \frac{1{,}302 \times 0.000802}{0.589 \times 3.35} = 0.529$$

This is higher than the assumed value of 0.45, and before proceeding it will be wise to correct h in order to avoid too low a temperature rise. As temperature rise varies inversely as the space factor, and inversely as h^2, the new h should be

$$3.35 \times \sqrt{\frac{0.45}{0.529}} = 3.09, \text{ say } 3\tfrac{1}{8} \text{ in.}$$

The new net winding length will be $3.125 - 0.102 = 3.023$ in., and the turns per layer will be

$$\frac{3.023}{0.0338} \times 0.95 - 1 = 84$$

The total turns will be $14 \times 84 = 1{,}176$, and the space factor

$$f = \frac{1{,}176 \times 0.000802}{0.589 \times 3.125} = 0.512$$

The coil resistance may now be computed as follows: The mean radius of a turn will be r_1 plus the winding thickness allowances $(a) + (b) + \tfrac{1}{2}$ the net coil winding thickness $= 0.631 + 0.047 + 0.254 = 0.932$ in. The total length of wire will be $2\pi \times 0.932 \times 1{,}176 = 6{,}880$ in., and the resistance at 90° C., applying the factor 0.8458 ohm per 1,000 in., and the temperature correction factor 1.2751, from Tables II and III, respectively, of Chapter VI, will be

$$6.88 \times 0.8458 \times 1.2751 = 7.42 \text{ ohms at } 90° \text{ C.}$$

The coil current will be

$$\frac{12.0}{7.42} = 1.62 \text{ amperes}$$

and the coil magnetomotive force will be

$$NI = 1{,}176 \times 1.62 = 1{,}904 \text{ ampere-turns}$$

Temperature Rise.

$$\theta_f = \frac{1 \times 0.865 \times 10^{-6}}{2 \times 0.0076 \times 0.512 \times 0.589} \left(\frac{1{,}904}{3.125}\right)^2 = 70.4° \text{ C.}$$

This is satisfactory.

Fixed Cylindrical Gap. Assume that the fixed cylindrical gap is to be designed to use 10 per cent of the available coil magnetomotive force = $1{,}904 \times 0.1 = 190$ ampere-turns. Then the permeance of this gap will be

$$P_c = \frac{\phi}{F} = \frac{150}{190} = 0.79 \text{ kmax. per ampere-turn}$$

where 150 kmax. is the maximum plunger flux based on the assumption made in the preliminary design. The gap permeance in terms of its dimensions will be

$$P_c = \frac{\mu S}{l} = \frac{\mu \pi (2r_1 + g) t_5}{g}$$

Substituting the known values we have

$$t_5 = \frac{0.79 \times 0.025}{0.00319 \times \pi \times 1.287} = 1.53 \text{ in.}$$

which may be called 1.5 in. as the fringing permeance around the edges of the gap will more than make up this difference.

Other Iron Dimensions. The end pieces may now be designed to have a substantially constant area equal to that of the shell. The shell area is $\pi(1.41^2 - 1.22^2) = 1.57$ sq. in.

$$t_2 = \frac{1.57}{2\pi r_2} = \frac{1.57}{2\pi \times 1.22} = 0.204 \text{ in.}$$

The area through t_4, as it does not carry the entire flux, can be made approximately equal to that of the plunger. Try $t_4 = 0.25$ in.; then area = $\pi(1.3125 + 0.25)0.25 = 1.23$ sq. in. This is sufficient. Then

$$t_3 = \frac{1.57}{2\pi(0.631 + 0.025 + 0.25)} = 0.275 \text{ in.}$$

Magnetic Circuit Calculations. At this point a sketch of the magnet should be made so that the mean lengths of the magnetic circuit may be obtained. This sketch is shown in Fig. 17.

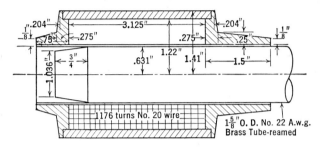

Fig. 17. Cross section through the final design of the tapered cylindrical plunger magnet, showing all dimensions.

The mean lengths and areas are tabulated below:

No.	Part	Length	Area	Material
1	Plunger.......................	4.0	1.25	S.A.E. 10–10
2	Shell.........................	3.25	1.57	"
3	Front end piece...............	1.25	1.57	"
4	Back end piece...............	1.00	1.57	"
5	Fixed cylindrical air gap.......	Permeance = 0.79 kmax. per a-t.		

The saturation curve of these parts may now be computed. In computing this curve all parts will be assumed to carry the same flux as was done in Art. 77. The tabular computation is shown in Table V, and the resulting saturation curve is plotted in Fig. 19.

Force-Stroke Curve. In this type of magnet, owing to the possibility of the force decreasing with the stroke, it is necessary to check the force over the entire stroke to make sure that it does not fall below the desired value at any point. The most convenient and accurate way to do this is to follow the method of Art. 77, Chapter VIII.

Following this scheme, we shall first calculate the permeance of the working gap P_w, and its rate of change at several points. We shall choose the points where $V = \frac{3}{4}, \frac{1}{2}, \frac{1}{4}$ and 0 in. In order to show the

TABLE V

CALCULATION OF SATURATION CURVE OF IRON PARTS AND FIXED CYLINDRICAL AIR GAP OF A TAPERED PLUNGER MAGNET

(See saturation curve of Fig. 11a, Chapter II, for sample 3)

Part	Length, in.	Area, sq. in.	ϕ kmax.	B kmax. per sq. in.	H a-t. per in.	F a-t.
1	4.00	1.25	156	125	410	1,640
2+3+4	5.50	1.57	156	99.5	31	171
5	$P = 0.79$ kmax. per a-t.		156	.....	......	197
					Total	2,008

Part	ϕ	B	H	F	ϕ	B	H	F
1	150	120	270	1,080	144	115	165	660
2+3+4	150	95.6	24	132	144	91.5	17	94
5	150	......	......	190	144	.....	......	182
			Total	1,402			Total	936
1	137.5	110	100	400	131	105	60	240
2+3+4	137.5	87.5	14	77	131	83.6	115	63
5	137.5	......	......	174	131	.....	......	166
			Total	651			Total	469
1	125	100	31	124	112.5	90	16	64
2+3+4	125	79.6	10	55	112.5	71.6	8	44
5	125	......	......	158	112.5	.....	......	142
			Total	337			Total	250

method of evaluating these permeances and their derivatives, the calculations at $V = \frac{1}{2}$ in. are given as an example.

Referring to Fig. 18

$$P_t(\text{taper gap}) = \pi\mu \left[\frac{2(r_1 + g)}{\theta} \log_\epsilon \left(\frac{\theta m + g}{\theta V + g} \right) + V - m \right]$$

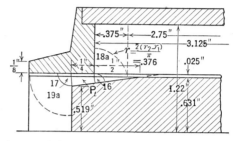

FIG. 18. Enlarged view of the tapered end of the magnet of Fig. 17 for calculating the permeance of the working gap.

from equation 13, Chapter VIII,

$$P_t = 3.19\pi \left[\frac{2 \times 0.656}{0.15} \log_\epsilon \left(\frac{0.137}{0.100} \right) + \frac{1}{2} - \frac{3}{4} \right] = 25.0 \text{ max. per ampere-turn}$$

$$P_{18a} = 4\mu \left[\bar{r}_1 + \sqrt{\bar{g}(t + \bar{g})} \right] \log_\epsilon \frac{t + g}{g}$$

where $t + \bar{g} = 0.376$, $\bar{r}_1 = r_1 + g - \bar{g} = 0.556$, and $\bar{g} = \theta V + g = 0.1$. See equation 18a, Chapter V.

$$P_{18a} = 4 \times 3.19 \left(0.556 \sqrt{0.1 \times 0.376} \right) \log_\epsilon 3.76 = 12.7 \text{ max. per ampere-turn}$$

$$P_{16a} = 3.3\mu \left(\bar{r}_1 + \frac{\bar{g}}{2} \right)$$

where $\bar{r}_1 = 0.556$, and $\bar{g} = 0.1$. See equation 16a, Chapter V.

$$P_{16a} = 3.3 \times 3.19 \times 0.616 = 6.5 \text{ max. per ampere-turn}$$

$$P_{17a} = 3.3\mu \left(\bar{r}_1 + \frac{\bar{g}}{2} \right)$$

where $\bar{r}_1 = 0.519$, and $\bar{g} = 0.137$.

$$P_{17a} = 6.2 \text{ max. per ampere-turn}$$

$$P_{19a} = 4\mu \left[\bar{r}_1 + \bar{g} - \sqrt{\bar{g}V} \right] \log_\epsilon \frac{V}{g}$$

where $\bar{g} = \theta m + g = 0.137$, $\bar{r}_1 = r_1 - \theta m = 0.519$. See equation 19a, Chapter V, V being substituted in those places where $\bar{r}_1 + \bar{g}$ does not apply.

$$P_{19a} = 4 \times 3.19 \,[0.519 + 0.137 - \sqrt{0.137 \times 0.50}] \log_\epsilon \frac{0.500}{0.137}$$

$$= 5.31 \text{ max. per ampere-turn}$$

$$P_w = P_t + P_{18a} + P_{16a} + P_{17a} + P_{19a} = 55.7 \text{ max. per ampere-turn}$$

The radial leakage permeance which must be added in parallel with P_w for determining the magnetomotive force across the working gap is, using formula 15b of Chapter V, and applying the corrections as used in Art. 77,

$$\text{Eff. } P_L = \frac{2.75}{3.12} \times \frac{2}{3} \times \frac{1}{2} \times \frac{2\pi \mu l}{\log_\epsilon \frac{r_2}{r_1}}$$

$$= 0.88 \times \frac{2}{3} \times \frac{1}{2} \times \frac{2\pi \times 3.19 \times 2.75}{\log_\epsilon \frac{1.22}{0.631}} = 0.294 \times 83.5$$

$$= 24.7 \text{ max. per ampere-turn}$$

The total permeance effective in determining the magnetomotive force across the working gap at $V = \frac{1}{2}$ in. is $P_w + P_L = 55.7 + 24.7 = 80.4$ max. per ampere-turn. In a similar manner the permeances at the other points can be calculated, and are given in columns 2, 3, and 4 of Table VI. With these values the four air-gap permeance lines of Fig. 19 have been drawn to determine the point of operation on the saturation curve for each plunger position, and the corresponding air-gap magnetomotive force, F_w, as tabulated in column 5.

Because the fringing permeances P_{19a}, P_{17a}, P_{16a}, and P_{18a}, are varying in a complicated manner with V, the rate of change of permeance with stroke can best be found graphically. This has been done by plotting the data of column 2 in Fig. 20 and determining the slope at each point desired. The results are tabulated in column 6.

The remainder of Table VI has been computed as indicated in the column headings, and follows Art. 77, Chapter VIII, exactly. The final net effective force of column 15 has been plotted in Fig. 21.

Actual Work and Optimum Work. The least force throughout the stroke is 12.35 lb., which is satisfactory as regards the specifications. The useful work will therefore be

$$12.35 \times \tfrac{3}{4} = 9.26 \text{ in-lb.}$$

The optimum cannot, as has been done in the previous designs, be determined from Fig. 19, by making the differential permeance of

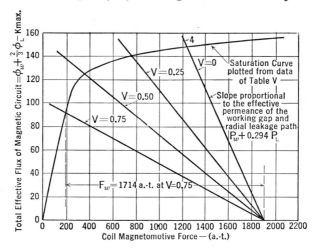

Fig. 19. Magnetization curve and graphical construction for determining the working air-gap magnetomotive force of the magnet of Fig. 17 as a function of the plunger displacement.

the iron circuit equal to that of the working gap. This is because the useful work, which is equal to the product of the least force by the stroke,

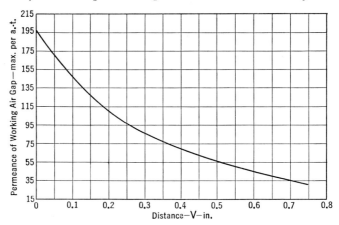

Fig. 20. Variation of working air-gap permeance with plunger displacement for the magnet of Fig. 17.

will not necessarily be the product of the force at the beginning of the stroke by the stroke.

TABLE VI

1	2	3	4	5	6	7	8
	Permeance of Working Gap	Permeance of Radial Leakage Path Which May Be Considered in Parallel with Main Working Gap	Total Air Path Permeance Effective in Determining M.m.f. across the Working Gap	M.m.f. of Working Gap	Rate of Change of Permeance of Working Air Gap	Force Derived from Working Gap	Radial Leakage Flux, ϕ_L
V	P_w	$0.293 P_L$	$2+3$	F_w	$\dfrac{dP_w}{dV}$	$4.43\times 10^{-8} \times F_w^2\, dP_w/dV$	$\tfrac{1}{2}\times 0.88\, P_L F_w$
in.	max. per a-t.	max. per a-t.	max. per a-t.	a-t.	max. per a-t. per in.	lb.	kmax.
3/4	32.0	24.7	56.7	1,714	69	9.0	63.0
1/2	55.7	↓	80.4	1,554	125	13.4	57.1
1/4	97.6		122.3	1,144	225	13.0	42.1
0	198.3		223.0	664	600	11.7	24.4

1	9	10	11	12	13	14	15
	Force due to ϕ_L, $H=\dfrac{1904}{3.125}=610$	Flux Through Working Gap, ϕ_w	Maximum Plunger Flux, ϕ_p	Maximum Plunger Flux Density, B_p	Loss in Work Due to Magnetizing 1 cu. in. of Plunger, W_L	Loss in Force Due to W_L	Net Magnetic Force
V	$8.86\times 10^{-5} H\phi_L$	$P_w F_w \times 10^{-3}$	$\phi_w + \phi_L$	ϕ_p/S	$\int_0^{B_p} H_p\, dB$	$8.86\, W_L S$	$7+9-14$
in.	lb.	kmax.	kmax.	kmax. per sq. in.	joules per cu. in.	lb.	lb.
3/4	3.41	54.8	117.8	94.2	0.005	0.06	12.35
1/2	3.09	86.6	143.7	115.0	0.019	0.21	16.28
1/4	2.28	111.8	153.9	123.0	0.038	0.42	14.86
0	1.32	131.5	155.9	124.6	0.043	0.48	12.54

The optimum can, however, be determined by reference to **Fig. 21**, by finding the largest rectangle which can be drawn under the force-stroke curve. This rectangle coincides so closely with the actual least force and required stroke as to be almost indistinguishable.

The total available work under the force-stroke curve, by measurement of the area, is 11.06 in-lb. The mechanical efficacy of the design will be 9.26 ÷ 11.06 = 83.8 per cent, which, though high, is not so high as to cause too sluggish an action.

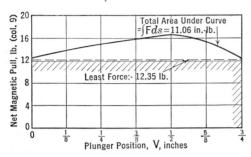

Fig. 21. Computed force-stroke curve for the magnet of Fig. 17 plotted from the data of column 15, Table VI.

The magnetic efficacy of the actual design is equal to the quotient of the total work under the force-stroke curve divided by the potential work ability of the magnet, $NI(\phi_m)$, where ϕ_m is the maximum value of the flux which may be considered to link with all the turns N. This flux will equal $\phi_w + \tfrac{2}{3}\phi_L$ as given by equation 3, Art. 46, Chapter IV, and may therefore be found from Fig. 19, point 4, as 148 kmax. Therefore

$$\eta = \frac{11.06}{8.86 \times 1904 \times 148 \times 10^{-5}} = 0.442$$

which is higher than the value of 0.4 assumed in the preliminary design, but checks very well with the statements made regarding the choice of η.

Weight. The weight of the copper of the coil will be

$$\frac{6{,}880}{12} \times \frac{1}{323.4} = 1.77 \text{ lb.}$$

where 323.4 is the number of feet per pound of No. 20 bare wire from Table II, Chapter VI. The insulation around the brass tube, end flanges, etc., may be estimated to bring the coil weight to 1.87 lb.

The volume of the brass tube will be

$$\pi \times 1.287 \times 0.025 \times 5.375 = 0.542 \text{ cu. in.}$$

and its weight

$$0.542 \times 0.32 = 0.174 \text{ lb.}$$

The volume of the iron pieces are computed as follows:

Plunger = $1.25 \times 4.625 + \frac{3}{4} [0.842 + (0.41/3)]$ = 6.51 cu. in.
Shell = 1.57×3.533 = 5.54
Front end piece   1.77
Back end piece   1.20
                                              Total volume   15.02

Weight of iron = 15.02×0.283 = 4.25 lb.

The total weight will be $1.87 + 0.17 + 4.25 = 6.29$ lb., and the weight economy will be $6.29 \div 9.26 = 0.68$ lb. per in-lb. of work. This checks very well with the weight economy of 0.78 given in Fig. 10, Chapter IX, for the same index number.

Final Results. The design is now finished. Figure 17 shows a complete sketch of the magnet with all dimensions. Below are tabulated all the principal data.

Least force = 12.35 lb.
Stroke = 0.75 in.
Voltage = 12 volts.
Current (hot) = 1.62 amperes.
Power (hot) = 19.44 watts.
Temperature rise (continuous excitation) = 70.4° C.
Weight = 6.29 lb.
Maximum useful work = 9.26 in-lb.
Weight per unit of maximum useful work = 0.68 lb. per in-lb.
Magnetic efficacy = 0.44.

Design with Magnetic Stop at End of Stroke. The use of a magnetic stop will make the force rise very sharply at the end of the stroke. Referring to Fig. 21 it will be seen that without a stop the force decreases with stroke as soon as the plunger saturates. This can be counteracted in the very last portion of the stroke by the use of a magnetic stop. The effective range of the stop will depend on the type of face employed. Thus, in the magnet designed, a flat stop would be effective for only a small portion of the stroke, say $\frac{1}{16}$ in., while a conical stop could be made effective over a much greater range.

When designing with a stop, the taper portion, using equation 13, can be designed for a shorter stroke by the effectiveness of the stop. The total force derived with the stop is computed exactly as without the stop, the effect of the stop in increasing the permeance of the working gap being taken into account when that permeance is computed.

90. Design of a Leakage-Flux Type of Magnet

1. General. The leakage-flux magnet is of a solenoid and plunger type and is used in all applications requiring a long stroke. It is quite different from the types previously discussed in that the force does not depend primarily on the attraction between magnetized surfaces, but on the reaction between a current-carrying conductor and a magnetic field. The design procedure is thus different from that outlined in Art. 84.

The strength of the magnetic field is limited by saturation, but the momentary current density of the conductor is limited only by vaporization of the copper, and hence from a practical point of view it is possible to obtain very high momentary forces. Because there is no limit to the stroke, the magnetic efficacy can be made very high.

As the coil generally is very long it is economical to wind it directly in the bobbin formed by the brass tube and iron end pieces. The length of the coil must be made equal to the stroke plus the initial distance the plunger is inserted. This length can be shortened somewhat if the latter part of the stroke is carried out by the taper plunger action of Art. 89. Experience shows that the maximum length of the taper end should not exceed one-third the distance the plunger is inserted into the coil at the start of the stroke. The angle of the taper should be about 0.15 radian, equal to 8.6°.

2. Design Data. For the purpose of illustration let it be required to design a magnet having given the following data:

Stroke = 4.0 in.
Force = 3.0 lb. = $\mathscr{F}$
Voltage = 120.0 volts.
Excitation = continuous.
Temperature rise = 70° C. = θ_f, average temperature rise (by change in resistance) of coil above surrounding ambient air temperature of 20° C.
Material = mild cold-rolled steel, S.A.E. 10–10, unannealed after machining, sample 3.
Coils = enameled wire with paper between layers wound in bobbin formed by brass tube and iron end pieces. Use phenolic insulation over brass tube and end pieces.
Plunger. Finish stroke by use of taper plunger force action.

The index number of the required design will be $\sqrt{\mathscr{F}}/s = \sqrt{3.0}/4 = 0.432$, which by reference to Fig. 10, Chapter IX, indicates the leakage-flux type. Figure 22 is a sketch of the conventional type of ironclad leakage-flux magnet, showing all the dimensions indicated in the formulas.

3. Design Equations.[21]

The four fundamental design equations of Art. 84, modified to suit the leakage-flux magnet, are as follows:

Force Equation. Equation 20e of Art. 76, Chapter VIII, applies directly to this type of magnet.

$$\text{Force} = KH\phi \text{ joules per inch}$$

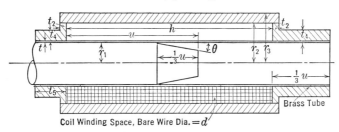

FIG. 22. Cross section through a cylindrical plunger leakage-flux magnet, illustrating the design symbols.

where K is a factor decreasing the force because of the loss due to energy stored in the plunger. If a good quality of soft iron is used for the iron parts the value of 0.85 as suggested in Art. 76 may be used. Where the material is quite different the factor should be evaluated as explained. Using 0.85 for this factor and expressing H and ϕ in terms of the magnet dimensions, we have

$$\text{Force} = \frac{8.86 \times 0.85 \; NI(10^{-5}B\pi r_1^2)}{h} = \frac{NIBr_1^2}{4{,}220h} \text{ lb.} \qquad (17)$$

Magnetic Circuit Equation. This equation is necessary in order to determine the distance u that the plunger must be inserted at the beginning of the stroke in order to produce the value of B required in equation 17. At insertions greater than u, the plunger will reach a saturation such that H of the plunger $+H$ of the shell will equal NI/h, the magnetic intensity of the coil.

The flux density B, with the plunger inserted a distance u (before the plunger saturates) can be approximated [22] as follows: The total magnetomotive force for that part of the flux crossing the winding at u is $(NI/h)u$ ampere-turns. Of this a portion, say, 10 per cent [23] is used

---

[21] The design equations and method of procedure in designing this magnet are similar to those developed by L. A. Hazeltine, for use in the Department of Electrical Engineering at Stevens Institute of Technology.

[22] When the dimensions of the magnet are known it is possible to calculate the maximum plunger flux density B with fair precision. The method will be explained in Sec. 6 when the final design is checked.

[23] Should the allowance for this gap be different, equation 18 should be changed accordingly.

up in the fixed cylindrical gap at the left, and a second portion is used up in the plunger and shell. As the flux density in the plunger decreases from its full value B at the cylindrical gap end to zero at u, we may estimate the average magnetic intensity in the plunger as $0.5H_p$, where H_p is the value corresponding to B, as obtained from the magnetization curve. The shell will not usually be so highly saturated as the plunger; estimating its average magnetic intensity as $0.3H_p$, we have for the magnetomotive force across the winding at the distance u

$$0.9\frac{NI}{h}u - 0.5\,H_p u - 0.3\,H_p u = 0.9\frac{NI}{h}u - 0.8H_p u \quad \text{ampere-turns}$$

At the cylindrical gap end, however, the magnetomotive force across the radial leakage path is the reluctance drop across the fixed cylindrical gap equal to $-0.1(NI/h)u$. Therefore the average magnetomotive force across the leakage path will be

$$F_{(\text{avg.})} = 0.4u\left[\frac{NI}{h} - H_p\right] \quad \text{ampere-turns, neglecting the taper}$$

The permeance of the path from the plunger to the shell is by equation 15a, Chapter V,

$$P = \frac{\mu\pi(r_1 + r_2)u}{r_2 - r_1}$$

The total flux of the plunger (that at the cylindrical gap end) is then

$$\phi = PF_{(\text{avg.})} = \frac{\mu\pi(r_1 + r_2)u}{r_2 - r_1}0.4u\left[\frac{NI}{h} - H_p\right]$$

and the flux density is

$$B = \frac{\phi}{\pi r_1^2} = \frac{0.4\mu(r_1 + r_2)u^2}{r_1^2(r_2 - r_1)}\left[\frac{NI}{h} - H_p\right] \tag{18}$$

Heating Equation. This remains unchanged.

$$\theta_f = \frac{q\rho}{2kf(r_2 - r_1)}\left[\frac{NI}{h}\right]^2 \tag{3}$$

Voltage Equation. This remains unchanged.

$$E = \frac{4\rho(r_2 + r_1)NI}{d^2} \tag{4}$$

4. Preliminary Design. Equations 17 and 3 may be solved for $[NI/h]$; thus from equation 17

$$\frac{NI}{h} = \frac{4{,}220 \text{ Force}}{Br_1^2}$$

and from equation 3

$$\frac{NI}{h} = \sqrt{\frac{2kf\theta_f(r_2 - r_1)}{\rho q}}$$

Equating these and solving for r_1, we have

$$r_1^5 = \frac{4{,}220^2 \, (\text{Force})^2 \, \rho q}{B^2 2kf\theta_f \left[\dfrac{r_2}{r_1} - 1\right]} \tag{19}$$

Equation 19 may now be solved by making the following substitutions:

$\theta_f = 70°$ C.

$q = 1$.

$\rho = 0.865 \times 10^{-6}$ ohm-inch at 90° C.[24]

$k = 0.0076$ watt per sq. in. per ° C. temperature difference from Fig. 7b, Chapter VII, for a temperature rise of 70° C. and for the coil in good thermal contact with iron parts. In this type of magnet the coil will generally be wound on the brass tube with the iron end pieces forming the bobbin flanges. Hence, the coil will be in good thermal contact with the iron parts.

$f = 0.45$, assumed.[24]

$\dfrac{r_2}{r_1} = 2.0$, in this type of magnet, gives the most economical proportions. However, under special conditions it might be desirable to change the ratio. Thus a light plunger will be obtained with a large value of r_2/r_1, as this will allow the magnetic intensity of the winding to be higher, resulting in an increase in copper weight and power consumption. Where a magnet is used for only a very small portion of the time and where power consumption is no object, a small value of r_2/r_1 will result in a design of small dimensions.

$B = 110$ kmax. per sq. in. The ultimate saturation in the plunger will reach a value such that H of the plunger $+ H$ of the shell will equal NI/h for the coil. As NI/h in this type of magnet will be between 500 and 1,000 ampere-turns per inch, the value of B will be determined by the saturation density of the iron at a value of H between, $\tfrac{1}{2}NI/h$ for the plunger area equal to the shell area, and higher as the shell area is increased. The maximum force during the stroke will be determined by this saturation density. Hence, B at the beginning of the stroke should have a value of about 90 per cent of this saturation value.

---

[24] See Art. 86, Sec. 4.

Making these substitutions, we have

$$r_1^5 = \frac{4{,}220^2 \times 3.0^2 \times 0.865 \times 10^{-6} \times 1}{110^2 \times 2 \times 0.0076 \times 0.45 \times 70 \times 1} = 0.0215$$

and
$$r_1 = 0.474 \text{ in.}$$
$$r_2 = 2r_1 = 0.948 \text{ in.}$$

Equation 17 may now be solved for NI/h.

$$\frac{NI}{h} = \frac{4{,}220 \times \text{Force}}{Br_1^2} = \frac{4{,}220 \times 3}{110 \times 0.474^2} = 514 \text{ ampere-turns per inch}$$

Referring now to Fig. 11a, Chapter II, H for $B_p = 110$ will be 104. Solving equation 18 for u we have

$$u = \sqrt{\frac{Br_1^2(r_2 - r_1)}{0.4\mu(r_1 + r_2)\left[\dfrac{NI}{h} - H\right]}}$$

and substituting the known values, we have

$$u = \sqrt{\frac{110 \times 0.474^2 \times 0.465}{0.4 \times 0.00319 \times 1.422(514 - 104)}} = \sqrt{15.44}$$

$$= 3.93 \text{ in.}$$

Axial Length of Coil h. In accordance with the statement made in Sec. 1 of this article, let the end of the stroke be carried through by a taper plunger action. Then the coil length can be made equal to

$$h = s + \tfrac{2}{3}u = 4.0 + 2.62 = 6.62 \text{ in.}$$

Wire Size. The total coil ampere-turns will be equal to

$$\frac{NI}{h} \times h = 514 \times 6.62 = 3{,}410 \text{ ampere-turns}$$

The bare wire diameter from equation 4 will be

$$d^2 = \frac{4 \times 0.865 \times 10^{-6} \times 1.422 \times 3{,}410}{120} = 0.1398$$
$$d = 11.83 \text{ mils}$$

Outside Steel Shell. Owing to the extremely high saturation at which the plunger will operate it is wise to make the flux density in the outer shell only about 80 per cent of that in the plunger; then

$$0.8\pi(r_3^2 - r_2^2) = \pi r_1^2$$

$$r_3^2 = \frac{r_1^2}{0.8} + r_2^2 = 1.180$$

$$r_3 = 1.087 \text{ inches}$$

5. Modification to Suit Stock Sizes. The preliminary design has now been finished; the next step will be to modify such dimensions as can be conveniently replaced by stock sizes.

Wire Size. From Table I, Chapter VI, 11.83 mils is between No. 29 and No. 28 wire. As No. 29 will produce less than the desired number of ampere-turns and hence make the force too low, it is best to use No. 28. This may make the temperature rise slightly high.

Iron Dimensions. Plunger Radius (r_1). This must be chosen in relation to the brass tube. Try a No. 24 A.w.g. brass tube, 1 in. in outside diameter. This will have an internal diameter of $1.000 - 0.0402 = 0.9598$ in. Allowing 0.0048 in. total clearance, the plunger diameter will be 0.955 in., and its area 0.716 sq. in. This is 1.5 per cent greater than the desired preliminary value, which should be sufficient.

Outside Steel Shell. Try a steel tube $2\frac{1}{4}$ in. in outside diameter, with a (0.134 in.) No. 10 B. & W. gauge wall. The internal diameter will be 1.982 in., and the area will be 0.89 sq. in., which is 24 per cent larger than that of the plunger.

Radial Thickness of the Iron End Piece (t_4). This should be thick enough to make its annular area equal to the area of the shell, thus

$$\pi(2r_1 + 2t + t_4)t_4 = 0.89$$

$$\pi t_4^2 + 2\pi(r_1 + t)t_4 - 0.89 = 0$$

$$t_4 = 0.231 \text{ in.}$$

Axial Thickness of the Iron End Piece (t_2). This should be made thick enough to make its circumferential area at a radius equal to ($r_1 + t + t_4$) equal to the area of the shell, thus

$$2\pi(r_1 + t + t_4)t_2 = 0.89$$

$$t_2 = \frac{0.89}{2\pi \times 0.731} = 0.194 \text{ in.}$$

Coil Length (h). Let this be equal to $6\frac{3}{4}$ in. This will allow a slightly greater length for u and tend to compensate for the reduction in permeance due to the taper end.

Length of taper section of plunger and length of corresponding iron end piece: say 1.25 in.

Initial Distance Plunger Is Inserted (u): This will be h + length of taper $- s = 6\frac{3}{4} + 1\frac{1}{4} - 4 = 4$ in.

6. Check of Preliminary Design Using Stock Sizes.

Final Sizes: $r_1 = 0.4775$ in.; $r_2 = 0.991$ in.; $r_3 = 1.125$ in.; $t = 0.0225$ in.; $t_2 = 0.194$ in.; $t_4 = 0.231$ in.; $h = 6\frac{3}{4}$ in.; $u = 4$ in.; length of taper $= 1\frac{1}{4}$ in.; plunger area $= 0.716$ sq. in.; shell area $= 0.89$ sq. in. No. 28 enamel-covered wire.

Coil Design—Space Factor. Following the exact same procedure as in Art. 87, Sec. 6, the gross winding depth $(r_2 - r_1) = 0.514$ in. will include the following allowances:

(a) 0.022 = thickness of brass tube and clearance to plunger = t
(b) 0.045 = thickness of phenolic tube between brass tube and coil
(c) 0.025 = insulation outside of coil
(d) 0.010 = allowance for coil irregularities

0.102 = total allowances

The net winding depth will be $0.514 - 0.102 = 0.412$ in. Allowing 1.5 mils of glassine paper between layers and 13.8 mils for the diameter of enameled-covered No. 28 wire, the thickness of a layer will be 0.0153 in. The number of layers will be $0.412 \div 0.0153 = 26.9$ layers. This may be increased to 28 layers to allow for embedding, which in a coil of this type may run as high as 10 per cent. This being an even number it will be possible to bring both leads out at one end through an insulated slot.

As the iron end pieces are to form the bobbin flange it will not be necessary to allow for bobbin flanges or for imperfect fit of the bobbin, in determining the net winding length of the coil. Then making an allowance of 45 mils at each end for a phenolic end washer, the net winding length will be $6.75 - 0.09 = 6.66$ in., and the turns per layer will be

$$\frac{6.66}{0.0138} \times 0.95 - 1 = 458$$

assuming that on account of the spread of the winding only 95 per cent as many turns can be wound, and that one turn is lost at the end of each layer.

The total turns will be 28 × 458 = 12,830. As the cross-sectional area of the bare copper wire is 0.000126 sq. in., the space factor will be

$$f = \frac{12{,}830 \times 0.000126}{0.514 \times 6.75} = 0.466$$

The coil resistance may be computed as follows: The mean radius of a turn will be r_1 + winding thickness allowances $(a) + (b) + \frac{1}{2}$ the net coil winding thickness = 0.4775 + 0.067 + 0.206 = 0.7505 in. The total length of wire will be $2\pi \times 0.7505 \times 12{,}830 = 60{,}500$ in., and the resistance at 90° C., applying the factor 0.00541 ohm per inch for No. 28 wire from Table III, Chapter VI, will be

$$60.5 \times 5.41 \times 1.2751 = 417 \text{ ohms at } 90° \text{ C.}$$

The coil current will be 120 ÷ 417 = 0.287 amperes, and the magnetomotive force developed by the coil will be

$$NI = 12{,}830 \times 0.287 = 3{,}680 \text{ ampere-turns}$$

Temperature Rise. Substituting the values so far obtained into equation 3, we have

$$\theta_f = \frac{1 \times 0.865 \times 10^{-6}}{2 \times 0.0076 \times 0.466 \times 0.514} \left[\frac{3{,}680}{6.75}\right]^2 = 70.5° \text{ C.}$$

This may be considered satisfactory.

Fixed Cylindrical Gap. The length of this gap can now be computed so that it will take about 10 per cent of the initial available magnetomotive force = $(NI/h)\,u$. Assuming, for this purpose, that the maximum plunger flux density is 110 kmax. per sq. in., the total plunger flux will be 110 × 0.716 = 78.7 kmax. The permeance of the fixed cylindrical gap will be

$$P_c = \frac{\phi}{0.1 \dfrac{NI}{h} u} = \frac{78.7}{218} = 0.361 \text{ kmax. per ampere-turn}$$

$$P_c = \frac{\mu S}{t} = \frac{\mu \pi (2r_1 + t) t_5}{t}$$

$$t_5 = \frac{0.361 \times 0.0225}{0.00319 \times \pi (0.977)} = 0.83 \text{ in.}$$

Let this be $\frac{7}{8}$ in.; then P_c will equal 0.361 × 875 ÷ 830 = 0.381 kmax. per ampere-turn.

316 DESIGN OF TRACTIVE MAGNETS [Chap. X

Force. In this type of magnet the force can be checked very easily, as follows:

$$\frac{NI}{h} = \frac{3{,}680}{6.75} = 545 \text{ ampere-turns per inch}$$

The flux density necessary to produce the required force of 3.0 lb. will be, by solving equation 17,

$$B = \frac{3.0 \times 4{,}220}{545 \times 0.4775^2} = 101.9 \text{ kmax. per sq. in.}$$

Whether this flux density will be produced may be checked by substituting a trial value of H_p in equation 18. Thus for a first trial let us assume that $B = 102$; then H_p will be 39; and substituting into equation 18, we have

$$B = \frac{0.4 \times 0.00319 \times 1.4685 \times 4^2}{0.4775 \times 0.514} (545 - 39)$$

$$B = 0.255 \times (545 - 39) = 129 \text{ kmax. per sq. in.}$$

Therefore the assumed value of B is too low. Let us try $B = 110.6$; then $H_p = 112$ and

$$B = 0.255 \times (545 - 112) = 110.6 \text{ kmax. per sq. in.}$$

When the plunger saturation is very high it is sometimes desirable to check this by a more exact method.[25] As the leakage flux path is entirely distributed, and the iron is highly saturated, it is very tedious to solve, accurately, for the exact flux. Therefore, in a manner like the approximate check by means of equation 18, a flux density must be assumed and the calculation carried through, step by step, to see whether the assumed value checks.

A sketch of the complete magnet is shown in Fig. 23. The diameter of the plunger at the end of the taper is equal to $0.955 - 2 \times 1.25 \times \tan 8.6° = 0.955 - 0.378 = 0.577$ in.

The first step is to divide the leakage flux path into short sections and compute the permeance of these sections. We shall assume five sections, each 0.8 in. long, and an end section to take care of the fringing flux at the end of the plunger, as illustrated in Fig. 23.

The permeance of the first three sections may be taken, using formula 15b of Chapter V, as

[25] This method is applicable to any case where the flux is distributed along a plunger which is highly saturated and could be applied, for instance, if one desired accurately to predetermine the force as the taper section enters the steel end piece.

$$P_{1-2} = P_{2-3} = P_{3-4} = \frac{2\pi\mu l}{\log_\epsilon \frac{r_2}{r_1}} = \frac{2\pi\mu 0.8}{\log_\epsilon \frac{0.991}{0.477}} = 0.0214 \text{ kmax. per ampere-turn}$$

For section 4–5, the average value of r_1 may be taken as 0.447 in., and

$$P_{4-5} = \frac{16}{\log_\epsilon \frac{0.991}{0.447}} = 0.0201 \text{ kmax. per ampere-turn}$$

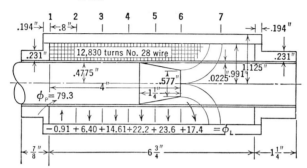

Fig. 23. Cross section through the final design of the cylindrical plunger leakage-flux magnet, showing all dimensions and the distributed leakage path divided into sections for checking the actual plunger flux density by "step-by-step" calculations. *Note:* The vertical and horizontal scales are not the same.

For section 5–6, the average value of r_1 may be taken as 0.349 in., and

$$P_{5-6} = \frac{16}{\log_\epsilon \frac{0.991}{0.349}} = 0.0154 \text{ kmax. per ampere-turn}$$

For section 6–7, we shall use permeance formulas 17 and 19a, of Chapter V; thus

$$P_{17} = 3.3\mu [r_1 + 0.575(r_2 - r_1)]$$

where $r_1 = 0.289$ in.

$$= 3.3\mu [0.289 + 0.575 \times 0.702] = 0.00728$$

$$P_{19a} = 4\mu [r_2 - \sqrt{r_2(r_2 - r_1)}] \log_\epsilon \frac{r_2}{r_2 - r_1}$$

$$= 4\mu [0.991 - \sqrt{0.991 \times 0.702}] \log_\epsilon \frac{0.991}{0.702} = 0.00068$$

$$P_{6-7} = P_{17} + P_{19a} = 0.008 \text{ kmax. per ampere-turn}$$

318 DESIGN OF TRACTIVE MAGNETS [Chap. X

The calculation is carried forward by assuming a flux density at 1, and calculating through, section after section, to the end of the plunger. If the calculations show that all the assumed flux will leak to the shell by the time the last section (6–7) is accounted for, the assumed B is correct; if less than the assumed flux, B has been assumed high; if more, B has been assumed low.

Let us assume B equal to 110.6 kmax. per sq. in. Then $\phi_1 = B_1 \pi r_1^2$ = 110.6 × 0.716 = 79.3 kmax. The magnetomotive force across the leakage path at 1, designated by F_1, will, for all practical purposes, be equal to that across the fixed cylindrical gap, as the average flux density in the iron end piece will be relatively low, and its reluctance drop may therefore be neglected.

$$F_1 = \frac{\phi_1}{P_c} = \frac{79.3}{0.381} = -\ 208 \text{ ampere-turns}$$

where the minus sign indicates a reluctance drop; i.e., the magnetic potential of the shell at section 1 is higher than that of the plunger.

The magnetomotive force at 2 may be found by equating to zero the sum of the magnetomotive forces around the closed loop formed by following a flux line from the shell to the plunger at section 1, through the plunger to section 2, from the plunger to the shell at section 2, and thence back through the shell to section 1, plus the coil ampere-turns included in the loop. Considering, as a general case, the plunger at both ends of the section to be at a higher magnetic potential than the shell, we may write:

$$+ F_1 - H_p l_{1-2} - H_s l_{1-2} - F_2 + \frac{NI}{h} l_{1-2} = 0$$

or

$$F_2 = \frac{NI}{h} l_{1-2} + F_1 - l_{1-2}(H_p + H_s)$$

where the F's will be positive when the flux goes from the plunger to the shell; H_p and H_s are the average magnetic intensities in the plunger and shell, respectively, over the section; and l is the length of the section.

For a first approximation assume the leakage flux in section 1–2 equal to 0; then the average $B_p = 110.6$, and the average $B_s = 89.2$. The corresponding magnetic intensities from curve 3, Fig. 11a, Chapter II, will be $H_p = 108$, $H_s = 15$. Then

$$F_2 = 0.8 \times 545 - [208 + 0.8(108 + 15)] = +\ 130 \text{ ampere-turns}$$

The average magnetomotive force available in this section designated by $F_{1\text{-}2}$ is $(-208 + 130)/2 = -39$ ampere-turns. The minus sign indicates that over this section the shell is at a higher magnetic potential than the plunger and the leakage flux in this section will be in the backwards sense, that is, the permeance of section 1–2 acts in parallel with the fixed cylindrical gap. The flux at section $2 = \phi_2$ will be

$$\phi_2 = \phi_1 - (F_{1\text{-}2})(P_{1\text{-}2}) = 79.3 + 39 \times 0.0214 = 79.3 + 0.84 = 80.14 \text{ kmax.}$$

and

$$B_2 = \frac{\phi_2}{\pi r_1^2} = \frac{80.14}{0.716} = 111.9 \text{ kmax. per sq. in.}$$

The average flux density in section 1–2 of the plunger, $B_{1\text{-}2}$, will be

$$\frac{110.6 + 111.9}{2} = 111.25$$

and that in the shell will be 89.8. The corresponding H's will be 115 and 16, respectively. Using these values a second approximation may be made; then

$$F_2 = 0.8 \times 545 - [208 + 0.8(115 + 16)] = +123 \text{ ampere-turns}$$

$$F_{1\text{-}2} = \frac{-208 + 123}{2} = -42.5 \text{ ampere-turns}$$

$$\phi_2 = 79.3 + 42.5 \times 0.0214 = 79.3 + 0.91 = 80.21$$

$$B_2 = \frac{80.21}{0.716} = 112.0 \text{ kmax. per sq. in.}$$

Continuing the calculation in this manner we arrive at the results tabulated on page 320.

The final value of $\phi_p = -4.0$ shows that the original estimate of $B_p = 110.6$ is slightly low, but that for all practical purposes the approximate estimate is close enough. The distribution of the leakage flux is shown in the lower half of Fig. 23.

The force,[26] at the starting position of the plunger, letting $B_p = 111$, will be

$$F = \frac{Br_1^2}{4{,}220} \frac{NI_m}{h} = \frac{111 \times 0.4774^2 \times 545}{4{,}220} = 3.26 \text{ lb.}$$

---

[26] Whereas the maximum flux density of the plunger occurs at position 2, the flux density at position 1 still determines the force. This is so because the force depends on the net leakage flux of the plunger that cuts across the coil, due regard being given to sign.

Position	B_p	ϕ_p	ϕ_L	F	P	Plunger Area
1	110.6	79.3	....	− 208	....	0.716
1–2	....	....	− 0.91	− 42.5	0.0214	0.716
2	112.0	80.21	....	+ 123.0	....	0.716
2–3	....	....	+ 6.40	+ 299.5	0.0214	0.716
3	103.0	73.81	....	+ 476	....	0.716
3–4	....	....	+14.61	+ 684	0.0214	0.716
4	82.5	59.2	....	+ 891	....	0.716
4–5	....	....	+22.2	+1104	0.0201	0.630
5	70.5	37.0	....	+1317	....	0.525
5–6	....	....	+23.6	+1531	0.0154	0.383
6	51.0	13.4	....	+1753	....	0.262
6–7	....	....	+17.4	+2180	0.0080	....
7	....	− 4.0	....	....	....	....

This is satisfactory. In this type of magnet there is no optimum flux density at the beginning of the stroke.

Weight. The weight of the copper of the coil will be

$$\frac{60{,}500}{12} \times \frac{1}{2{,}067} = 2.44 \text{ lb.}$$

where 2,067 is the number of feet of No. 28 wire per pound from Table II, Chapter VI.

The iron weight can be estimated by calculating the volume of the parts:

Outer shell = 0.89×7.14 = 6.35
Plunger = $0.716 \times 7.63 + 0.262 \times 1.25 + \frac{1}{3}(0.716 - 0.262) \times 1.25$ = 5.99
Ends = $0.89 \times 2.125 + \pi \times 1.72 \times 0.260 \times 2 \times 0.194$ = 2.44

Total = 14.78

Weight iron = 14.78 × 0.283 = 4.18 lb.
Brass tube = π × 0.98 × 0.0201 × 0.32 = 0.02 lb.
Total weight = 2.44 + 4.18 + 0.02 = 6.64 lb.

Weight Economy. The weight per unit of work will be

$$\frac{6.64}{3.26 \times 4} = 0.51 \text{ lb. per in-lb. of work}$$

This checks very well with the data of Fig. 10, Chapter IX, which shows a weight economy of 0.53 for an index number 0.43.

Final Results. The design is now finished. Figure 23 shows a complete sketch of the magnet with all dimensions. Below are tabulated the principal data:

Force = 3.26 lb.
Stroke = 4.0 in.
Voltage = 120 volts.
Current (hot) = 0.287 amperes.
Power (hot) = 34.5 watts.
Final temperature rise = 70.5° C.
Excitation = continuous.
Weight = 6.64 lb.
Useful work = 13.04 in-lb.
Weight economy = 0.51 lb. per in-lb. of work.

91. Design of a Horseshoe Type of Magnet

1. General. The horseshoe type of magnet is used primarily as a general-purpose magnet for relatively small amounts of work. Thus it is very often used in machines to perform light work where weight economy is not an object, but where ease and cheapness of construction and freedom from mechanical friction of sliding plungers are desired. For this reason it is used over a very wide range of index numbers.

Where the designer is free to choose the shape of magnet (ratio of pole length to distance between pole centers) it is possible to standardize the horseshoe design as was done for the other types. However, if the shape has already been determined by factors outside of the control of the designer, the design data given in this article will not necessarily apply; the designer must then resort to cut-and-try methods, following the general procedure outlined in this article.

In the design of a horseshoe magnet it is often necessary to consider the residual force in the closed-gap position. This is particularly so when the force is small and the stroke large. Where the residual force must be kept very low it will generally be necessary to insert a nonmagnetic stop or spacer between the armature and pole faces. In designing the magnet this can be taken care of by adding a nominal amount to the required stroke. Usually 0.010 in. is sufficient. As regards the index number, calculation, etc., the additional amount should be considered part of the required stroke.

2. **Design Data.** For the purpose of illustration let it be required to design a magnet having given the following data:

Stroke = 0.09 in.
Force = 10 lb. = $\mathscr{F}$
Voltage = 120 volts.
Excitation = continuous.
Temperature rise. 70° C. = θ_f, average temperature rise (by change of resistance) of coil above surrounding ambient air temperature of 20° C.
Material [27] = mild cold-rolled steel, S.A.E. 10–10, unannealed after machining, sample 3.
Residual force. Use a 0.010 in. non-magnetic spacer between armature and pole faces to prevent sticking.
Shape. No restriction.
Coils. Paper-section enameled copper.

The index number of this required design, allowing 0.010 in. for a non-magnetic spacer, will be $\sqrt{\mathscr{F}/s} = \sqrt{10/0.10} = 31.6$. Figure 24 is a sketch of the conventional type of horseshoe magnet, showing all the dimensions indicated in the formulas. It is nearly always necessary to provide polar enlargements in order that the pole cores and yoke can be operated at an optimum flux density. We shall assume the polar enlargements to be square.

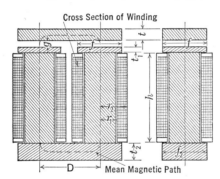

FIG. 24. Cross section through a horseshoe magnet, illustrating the design symbols.

DESIGN LIMITATIONS

$D = 2r_2 + \frac{1}{8}''$, but $\geq f + \frac{1}{4}''$;
$f \geq 2r_1$; $t_1 = \frac{1}{8}''$.

3. **Design Equations.** The four fundamental design equations of Art. 84, modified to suit the horseshoe magnet, are as follows:

Force Equation. The force equation is

$$\text{Force} = \frac{B_g^2 f^2}{36} \text{ lb.} \qquad (20)$$

[27] If a steel having a high coercive intensity, such as sample 4, is used particular attention must be paid to the residual force. For the method of calculating the residual force see Article 49, Chapter IV, or the relay example of Secs. 16 and 17, Art. 135, Chapter XIV.

BIPOLAR OR HORSESHOE TYPE OF MAGNET

from equation 8b, Chapter VIII, remembering that the denominator must be reduced to one-half as the horseshoe magnet has two working faces.

Magnetic Circuit Equation. In the horseshoe magnet the optimum magnetomotive force to use in the iron parts depends not only on the kind of steel used but also on the index number. The reason for this is that the total magnet ampere-turns generally drop at the high index numbers owing to the short air gap. Between index numbers of 3 and 100, the quality of the iron is the main factor. In this range the

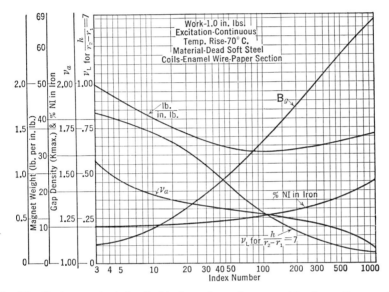

FIG. 25. Design constants for the bipolar or horseshoe magnet having optimum polar enlargements.

magnetomotive force used in the iron parts may be provisionally estimated as between 10 and 20 per cent of that developed by the coils. The low value is about right for the very soft magnet steels, such as Swedish charcoal iron; the high value is about right for the harder steels such as sample 4 machinery steel $\frac{1}{2}$ hard. Figure 25 gives the variation in this quantity over the entire range of index numbers for dead soft steel.

Referring to Fig. 25, for the index number of 31.6, we shall assume 11 per cent. Making this substitution in equation 2 for $\Sigma H_i l_i$, and noting that there are two air gaps in series, we have

$$NI = \frac{2B_g g}{\mu} + 0.11 NI$$

or
$$NI = \frac{2.25 B_g g}{\mu} \tag{21}$$

Heating Equation. This remains unchanged,

$$\theta_f = \frac{q\rho}{2kf(r_2 - r_1)} \left[\frac{NI}{h}\right]^2 \tag{3}$$

provided that (NI/h) is interpreted as the magnetic intensity of the coils, i.e., the ampere-turns of one coil divided by its length.

Voltage Equation. This remains unchanged,

$$E = \frac{4\rho(r_2 + r_1)NI}{d^2} \tag{4}$$

provided that the coils are connected in series, NI is the total magnetomotive force of both coils, and E is the supply voltage.

4. Preliminary Design. Equation 20 can now be solved for f if a value of B_g can be assumed. Referring to Fig. 25, data are given for B_g as a function of the index number for medium-sized horseshoe magnets (1 in-lb. of work).[28] Using these data, a preliminary value of 19 kmax. per sq. in. is obtained for B_g corresponding to an index number of 31.6.

Substituting into 20, we have, for f,

$$f = \sqrt{\frac{10 \times 36}{19^2}} = 0.998, \text{ say } 1.0 \text{ in.}$$

This is merely the size of the square polar enlargement. To get the radius r_1 of the pole core the leakage coefficients ν_a, equal to the ratio of the flux leaving the polar enlargements and passing through the armature to the useful flux in the working gap, and ν_L, equal to ratio of the leakage flux between pole cores to the useful flux in the working gap, must be known. Using data of Fig. 25 these leakage coefficients can be provisionally estimated as

$$\nu_a = 1.31$$
$$\nu_L = 0.54$$

The flux through the base of the pole cores and the yoke may now be computed as

$$\phi_y = B_g f^2 (\nu_a + \nu_L)$$
$$= 19 \times 1.0^2 \times 1.85 = 35.2 \text{ kmax.}$$

[28] Smaller magnets will have larger leakage coefficients, and therefore the value of B_g should be chosen somewhat smaller than that given in Fig. 25.

Letting B_y be the flux density at the base of pole cores or in the yoke, r_1 will be

$$r_1 = \sqrt{\frac{\phi_y}{B_y \pi}}$$

B_y, the highest flux density in the magnet, may be assumed on the basis of the saturation of the iron. It should be between 90 and 110 kmax. per sq. in., the low value being suitable for magnets having poor-quality iron and the higher values for large magnets which are excited only intermittently. For magnets having a rated work of about 1 in-lb., made of soft steel, a value of 100 is about optimum. Using this, we have

$$r_1 = \sqrt{\frac{35.2}{100 \times \pi}} = 0.335 \text{ in.}$$

The total coil magnetomotive force will be, from equation 21,

$$NI = \frac{2.25 \times 19 \times 0.1}{0.00319} = 1{,}342 \text{ ampere-turns}$$

Equation 3 may now be solved by making the following substitutions:

$\theta_f = 70°$ C.
$q = 1$.
$\rho = 0.865 \times 10^{-6}$ ohm-inch at 90° C.[29]
$k = 0.0067$ watt per sq. in. per ° C. temperature difference from Fig. 7b, Chapter VII, at $\theta_f = 70°$ C., for the coil in poor thermal contact with the iron parts. In the horseshoe magnet a paper-wound coil should always be considered in poor contact with the iron pole core. A snugly fitting brass bobbin coil may be considered to be in good thermal contact.
$f = 0.4$, assumed; estimated from Table V, Chapter VI, by assuming a wire size and making allowance for the paper margins and core tube.
$\dfrac{h}{r_2 - r_1} = 7$. Actual trial calculations made in computing the data of Fig. 25 show that this ratio does not seem to depend on the index number, and may be taken anywhere between 6 and 8, with a relatively flat optimum at 7.

Solving equation 3 for h and substituting the above values, we have

$$h = \sqrt[3]{\frac{3.5 q \rho (NI)^2}{k f \theta_f}} = \sqrt[3]{\frac{3.5 \times 1 \times 0.865 \times 10^{-6} \times \left(\dfrac{1{,}342}{2}\right)^2}{0.0067 \times 0.4 \times 70}} = 1.94 \text{ in.}$$

[29] See Art. 86, Sec. 4.

and $r_2 - r_1 = h/7 = 0.277$ in.

$$r_2 = 0.277 + 0.335 = 0.612 \text{ in.}$$

Equation 4 may now be solved for d:

$$d = \sqrt{\frac{4\rho(r_2 + r_1)NI}{E}} = \sqrt{\frac{4 \times 0.865 \times 10^{-6} \times 0.947 \times 1{,}342}{120}} = 0.00605$$

Assuming the width of the yoke to be equal to f, its thickness will be

$$t_2 = \frac{\pi r_1^2}{f} = \frac{\pi \times 0.335^2}{1.0} = 0.352 \text{ in.}$$

The thickness, t, of the armature, assuming its width equal to f, and its maximum flux density equal to that of the yoke, will be

$$t = \frac{\phi_a}{B_a f} = \frac{\nu_a \phi_g}{B_a f} = \frac{\nu_a B_g f}{B_a} = \frac{1.31 \times 19 \times 1.0}{100} = 0.249 \text{ in.}$$

5. Modification to Suit Stock Sizes. The preliminary design has now been finished and the next step will be to modify such dimensions as can conveniently be replaced by stock sizes.

Wire Size. From Table I, Chapter VI, the diameter of 6.05 mils corresponds very closely to No. 34.

The distance between pole centers can be chosen as soon as the coil diameter is fixed. This may be taken as 1.25 in., which will make $r_2 = 0.625$ in. instead of 0.612 in. as computed. This will make the mean length of turn slightly longer and hence decrease the ampere-turns slightly. The distance between pole centers, D, can then be taken as $1\frac{3}{8}$ in., allowing $\frac{1}{8}$-in. clearance between coils, as specified in Fig. 24.

Iron Dimensions. The pole core diameter $2r_1$ comes out almost exactly equal to $\frac{11}{16}$ in., and its length may be taken as $1\frac{15}{16}$ in. The armature width, f, can be made equal to 1.0 in. and its thickness t, $\frac{1}{4}$ in. The yoke width, f_2, can be made 1 in. and its thickness t_2, $\frac{3}{8}$ in. The thickness of the polar enlargement, t_1, can be taken equal to $\frac{1}{8}$ in.

6. Check of Preliminary Design Using Stock Sizes.

Final Sizes: $r_1 = \frac{11}{32}$ in.; $h = 1\frac{15}{16}$ in.; $r_2 = \frac{5}{8}$ in.; $D = 1\frac{3}{8}$ in.; $f = 1$ in.; $t = \frac{1}{4}$ in.; $f_2 = 1$ in.; $t_2 = \frac{3}{8}$ in.; $t_1 = \frac{1}{8}$ in.

Coil Design—Space Factor. The gross winding depth, $(r_2 - r_1) = 0.281$ in., will include, besides the thickness of the winding itself: (a) the thickness of the core tube; (b) the cover over the coil. Referring to Art. 59, Chapter VI, the core tube should consist of 7 wraps of 0.005 in.

Art. 91]   BIPOLAR OR HORSESHOE TYPE OF MAGNET   327

Kraft paper and the cover of 0.005 in. Kraft paper. The net winding depth will therefore be 0.281−0.040 = 0.241 in. Referring to Table IV, Chapter VI, the thickness of 1 layer will be 7.1 mils for the enameled wire plus 1.0 mil for the interlayer paper, making 8.1 mils. The number of layers will be 0.241 ÷ 0.0081 = 29.8. As the layers will compress slightly this can safely be called 30.

Besides the insulated wire the axial length of the coil includes two paper margins which from Table IV may be taken as $\frac{3}{32}$ in. each. The net winding length of the coil will then be $1\frac{15}{16} - 2 \times \frac{3}{32}$ in. = $1\frac{3}{4}$ in. As the maximum turns per inch are given as 130 for No. 34 enamel wire in Table IV the turns per layer will be $130 \times 1\frac{3}{4}$ = 228. Total turns of one coil will be $30 \times 228 = 6,840$. The space factor of the coil will therefore be

$$f = \frac{6{,}840 \times 0.0312 \times 10^{-3}}{0.281 \times 1.9375} = 0.392$$

where 0.0312×10^{-3} is the copper area of No. 34 wire from Table II Chapter VI.

The coil resistance may now be computed as follows: The mean radius of a turn will be r_1 + the thickness of the core tube + $\frac{1}{2}$ the net coil winding thickness = $0.344 + 0.035 + 0.120 = 0.499$ in. The total length of wire of one coil will be $2\pi \times 0.499 \times 6{,}840 = 21{,}400$ in., and the resistance at 90° C. of one coil, applying the data of Tables II and III, Chapter VI, is

$$21.4 \times 21.74 \times 1.2751 = 594 \text{ ohms}$$

The coil current will be

$$\frac{120}{2 \times 594} = 0.101 \text{ ampere (hot)}$$

and the magnetomotive force developed by both coils is

$$NI = 2 \times 6{,}840 \times 0.101 = 1{,}380 \text{ ampere-turns}$$

Temperature Rise. Substituting the values so far obtained into equation 3, we have

$$\theta_f = \frac{1 \times 0.865 \times 10^{-6}}{2 \times 0.0067 \times 0.392 \times 0.281} \left(\frac{690}{1.9375}\right)^2 = 74.3° \text{ C.}$$

This is too high and should be lowered before proceeding to check the force. It can be lowered without changing the ampere-turns of the coil by increasing h. The new value of h should be equal to $1.9375 \times \sqrt{74.3 \div 70.0} = 2.0$.

Coil Redesign with New h. The new h will change the turns per layer, which will now be

$$130 \left[2 - 2 \times \tfrac{3}{32}\right] = 236$$

The turns per coil will now be $30 \times 236 = 7,080$. As the mean length of turn remains unaltered the new resistance will be

$$594 \times \frac{7,080}{6,840} = 615 \text{ ohms}$$

and the coil current 0.0976 ampere. The total magnetomotive force of both coils will be

$$NI = 2 \times 7,080 \times 0.0975 = 1,380 \text{ ampere-turns}$$

and their space factor

$$f = \frac{7,080 \times 0.0312 \times 10^{-3}}{0.281 \times 2.0} = 0.393$$

The new temperature rise will be

$$\theta_f = \frac{0.865 \times 10^{-6}}{2 \times 0.0067 \times 0.393 \times 0.281} \left(\frac{690}{2.0}\right)^2 = 69.6° \text{ C.}$$

This is satisfactory.

Magnetic Circuit Calculation. At this point a sketch of the magnet should be made so that the mean lengths of the magnetic circuit may be obtained. The sketch is shown in Fig. 26, drawn to scale. The mean lengths of the magnetic circuit as obtained from this sketch are:

	Length, in.	Area, sq. in.
Armature	1.63	0.250
Pole cores	2.0	0.371
Yoke	1.75	0.375

Before the saturation curve of the magnet can be calculated it is necessary to determine the leakage coefficients so that the distribution of flux in the various parts will be known. Referring now to Sec. 2, Art. 54, Chapter V, it will be seen that the leakage coefficient derived there for the horseshoe magnet cannot be applied directly as the distance between polar enlargements equal to $D - f = \tfrac{3}{8}$ in. is less than $\pi(g + t) = 0.9$ in.

The leakage coefficient ν_a, as given by formula 24, Chapter V, must be modified by dropping out the term $1.47 \log_{10}(1 + t/g)$ corresponding to path 12b of Fig. 18, Chapter V, giving

$$\nu_a = 1 + \frac{g}{f^2}\left[1.5t + 0.46g + 1.3f + 2.2f \log_{10}\left(1 + \frac{2t}{g}\right)\right]$$

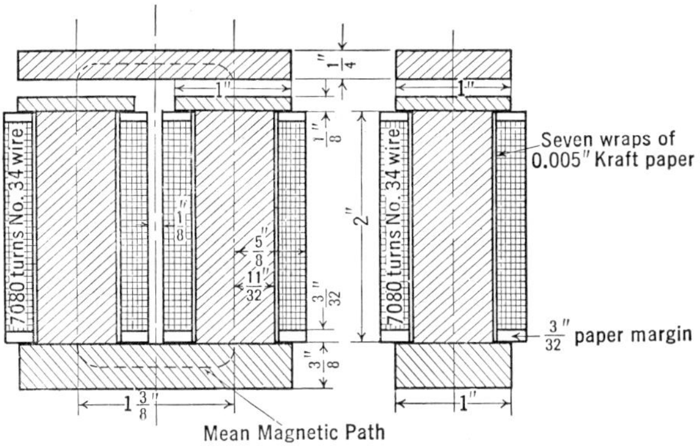

Fig. 26. Cross section through the final design of the bipolar or horseshoe magnet, showing all dimensions.

Likewise formula 25 for ν_L must be modified by adding paths 1 and 7 between the opposite faces and adjacent edges of the polar enlargements, giving:

$$\nu_L = \frac{g}{f^2}\left[\frac{\pi h'}{\log_\epsilon \frac{D}{r_1}} + \frac{ft_1}{D - f} + 0.52t_1\right]$$

$g = 0.10$ in. $r_1 = 0.344$ in.

avg t for $\nu_a = \dfrac{\frac{1}{4} + \frac{1}{8}}{2} = 0.1875$ in. $D = 1.375$ in.

$f = 1.0$ in. $h' = 2.0 + \frac{3}{32} = 2.094$ in.

Making these substitutions, we have

$$\nu_a = 1.31$$
$$\nu_L = 0.51$$

which checks very well with the provisional estimate.

The useful air-gap permeance

$$P_u = \frac{\mu f^2}{2g} = \frac{3.19 \times 1.0^2}{2 \times 0.1} = 15.95 \text{ max. per ampere-turn}$$

The saturation curve of the iron parts for the useful flux, ϕ_g, plotted against magnetomotive force for a constant total gap length of 0.10 in., may now be computed. Using ϕ_g as a reference, and assuming the permeance of the armature to be infinite,[30] the flux of the armature will be

$$\phi_a = \phi_g \nu_a = 1.31\phi_g$$

and the flux in the yoke will be

$$\phi_y = \phi_g[\nu_a + \nu_L] = 1.82\phi_g$$

The flux in the pole cores will be equal to ϕ_g at the armature end and ϕ_y at the yoke end. The average flux density in the pole cores, as outlined in Art. 46, Chapter IV, may be taken as $\phi_a + \tfrac{2}{3}\phi_L$; therefore

$$\phi_p = \phi_g[\nu_a + \tfrac{2}{3}\nu_L] = 1.65\phi_g$$

The actual calculation of the saturation curve will be carried out in tabular form; it is started by assuming the value of flux density in the yoke.[30] This is a logical procedure as this is where the magnet will saturate first.

Actual Work and Optimum Work of Design. In Fig. 27 is shown

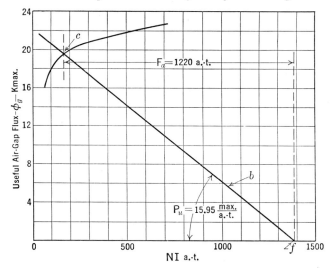

Fig. 27. Magnetization curve and graphical construction for determining the operating point of the magnet of Fig. 26.

plotted the saturation curve of the iron parts from Table VII. From point f at 1,380 ampere-turns (actual coil NI) on the axis of abscissas

[30] See footnote pages 262 and 263.

Art. 91]   BIPOLAR OR HORSESHOE TYPE OF MAGNET   331

TABLE VII

Calculation of Saturation Curve of Iron Parts of Horseshoe Magnet (Air Gap Excluded) for a Useful Air-Gap Length of 0.10 In.

(See saturation curve of Fig. 11a, Chapter II, for sample 3)

Part	Length, in.	Area, sq. in.	Flux, ϕ		B kmax. per sq. in.	H a-t. per in.	F a-t.
				kmax.			
1. Yoke.........	1.75	0.375	$1.82\phi_g$	41.3	110	102	179
2. Pole cores.....	2.0 × 2	0.371	$1.65\phi_g$	37.4	101	35	140
3. Useful air gap..	$P_u = 15.95$ max. per a-t.		ϕ_g	22.7	22.7	....	....
4. Armature.....	1.63	0.250	$1.31\phi_g$	29.7	119	245	400

Total 719

Part	ϕ	B	H	F	ϕ	B	H	F
1	39.4	105	60	105	37.5	100	31	54
2	35.7	96.2	23	92	34.0	91.6	16	64
3	21.6	21.6	....	....	20.6	20.6	....	....
4	28.4	113.5	148	242	27.0	108	83	136
			Total 439				Total 254	

Part	ϕ	B	H	F	ϕ	B	H	F
1	35.6	95	21	37	33.8	90	16	28
2	32.3	87	13.5	54	30.6	82.5	11.3	45
3	19.6	19.6	....	....	18.5	18.5	....	....
4	25.6	102.5	43	70	24.3	97	24	39
			Total 161				Total 112	

there is drawn the working air-gap permeance line b, with a negative slope of 0.01595 kmax. per ampere-turn. The intersection of line b with the saturation curve at c gives the operating point of the magnet with its rated air gap and current. The optimum operating point for this particular magnet with an excitation of 1,380 ampere-turns (as explained in Art. 85) is also at this point.

The energy actually available in the air gap at 0.1-in. gap corresponding to the useful work of the magnet (full 0.1-in. stroke) is equal to area fce, which is

$$\tfrac{1}{2} \times 19.6 \times 10^{-5} \times 1{,}220 \times 8.86 = 1.06 \text{ in-lb.}$$

Force. The only factor left to be checked is the force, which according to the stored energy of the air gap should be $1.06 \div 0.1 = 10.6$ lb.

The flux density in the working gap B_g will be $\phi_g/f^2 = 19.6/1.0 = 19.6$ kmax. per sq. in. The force by equation 20 will be

$$\text{Force} = \frac{B_g^2 f^2}{36} = \frac{19.6^2 \times 1.0}{36} = 10.6 \text{ lb.}$$

This force is slightly higher than that designed for. Some excess force is, of course, desirable as a factor of safety. Generally 5 to 10 per cent is sufficient.

Weight. The weight of the copper of the coils will be

$$2 \times \frac{22{,}160}{12} \times \frac{1}{8{,}310} = 0.444 \text{ lb.}$$

where 8,310 is the number of feet per pound of No. 34 enameled wire from Table II, Chapter VI. The paper and other insulation on the coils may be estimated to bring the total coil weight up to 0.5 lb.

The volume of the iron parts are computed as follows:

$$
\begin{array}{ll}
\text{Armature} = 0.250 \times 2\tfrac{3}{8} = & 0.59 \\
\text{Pole cores} = 0.371 \times 4 = & 1.48 \\
\text{Polar enlargements} = 2 \times \tfrac{1}{8} \times 1.0 = & 0.25 \\
\text{Yoke} = 0.375 \times 2\tfrac{3}{8} = & 0.89 \\
\hline
\text{Total} & 3.21 \text{ cu. in.}
\end{array}
$$

and its weight will be

$$3.21 \times 0.283 = 0.91 \text{ lb.}$$

The total weight will be $0.91 + 0.50 = 1.41$ lb., and the weight economy will be

$$\frac{1.41}{1.06} = 1.33 \text{ lb. per in-lb. of work}$$

considering that without the non-magnetic stop a stroke of 0.1 in. is possible. This checks with the value of 1.34 given in Fig. 25 for an index number = 31.6.

Final Results. The design is now finished. Figure 26 shows a complete sketch of the magnet with all dimensions. Below are tabulated all the principal data.

Force = 10.6.
Stroke (useful) = 0.09 in.; (maximum possible) = 0.1 in.
Voltage = 120 volts.
Current (hot) = 0.0975 ampere.
Power (hot) = 11.7 watts.
Temperature rise = 69.6° C.
Weight = 1.41 lb.
Maximum useful work (no non-magnetic stop) = 1.06 in-lb.
Weight per unit of maximum work = 1.33 lb. per in-lb. of work.

PROBLEMS

1. Make a preliminary design of a direct-current lifting magnet to hold a load of 4,000 lb. Assume the following:

  a. 65 per cent of continuous excitation.
  b. 50° C. final temperature rise.
  c. The force of 4,000 lb. must be exerted over a stroke of $\frac{1}{32}$ in.
  d. Soft iron such as S.A.E. 10–10.
  e. Voltage = 110 volts.

Make a sketch, to scale, of the design and tabulate all the principal data, using stock sizes wherever possible.

2. It is proposed to build a demonstration lifting magnet having the following features:

  a. Smallest possible size.
  b. Capable of supporting a man (assume 200 lb.) with its armature in contact with the pole face. (See footnote 4, Chapter IV.)
  c. Operable from a large flashlight cell (Eveready No. 950 size D) for 15 seconds at a time, in intervals of 5 minutes.

There are no restrictions as to material or cost. A new No. 950 cell will deliver not less than 5 amperes on short circuit for the first few intervals as described above. Make a complete design, giving all data, and a sketch, to scale, with dimensions.

3. Design a magnet suitable for a 50-lb. pull through a $\frac{1}{16}$ in. for operation from a 12-volt battery. Maximum allowable temperature rise with continuous operation, 70° C. Make a sketch, to scale, of the final design and tabulate all the principal data.

4. Design a magnet to meet the following specifications:

  1. Minimum force throughout stroke = 25 lb.
  2. Useful stroke = $\frac{23}{64}$ in.
  3. Time of excitation = $\frac{1}{4}$.
  4. Maximum temperature rise = 100° C.
  5. Supply voltage = 12 volts.

This magnet is to be designed with no restrictions as to cost, and everything possible, including the use of minimum plunger eccentricity tolerances and a high permeance fixed cylindrical gap, should be done to reduce its size. Make a sketch, to scale, of the final design, and tabulate all the principal data, including the weight.

5. Compute, for the magnet of Problem 4, the time of continuous excitation necessary to raise the average coil temperature to the rated temperature rise from an ambient temperature of 20° C. To simplify the calculation assume the magnet constants and power input to stay constant at the final value.

6. Calculate, from the allowed tolerance between the plunger and tube of your magnet of Problem 4, the maximum side pull that can occur in (a) the fixed cylindrical gap, (b) the working gap.

Note: The side pull in the working gap of a conical-type magnet can be estimated by dividing the gap into a number of sections and computing an equivalent cylindrical-faced gap for each. Formula 16a of Sec. 8, Art. 72, may be applied to each equivalent gap section. Referring to the conical gap illustrated in Fig. 28, the conical gap has been divided into four sections. The cylindrical gap, equivalent to section 3, is shown at A, where r is the average radius of 3, and the other dimensions are as indicated.

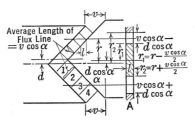

FIG. 28. Diagram for estimating the side pull in the working gap of a full-conical cylindrical plunger magnet.

7. Assuming a coefficient of friction between brass and steel of 0.15, compute for the magnet of Problem 4 the percentage reduction in available force at the rated stroke due to friction.

8. It is proposed, for commercial reasons, to redesign the magnet of Problem 4 so as to reduce its cost of manufacture. To obtain this result it is desired to use standard rods and tubes for the plunger, shell, and plunger tube, and to machine only the end pieces. Because of the tolerances necessary with this type of construction, the eccentricity of the plunger must be considered with respect to the side pull that will be produced, and the consequent increase of sliding friction. Redesign the magnet, as suggested, keeping its net force (exclusive of friction) at rated stroke the same. Draw a sketch of the final design and tabulate its principal data, including weight, in comparison with that of the magnet of Problem 4.

9. Devise a mechanical arrangement that will eliminate any detrimental eccentricity in the working gap due to necessary tolerance between the plunger and brass tube when a magnet is built as suggested in Problem 8 for the following magnet types:

1. Conical faced   as illustrated in Fig. 4, Chapter IX
2. Cylindrical faced   "   "   "   " 5,   "   "
3. Stepped cylindrical faced "   "   "   " 6,   "   "
4. Taper   "   "   "   " 7,   "   "
5. Stepped cylindrical faced "   "   "   " 29, below

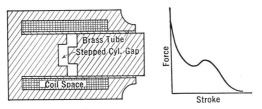

FIG. 29. Stepped cylindrical-faced plunger magnet.

10. Design a magnet to meet the following specifications:

Stroke = $\frac{1}{2}$ in.
Force at the beginning of stroke = 1 lb.
Maximum length = $1\frac{3}{4}$ in.
Maximum diameter = $\frac{3}{4}$ in.
Minimum voltage = 10 volts.
Current to be kept as low as possible, but not to exceed 5 amperes at 12 volts.
Excitation sufficiently short that heating is not a limitation.
Force must gradually rise toward end of stroke.
Draw, to scale, a sketch of the final design, and tabulate all the principal data.

11. Design a magnet to meet the following specifications:

1. Minimum force throughout stroke = $\frac{1}{2}$ lb.
2. Stroke = 2 in.
3. Temperature rise = 50° C.
4. Time of excitation = $\frac{1}{4}$.
5. Supply voltage = 120 volts.

12. Compute for the magnet designed in Art. 90 the force-stroke characteristic for the tapered end section (the last $1\frac{1}{2}$ in. of stroke). Use the method as developed in that article for handling the distributed leakage permeance.

13. Design a horseshoe magnet to meet the following specifications:

1. Minimum force throughout stroke = 8 lb.
2. Useful stroke = $\frac{1}{8}$ in.
3. Temperature rise = 50° C.
4. Time of excitation = $\frac{1}{2}$.
5. Supply voltage = 24 volts.
6. Allow for $\frac{1}{64}$-in. non-magnetic spacer between armature and pole faces at the end of the stroke.

Draw, to scale, a sketch of the final design, and tabulate all the principal data.

14. Design a horseshoe magnet to meet the following specifications:

1. Minimum force throughout the stroke = 2 lb.
2. Useful stroke = $\frac{1}{2}$ in.
3. Temperature rise = 70° C.
4. Excitation = continuous.
5. Supply voltage = 120 volts.

Draw, to scale, a sketch of the final design, and tabulate all the principal data.

15. Calculate for the magnet of Problem 14 the following:

1. The force at the end of the stroke with full excitation (see footnote 2, page 86).
2. The residual force following full excitation.

Assume that there is no non-magnetic spacer between the armature and pole faces.

CHAPTER XI

TIME-DELAYED MAGNETS

92. General

In many applications it is desirable to make the action of a magnet occur sometime after the closing or opening of the switch which initiates the action. Such a magnet system is called time-delayed. Its practical application generally occurs in a sequence or selecting mechanism. Thus, in automatic telephone equipment there often is a bank of relays in parallel, which must not operate simultaneously in response to the same voltage. This calls for time-delayed action in some of the relays. Likewise in automatic elevator equipment, the pressing of the start button not only starts the elevator but also closes the doors. The starting of the elevator, obviously, must be delayed to occur later than the closing of the doors. In some circuits instability results if the actuating relay is not allowed to complete its stroke before the action it initiates commences. These and many other devices require a time-delay action.

93. Methods of Producing Time Delay

Delayed action can be produced in two essentially different ways, or by a combination of them:

1. In the magnet itself, by causing the change in flux to lag behind the change in current. Such a magnet can be properly termed a time-delay magnet.

2. By using a non-time-delayed magnet in conjunction with (a) a time-delay circuit or (b) a time-delay mechanism. The time-delay circuit retards the change of current through the magnet, while the time-delay mechanism uses an auxiliary device in conjunction with the magnet.

Method 1 is always accomplished by means of a secondary winding on the magnet which is short-circuited or closed on itself, as illustrated in Fig. 1. Such a winding, called a lag coil, may consist of a copper cylinder of one turn, or a regular winding with insulated turns closed on itself. This method of time delay is suitable for producing time delays of the order of the electrical time constant of the primary circuit of the magnet only.

Method 2a consists of introducing a series-connected energy-storing device to delay the building up of the current through the magnet coil when the circuit is closed, as illustrated by the series inductance of Fig. 2a, or a parallel-connected energy-storing device to delay the fall of current through the magnet coil when the circuit is opened, as illustrated by the condenser of Fig. 2b in parallel with the magnet coil. The time

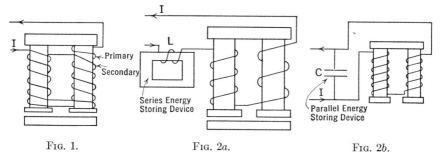

FIG. 1.   FIG. 2a.   FIG. 2b.
Methods of producing time delay.

delay that can be produced by this method depends entirely on the ratio of the amount of energy stored in the auxiliary device to the normal rate of energy consumption of the magnet. When the condenser is used the circuit under some conditions may become oscillatory; the time delay then depends on the natural period of oscillation. This method can be conveniently combined with method 1, and will then give time delays greater than can be obtained with method 1 alone.

Method 2b, consisting of the interposition of an auxiliary device between the magnet and the initiating switch, does not depend in any way on the characteristics of the magnet, but only on the nature of the auxiliary device. The auxiliary devices used for this purpose may be divided into two kinds: mechanical and electrical.

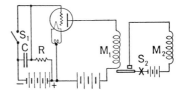

FIG. 3. An electronic method of producing time delay.

Mechanical devices are usually a dashpot or clockwork mechanism which, at the end of a predetermined time interval, opens or closes the magnet circuit.

An electrical device which is electronic is illustrated in Fig. 3. In this circuit when switch S_1 is closed the three-electrode tube will be biased to cut-off and no plate current will flow to actuate the relay magnet M_1. Hence, the relay switch S_2 will remain open and the power magnet M_2 will remain unenergized. When S_1 is opened the potential of the grid will rise exponentially from a high negative value to zero as the condenser C charges through the resistance R. As the grid becomes more positive, plate current will commence to flow through the relay magnet M_1, until it operates at some predetermined value. Relay M_1 will then close the circuit to the power magnet. It is not necessary to use the intermediate relay M_1 if the plate current of the tube is sufficient to operate M_2 directly. The time elapsing between the opening of S_1 and the operation of M_1 will depend on the time constant of the condenser and resistance in the grid circuit. This time constant is equal to CR seconds, where C is in farads and R in ohms. Thus a 1-microfarad condenser and a 1-megohm resistance would be suitable for producing a time delay of the order of 1 second.

The clockwork mechanism is suitable for producing relatively long time delays of the order of a fraction of a minute and up. The electronic device is useful in the range of several seconds and down. It will accurately produce and repeat not only long time delays but also short ones of the order of a fraction of a second.

94. Method of Computing the Time-Delay for Various Types of Delay

The generalized scheme of computing time delay is to compute the flux of the magnet, and hence its force, as a function of time. With this information the amount of time delay can be determined if the force against which the armature acts is known.

The flux of the magnet as a function of time can be determined from an equation expressing the relationship between the instantaneous values of flux, time, and impressed voltage. This equation takes the form of Kirchhoff's second law, equating the algebraic sum of the voltages around the complete circuit to zero. Thus, in general,

$$IR + \frac{d(N\phi)}{dt} = E,$$

where E is the impressed voltage, $d(N\phi)/dt$ is the voltage induced in the magnet coil, and IR is the resistance drop in the magnet coil. This equation is perfectly general and applies to any magnet regardless of saturation, motion of the armature, or any other change that may be taking place. It is not usually possible to solve the equation analytically when saturation occurs, because then ϕ cannot be expressed as an analytical function of I. In such cases the solution must be obtained by an integration performed by a step-by-step method. When motion occurs another variable, the distance s, locating the position of the armature, is introduced. In this case ϕ becomes a function of both

I and s, which relationship can be expressed only by a family of curves. This problem, also, must be solved by step-by-step integration.[1]

Fortunately in time-delayed magnets such methods are seldom necessary, as the calculations involve only incipient motion with an air gap in the magnetic circuit. The air gap will, in general, prevent saturation and so minimize any change in permeance of the over-all magnetic circuit due to change of flux density in the iron. Likewise the absence of motion prevents any change in the air-gap permeance. These two factors make valid the assumption of constant inductance, and hence the method of analytical solution employing linear differential equations.

95. Simple Magnet—Build-Up of Current

In the ordinary magnet the build-up of current lags behind the impressed voltage, owing to the self-inductance of the winding. Hence time delay is introduced into the operation of every magnet. The magnitude of this time delay depends on the relative magnitudes of the self-inductance and the resistance, and the percentage of the final current [2] necessary to initiate action.

Figure 4 illustrates the simple magnet, where L_1 and R_1 are the self-inductance and resistance, respectively. The relationship between current and impressed voltage is given by the following equation:

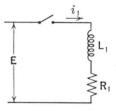

Fig. 4. The electrical circuit of a simple magnet.

$$E = i_1 R_1 + L_1 \frac{di_1}{dt} \quad (1)$$

This equation expresses Kirchhoff's law and states that the algebraic sum of the impressed voltage E, the self-induced voltage of the magnet $L(di_1/dt)$, and the resistance voltage of the magnet $i_1 R_1$ must add to zero. It is a differential equation of the first order, and if R_1 and L_1 can be assumed to remain constant, it has constant coefficients. We shall solve [3] this equation by the classical method of breaking the solution

---

[1] This type of problem occurs when the time of action of a magnet is computed. This is taken up in Chapter XII in connection with high-speed magnets.

[2] When the magnet is laminated, or the rise of flux is sufficiently slow, so that the effect of eddy currents can be neglected, it may be assumed that the static magnetization curve between flux and current will apply. Hence the time delay can be determined by the time required for the current to reach a predetermined value.

[3] For a detailed discussion of the solution of differential equations for electrical circuits of the type discussed in this chapter the reader is referred to the following Books: Woodruff, "Electric Power Transmission and Distribution," Chapter XV, 1925; Guillemin, "Communication Networks," Vol. I; Bush, "Operational Circuit Analysis." All published by John Wiley & Sons.

into two parts, one, i_s, called the steady state or particular integral, and the other, i_t, the transient or complementary function. Thus

$$i_1 = i_s + i_t \tag{2}$$

The steady-state solution is the so-called "forced response" of the circuit, and is the value of current produced by the steady impressed supply voltage E. Its value can be obtained in three ways:

    a. Ordinary steady state theory.
    b. Solution of the circuit differential equation 1, assuming for the form of the solution the same shape as the impressed voltage.
    c. By inspection.

This problem is so simple that the steady-state solution can be determined directly by inspection, as

$$i_s = \frac{E}{R_1} \tag{3}$$

The transient part of the solution is the surge produced in the system by the sudden application or removal of the supply voltage, or by any sudden change in applied voltage. This is the so-called "free response" or "normal behavior" of the circuit in response to an impulse, or shock excitation. It is due to the fact that a change in current through an inductive element or change in voltage across a capacitive element is associated with a change in stored energy of the circuit. A change in stored energy can be brought about only by a finite power acting for a finite time. This transient power necessary to effect the readjustment in energy comprises the transient current and voltage of the circuit, which disappears when the energy adjustment has been effected. Consequently, the transient current is not associated with any impressed voltage, and its equation may be written directly from the circuit differential equation 1 by letting the impressed voltage equal zero, thus

$$0 = i_t R_1 + L_1 \frac{di_t}{dt} \tag{4}$$

In linear electrical circuits, the transient currents and voltages are always of the exponential form, and hence the exponential function is called the normal function of the electric circuit. It is the only mathematical function which will express the normal behavior or relationship which exists between current and voltage in a freely responding or oscillating circuit. The solution of equation 4, therefore, is always obtained by assuming an exponential form for the transient current,

$$i_t = I_1 \epsilon^{a_1 t} \tag{5}$$

where I_1 and a_1 are constants which must be determined to fit the particular problem.

Differentiating equation 5 and substituting into 4, we have

$$0 = I_1 \epsilon^{a_1 t}(R_1 + a_1 L_1) \qquad (6)$$

Inasmuch as $i_t = I_1 \epsilon^{a_1 t}$ cannot be zero, without producing a trivial solution, we can equate

$$R_1 + a_1 L_1 = 0 \qquad (7)$$

giving

$$a_1 = -\frac{R_1}{L_1}$$

Equation 7 is known as the determinantal equation of the circuit, and determines the values of a, or the normal modes of the circuit.

The transient current will therefore be

$$i_t = I_1 \epsilon^{\frac{-R_1}{L_1} t} \qquad (8)$$

where the constant I_1 must still be determined.

The most systematic method of determining such constants is by substituting known values, or boundary conditions, into equation 2.

In the circuit of Fig. 4, it is known that, when the switch is closed, the initial current through the magnet is zero. Thus when $t = 0$, $i_1 = 0$; therefore

$$0 = i_s + i_t = \frac{E}{R_1} + I_1 \epsilon^{\frac{-R_1}{L_1} t} \qquad (9)$$

which, on substituting 0 for t, gives

$$I_1 = -\frac{E}{R_1} \qquad (10)$$

Therefore the final solution is

$$i_1 = i_s + i_t = \frac{E}{R_1} - \frac{E}{R_1} \epsilon^{\frac{-R_1}{L_1} t}$$

or

$$i_1 = \frac{E}{R_1}\left(1 - \epsilon^{\frac{-R_1}{L_1} t}\right) \qquad (11)$$

This equation is shown plotted in Fig. 5.

The exponential current i_1 starts with an initial slope of E/L_1 amperes per second and approaches the steady-state value i_s, shown by line b as an asymptote. The transient current i_t is the difference between the actual current i_1 and the steady-state current i_s, and gradually disappears as time goes on. The time required for the current to reach its

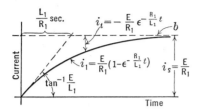

FIG. 5. Build-up of current in a simple magnet.

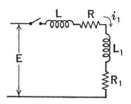

FIG. 6. The electrical circuit of a simple magnet with a series choke coil like that illustrated in Fig. 2a.

final value, were it to continue to change at its initial rate, is known as the time constant of the circuit, and is equal to

$$\frac{1}{a_1}, \text{ or } \frac{L_1}{R_1} \text{ seconds}$$

The actual time delay which can be produced on closing the switch is determined by the time required for the flux to reach a value sufficient to cause the magnet to develop just enough force to overcome its armature load.

The time delay of the circuit of Fig. 4 can be increased by adding a choke coil, having a time constant considerably longer than that of the magnet, in series. This is shown in Fig. 6. The equation for the current i_1 on closing the switch will be

$$i_1 = \frac{E}{R + R_1} \left[1 - \epsilon^{-\frac{R+R_1}{L+L_1}t} \right] \tag{12}$$

96. Simple Magnet with Resistance Shunt—Build-Down of Current

The simplest circuit for producing time delay when the circuit to a magnet is broken consists in shunting the magnet with a resistance as illustrated in Fig. 7. This circuit will not only produce time-delay but will also limit the surge voltage across the coil when the circuit is opened. Remembering that time starts at the instant the switch is opened, the supply voltage will be zero and the voltage equation will be

$$0 = i_1(R + R_1) + L_1 \frac{di_1}{dt} \tag{13}$$

The steady-state solution will be by inspection

$$i_s = 0 \tag{14}$$

therefore $i_1 = i_t$. Making the substitution as in equation 6, the determinantal equation will be

giving
$$R + R_1 + a_1 L_1 = 0 \tag{15}$$

$$a_1 = -\frac{R + R_1}{L_1} \tag{16}$$

The transient current will therefore be

$$i_t = I_1 \epsilon^{-\frac{R+R_1}{L_1} t}$$

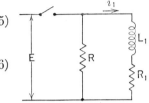

Fig. 7. The electrical circuit of a simple magnet with a resistance shunt.

The constant I_1 may be determined from the condition that the current i_1 at the instant the switch is opened ($t = 0$) will be E/R_1. Therefore

$$\frac{E}{R_1} = i_s + i_t = 0 + I_1 \epsilon^{-\frac{R+R_1}{L_1} t} \tag{17}$$

and

$$I_1 = \frac{E}{R_1} \tag{18}$$

The final solution is

$$i_1 = i_s + i_t = \frac{E}{R_1} \epsilon^{-\frac{R+R_1}{L_1} t} \tag{19}$$

This equation is shown plotted in Fig. 8.

In order to test this circuit experimentally, the magnet of Fig. 2, Chapter IX, was set up so as to support the weight of its armature, 9 oz., with an air-gap length $\frac{9}{64}$ in. The inductance of the magnet had been measured previously and was found to be substantially independent of current at this gap length. The circuit arrangement, and constants, together with the computed and test results, are given in Fig. 9. Referring to the figure it will be noticed that the current by test drops more slowly than the computed results. This probably is due to an error in the value of inductance used. The

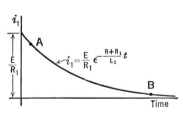

Fig. 8. Build-down of current in a simple magnet with a resistance shunt.

measured inductance represents a normal value determined by reversing the flux linkage through distributed exploring coils. The actual inductance effective is a transient inductance determined by the falling branch of the hysteresis loop.

In a time-delay circuit of this type the actual delay produced is dependent on the air-gap length and the load, as these two items determine the ampere-turns at which the armature will release. Referring to Fig. 8, a long air gap and a heavy load will cause the armature to release

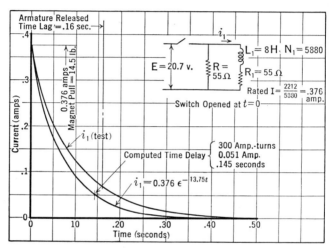

Fig. 9. Experimental and computed curves of the build-down of current in a simple magnet with a resistance shunt. Horseshoe magnet of Fig. 2, Chapter IX, supporting armature weighing 9 oz. through a $\frac{9}{64}$-in. air gap. Actual armature release occurs at 300 ampere-turns or 0.0511 ampere.

with a short time delay as at A, while a short air gap and light load will cause a very long time delay before release, as at B. When the air gap is made too short, the time delay becomes uncertain and the action unreliable, with possibility of non-release due to the residual force. In such cases [4] it is common practice to use a second coil on the magnet which is separately and continually energized to produce a relatively small magnetomotive force opposing the main winding. This will cause the curve of effective magnetomotive force to cross the axis, instead of being asymptotic as shown in Fig. 8, and results in a definite release time.

[4] See Fig. 14c, of Art. 109, Chapter XII, and the associated text.

97. Magnet with Lag Coil—Build-Up of Effective Magnetomotive Force

A lag coil is a short-circuited coil surrounding the magnetic circuit. A magnet so arranged is the same as a short-circuited transformer, and its solution can be handled in a similar manner. Figure 10a represents such a magnet with a lag coil. The lag coil winding is so arranged with respect to the main exciting winding that the same flux may be assumed to link with both windings.[5]

In Fig. 10b, let L_1 and R_1 represent the self-inductance and resistance

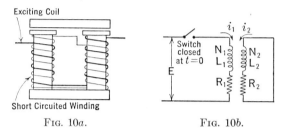

FIG. 10a.   FIG. 10b.

Magnet with lag coil and its equivalent electric circuit.

of the exciting circuit, and L_2 and R_2 the self-inductance of the lag coil, of the magnet of Fig. 10a. L_{12} represents the mutual inductance between the primary and the secondary. The current i_2 of the lag coil will always flow in such a direction as to tend to prevent a change in flux. If, as shown in Fig. 10b, the positive directions of i_1 and i_2 are taken so that both produce flux in the same direction in the common core, then the effective magnetomotive force producing flux will be the algebraic sum of the coil magnetomotive forces.

$$F_e = N_1 i_1 + N_2 i_2 \tag{20}$$

The voltage equations for this circuit will be

$$\text{Primary: } E = i_1 R_1 + L_1 \frac{di_1}{dt} + L_{12} \frac{di_2}{dt} \tag{21}$$

$$\text{Secondary: } 0 = i_2 R_2 + L_2 \frac{di_2}{dt} + L_{12} \frac{di_1}{dt} \tag{22}$$

---

[5] This is accomplished most perfectly by winding the two coils simultaneously by means of a doubled feed wire. The next best scheme is to put the lag coil on the magnet next to the pole cores under the exciting coil, and covering the same length of pole core as the exciting coils. With this construction it is advantageous to use a lag coil made from a seamless copper tube. Although tight coupling is generally desired, the lag coil is sometimes made much shorter than the pole cores and placed next to the polar enlargements or yoke, to obtain special effects.

If equation 21 is solved for di_2/dt and d^2i_2/dt^2, and these substituted into the derivative of 22, we will obtain the following:

$$\left[\frac{L_1L_2 - L_{12}^2}{R_2}\right]\frac{d^2i_1}{dt^2} + \left[L_1 + \frac{R_1}{R_2}L_2\right]\frac{di_1}{dt} + R_1i_1 = E \quad (23)$$

The steady-state solution of this circuit will be, by inspection,

$$(i_1)_s = \frac{E}{R_1} \quad (24)$$

$$(i_2)_s = 0 \quad (25)$$

The equation of the transient is obtained by writing one like (23) with zero impressed voltage, thus:

$$(L_1L_2 - L_{12}^2)\frac{d^2(i_1)_t}{dt^2} + (L_1R_2 + R_1L_2)\frac{d(i_1)_t}{dt} + R_1R_2(i_1)_t = 0 \quad (26)$$

Making the substitution as in equation 6, the determinantal equation will be

$$(L_1L_2 - L_{12}^2)a^2 + (L_1R_2 + R_1L_2)a + R_1R_2 = 0 \quad (27)$$

Solving this by the quadratic formula, the normal modes will be

$$a_1;\ a_2 = -\frac{R_1L_2 + R_2L_1}{2(L_1L_2 - L_{12}^2)} \pm \frac{1}{2}\sqrt{\frac{(R_1L_2 + R_2L_1)^2}{(L_1L_2 - L_{12}^2)^2} - \frac{4R_1R_2}{L_1L_2 - L_{12}^2}} \quad (28)$$

The transient current in the primary will therefore be

$$(i_1)_t = I_{11}\epsilon^{a_1t} + I_{12}\epsilon^{a_2t} \quad (29)$$

where I_{11} and I_{12} are the amplitudes of the modes a_1 and a_2, respectively. The transient currents in the secondary will be of the same form as in the primary and will differ only in the magnitude of each mode

$$(i_2)_t = I_{21}\epsilon^{a_1t} + I_{22}\epsilon^{a_2t} \quad (30)$$

where I_{21} and I_{22} are the amplitudes of the modes a_1 and a_2, respectively.

When $t = 0$, $i_1 = 0$; therefore

$$i_1 = (i_1)_s + (i_1)_t = \frac{E}{R_1} + I_{11}\epsilon^{a_1t} + I_{12}\epsilon^{a_2t} = 0$$

or

$$I_{11} + I_{12} = -\frac{E}{R_1} \quad (31)$$

Also at $t = 0$, $i_2 = 0$; therefore

$$i_2 = (i_2)_s + (i_2)_t = 0 + I_{21}\epsilon^{a_1 t} + I_{22}\epsilon^{a_2 t}$$

or

$$I_{21} + I_{22} = 0 \tag{32}$$

I_{21} and I_{22} can be expressed in terms of I_{11} and I_{12}; thus the voltage induced in the secondary due to mode a_1 will be, by equation 22,

$$e_{21} = -L_{12}\frac{d}{dt}(I_{11}\epsilon^{a_1 t}) = R_2 I_{21}\epsilon^{a_1 t} + L_2\frac{d}{dt}(I_{21}\epsilon^{a_1 t})$$

or

$$-L_{12}a_1 I_{11}\epsilon^{a_1 t} = (R_2 + L_2 a_1)I_{21}\epsilon^{a_1 t}$$

which gives

$$I_{21} = -\frac{L_{12}a_1 I_{11}}{R_2 + L_2 a_1} \tag{33}$$

Likewise

$$I_{22} = -\frac{L_{12}a_2 I_{12}}{R_2 + L_2 a_2} \tag{34}$$

Substituting equations 33 and 34 into 32, we have

$$\frac{a_1 I_{11}}{R_2 + L_2 a_1} + \frac{a_2 I_{12}}{R_2 + L_2 a_2} = 0 \tag{35}$$

which on substituting into 31 gives

$$I_{11} - \frac{a_1}{a_2} \times \frac{R_2 + L_2 a_2}{R_2 + L_2 a_1} I_{11} = -\frac{E}{R_1}$$

solving for I_{11}, we have

$$I_{11} = \frac{E}{R_1 \left[\dfrac{a_1}{a_2} \times \dfrac{R_2 + L_2 a_2}{R_2 + L_2 a_1} - 1\right]} \tag{36}$$

In a similar manner

$$I_{12} = \frac{E}{R_1 \left[\dfrac{a_2}{a_1} \times \dfrac{R_2 + L_2 a_1}{R_2 + L_2 a_2} - 1\right]} \tag{37}$$

The primary current will be

$$i_1 = \frac{E}{R_1} + I_{11}\epsilon^{a_1 t} + I_{12}\epsilon^{a_2 t} \tag{38}$$

and the secondary current

$$i_2 = I_{21}\epsilon^{a_1 t} + I_{22}\epsilon^{a_2 t} \tag{39}$$

where the coefficients I_{11}, I_{12}, I_{21}, and I_{22} are as defined above. The effective magnetomotive force acting on the magnet will be given by the sum of the primary and secondary magnetomotive force defined in equation 20.

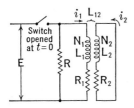

Fig. 11. The electrical equivalent circuit of a magnet with a lag coil and resistance shunt on the primary.

98. Magnet with Lag Coil—Build-Down of Effective Magnetomotive Force

The build-down of effective magnetomotive force with a lag coil is the same problem as that of Art. 97, except that the switch is opened at $t = 0$ instead of being closed. In order to decrease the sparking when the switch is opened, and also to increase the time delay, the primary is shunted by a resistance R as illustrated in Fig. 11. If this resistance is made equal to R_1, there will be no rise in voltage across the primary at the instant of opening the switch.

The voltage equations for the circuit of Fig. 11 are

$$\text{Primary:} \quad 0 = i_1(R + R_1) + L_1 \frac{di_1}{dt} + L_{12} \frac{di_2}{dt} \tag{40}$$

$$\text{Secondary:} \quad 0 = i_2 R_2 + L_2 \frac{di_2}{dt} + L_{12} \frac{di_1}{dt} \tag{41}$$

Solving these equations in the same manner as 21 and 22 of Art. 97, we obtain for the modes

$$a_1; a_2 = -\frac{R_2 L_1 + (R + R_1) L_2}{2(L_1 L_2 - L_{12}^2)}$$

$$\pm \left\{ \frac{(R_2 L_1 + (R + R_1) L_2)^2}{2(L_1 L_2 - L_{12}^2)} - \frac{(R + R_1) R_2}{L_1 L_2 - L_{12}^2} \right\}^{\frac{1}{2}} \tag{42}$$

and for the amplitudes of the modes,

$$I_{11} = -\frac{E}{R_1 \left[\dfrac{a_1}{a_2} \times \dfrac{R_2 + L_2 a_2}{R_2 + L_2 a_1} - 1 \right]} \tag{43} \qquad I_{21} = -\frac{a_1 L_{12} I_{11}}{R_2 + L_2 a_1} \tag{44}$$

$$I_{12} = -\frac{E}{R_1 \left[\dfrac{a_2}{a_1} \times \dfrac{R_2 + L_2 a_1}{R_2 + L_2 a_2} - 1 \right]} \tag{45} \qquad I_{22} = -\frac{a_2 L_{12} I_{12}}{R_2 + L_2 a_2} \tag{46}$$

which are determined from the following initial conditions: at $t = 0$

$$i_1 = I_{11} + I_{12} = \frac{E}{R_1} \tag{47}$$

$$i_2 = I_{21} + I_{22} = 0 \tag{48}$$

$$\frac{di_1}{dt} = a_1 I_{11} + a_2 I_{12} = -\frac{E(R + R_1)}{R_1 \left(L_1 - \frac{L_{12}^2}{L_2}\right)} \tag{49}$$

The primary current will be

$$i_1 = I_{11} \epsilon^{a_1 t} + I_{12} \epsilon^{a_2 t} \tag{50}$$

and the secondary current

$$i_2 = I_{21} \epsilon^{a_1 t} + I_{22} \epsilon^{a_2 t} \tag{51}$$

The effective magnetomotive force acting on the magnet will be

$$F_{\text{eff.}} = N_1 i_1 + N_2 i_2 \tag{52}$$

In order to test the time delay of this circuit experimentally, the magnet of Fig. 2, Chapter IX, was set up as in Art. 96, to support the weight of its armature (9 oz.), with an air-gap length of $\frac{9}{64}$ in. The magnet was fitted with solid copper lag coils of $\frac{1}{4}$-in. radial thickness and 3-in. axial length, placed directly around each pole core. The turns and outside diameter of the exciting coil was kept the same as indicated in Fig. 2, Chapter IX. Owing to the greater mean length of turn, and the necessary decrease in wire size, the resistance of the primary winding increased from 55 to 126 ohms. The maximum excitation was held substantially the same as that of the tests shown in Fig. 9. For the purpose of computing the primary and secondary currents the primary inductance was measured and was found to remain substantially constant for the gap length used. The resistance of the lag coils was computed from their dimensions, and their inductance was computed by assuming that the permeance of their magnetic path was the same as that of the primary. The coefficient of coupling between the primary and secondary was taken as 0.95. Test results show that for the manner in which the coils are arranged this coefficient is close to unity and slightly higher than 0.95. Test results for the primary current were obtained by means of an oscillograph. Because a solid lag coil was used it was impossible to record the current of the secondary.

In Fig. 12, computed results for the build-up of the effective magnetomotive force for the magnet described above are given. The circuit and its constants are shown directly on the figure.

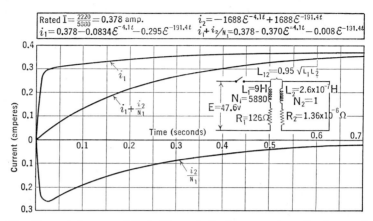

Fig. 12. Computed curves of the build-up of effective magnetomotive force and current in a magnet with lag coils. Horseshoe magnet of Fig. 2, Chapter IX, with a $\frac{9}{64}$-in. air gap between armature and pole faces.

Figure 13 shows the computed results for the build-down of the effective magnetomotive force for the same magnet with a resistance of

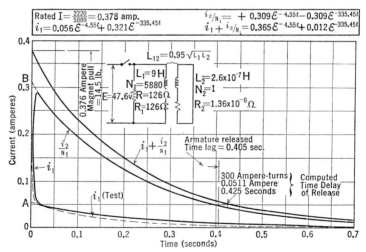

Fig. 13. Experimental and computed curves of the build-down of effective magnetomotive force in a magnet with lag coils and a resistance shunt on the primary. Horseshoe magnet of Fig. 2, Chapter IX, supporting armature weighing 9 oz. through a $\frac{9}{64}$-in. air gap. Actual armature release occurs at 300 ampere-turns or 0.0511 ampere.

126 ohms shunted across the primary. The primary current as recorded on an oscillograph and the time delay as measured on the oscillograph

ART. 99]   MAGNET WITH UNITY COUPLED LAG COIL   351

are shown on the figure. The computed time delay was determined by finding, experimentally, the ampere-turns at which the armature was released. This number of effective ampere-turns determined the release point on the effective magnetomotive force curve. By referring to the curve it will be noticed that the computed and test values of primary current and release time compare quite well. It will also be observed that the contribution of the primary to the time delay, obtained by shunting the resistance R across the primary, though appreciable, is small compared to that of the lag coil. This is because the resistance of the primary, including its added resistance R, is relatively large compared to the equivalent primary resistance of the solid lag coils.

If the resistance R is omitted the primary current would cease at the instant of opening the switch, provided that an ideal break could be assumed. In this event a secondary current equal to $i_1 \times N_1/N_2$ would immediately flow in order that the circuit magnetomotive force, and hence flux, could remain constant at the instant of opening the switch. Actually this could not be entirely realized in practice unless the coupling coefficient were unity or very close to unity. Assuming the coupling coefficient as unity, the time delay produced on opening the switch would be that produced by the constants of the secondary circuit only. The solution of this circuit will then be identical to that of Fig. 7, Art. 96, replacing $(R + R_1)$ by R_2, and L_1 by L_2. The initial secondary current at the instant of opening the switch may be taken as $i_1 \times N_1/N_2$.

99. Magnet with Unity Coupled Lag Coil—Build-Down and Build-Up of Effective Magnetomotive Force

Although unity coupling between the primary and secondary of a transformer, or between the exciting winding and the lag coil of a magnet, is an ideal condition which cannot be realized in practice, it is possible to attain coupling coefficients close to unity. The solutions for unity coupling are much simpler than those presented in Arts. 97 and 98, and may be used as a convenient approximation when the coupling is reasonably tight, say over 0.95.

Referring to Fig. 13, the initial steep portion of i_1 and i_2 are due to the mode a_2, which is large in value. As the coupling is tightened a_2 will increase, and the initial slopes of the currents become steeper until, when $k = 1$, the initial slope is infinity, and the current i_1 will instantaneously drop to a value approximately equal to A, and i_2 will instantaneously rise to a value approximately equal to B. A and B are determined by extending the gradually sloped portions of the currents i_1 and i_2 to the axis. Thus, when $k = 1$, mode a_2 will become minus infinity and the currents of this mode will disappear, while the currents of mode a_1 remain. The value of mode a_1 does not change much as k approaches unity. Evaluation of the indeterminate form of equation 42 for the normal modes at $k = 1$ gives

$$a_1 = -\frac{(R + R_1)R_2}{R_2 L_1 + (R + R_1)L_2} \quad \text{and} \quad a_2 = -\text{infinity} \quad (53)$$

Thus, when the coefficient of coupling is tight, the two coils may be treated as a single winding having a time constant of $1/a_1$ seconds. For the circuit of Fig. 11, the primary and secondary currents will be

$$i_1 = I_{11} \epsilon^{a_1 t} \quad \text{and} \quad i_2 = I_{21} \epsilon^{a_1 t}, \text{ respectively} \quad (54)$$

The effective magnetomotive force of the magnet will be

$$F_e = N_1 \frac{E}{R_1} \epsilon^{a_1 t} \quad (55)$$

where the coefficient $N_1(E/R_1)$ can be written by inspection on the basis that, at the instant of opening the switch, the effective magnetomotive force of the combined primary and secondary currents must remain constant at the value before the switch was opened.

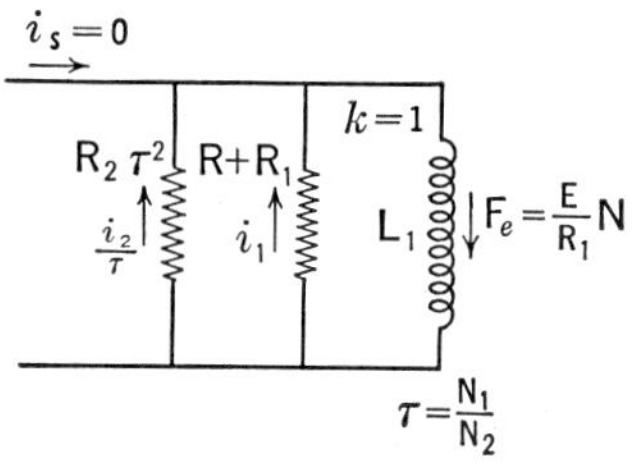

The coefficients I_{11} and I_{21} of equation 54 can be evaluated on the basis of the following logic: Because there is no leakage flux between the primary and the secondary, the induced voltage of both coils will be proportional to their turns, and the transient currents proportional to these voltages and inversely proportional to their respective resistances. Thus the secondary may be considered a resistance of $R_2\tau^2$ ohms, shunted across the primary, where τ is the ratio of the primary to the secondary turns N_1/N_2. Figure 14 shows the equivalent circuit of Fig. 11 when $k = 1$, at $t = 0$. On the basis of this circuit, I_{11} and I_{21} will be

FIG. 14. Equivalent circuit of Fig. 11, for $k = 1$ at the time $t = 0$.

$$I_{11} = \left(\frac{E}{R_1}\right)\left[\frac{R_2\tau^2}{R_2\tau^2 + (R + R_1)}\right] \quad (56)$$

$$I_{21} = \left(\tau \frac{E}{R_1}\right)\left[\frac{R + R_1}{R_2\tau^2 + (R + R_1)}\right] \quad (57)$$

In a similar manner, for the circuit of Fig. 10b, the mode a_1 when $k = 1$ will be

$$a_1 = -\frac{R_1 R_2}{R_2 L_1 + R_1 L_2} \tag{58}$$

and the primary and secondary currents will be

$$i_1 = \frac{E}{R_1} + I_{11} \epsilon^{a_1 t}, \text{ and } i_2 = I_{21} \epsilon^{a_1 t} \tag{59}$$

The effective magnetomotive force of the magnet will be

$$F_e = N_1 \frac{E}{R_1} (1 - \epsilon^{a_1 t}) \tag{60}$$

which can be written by inspection on the basis that the steady-state magnetomotive force is $(E/R_1)N$, and that its initial value is zero.

The equivalent circuit for Fig. 10b at the instant of closing the switch is shown in Fig. 15, and the coefficients I_{11} and I_{21} will be

$$I_{11} = -\frac{E}{R_1} \left[\frac{R_2 \tau^2}{R_2 \tau^2 + R_1} \right] \tag{61}$$

$$I_{21} = -\left(\tau \frac{E}{R_1} \right) \left[\frac{R_1}{R_2 \tau^2 + R_1} \right] \tag{62}$$

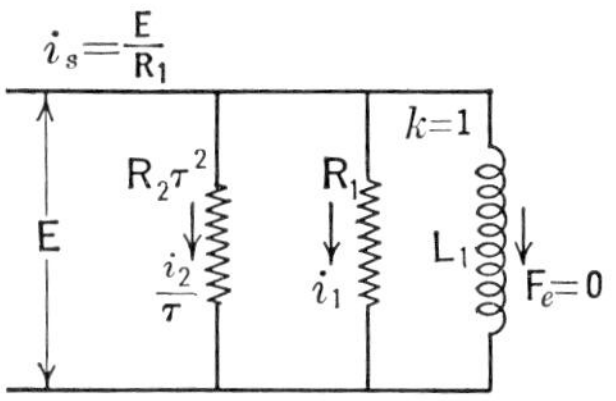

Fig. 15. Equivalent circuit of Fig. 10b, for $k = 1$ at the time $t = 0$.

100. Magnet with Shunt Condenser—Build-Down of Effective Magnetomotive Force

This circuit arrangement is illustrated in Fig. 16, where the magnet having a self-inductance and resistance of L_1 and R_1, respectively, is shunted by a series group consisting of a condenser C and a resistance R. This method of time delay can be applied only to delay the release of a magnet after interruption of the exciting circuit. It can easily be applied to a magnet which has been built, and it can be made as effective as a lag coil when the supply voltage E is sufficiently high. Its effectiveness depends upon the ability of the condenser to store energy, which energy varies as the square of the voltage and directly with the capacity. When the voltage is very low electrolytic condensers can be profitably employed. The resistance added in series with the condenser serves two purposes: (1) to prevent a bad spark at the contacts when the switch is closed, and (2) to prevent the system from becoming oscillatory when the switch is opened. A non-oscillatory circuit will give a longer time delay than the oscillatory one if the release current of the magnet is a small fraction of its normal current.

The voltage equation for the circuit of Fig. 16, after the switch has been opened, is

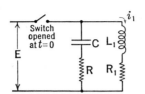

FIG. 16. Equivalent electrical circuit of a magnet with a shunt, consisting of a condenser and resistance in series, across the exciting winding.

$$i_1 R_1 + L_1 \frac{di_1}{dt} + i_1 R + \frac{1}{C} \int_0^t i_1 dt = 0 \quad (63)$$

where the last term represents the voltage across the condenser. Differentiating equation 63, we have

$$L_1 \frac{d^2 i_1}{dt^2} + (R_1 + R) \frac{di_1}{dt} + \frac{i_1}{C} = 0 \quad (64)$$

which is a second-order linear differential equation. Solving this equation in the manner indicated in Arts. 96 and 97, we will obtain for the normal modes

$$a_1, a_2 \quad \text{or} \quad \alpha + j\beta = -\frac{R + R_1}{2L_1} \pm \sqrt{\left(\frac{R + R_1}{2L_1}\right)^2 - \frac{1}{L_1 C}} \quad (65)$$

If the radical is real, the circuit will be non-oscillatory and the current will be made up of two damped exponentials. The modes will then be designated by a_1 and a_2 as in the previous cases. As the steady-state current of this circuit is zero, the final solution will be of the form

$$i_1 = I_1 \epsilon^{a_1 t} + I_2 \epsilon^{a_2 t} \quad (66)$$

The constants I_1 and I_2 will be determined from the following boundary conditions:

$$i_1 = I_1 + I_2 = \frac{E}{R_1} \quad \text{at } t = 0 \quad (67)$$

$$\frac{di_1}{dt} = a_1 I_1 + a_2 I_2 = -\frac{ER}{LR_1} \quad \text{at } t = 0 \quad (68)$$

$$\int_0^\infty i_1 dt = -\frac{I_1}{a_1} - \frac{I_2}{a_2} = CE \quad \text{at } t = \infty \quad (69)$$

Their numerical values can be found by solving any two of the above equations simultaneously after substituting for a_1 and a_2.

If the radical of equation 65 is imaginary, the circuit will be oscillatory and the current will be a damped sinusoidal wave. The modes will then be conjugate complex quantities and will be designated by $\alpha \pm j\beta$

$$i_1 = I\epsilon^{\alpha t} \cos(\beta t + \theta) \tag{70}$$

The constants I and θ will be determined from the same boundary conditions as before, rewritten in a more suitable form. Thus equations 67 and 68 will now be

$$i_1 = I \cos \theta = \frac{E}{R_1} \text{ at } t = 0 \tag{71}$$

$$\frac{di_1}{dt} = +I(\alpha \cos \theta - \beta \sin \theta) = -\frac{ER}{L_1 R_1} \text{ at } t = 0 \tag{72}$$

In order to test the time delay of this circuit experimentally, the magnet of Fig. 2, Chapter IX, was again used. To decrease the amount of capacity necessary to make the time delay effective, the coils were

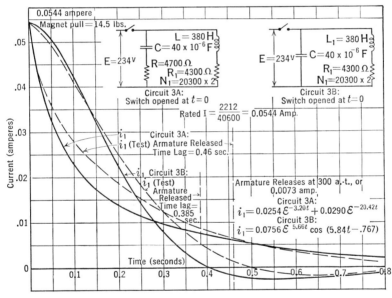

Fig. 17. Computed and experimental curves for the build-down of current in a magnet with a shunt condenser. Horseshoe magnet of Fig. 2, Chapter IX, supporting armature weighing 9 oz. through an air-gap length of $\frac{9}{64}$ in. Armature released at 300 a-t., or 0.0073 amp.; switch opened at $t = 0$.

rewound with 20,300 turns each, for operation at 248 volts. The overall dimensions of the coils, and their maximum ampere-turns, were retained the same as in the previous examples. Computations were made and data taken for two cases: (a) with sufficient resistance R in series with the condenser to prevent oscillation, and (b) with no resistance in series with the condenser. The circuit data and results are shown in Fig. 17 for both circuits. Although the experimental and computed results check well over a large portion of the characteristic, there are ranges where there is a considerable deviation. This error is most marked where the current is changing rapidly and is probably due to eddy currents in the solid iron magnetic circuit of the magnet. This would also account for the measured time delay being longer than the computed value. In the previous case with the lag coil, the effect of eddy currents would not be particularly noticeable, as they merely tend to supplement the action of the lag coils. It is interesting to note that the introduction of resistance, which consumes energy, actually makes the time delay longer. Whether or not this will occur depends on the portion of the curve at which release takes place. In a magnet where the excitation must be kept high to prevent release, the oscillatory condition is far superior as it causes the current to drop less rapidly in the beginning.

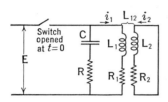

Fig. 18a. Equivalent electrical circuit of a magnet with lag coils and a shunt, consisting of a condenser and resistance in series, across the exciting winding.

101. Magnet with Shunt Condenser and Lag Coils—Build-Down of Effective Magnetomotive Force

This circuit arrangement, illustrated in Fig. 18a, represents a combination of the circuits of Arts. 98 and 100. The lag coils prevent the current from dropping so rapidly, with the result that the time delay can be made longer than with the circuits of either Art. 98 or 99.

The voltage equations for the circuit of Fig. 18a are

$$\text{Primary: } i_1(R + R_1) + L_1 \frac{di_1}{dt} + L_{12} \frac{di_2}{dt} + \frac{1}{C} \int_0^t i_1 dt = 0 \quad (73)$$

$$\text{Secondary: } i_2 R_2 + L_2 \frac{di_2}{dt} + L_{12} \frac{di_1}{dt} = 0 \quad (74)$$

Solving these equations in the manner of Art. 97, the determinantal equation will be

ART. 101] MAGNET WITH SHUNT CONDENSER AND LAG COILS 357

$$(L_1L_2 - L_{12}^2)a^3 + [(R + R_1)L_2 + R_2L_1]a^2$$
$$+ \left[(R + R_1)R_2 + \frac{L_2}{C}\right]a + \frac{R_2}{C} = 0 \quad (75)$$

This equation is most easily solved for the three modes by making numerical substitution into it, and solving the resulting cubic by some standard method. There are two possibilities: (a) three real negative roots, giving rise to three damped exponential terms; or (b) a real negative root and one pair of conjugate complex roots, giving rise to one damped exponential and one damped sinusoid.

As the steady-state current will be zero, the resulting current equations will have the following forms:

Non-oscillatory: $i_1 = I_{11}\epsilon^{a_1t} + I_{12}\epsilon^{a_2t} + I_{13}\epsilon^{a_3t}$ (76)

$\qquad\qquad i_2 = I_{21}\epsilon^{a_1t} + I_{22}\epsilon^{a_2t} + I_{23}\epsilon^{a_3t}$ (77)

Oscillatory: $i_1 = I_{11}\epsilon^{a_1t} + I_1\epsilon^{\alpha t}\cos(\beta t + \theta_1)$ (78)

$\qquad\qquad i_2 = I_{21}\epsilon^{a_1t} + I_2\epsilon^{\alpha t}\cos(\beta t + \theta_2)$ (79)

The coefficients can be evaluated from the following boundary conditions:

$$i_1 = I_{11} + I_{12} + I_{13} = \frac{E}{R_1} \quad \text{at } t = 0 \quad (80)$$

$$\frac{di_1}{dt} = a_1I_{11} + a_2I_{12} + a_3I_{13} = -\frac{ER}{R_1\left(L_1 - \frac{L_{12}^2}{L_2}\right)} \quad \text{at } t = 0 \quad (81)$$

$$\int_0^\infty i_1 dt = -\frac{I_{11}}{a_1} - \frac{I_{12}}{a_2} - \frac{I_{13}}{a_3} = CE \quad \text{at } t = \infty \quad (82)$$

$$i_2 = I_{21} + I_{22} + I_{23} = 0 \quad \text{at } t = 0 \quad (83)$$

This last equation may be rewritten in terms of the primary constants as was done in Art. 97, as follows:

$$i_2 = -\frac{a_1I_{11}}{R_2 + L_2a_1} - \frac{a_2I_{12}}{R_2 + L_2a_2} - \frac{a_3I_{13}}{R_2 + L_2a_3} = 0 \quad \text{at } t = 0 \quad (84)$$

If the circuit is oscillatory, the equations may be written in the other form, using 78 and 79.

$$i_1 = I_{11} + I_1 \cos \theta_1 = \frac{E}{R_1} \quad \text{at } t = 0 \tag{85}$$

$$\frac{di_1}{dt} = a_1 I_{11} + I_1(\alpha \cos \theta_1 - \beta \sin \theta_1) = -\frac{ER}{R_1 \left(L_1 - \frac{L_{12}^2}{L_2}\right)} \quad \text{at } t = 0 \tag{86}$$

$$i_2 = I_{21} + I_2 \cos \theta_2 = 0 \quad \text{at } t = 0, \text{ etc.} \tag{87}$$

The effective magnetomotive force will be given by the algebraic sum of the magnetomotive forces of the primary and secondary,

$$F_e = N_1 i_1 + N_2 i_2 \tag{88}$$

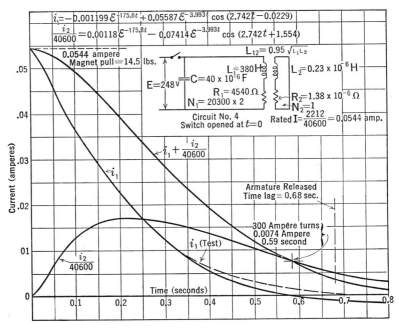

FIG. 18. Computed and experimental curves for the build-down of current in a magnet with $\frac{1}{4}$-in. lag coils and a shunt condenser. Horseshoe magnet of Fig. 2, Chapter IX, supporting armature weighing 9 oz. through an air-gap length of $\frac{9}{64}$ in. Armature released at 300 a-t., or 0.0074 amp.; switch opened at $t = 0$.

In order to test the time delay of this circuit experimentally the lag coils of Art. 98 and the condenser and coils of Art. 100 were used in conjunction with the magnet of Fig. 2, Chapter IX. The circuit data and the results of both computation and experiment are shown in Fig.

Art. 101] MAGNET WITH SHUNT CONDENSER AND LAG COILS 359

18. Computations based on the circuit constants indicated show the circuit to be slightly oscillatory.

The experimental results check perfectly with the computation, except at the small values of current, where the difference is sufficient to make the actual circuit just non-oscillatory.

The method outlined for solving for the constants is particularly laborious when the circuit is oscillatory, owing to the simultaneous equations with complex numbers. In this case, operational methods afford an easier method of solution. When the coefficient of coupling is tight, the magnet and its lag coil can be replaced by a primary equivalent as was done in Art. 99, and then the solution for the effective magnetomotive force of the magnet can be carried out by the method of Art. 100.

PROBLEMS

1. The magnet of Fig. 9 (with R omitted) is arranged so as to lift a total load of 12 lb. through a stroke of $\frac{9}{64}$ in. Its force-current characteristic is given in Fig. 19. Assuming that the maximum (at $t = \infty$) magnetomotive force is to be held constant at its rated value of 2,220 ampere-turns, calculate the necessary time constant of a choke coil which when connected in series with the magnet will cause the load to be picked up 1 second after the circuit is closed. Assume that the supply voltage available is 120 volts.

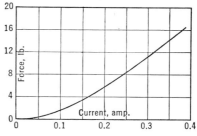

Fig. 19. Force-stroke characteristic of the magnet of Fig. 2, Chapter IX, at a constant air-gap length of $\frac{9}{64}$ in. $N = 5880$ turns.

2. It is proposed to speed up the operation of the magnet of Problem 1 by adding resistance in series with it, and raising the supply voltage so as to keep the maximum magnetomotive force constant. Compute the necessary series resistance and the supply voltage so that the load of 12 lb. will be lifted 0.05 second after the switch is closed.

3. Design a choke coil which will be suitable for Problem 1. *Note*: In a well designed choke coil, to be used as above, the flux density of the iron should be chosen so as to make the stored energy of the air gap a maximum at the current corresponding to operation of the magnet.

4. The statement has been made that the time delay that can be produced in a magnet depends on its time constant. By comparing the time constants, and the computed results for the time lag of the circuits of Figs. 9 and 13, determine the accuracy of this statement.

5. Compute the time delay for the circuit of Fig. 13 using the approximation of equation 55. What is the percentage error of your answer, assuming that the value as computed by the coupled-circuit theory and shown on Fig. 13 as 0.425 second is correct?

6. The magnet of Fig. 12 has the force-current characteristic shown in Fig. 19. Compute from the data of Fig. 12 the time delays that will be produced by loads of 4, 8, 12, and 15 lb.

7. Recompute Problem 6 with the following changes:

(a) The lag coils, which were $\frac{7}{8}$ in. in inside diameter, $\frac{1}{4}$ in. in radial thickness, and 3 in. long, have been changed to $\frac{7}{8}$ in. in inside diameter, $\frac{1}{8}$ in. in radial thickness, and 3 in. long.

(b) The mean diameter of the exciting coils has been reduced by $\frac{1}{4}$ in., from $1\frac{3}{4}$ to $1\frac{1}{2}$ in., the wire size and number of turns remaining constant.

(c) The voltage is changed to compensate for the decreased winding resistance so that the maximum magnetomotive force will be the same as in Problem 6.

Assume that the above changes will affect the resistance of the coils but not their inductance. Use the approximate method of solution as indicated in Art. 99.

8. Show that NBS/E seconds is an approximate expression for the time constant of a magnet with a single exciting winding, where S is the core cross section, B the maximum core flux density, N the winding turns, and E the supply voltage. Explain why this formula takes into account the coil space factor, and whether it over or underestimates the time constant.

9. Using the formula of Problem 8, estimate the time constant of the magnet of Fig. 9 (without the shunt resistance), using the dimensional data shown in Fig. 2, Chapter IX. Compare this estimate with the value computed from the winding resistance and inductance shown on Fig. 9.

10. Show that $\dfrac{N_1 B S R_1}{E} \times \dfrac{R_1 + \tau^2 R_2}{R_1 \times \tau^2 R_2}$ is an approximate expression for the effective time constant of a magnet, like that of Fig. 10b, with a lag coil, where N_1 and R_1 are the turns and resistance of the primary respectively, R_2 is the resistance of the secondary, τ is the ratio N_1/N_2, B the maximum core flux density, S the core cross section, and E the primary supply voltage.

11. Derive a formula, like that of Problem 10, for the approximate value of the time constant, effective on opening the circuit, for a magnet arranged like that of Fig. 13.

12. Using the formula of Problem 10, estimate the time constant of the magnet of Fig. 12, using the dimensional data of Fig. 2, Chapter IX. Compare the estimate with the value computed from equation 58, using the winding resistance and inductance shown in Fig. 12.

13. Compute the time delay of the magnet designed in Art. 91, Chapter X, if the armature is to lift its rated load of 10.6 lb. through a distance of 0.1 in. Assume all the data the same as tabulated for the final design, except that winding is to be assumed to be at a temperature of 20° C.

14. Repeat Problem 13, with the following changes in design:

(a) A solid copper lag coil $\frac{1}{8}$ in. thick and 2 in. long is slipped over each field pole.

(b) The turns and wire size of each exciting coil are retained constant, but the mean diameter is increased sufficiently so that the coil can be slipped over the lag coil.

(c) The poles are moved apart to permit the larger-diameter exciting coils to be used.

(d) The voltage is increased sufficiently to compensate for the increased coil resistance so that the maximum magnetomotive force is kept the same as in Problem 13.

Assume that the change of pole spacing does not affect the leakage coefficients or otherwise alter the magnetic circuit.

15. Compute the size of a condenser which when used in conjunction with the magnet designed in Art. 91, Chapter X, will allow the coil current to fall to zero 1 second after the switch is opened. Assume the circuit of Fig. 16, neglect the resistance of the exciting coil, and let $R = 0$.

16. Using the condenser as computed in Problem 15, and the magnet designed in Art. 91, Chapter X, connected as per circuit B of Fig. 17, compute the time delay which will be produced on opening the circuit when the magnet is supporting a load of 3.0 lb. through an air gap of 0.1 in. Assume the coils to be at 20° C., and use whatever data are necessary from Art. 91 to determine the winding inductance and required magnetomotive force.

17. Figure 20 shows the magnet of Fig. 2, Chapter IX, arranged to operate in conjunction with a radio code receiver so as to select between desired long-duration signals and short interfering signals. The magnet is controlled by a fast relay (one which responds rapidly to current variations) connected in the plate circuit.

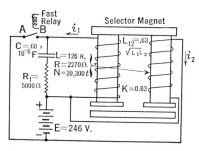

Fig. 20. Circuit of magnet which selects between desired long duration signals, and short duration interfering signals.

Part I. If the selector magnet will pick up its armature when the effective magnetomotive force is 2,000 ampere-turns, compute the duration of a radio signal which will just commence to move the selector armature.

Part II. If the relay circuit has been open a long time, compute how long it must remain closed to destroy the force-producing flux. Assume that this flux will be inappreciable when the magnetomotive force has dropped to 100 ampere-turns. This time will be the minimum time between radio signals which will prevent the signals from acting cumulatively.

Carry out your calculations in the following order:

1. Assuming the circuit non-oscillatory, sketch, with proper regard to their magnetizing effect, approximate curves of the currents in the two branches as a function of time.

2. Write the differential equations for the circuit, and solve them for the normal modes.

3. Write the equations from which the constants of integration are to be determined, and write a short explanation of each equation, explaining its physical significance.

4. Find the constants of integration.

5. Plot the curve of effective magnetomotive force and thence determine the answer to the question.

Answer. Considering the magnetomotive force produced by the steady-state value of i_2 as positive:

Part I:

$$i_1 = -0.0325\epsilon^{-2.45t} - 0.0133\epsilon^{-15.18t} - 0.0637\epsilon^{-108.4t}$$

$$i_2 = +0.1095 - 0.003\epsilon^{-2.45t} - 0.045\epsilon^{-15.18t} + 0.048\epsilon^{-108.4t}$$

$$F_e = N(i_1 + i_2) = N(0.1095 - 0.0355\epsilon^{-2.45t} - 0.0583\epsilon^{-15.18t} - 0.0157\epsilon^{-108.4t})$$

Part II:

$i_1 = -0.1095 + 0.0548\epsilon^{-11.05t} + 0.0548\epsilon^{-48.6t}$

$i_2 = +0.1095 + 0.0548\epsilon^{-11.05t} - 0.0548\epsilon^{-48.6t}$

$F_e = N(i_1 + i_2) = N(0.1095\epsilon^{-11.05t})$

These equations are shown plotted in Fig. 21.

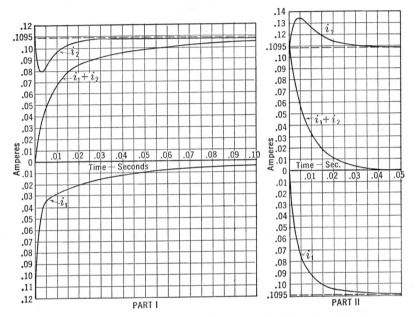

Fig. 21. Response curves for the selector circuit of Fig. 20, and answers to problem 17.

CHAPTER XII

HIGH-SPEED MAGNETS

102. General

High-speed or quick-acting magnets include, as a classification, all those applications where the time of action of the magnet must be accurately predetermined, or where the operation must take place in less than a specified maximum time interval. This would automatically include many relay applications, where quick action is often a prime requisite. However, in this chapter we are mainly interested in quick-acting tractive magnets. Some noteworthy applications in this field have been the electromagnetic camera shutter and electromagnetically operated control mechanisms for airplanes, such as bomb releases.

The design of a tractive magnet when the speed of action is stipulated becomes rather involved, first, because of the difficult mathematical interrelationships between the dynamic and electromagnetic laws controlling the operation, and second, merely because three independent variables are involved, namely, force, stroke, and time. The mathematical relationships involve linear differential equations with variable parameters. Some of the parameters are functions of two variables. Generally these equations can be solved only by a step-by-step method.

These difficulties, combined with the range of designs required because of the three possible variables, make it impossible to design such a magnet in a rational manner. The usual approach is by the cut-and-try method aided by the experience of the designer. However, the actual "try" of a proposed design may take several days to execute because of the tedious and involved graphical work necessary to pre-determine its operating characteristics. In order to facilitate such design work the author with the assistance of a student worker [1] has developed a scheme which is of material assistance in making a rational approach to the "try" design. This will be presented in Art. 112 of this chapter.

---

[1] See "A Rational Approach to the Design of Quick-Acting Electromagnets," a master's thesis by Joseph C. Boyle at the Stevens Institute of Technology in 1936.

From the point of view of execution, the converse problem is much simpler, and therefore we shall solve the problem of computing the time of action of a magnet whose complete design data are known, before attempting to design a magnet to specific time of action requirements.

103. Eddy Currents in Quick-Acting Magnets

One factor which causes complication, from both a practical and a theoretical point of view, is the possibility of eddy currents in the solid steel parts of the magnet. Practically, unless precautions are taken to avoid eddy currents, it is impossible to produce a high-speed magnet. Theoretically, if eddy currents are assumed to exist, the accurate predetermination of the stroke-time characteristic of the magnet is impossible. This is because the eddy-current density at any point in the iron is a complicated time function of the iron resistivity, the permeability, and the manner of variation of the exciting magnetomotive force.

Consequently, extreme precautions must be taken to laminate the iron parts thoroughly, and to pay particular attention to rivets passing through or around the iron circuit so that no paths for eddy currents are produced. Solid iron can often be used if it is slotted sufficiently. It is wise, however, to use a high-silicon bar stock if obtainable, as its high resistivity will greatly assist in the reduction of eddy currents and minimize the amount of slotting necessary.

Some idea of the effect of eddy currents, and what degree of slotting is necessary when using solid iron, may be obtained by considering the transient magnetization cycles to occur periodically in a continuous succession. This concept immediately establishes an equivalent sine wave of flux variation and establishes an equivalent frequency. Steinmetz [2] states that in a piece of solid iron the depth of penetration of an alternating flux is:

$$\text{Depth of penetration} = 2{,}240 \sqrt{\frac{\rho}{\mu f}} \text{ inches}$$

where ρ is the resistivity in ohm-inches, μ the relative permeability, and f the frequency in cycles per second. The depth of penetration is defined as the thickness of the surface layer of iron which at a uniform flux density, equal to that at the outside surface, will give the same total flux as exists in the entire piece of iron.

[2] See Sec. II, Chapter VI, of "Theory and Calculation of Transient Electric Phenomena and Oscillations," Steinmetz, third edition, McGraw-Hill Book Co.

The following table gives the depth of penetration for various materials at various frequencies, and at a permeability which is the average value of the normal permeability between flux densities of 1 to 100 kmax. per sq. in. for high-quality, carefully annealed, present-day commercial steels.

Material			Frequency, cycles per second			
			25	60	120	240
Soft iron		$\mu = 3{,}500, \rho = 4 \times 10^{-6}$	0.0152	0.0098	0.0069	0.0049
1.0% silicon steel		$\mu = 3{,}500, \rho = 9.5 \times 10^{-6}$	0.0234	0.0151	0.0106	0.0076
2.5%	" "	$\mu = 3{,}500, \rho = 16 \times 10^{-6}$	0.0304	0.0196	0.0138	0.0098
3.25%	" "	$\mu = 3{,}500, \rho = 19 \times 10^{-6}$	0.0331	0.0214	0.0150	0.0107
4.25%	" "	$\mu = 3{,}500, \rho = 24 \times 10^{-6}$	0.0372	0.0240	0.0169	0.0120

Thus, a soft iron rod $\frac{1}{2}$ in. in diameter, whose average permeability is 3,500, will carry just as much flux at 25 c.p.s. as a tube, $\frac{1}{2}$ in. in outside diameter and of 0.0152-in. wall thickness, which has a uniform flux density equal to that at the surface of the rod.

Actually, the permeability which is effective at any instant is the differential permeability obtained by drawing a tangent to the normal cyclic hysteresis loop at the instantaneous flux density. If a curve of this instantaneous effective permeability is plotted as a function of time a more representative value of the μ determining the depth of penetration may be found. At the higher cyclic flux densities the flux wave will, for a considerable portion of the cycle, stay in the flat region of the hysteresis loop where the permeability is low, and thus give a low average permeability resulting in a greater depth of penetration.[3] For this reason, the depths of penetration shown in the preceding tabulation may be considered low, because they are based on the average normal permeability.

In a high-speed magnet it is desirable that the thickness of the laminations should not exceed about twice the depth of penetration. However, where it is necessary for mechanical reasons to make the laminations thicker than this, a greater depth of penetration can be secured by decreasing the iron permeability by using very high iron saturations.

When regular laminations are used they should be held together by clamping plates if possible. In many instances this is not possible, and

---

[3] For a further discussion and application of this point see Problem 1 at the end of this chapter.

then rivets must be used. It is common practice to rivet a stack of thin laminations together between two heavier steel plates in order to maintain the shape of the stack better and to allow for finishing. In order not to nullify the benefit of using the laminations, particular attention must be paid to the placement of the rivets. This is best done by considering the magnet to be the core of an alternating-current transformer, and placing them so that as few lines of flux link with the metallic circuit formed by any two rivets as is possible. For instance, referring to the magnet illustrated in Fig. 5a, the plunger rivets can be placed only along the center line of the plunger, and those of the stationary section should be placed as near the outer edge of the lamination as is mechanically feasible.

Throughout this chapter it will be assumed that all the magnets considered have been laminated, and that the effect of eddy currents is negligible.

104. Flux-Time Characteristics—No Armature Motion

The simplest case to consider is the build-up of flux, or the related quantities of current and force, of an iron-core magnet with the armature stationary. The limitation removing armature motion simplifies the problem greatly, because it eliminates the space variable and its attendant dynamic relationships. The practical application of this case occurs when it is desired to calculate the following:

1. The time required for a lifting magnet or a brake magnet to develop a particular static force.
2. The time required for the current to rise, or the stored energy to build up to a specified amount in a choke coil.
3. The time required for the current of a high-speed tractive magnet to build up to the point where armature motion is incipient.

In all these cases the current is related to the constant impressed voltage E by the following equation:

$$E = N\frac{d\phi}{dt} + IR \tag{1}$$

where the relationship between the flux ϕ and the current I is non-linear and must be expressed graphically.

Let Fig. 1a represent an iron-core choke coil, and Fig. 1b the magnetization curve for the iron parts only. Strictly speaking, the induced voltage of equation 1 should be written as $d(N\phi)/dt$, and the magnetization curve should have $N\phi$ plotted against I. However, it is more convenient for our purpose to factor out N. This is accurate only if all the lines of flux ϕ link with all the turns N. Consequently, when using the form of equation 1 and Fig. 1b, care must be taken to see that the ϕ

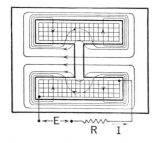

Fig. 1a. Iron-core choke.

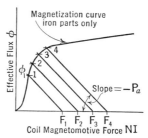

Fig. 1b. Graphical construction necessary to obtain the required coil magnetomotive force for any value of flux for the magnetic circuit of Fig. 1a.

used is an effective value which may be considered to link with all the turns. See Art. 46, Chapter IV.

The two variables, ϕ and I, of equation 1, are functions of the time t, which is the independent variable. This equation, because of the non-linear relationship between ϕ and I which cannot be expressed analytically, must be solved by the step-by-step method. It can be solved in terms of either ϕ or I, but because it is more direct we shall develop the solution in terms of ϕ first. Let the curve of Fig. 2 represent the form of solution, and the ordinates 0, 1, 2, 3, etc., the boundaries of the various intervals or steps of the independent variable t. The method of solution is to determine the value of the dependent variable at the end of the step in terms of its value, and that of its derivatives, at the beginning of the step. Having obtained its value at the end of the step, its derivatives may be determined at that point, and progress may then be made from that point to the end of a new step.

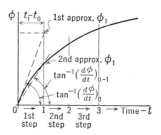

Fig. 2. Curve illustrating the step-by-step solution for the flux-time characteristic of the choke of Fig. 1a.

In order to start the process and again to know when to stop it, the boundary conditions of the equation must be known. These boundary conditions are the physical limitations which the solution must satisfy and are determined by the governing laws and any initial conditions which may be arbitrarily imposed. In our illustrative problem the boundary conditions, at $t = 0$, which determine the start are: $\phi = 0$ and $I = 0$. The boundary conditions at $t = \infty$ are: $d\phi/dt = 0$, and $I = E/R$.

Commencing with the initial boundary conditions we can solve equation 1 at $t = 0$ for

$$\left(\frac{d\phi}{dt}\right)_0 = \frac{E}{N} \qquad (2)$$

where the subscript 0 denotes the value at $t = 0$. The value of ϕ at the end of the first step will be

$$\phi_1 = \phi_0 + \left(\frac{d\phi}{dt}\right)_0 (t_1 - t_0) \qquad (3)$$

The construction of this equation is shown graphically in Fig. 2, and, as can be seen, the value of ϕ_1 so obtained is too large.

Referring this value of ϕ_1 to Fig. 1b, it is seen that the coil magnetomotive force F_1 is required to produce the flux ϕ_1. Point F_1 is obtained by drawing from the value ϕ_1 on the magnetization curve the line $1 - F_1$, having a negative slope of P_a, where P_a is the permeance of the air gap through which the flux ϕ_1 passes. The current at 1 will therefore be $I_1 = F_1/N$. This value of I_1 will be high because ϕ_1 is high.

Equation 1 may now be solved for $(d\phi/dt)_1$, giving

$$\left(\frac{d\phi}{dt}\right)_1 = \frac{E - I_1 R}{N} \qquad (4)$$

which will be slightly low. The average rate of change of flux during the first interval will be

$$\left(\frac{d\phi}{dt}\right)_{0-1} = \frac{1}{2}\left[\left(\frac{d\phi}{dt}\right)_0 + \left(\frac{d\phi}{dt}\right)_1\right] \qquad (4a)$$

Using this, a second approximation to ϕ_1 can be made, as:

$$\phi_1 = \phi_0 + \left(\frac{d\phi}{dt}\right)_{0-1} (t_1 - t_0) \qquad (5)$$

Using this value of ϕ_1, a new value of I_1 may be obtained in the same manner as before. For most purposes the second approximation is sufficient, and the first step may be considered finished.

CURRENT-TIME CHARACTERISTIC

The values obtained from 5 may now be used as the initial ones for the second step, and the process repeated until finally the flux, or current, reaches the desired value. In general, when working graphically, higher accuracy may be obtained if the first approximation for the value of ϕ at the end of an interval is obtained by extrapolating the curve of ϕ already obtained, instead of using the method of equation 3. This virtually gives the effect of one additional approximation without any additional labor. The actual computations should be carried out in tabular form as illustrated in Table I. The final results can be plotted as either flux

TABLE I

STEP-BY-STEP TABULATION FOR THE COMPUTATION OF THE FLUX-TIME CHARACTERISTIC OF A MAGNET—NO ARMATURE MOTION

Point or Interval	Degree of Approx.	Time or Interval	Flux ϕ	Coil M.M.F. NI	Current I	$\dfrac{d\phi}{dt}$
0	Correct	t_0	0	0	0	Eq. 2
0–1	First	$t_1 - t_0$	......	....	....	Eq. 4a
1	First	t_1	Eq. 3 → from curve	→ I_1	→ Eq. 4	
1	Second	t_1	Eq. 5 → from curve	→ I_1	→ $\left(\dfrac{d\phi}{dt}\right)_1$	
2	First	t_2	Eq. 3 → from curve	→ I_2	→ Eq. 4	

or current against time. If desired to plot force as a function of time the effective force may be determined for each value of flux. A completely worked out numerical example of this method of solution will be found in Art. 108.

105. Current-Time Characteristic—No Armature Motion

If desired, the current-time characteristic can be worked out directly, using current for the dependent variable instead of flux as in the last article. To do this the rate of change of flux of equation 1

$$E = N \frac{d\phi}{dt} + IR \tag{1}$$

must be expressed in terms of the rate of change of current. Referring to Fig. 3, let curve a represent the magnetization curve for the iron parts of a magnet plotted in terms of the magnetomotive force required in the entire magnetic circuit (exclusive of the main gap) to establish various values of flux in the main gap of permeance P_a, where P_a includes fringing and the effective leakage permeance.

Then for any value of NI developed in the coil we may write

$$\phi = f(NI - F_a) = F_a P_a \qquad (6)$$

Differentiating this with respect to I, we have

$$\frac{d\phi}{dI} = f'(NI - F_a)\left(N - \frac{dF_a}{dI}\right) = P_a \frac{dF_a}{dI}$$

where $f'(NI - F_a)$ is the slope of the magnetization curve of Fig. 3.

Solving for dF_a/dI, we obtain

$$\frac{dF_a}{dI} = N \frac{f'(NI - F_a)}{P_a + f'(NI - F_a)}$$

and

$$\frac{d\phi}{dI} = \frac{f'(NI - F_a)}{P_a + f'(NI - F_a)}(NP_a)$$

as

$$\frac{d\phi}{dt} = \frac{d\phi}{dI} \times \frac{dI}{dt}$$

$$\frac{d\phi}{dt} = \frac{f'(NI - F_a)}{P_a + f'(NI - F_a)} NP_a \frac{dI}{dt} \qquad (7)$$

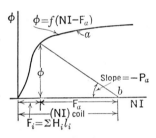

FIG. 3. Graphical construction necessary to express the rate of change of flux in a saturated iron circuit in terms of the rate of change of current.

Substituting this into equation 1, we have

$$E = (N^2 P_a) \frac{f'(NI - F_a)}{P_a + f'(NI - F_a)} \frac{dI}{dt} + IR \qquad (8)$$

It is interesting to note that the coefficient of dI/dt is the effective value of the inductance of the magnet winding. The evaluation of this equation is carried out in the same manner as for equation 1. When $t = 0$, I will be 0, and therefore NI also will be 0. By drawing a tangent to the magnetization curve of Fig. 3, at $F_i = 0$, the slope equal to $f'(NI - F_a)$

may be found. Substituting these values into equation 8, $(dI/dt)_0$ may be found. Then a first approximation for I will be

$$I_1 = I_0 + \left(\frac{dI}{dt}\right)_0 (t_1 - t_0) \tag{9}$$

Using the slope of the magnetization curve at $(F_i)_1$, corresponding to I_1, $(dI/dt)_1$ may be determined from equation 8. Averaging $(dI/dt)_0$ and $(dI/dt)_1$, the average value of that derivative, $(dI/dt)_{0-1}$, of the interval from t_0 to t_1 can be determined. Then a second approximation for I_1 will be

$$I_1 = I_0 + \left(\frac{dI}{dt}\right)_{0-1} (t_1 - t_0) \tag{10}$$

This process can now be continued in the same manner as was done in Art. 104. Table II shows a convenient way of arranging the computations.

TABLE II

Step-by-Step Tabulation for the Computation of the Current-Time Characteristic of a Magnet—No Armature Motion

Point or Interval	Degree of Approx.	Time or Interval	Current I	Coil M.M.F. NI	Slope of Mag. Curve	$\frac{dI}{dt}$
0	Correct	t_0	0	0	at $I = 0$	Eq. 8
0–1	......	$t_1 - t_0$	....	....	....	$\left(\frac{dI}{dt}\right)_{0-1}$
1	First	t_1	Eq. 9 →	NI_1 →	at I_1	Eq. 8
1	Second	t_1	Eq. 10 →	NI_1 →	at I_1 →	$\left(\frac{dI}{dt}\right)_1$
1–2	......	$t_2 - t_1$	....	....	....	$\left(\frac{dI}{dt}\right)_{1-2}$
2	First	t_2	Eq. 9 →	NI_2 →	at I_2	Eq. 8

Although the method of Art. 104 using ϕ as the dependent variable, and that above using I as the dependent variable, give the same result, there is one important difference: to find $d\phi/dt$ of Art. 104 it is only necessary to read I from the magnetization curve, whereas to obtain

dI/dt of this article it is necessary to draw a tangent to the magnetization curve. The latter process is less accurate than the former, and hence the method of Art. 104 is to be preferred.

106. Determination of the Time Required for the Flux of a Magnet to Rise from 0 to Some Value ϕ_a

For a constant impressed voltage E, the current and voltage will be related as given in equation 1:

$$E = N\frac{d\phi}{dt} + IR \tag{1}$$

Rearranging, we have

$$\left(\frac{E}{R} - I\right) = \frac{N}{R} \cdot \frac{d\phi}{dt}$$

If, at $t = 0$, both ϕ and I are zero, this equation may be solved for t in the form of the following definite integral:

$$t = \frac{N}{R} \int_0^{\phi_a} \frac{d\phi}{\left(\frac{E}{R} - I\right)} \tag{11}$$

This equation must be solved graphically by plotting ϕ as a function of $1/(E/R - I)$ and evaluating the area back of the curve. This curve may be plotted as follows: Let Fig. 4a represent the magnetization

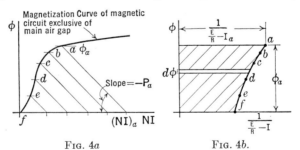

Fig. 4a.   Fig. 4b.

Graphical construction for directly determining the time required for the flux of a saturated iron circuit to rise from 0 to some value ϕa.

curve of all the iron parts and air gaps exclusive of the main air gap which has a permeance P_a. Then the value of NI_a required to produce ϕ_a may be determined by the graphical construction shown; I_a will then equal $(NI)_a/N$. This value of ϕ_a may then be plotted against $1/(E/R - I_a)$, and is shown as point a on Fig. 4b. Other points from the magnetization curve may be transferred in a similar manner, giving the curve a–b–c–d–e–f of Fig. 4b. The shaded area behind this curve is evidently the desired integral, which when multiplied by N/R will give the time for the flux to rise from zero to the value ϕ_a. An illustration of this method, and a comparison of its results with those of the method of Art. 104, will be given in the problem solved in Art. 108.

If desired the flux-time curve can be plotted, using this method, by evaluating the total area in steps and calculating the time corresponding to the end of each step.

107. Space-Time Characteristic of a Magnet

1. General Explanation of Problem. By space-time characteristic is meant the manner of variation of the space covered by the armature as a function of time. There are two portions to this characteristic: (a) the portion prior to motion where the current, flux, and magnetic force rise to the point where the restraining forces are just overcome and motion is incipient; and (b) the portion covering the actual motion of the armature, starting with the end of part (a), and ending when the armature completes its travel. The method of computing the flux-time or current-time characteristic for portion (a) has already been discussed in Arts. 104, 105, and 106.

Let Fig. 5a illustrate a magnetic system which is

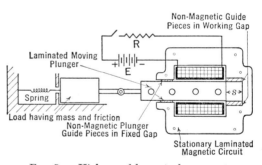

Fig. 5a. High-speed long-stroke magnet.

designed to move an external load having mass against the action of a constant load force, friction, and a spring. The particular magnet shown is a long-stroke one, having a square laminated plunger, and a return magnetic circuit formed of laminations which provide a stop for the plunger at the end of its stroke. The plunger is guided in the fixed gap and the working gap by means of non-magnetic, and preferably non-conducting, guide pieces. The entire assembly is very similar to that of a shell-type transformer.

2. Dynamic and Electric Equations of Motion. Figure 5b illustrates a free body of the dynamic system showing all the forces effective. Let

- x = displacement in inches of the plunger from its initial position.
- $\mathscr{F}$ = the magnetic pull on the plunger in pounds.
- t = the time in seconds from the start of motion.
- M = the combined mass of the plunger and all other moving parts in pounds per inch per second^2.
- k_1 = friction factor in pounds per inch per second.
- k_2 = spring constant in pounds per inch.
- $\mathscr{F}_0$ = the initial load, in pounds, to be overcome before motion can commence = f_0, the initial spring tension, plus f_1, the constant load force.
- ϕ = the total flux, in webers, of the magnet which may be considered as effectively linking with all the turns of the coil N (as discussed in the beginning of Art. 104).
- I = the coil current in amperes.
- R = the total resistance of the coil circuit in ohms.
- E = the constant impressed voltage in volts.

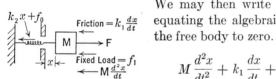

Fig. 5b. Free body of the dynamic system of Fig. 5a.

We may then write the following equation, equating the algebraic sum of the forces on the free body to zero.

$$M \frac{d^2x}{dt^2} + k_1 \frac{dx}{dt} + k_2 x = \mathscr{F} - \mathscr{F}_0 \qquad (12)$$

where d^2x/dt^2 is the acceleration of the plunger in inches per second^2, and dx/dt is the velocity of the plunger in inches per second. This equation is valid when $\mathscr{F} \geq \mathscr{F}_0$.

Equation 1 will still apply to the electric circuit.

$$E = N \frac{d\phi}{dt} + IR \qquad (1)$$

These equations have four variables, x, $\mathscr{F}$, ϕ, and I, which are functions of time. In order to solve them, two more relationships are necessary: one between current and flux, and another between the magnetic force and flux.

3. Coil Current for Any Plunger Position and Flux. The current, however, not only is dependent on flux, but also varies with the plunger position x, thus

$$I = f(\phi, x) \qquad (13)$$

If the length of the magnetic circuit through the iron parts and fixed gaps can be assumed to remain substantially constant with x, the magnetomotive force for these iron parts and gaps, F_i, will be a function of the effective flux only [4] and may be expressed by a magnetization curve exactly like that of Fig. 1a, or Fig. 4a, which is reproduced in Fig. 6. Then, for any value of flux ϕ_m at any position of the plunger x_m, the total coil magnetomotive force required can be found as

$$(NI)_m = F_i + \frac{\phi_m}{(P_w)_m} = F_i + F_w \qquad (14)$$

where $(P_w)_m$ is the permeance, at plunger position x_m, of the working gap and associated leakage paths through which the flux ϕ_m passes, and where F_w is the magnetomotive force across the working gap. Referring

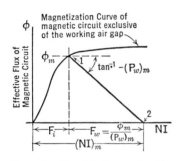

Fig. 6. Graphical construction necessary to obtain the required coil magnetomotive force for a flux ϕ_m, in the generalized case of a magnetic circuit consisting of a non-linear part in series with a working air gap of permeance P_w.

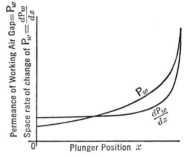

Fig. 7. Curves of working air-gap permeance and its space rate of change plotted as a function of the plunger position or air-gap length.

to Fig. 6, if a line is drawn through the magnetization curve at point 1, $\phi = \phi_m$, with a negative slope of $(P_w)_m$, the intersection of this line with the abscissa at point 2 will give the required coil ampere-turns. Dividing this by the coil turns the required current may be found. Inasmuch as P_w varies with x, a curve of P_w plotted against x will greatly assist the evaluation of coil ampere-turns from Fig. 6. Such a curve is shown in Fig. 7 along with its derivative dP_w/dx. The magnetization curve of

---

[4] In the leakage-flux magnet the effective length of the magnetic circuit changes greatly with motion, and this assumption is not valid. Likewise the effective number of turns linking with the flux depends on x, and the assumption that N can be factored out of equation 1 is fallacious. This problem must be handled by making $N\phi$ the variable instead of ϕ.

Fig. 6 is exactly similar to that of Fig. 16a, Chapter VIII, and is computed in the same manner, while the curves of Fig. 7 are similar to those of Fig. 16b except that x has been taken oppositely.

4. Magnet Force for Any Plunger Position and Flux. The relationship between the magnetic force and current is also complicated because it depends on x; thus

$$\mathscr{F} = f(\phi, x) \qquad (15)$$

This force is most easily evaluated graphically by the method of Art. 77, Chapter VIII, using the general magnetic force formula

$$\mathscr{F} = 4.43 \, F_w^2 \frac{dP_w}{dx} \quad \text{lb.} \qquad (16)$$

For any value of flux ϕ_m, at any plunger displacement x_m, the construction of Fig. 6 will give the value of F_w. The space rate of change of the permeance of the working gap, dP_w/dx, may be found by taking the slope of curve P_w of Fig. 7 at the required plunger position x_m. As mentioned before, it is convenient to plot a curve of dP_w/dx against x as shown in Fig. 7. This curve may be derived from that of P_w by graphical differentiation, or by the method suggested in Art. 77.

5. Step-by-Step Procedure. The four variables whose characteristics are to be plotted are: the effective magnet flux ϕ, the coil current I, the plunger velocity dx/dt, and the plunger displacement x. Let these characteristics be represented by the curves shown in Fig. 8, where t_0 is the zero of time when motion is just about to commence, and the other subscripts denote succeeding points in time as shown. Suppose that the characteristics have been evaluated to $t = t_m$, and that it is desired to carry the process forward to $t = t_n$. The procedure is as follows:

1. Project the curves ahead graphically from t_m to t_n, as shown by the dashed lines. This will give approximate values for the four variables at t_n.

2. Substitute the estimated value of I_n in equation 1, and solve for the rate of change of flux at n, thus

$$\left(\frac{d\phi}{dt}\right)_n = \frac{E - I_n R}{N} \qquad (17)$$

3. Average this value of $(d\phi/dt)_n$ with the final value at m, $(d\phi/dt)_m$, and designate this average value over the interval from m to n, by $(d\phi/dt)_{m-n}$.

4. The value of ϕ at n may then be obtained as

$$\phi_n = \phi_m + \left(\frac{d\phi}{dt}\right)_{m-n} (t_n - t_m) \qquad (18)$$

5. Using the estimated value of x_n from step 1 find the values of $(P_w)_n$ and $(dP_w/dx)_n$ by means of the data of Fig. 7.

6. Using ϕ_n from step 4, and $(P_w)_n$ from step 5, enter the magnetization curve of Fig. 6, and by means of the construction shown find $(NI)_n$ and $(F_w)_n$.

7. Using $(F_w)_n$ from step 6, and $(dP_w/dx)_n$ from step 5, calculate the magnetic force at n, as

$$\mathscr{F}_n = 4.43(F_w)_n^2 \left(\frac{dP_w}{dx}\right)_n \qquad (16a)$$

8. Using $\mathscr{F}_n$ from step 7, and $(dx/dt)_n$ and $(x)_n$ from step 1, substitute into equation 12 and solve for the plunger acceleration at n, as

$$\left(\frac{d^2x}{dt^2}\right)_n = \frac{1}{M}\left(\mathscr{F}_n - \mathscr{F}_0 - k_1\left(\frac{dx}{dt}\right)_n - k_2 x_n\right) \qquad (19)$$

9. Average this value of $(d^2x/dt^2)_n$ with that at m, $(d^2x/dt^2)_m$, and, designating this average value over the interval from m to n by $(d^2x/dt^2)_{m-n}$, find the plunger velocity at n, as

$$\left(\frac{dx}{dt}\right)_n = \left(\frac{dx}{dt}\right)_m + \left(\frac{d^2x}{dt^2}\right)_{m-n} (t_n - t_m) \qquad (20)$$

10. Average the value of $(dx/dt)_n$ with that at m, $(dx/dt)_m$, and, designating this average value over the interval from m to n by $(dx/dt)_{m-n}$, find the plunger displacement at n, as

$$x_n = x_m + \left(\frac{dx}{dt}\right)_{m-n} (t_n - t_m) \qquad (21)$$

11. This completes the computations for the step from m to n, but in order to carry out the next step, a corrected value of I_n should be computed using ϕ_n from step 4 and x_n from step 10, in the manner explained in connection with Sec. 3. By means of equation 17 of step 2, and the corrected value of I_n, a corrected value for $(d\phi/dt)_n$ may be obtained. Finally a corrected value of ϕ_n should be obtained by repeating steps 3 and 4 substituting the corrected values of $(d\phi/dt)_n$. If this value of ϕ_n differs materially from the approximate value obtained in step 4, the computations should be repeated. The final computed values of I_n

and ϕ_n, and the values of $(dx/dt)_n$ and x_n obtained from steps 9 and 10, respectively, should now be plotted in Fig. 8 and the process repeated.

6. Method of Starting Step-by-Step Procedure at $t = 0$. In the beginning, when t is small, the ordinary averaging described in Sec. 5 will not be very accurate. The reason for this is that these averages are based upon the premise of constant acceleration, while when t is small the acceleration is proportional to t. The following development will explain this point.

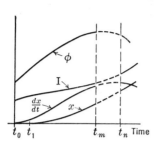

FIG. 8. Method of making the first approximation for the value of a function at a time t_n by graphically extrapolating the function from its last known value at t_m.

When t is small the force producing acceleration is

$$\text{Accelerating force} =$$

$$M \frac{d^2x}{dt^2} = 4.43\, F_w^2 \frac{dP_w}{dx} - \mathscr{F}_0 \qquad (22)$$

This is derived from equation 12 by noting that $k_1(dx/dt)$ and $k_2 x$ may be neglected when t is small. Expressing $\mathscr{F}_0$ in terms of the magnetic force equation, 22 may be written as

$$M \frac{d^2x}{dt^2} = 4.43 \left[F_w^2 - (F_w)_0^2 \right] \left(\frac{dP_w}{dx} \right)_0 \qquad (23)$$

where (dP_w/dx) is assumed to remain constant at the value for $x = 0$, because x during this interval will be very small. Now,

$$F_w^2 - (F_w)_0^2 = [F_w - (F_w)_0][F_w + (F_w)_0] \approx [F_w - (F_w)_0]2(F_w)_0$$

when $[F_w - (F_w)_0]$ is small. Substituting this into 23, we have

$$M \frac{d^2x}{dt^2} = 8.86 (F_w)_0 [F_w - (F_w)_0] \left(\frac{dP_w}{dx} \right)_0 \qquad (24)$$

The increment in air-gap magnetomotive force may be expressed in terms of the rate of change of flux at $t = 0$, as follows:

$$F_w - (F_w)_0 = \frac{\phi - \phi_0}{(P_w)_0} = \frac{1}{(P_w)_0} \frac{d\phi}{dt} t \qquad (25)$$

Substituting equation 25 into 24 and solving for the acceleration, we get

$$\frac{d^2x}{dt^2} = 8.86 \left[\frac{(F_w)_0}{M(P_w)_0} \left(\frac{dP_w}{dx} \right)_0 \left(\frac{d\phi}{dt} \right)_0 \right] t \qquad (26)$$

The velocity will be

$$\frac{dx}{dt} = \int_0^t \left(\frac{d^2x}{dt^2}\right) dt = \frac{8.86}{2} \left[\frac{(F_w)_0}{M(P_w)_0} \left(\frac{dP_w}{dx}\right)_0 \left(\frac{d\phi}{dt}\right)_0\right] t^2 \qquad (27)$$

and the space covered will be

$$x = \int_0^t \left(\frac{dx}{dt}\right) dt = \frac{8.86}{3!} \left[\frac{(F_w)_0}{M(P_w)_0} \left(\frac{dP_w}{dx}\right)_0 \left(\frac{d\phi}{dt}\right)_0\right] t^3 \qquad (28)$$

In these equations $(F_w)_0$ is the air-gap magnetomotive force necessary to produce the initial force $\mathscr{F}_0$ at $x = 0$; it may be determined by solving equation 16. $(d\phi/dt)_0$ may be evaluated by equation 17, where I_0 may be found by the construction of Sec. 3, and ϕ_0 will equal $(F_w)_0 (P_w)_0$. These equations make it very easy to get started. The duration of the first interval, $t_1 - t_0 = t$, over which these equations should be applied must be such that $d\phi/dt$ is substantially constant and x is small. ϕ_1 may be taken as $\phi_0 + (d\phi/dt)_0 t_1$, and I_1 found corresponding to ϕ_1 and x_1. $(d\phi/dt)_1$ should now be found from I_1. If it is very different from $(d\phi/dt)_0$, the first interval has been taken too long.

7. Tabulation of Step-by-Step Solution of Dynamic Characteristics. This tabulation starts at the end of step 1, because it is assumed that the values of x_1, $(dx/dt)_1$, $(d^2x/dt^2)_1$, I_1, $(d\phi/dt)_1$, and ϕ_1 have been found by the method of Sec. 6. The complete tabulation is shown in Table III. For the purpose of conserving space the columns of the table have been split into two sections, which are arranged one above the other. When computing, these columns should be placed in numerical order as is done in Table IX.

108. Sample Calculation of the Dynamic Characteristics of a High-Speed Magnet

1. Statement of Problem. In order to illustrate the methods developed in the last four articles a complete calculation of the dynamic characteristics for a fully laminated high-speed magnet will be carried out. The actual magnet is similar to that illustrated in Fig. 5a. Complete magnetization data are given in Fig. 10; Fig. 9 shows the effective air-gap permeance (including leakage permeances) P_w, and its derivative dP_w/dx, both as functions of the plunger displacement x. These curves are determined in the same manner as the similar ones of Art. 77, Chapter VIII.

TABLE III

Step-by-Step Tabulation for the Computation of the Dynamic Characteristics of a High-Speed Magnet

Column	1	2	3	4	5	6	7	8	9
Point or Interval	Time	Estimated from Curves		Approx. $\dfrac{d\phi}{dt}$	Approx. ϕ	P_w	$\dfrac{dP_w}{dx}$	$(F_w)_n$	Force
		I	x						
1	t_1	.....	.....	....	.....	$(P_w)_0$	$\left(\dfrac{dP_w}{dx}\right)_1$	$(F_w)_1$	...
1–2	...	.....	.....	$\left(\dfrac{d\phi}{dt}\right)_{1-2}$	.....	....	....	....	...
2	...	I_2	x_2	$\left(\dfrac{d\phi}{dt}\right)_2$	ϕ_2	$(P_w)_2$	$\left(\dfrac{dP_w}{dx}\right)_2$	$(F_w)_2$	$\mathscr{F}_2$

Column	10	11	12	13	14	15
Point or Interval	$\dfrac{d^2x}{dt^2}$	$\dfrac{dx}{dt}$	Final Values x	I	$\dfrac{d\phi}{dt}$	ϕ
1	$\left(\dfrac{d^2x}{dt^2}\right)_1$	$\left(\dfrac{dx}{dt}\right)_1$	x_1	I_1	$\left(\dfrac{d\phi}{dt}\right)_1$	ϕ_1
1–2	$\left(\dfrac{d^2x}{dt^2}\right)_{1-2}$	$\left(\dfrac{dx}{dt}\right)_{1-2}$	....	....	$\left(\dfrac{d\phi}{dt}\right)_{1-2}$	....
2	$\left(\dfrac{d^2x}{dt^2}\right)_2$	$\left(\dfrac{dx}{dt}\right)_2$	x_2	I_2	$\left(\dfrac{d\phi}{dt}\right)_2$	ϕ_2

The data for the magnet and its load are as follows:

E = supply voltage, constant at 12 volts.
s = required stroke, 1.0 in.
R = coil resistance, 0.311 ohm at 20° C.
N = coil turns, 160.
M = mass of plunger and load = 1.54 lb. = 0.004 lb. per in. per sec.^{2}
k_1 = friction coefficient, neglected.
k_2 = spring constant, 6.0 lb. per in.
$\mathscr{F}_0$ = initial load to be overcome before motion occurs, 10 lb.

ART. 108] SAMPLE CALCULATION—DYNAMIC CHARACTERISTICS

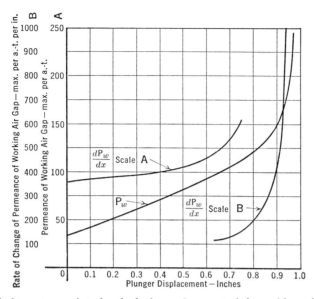

Fig. 9. Working air-gap data for the high-speed magnet of the problem of Art. 108.

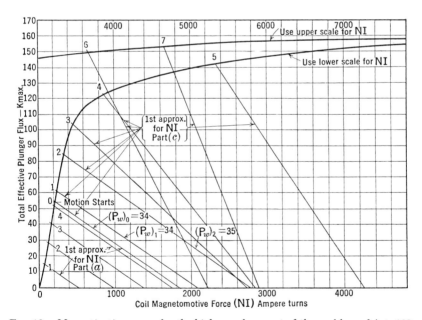

Fig. 10. Magnetization curve for the high-speed magnet of the problem of Art. 108, and the graphical construction necessary to determine the coil magnetomotive force for the various steps of the step-by-step solution.

The problem will be divided into three parts.

(a) Determination of characteristics from the time the circuit is closed until motion is just about to commence, using the method of Art. 104.

(b) Check of the total time of part (a), following the method of Art. 106.

(c) Determination of the dynamic characteristics from the time motion starts until the stroke has been completed, by the method of Art. 107.

2. **Part (a). Calculation of Magnet Characteristics from Time of Closing Circuit until Motion Is just about to Commence.** Motion will commence when the magnetic force $\mathscr{F} = \mathscr{F}_0 = 10$ lb. Let us first calculate the current I and the flux ϕ corresponding to this condition before beginning the step-by-step calculation of Art. 104. This will enable us to proportion the steps better. From Fig. 9 the working air-gap permeance and its derivative, at $x = 0$, are $(P_w)_0 = 34$ max. per ampere-turn, and $(dP_w/dx)_0 = 89$ max. per ampere-turn per inch, respectively. Substituting $(dP_w/dx)_0$ and $\mathscr{F} = 10$ lb. into equation 16, and solving for F_w, we have

$$F_w = \sqrt{\frac{\mathscr{F}}{4.43 dP_w/dx}} = \sqrt{\frac{10}{4.43 \times 89 \times 10^{-8}}} = 1{,}610 \text{ ampere-turns}$$

Then the effective flux of the plunger will be

$$\phi = P_w F_w = 34 \times 1{,}610 \times 10^{-3} = 54.7 \text{ kmax.}$$

Graphical construction on the magnetization curve of Fig. 10, point 0, gives NI equal to 1,825 ampere-turns and a current of $1{,}825/160 = 11.4$ amperes.

Starting at $t = 0$, ϕ will be zero, and $d\phi/dt$, by equation 2, will be

$$\left(\frac{d\phi}{dt}\right)_0 = \frac{E}{N} = \frac{12}{160} = 0.075 \text{ weber per second}$$

If we break up the entire computation into four intervals, $\Delta\phi$ for the first interval will be about 14 kmax. and

$$\Delta t = \frac{14 \times 10^{-5}}{0.075} = 0.001868 \text{ second}$$

Let $t_1 = 0.002$ second, and following the procedure of Art. 104, we have

$$\phi_1 = 7{,}500 \times 0.002 = 15 \text{ kmax.}$$

and I_1 from Fig. 10 will be $510/160 = 3.190$ amperes. Solving for $(d\phi/dt)_1$, we have

First approx. $\left(\dfrac{d\phi}{dt}\right)_1 = \dfrac{E - IR}{N} = \dfrac{12 - 3.190 \times 0.311}{160} = \dfrac{11.010}{160} = 0.0690$

ART. 108] SAMPLE CALCULATION—DYNAMIC CHARACTERISTICS

Then $(d\phi/dt)_{0-1}$ will be $(0.075 + 0.0690)/2 = 0.0720$. The second approximation for ϕ_1 will be, by equation 5,

Second approx. $\phi_1 = 0.0720 \times 10^{-5} \times 0.002 = 14.40$ kmax.

and I_1 will be $490/160 = 3.06$ amperes, and

$$\text{Second approx.} \left(\frac{d\phi}{dt}\right)_1 = \frac{12 - 3.06 \times 0.311}{160} = \frac{11.05}{160} = 0.0693$$

This completes the first step. The second step is started by making the first approximation for ϕ_2, as

$$\text{First approx. } \phi_2 = \phi_1 + \left(\frac{d\phi}{dt}\right)_1 (t_2 - t_1)$$

$$= 14.40 + 0.0693 \times 10^{-5} \times 0.002 = 14.40 + 13.90$$

$$= 28.3 \text{ kmax.}$$

These and the remainder of the computations are listed in Table IV, and

TABLE IV

Point	Degree of Approximation	Time, sec.	Flux, ϕ, kmax.	Coil, NI a-t.	Current, I, amperes	$\frac{d\phi}{dt}$ webers per sec.
0	Correct	0	0	0	0	0.0750
0–1	....	0.002	......	......	......	0.0720
1	First	0.002	15.00	510	3.19	0.0690
1	Second	0.002	14.40	490	3.06	0.0693
1–2	....	0.002	......	......	......	0.06655
2	First	0.004	28.30	950	5.94	0.0635
2	Second	0.004	27.77	925	5.78	0.0638
2–3	....	0.002	......	......	......	0.0612
3	First	0.006	40.53	1,355	8.47	0.0586
3	Second	0.006	40.01	1,335	8.34	0.0588
3–4	....	0.002	......	......	......	0.0564
4	First	0.008	51.77	1,725	10.79	0.0541
4	Second	0.008	51.29	1,710	10.69	0.0542
5	....	Assume $\phi =$	54.70	1,825	11.40	0.0529
4–5	....	0.00064	......	......	......	0.0535
5	....	0.00864	......	......	......	....

the graphical construction for the first approximations are shown in Fig. 10. The final computed results are plotted in Fig. 11.

In this particular case the saturation curve, up until the time of motion, is practically a straight line, and the computation could have been carried out with greater ease by the method of Art. 95, using the effective inductance of the coil.

3. **Part (b). Calculation of Total Time Required for Flux to Rise from Time of Closing Circuit until Motion Is just about to Commence.**

As in the last computation the first step is to find the values of ϕ and I when motion just starts. As this computation is the same as that of Sec. 2 we shall copy the results:

$\phi = 54.7$ kmax.

$NI = 1,825$

$I = 11.4$ amperes

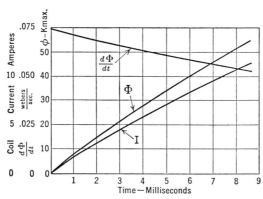

Fig. 11. Characteristics of the high-speed magnet of the problem of Art. 108, from the time of closing the switch until motion is just about to commence, computed by the method of Art. 104, and plotted from the tabulated results of Table IV.

In order to use the method of Art. 106 we must plot a curve of ϕ against $1/(E/R - I)$, between the limits of $\phi = 54.7$ and 0. This computation is carried out most easily in tabular form and is shown arranged in Table V. The values of ϕ are taken arbitrarily between the given limits

TABLE V

ϕ, kmax.	NI, a-t.	I, amperes	E/R, amperes	$E/R - I$, amperes	$\dfrac{1}{E/R - I}$
54.70	1,825	11.40	38.6	27.2	0.0368
51.29	1,710	10.69		27.9	0.0359
40.01	1,335	8.34		30.3	0.0330
27.77	925	5.78		32.8	0.0305
14.40	490	3.06		35.5	0.0282
0.00	0	0.00	↓	38.6	0.0259

Art. 108] SAMPLE CALCULATION—DYNAMIC CHARACTERISTICS

so as to get a good distribution of points. For each value of ϕ, NI is obtained directly from the magnetization curve of Fig. 10. The method of computation will be apparent from an inspection of the table.

The tabular results of ϕ against I are shown plotted in Fig. 12. Evaluation of the area behind the curve by graphical integration gives 1.68×10^{-5} ohm-second per turn. Then the time for the flux to rise from 0 to 54.7 kmax. will be, by equation 11,

$$t = \frac{N}{R} \int_0^{\phi = 54.7} \frac{d\phi}{(E/R - I)} \quad (11)$$

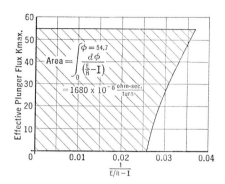

Fig. 12. Graphical construction for evaluating the time required for the flux to build up from zero to the value where motion is incipient, for the high-speed magnet of the problem of Art. 108, following the method of Art. 106.

Which, on substituting, gives

$$t = \frac{160}{0.311} \times 1{,}680 \times 10^{-8} = 0.00865 \text{ second}$$

This, as can be seen, checks perfectly with the result of Sec. 2. If just the total time is required, this method is much simpler than that of Sec. 2 because no approximations or step-by-step calculations are required.

4. Part (c). Calculation of Dynamic Characteristics from Time of Start of Motion until Stroke Is Completed. At the start of motion, t will equal 0.00865 second, reckoned from the time of closing the switch. Let this time $= t_0$, and designate all other quantities occurring at this instant with the subscript 0. We then have, from the computations of Sec. 2, the following:

$$\phi_0 = 54.7$$

$$(NI)_0 = 1{,}825$$

$$(F_w)_0 = 1{,}600$$

$$I_0 = 11.4$$

$$\left(\frac{d\phi}{dt}\right)_0 = 0.0529$$

$$(P_w)_0 = 34$$

$$\left(\frac{dP_w}{dx}\right)_0 = 89$$

In starting the computations let us follow the method of Art. 107, Sec. 6, and substitute into equations 26, 27, and 28, using for the first interval $t = (t_1 - t_0) = (0.010 - 0.00865) = 0.00135$ second.

Then

$$\left(\frac{d^2x}{dt^2}\right)_1 = 8.86 \left[\frac{(F_w)_0}{M(P_w)_0}\left(\frac{dP_x}{dx}\right)_0\left(\frac{d\phi}{dt}\right)_0\right] t \qquad (26)$$

$$= 8.86 \left[\frac{1{,}600 \times 89 \times 10^{-8} \times 0.0529}{0.004 \times 34 \times 10^{-8}}\right] 0.00135$$

$$= 8.86 \times 55{,}300 \times 0.00135 = 662 \text{ in. per sec. per sec.}$$

$$\left(\frac{dx}{dt}\right)_1 = 4.43 \times 55{,}300 \times (0.00135)^2 \qquad (27)$$

$$= 0.447 \text{ in. per sec.}$$

$$x_1 = \frac{4.43}{3} \times 55{,}300 \times (0.00135)^3 \qquad (28)$$

$$= 0.000201 \text{ in.}$$

$$\phi_1 = \phi_0 + \left(\frac{d\phi}{dt}\right)_0 t = 54.7 + 0.0529 \times 10^5 \times 0.00135$$

$$= 54.7 + 7.14 = 61.84 \text{ kmax.}$$

$(NI)_1$, at $\phi = 61.84$, $(P_w)_1 = 34$, $x = 0.0002$, is 2,060. Hence

$$I_1 = \frac{2{,}060}{160} = 12.88 \text{ amperes}$$

$$\left(\frac{d\phi}{dt}\right)_1 = \frac{E - IR}{N} = \frac{12 - 12.88 \times 0.311}{160} = 0.050 \text{ weber per second}$$

Then

$$\left(\frac{d\phi}{dt}\right)_{0-1} = \frac{0.0529 + 0.050}{2} = 0.05145$$

and a second approximation for ϕ_1 will be

$$\phi_1 = 54.7 + 0.05145 \times 10^5 \times 0.00135$$

$$= 54.7 + 6.95 = 61.65 \text{ kmax.}$$

This completes the first step, and we are now ready to start the tabular computation following the form of Table III. The results of each step are plotted directly in Fig. 13 so that the data for each step and for each succeeding point may be extrapolated from those preceding. The tabular computations are shown in Table VI.

For the purpose of illustration we shall go through one complete step, from $t_1 = 0.01$ to $t_2 = 0.015$, following exactly the procedure of Art. 107, Sec. 5.

Art. 108] SAMPLE CALCULATION—DYNAMIC CHARACTERISTICS

1. Extrapolating the curves of I and x of Fig. 13 from t_1 to t_2 we obtain the estimated values of 17.3 amperes and 0.01 inch, respectively, at point 2. These values are now entered into Table VI.

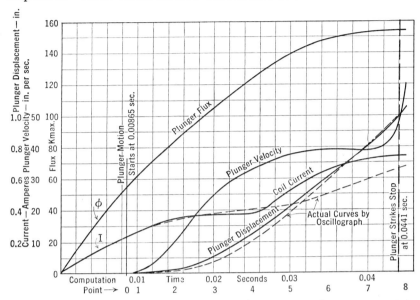

FIG. 13. Computed and experimental curves of the dynamic characteristics of the high-speed magnet of the problem of Art. 108. Computed results are plotted from the tabulation of Table VI.

2. The first approximation for $d\phi/dt$ at 2, will then be, by equation 17,

$$\left(\frac{d\phi}{dt}\right)_2 = \frac{E - IR}{N} = \frac{12 - 17.3 \times 0.311}{160} = 0.0413 \text{ weber per second}$$

3. Average values of $(d\phi/dt)$ during the interval from 1 to 2 will be

$$\left(\frac{d\phi}{dt}\right)_{1-2} = \frac{0.050 + 0.0413}{2} = 0.0456 \text{ weber per second}$$

4. The value of ϕ at 2 will be

$$\phi_2 = 61.65 + 0.0456 \times 10^5 \times 0.005$$
$$= 61.65 + 22.8 = 84.45 \text{ kmax.}$$

5. At the estimated value of $x_2 = 0.01$ in., $(P_w)_2$ will be 35 max. per ampere-turn and $(dP_w/dx)_2 = 90$ max. per ampere-turn per in., from Fig. 9.

6. From Fig. 10, at $\phi_2 = 84.45$ and $(P_x)_2 = 35$, $(NI)_2$ will be 2,750, and $(F_w)_2$ will be 2,410 ampere-turns.

7. The magnetic force at 2 will be, by equation 16a,
$$\mathscr{F}_2 = 4.43 \times 2{,}410^2 \times 90 \times 10^{-8} = 22.7 \text{ lb.}$$

8. The acceleration at point 2, by equation 19, will be
$$\left(\frac{d^2x}{dt^2}\right)_2 = \frac{1}{0.004}[22.7 - 10 - 0 - 6.0 \times 0.01]$$
$$= \frac{12.64}{0.004} = 3{,}160 \text{ in. per sec. per sec.}$$

9. The average acceleration during the interval 1 to 2 will be
$$\left(\frac{d^2x}{dt^2}\right)_{1-2} = \frac{662 + 3{,}160}{2} = 1{,}911$$

and the plunger velocity at 2 will be
$$\left(\frac{dx}{dt}\right)_2 = 0.447 + 1{,}911 \times 0.005$$
$$= 0.447 + 9.555 = 10.002 \text{ in. per sec.}$$

10. The average value of the plunger velocity over the interval from 1 to 2 will be
$$\left(\frac{dx}{dt}\right)_{1-2} = \left(\frac{dx}{dt}\right)_1 + \frac{1}{3}\left[\left(\frac{dx}{dt}\right)_2 - \left(\frac{dx}{dt}\right)_1\right]$$
$$= 0.447 + 3.18 = 3.63 \text{ in. per sec.}$$

This is taken as a parabolic average in line with equation 28, because $(dx/dt)_1$ is so small compared to $(dx/dt)_2$ that an ordinary average would be inaccurate. The plunger displacement at x_2 will be, from equation 21,
$$x_2 = 0.0002 + 3.63 \times 0.005$$
$$= 0.0002 + 0.0162 = 0.0167 \text{ in.}$$

11. The final value of I_2 may be computed directly from $(NI)_2$ found in step 6, because the difference in NI between $x_2 = 0.01$, the estimated value, and the final value of 0.0167 is negligible.[5] Then
$$I_2 = \frac{2{,}750}{160} = 17.2 \text{ amperes}$$

[5] This will not in general be true unless the actual value of the difference between the x's is negligible.

ART. 108] SAMPLE CALCULATION—DYNAMIC CHARACTERISTICS 389

TABLE VI

Point or Interval	Time, sec.	Estimated from Curves		First Approximations								Final Values					
		I, amp.	x, in.	$\frac{d\phi}{dt}$, webers per sec.	ϕ, kmax.	P_w, max. per a-t.	$\frac{dP_w}{dx}$, max. per a-t. per in.	NI, a-t.	F_w, a-t.	Force, lb.	$\frac{d^2x}{dt^2}$, in. per sec. per sec.	$\frac{dx}{dt}$, in. per sec.	x, in.	I, amp.	$\frac{d\phi}{dt}$, webers per sec.	ϕ, kmax.
1	0.010	....	....	0.0529	61.84	34.0	89.0	2,060	....	....	662	0.447	0.000201	12.88	0.050	61.65
1–2	0.005	....	....	0.0456	....	....	....	....	....	....	1,911	3.63*	....	....	0.0458	....
2	0.015	17.3	0.01	0.0413	84.45	35.0	90.0	2,750	2,410	22.70	3,160	10.00	0.0167	17.2	0.0416	84.55
2–3	0.005	....	....	0.0385	....	....	....	....	....	....	2,815	17.05	....	....	0.0410	....
3	0.020	20.2	0.13	0.0357	103.90	46.5	93.5	2,700	2,230	20.65	2,470	24.09	0.102	17.9	0.0403	105.05
3–4	0.005	....	....	0.0352	....	....	....	....	....	....	1,772	28.52	....	....	0.0393	....
4	0.025	20.7	0.31	0.0347	122.65	63.0	97.0	2,810	1,940	16.16	1,074	32.96	0.245	18.9	0.0383	124.7
4–5	0.005	....	....	0.0343	....	....	....	....	....	....	978	35.40	....	....	0.0309	....
5	0.030	21.2	0.43	0.0338	141.90	75.0	101.5	4,225	1,890	16.10	881	37.85	0.422	26.4	0.0236	140.2
5–6	0.005	....	....	0.0196	....	....	....	....	....	....	260	38.50	....	....	0.0174	....
6	0.035	30.5	0.642	0.0156	150.00	98.5	120.5	5,200	1,525	12.40	–362	39.15	0.615	32.8	0.01124	148.9
6–7	0.005	....	....	0.0085	....	....	....	....	....	....	–82	38.95	....	....	0.00725	....
7	0.040	37.8	0.829	0.00144	153.15	127.0	245	5,870	1,210	15.76	+198	38.74	0.810	36.9	0.00325	152.5
7–8	0.005	....	....	0.00265	....	....	....	....	....	....	4,349	49.62	....	....	....	....
8	0.045	37.5	1.00	0.00206	153.82	....	....	....	....	50.	8,500	60.49	1.058	....	....	....

* Parabolic average.

The final corrected value for $(d\phi/dt)_2$ will be

$$\left(\frac{d\phi}{dt}\right)_2 = \frac{12 - 17.2 \times 0.311}{160} = 0.0416 \text{ weber per second}$$

and the corrected average value of $(d\phi/dt)_{1-2}$ will be

$$\left(\frac{d\phi}{dt}\right)_{1-2} = \frac{0.050 + 0.0416}{2} = 0.0458 \text{ weber per second}$$

and the final value of ϕ_2 will be

$$\phi_2 = 61.65 + 0.0458 \times 0.005$$
$$= 61.65 + 22.90 = 84.55 \text{ kmax.}$$

In this manner we may proceed, step by step, until the stroke is completed. The tabular computations for each point are shown in Table VI, and the graphical construction required for the first approximation of NI and F_w are shown in Fig. 10. In computing the last step to point 8 at the completion of the stroke, the force cannot be determined as in step 7, because at this point the plunger touches its stop and the air-gap permeance becomes infinite. It can, however, be readily estimated on the basis of the areas of contact and the saturation area of the flux. The value of 50 lb. was obtained in this way and accounts for the sudden acceleration at the end.

The magnet described has actually been built and subjected to test. The oscillographic test curves of plunger displacement and coil current are shown plotted in Fig. 13, by means of the dashed lines. The close check between the computed and actual plunger displacement curves is remarkable. The computed time of motion is practically identical with the measured value. The computed current curve checks very closely up until about 0.026 sec. and then shows a deviation to the high side. This probably is caused by an error in the computed magnetization curve at high densities. A small error in saturation, though it will affect the force only slightly, will produce a large change in necessary magnetomotive force and hence a large error in current. It should also be noted that the computed curve is subject to a sudden change at about 0.026 sec. This is because the iron is saturating at this point, and the current which is derived from the first approximation of the flux is subject to a large error as explained before. If the current values are recomputed using the final flux and displacement values, a much smoother curve, more closely approaching the experimental curve, will be obtained.

If when the plunger approaches its stop, the saturation is low, most of the coil magnetomotive force will be across the air gap, and then as the air gap closes the necessary magnetomotive force will rapidly drop, resulting in a drop in coil current. Magnets of the flat-faced lifting type and flat-faced or coned plunger types have, generally, a very pronounced dip in current just before striking the stop. Computed characteristics for such a magnet are given in Art. 112.

109. Quick-Release Magnets

In certain types of holding magnets it is necessary that the magnet be capable of releasing its load in a very short time, of the order of a few thousandths of a second. When this is the case special attention must be paid to laminating or slotting the magnet, as the effect of eddy currents becomes increasingly important. Another feature which greatly influences the release is the manner in which the circuit is broken. When the release time is to be very short, it is not feasible to break the circuit in the ordinary manner because the arc formed at the switch contacts

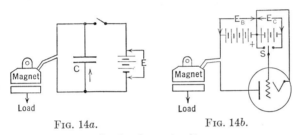

Fig. 14a. Fig. 14b.
Quick-release circuits.

will allow the current to persist far beyond the desired limit. In these cases some method must be used which will quickly dissipate the stored energy of the magnet. In Fig. 14 are illustrated two circuit arrangements which will accomplish this purpose.

In Fig. 14a, the magnet is shunted by a condenser, C. This condenser serves many purposes: (a) at the instant of opening the switch it provides a zero impedance path for the magnet current and thus prevents an arc at the switch; (b) by adjusting its capacitance in relation to the average inductance of the magnet the natural oscillation frequency can be adjusted to make the flux, and hence the force, fall to zero in the desired release time; (c) the reverse magnetizing effect produced by the negative lobe of the current oscillation can be made to remove any residual effect in the iron and also tend to counteract the effect of any eddy currents which may be present in the iron as the result of imperfect laminating. When using this system for very short release times, the mechanical design of the armature mechanism must be such that the armature will move away from the magnet pole faces sufficiently fast to prevent reattachment during the negative lobe.

In Fig. 14b, the excitation current is supplied from the plate circuit of a high-vacuum tube. When the magnet is to be excited switch S is thrown to left, placing zero bias on the grid, and allowing a large plate current to flow. When it is desired to release the load the switch is thrown to right, placing a large negative bias on the grid. This increases the resistance of the plate circuit to a very high value and very quickly dissipates the stored energy of the magnet. This system can be used only when the available voltage is sufficiently high for vacuum tube use.

In Fig. 14c is shown a circuit arrangement for producing a constant demagnetizing effect to offset the effects of residual magnetism and eddy currents. The full line winding is the main exciting winding; it produces enough excess excitation to balance out the magnetomotive force of the coil shown in dashed lines. The dashed winding is an auxiliary winding of relatively light wire; it produces a small magnetomotive force acting opposite to that of the main winding. When the circuit through the main winding is interrupted the negative magnetomotive force of the auxiliary coil will not only prevent sticking of the armature due to residual magnetism but will also make the effective magnetomotive force due to eddy currents come to zero quickly. Some arrangement, like that of Fig. 14b, must be employed to interrupt the main circuit quickly.

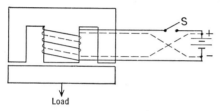

Fig. 14c. Circuit for minimizing the effects of residual magnetism and eddy currents.

In any magnet designed for quick release it is obvious that the stored energy of the magnet should be made as small as possible, as it is this energy which must be dissipated before release can be effected. Consequently, fixed air gaps should be eliminated, and the working air gap in the holding position should be made as small as possible. To mitigate any detrimental effects due to residual magnetism produced by a short air gap, the iron parts should be made of a low coercive intensity iron carefully annealed after forming. An arrangement like that of Fig. 14a or c for producing a demagnetizing effect will also be useful.

110. Method of Computing the Release Time for a Magnet Shunted by a Condenser

The voltage equation for the magnet circuit is

$$N\frac{d\phi}{dt} + IR = E - \frac{1}{C}\int_0^t I\,dt \qquad (29)$$

where E is the voltage existing across the condenser at the instant of opening the switch. If the magnet has a large air gap and is undersaturated, the saturation curve will be a straight line, and this equation may be solved by the method of Art. 100. However, in the case under consideration, where a short time of release is desired, the air gaps must necessarily be short, with the result that the iron is generally highly saturated. Under these conditions equation 29 cannot be solved analytically, and recourse to a step-by-step solution must be made.

Designating the instant of opening the switch as the zero of time, then, at this instant:

$$t = 0 = t_0, \quad I = I_0 = E/R, \quad \text{and the flux } \phi = \phi_0$$

may be read directly from the magnetization curve of the magnet corresponding to the value I_0, and the

$$\int_0^t I\,dt = \int_{t=0}^{t=0} I\,dt = 0$$

The current direction and relative polarity will be as shown in Fig. 15. The rate of change of flux at t_0 will be zero, because the IR drop of the coil will be exactly counter balanced by the condenser voltage, and hence cannot be used to estimate ϕ_1. Its derivative, however, will not be zero at t_0 and may be found by differentiating equation 29 and solving:

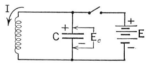

Fig. 15. Magnet with shunt condenser—switch opened at $t = 0$—build-down of current.

$$\left(\frac{d^2\phi}{dt^2}\right)_0 = -\frac{I_0}{CN} \qquad (30)$$

Using this, a first approximation of the flux at t_1 may be computed as

$$\phi_1 = \phi_0 - \frac{1}{2}\left(\frac{d^2\phi}{dt^2}\right)_0 (t_1 - t_0)^2 \qquad (31)$$

With this value of ϕ_1, the corresponding value of I_1 may be obtained by means of the magnetization curve. The value of $(d\phi/dt)_1$ may now be obtained by substituting into equation 29

$$\left(\frac{d\phi}{dt}\right)_1 = \frac{(E_c)_1 - I_1 R}{N} \tag{32}$$

where $(E_c)_1$, the voltage of the condenser at t_1, will be

$$(E_c)_1 = (E_c)_0 - (\Delta E_c)_{0-1} \tag{33}$$

The change in condenser voltage over the interval may be estimated as

$$\frac{1}{C} \int_{t=0}^{t=1} I\,dt = (\Delta E_c)_{0-1} = \frac{1}{C}[(I)_{0-1}(t_1 - t_0)] \tag{34}$$

where $(I)_{0-1}$ is the average current over the interval equal to $\frac{1}{2}(I_0 + I_1)$. The average value of the rate of change of flux over the interval will then be

$$\left(\frac{d\phi}{dt}\right)_{0-1} = \frac{1}{2}\left[\left(\frac{d\phi}{dt}\right)_0 + \left(\frac{d\phi}{dt}\right)_1\right] \tag{35}$$

and the second approximation for ϕ_1 will be

$$\phi_1 = \phi_0 - \left(\frac{d\phi}{dt}\right)_{0-1}(t_1 - t_0) \tag{36}$$

With this value of ϕ_1, a new value of I_1, $(I)_{0-1}$, $(E_c)_1$, and $(d\phi/dt)_1$ may be obtained. The final value of ϕ_1 can now be computed by repeating the evaluation of equation 35, and should be plotted. It should be noted that this value of ϕ is a third approximation.

This completes the first step. In the subsequent steps the first approximation of ϕ is preferably found by extrapolating the curve of flux plotted from the preceding steps. In Table VII the complete tabulation of the method of computation is set forth. The subscripts of equations 32 to 36 must be changed to correspond to the steps. Care must be taken to watch the signs of E_c and IR if the computation is carried beyond the first positive lobe. If, in following the above procedure, it is found that the third approximation of ϕ is practically identical with the second approximation, the calculations following the second approximation of equation 36 should be omitted.

The above method is accurate except in so far as it neglects iron loss. This could be taken into account by adding, in series with the coil resistance, a fictitious equivalent iron loss resistance equal to P_c/I^2, where P_c will be a function of ϕ, and the natural oscillation frequency. This resistance will therefore be different at each step.

TABLE VII

Step-by-Step Tabulation for the Computation of the Flux-Time Release Characteristics of a Magnet Shunted by a Condenser—No Armature Motion

Point or Interval	Degree of Approx.	Time or Interval	Flux, ϕ	Current, I	Condenser Voltage, E_c	$\left(\dfrac{d\phi}{dt}\right)$	Final ϕ
0	Correct	t_0	ϕ_0 from mag. curve	E/R	$+E$	0	ϕ_0
0–1	First	$t_1 - t_0$	....	$(I)_{0-1}$ → Eq. 34		Eq. 35	.....
1	First	t_1	Eq. 31 → I_1		Eq. 33 → Eq. 32		.....
0–1	Second	$t_1 - t_0$	....	$(I)_{0-1}$ → $(\Delta E_c)_{0-1}$		$\left(\dfrac{d\phi}{dt}\right)_{0-1}$	.....
1	Second	t_1	Eq. 36 → I_1		$(E_c)_1$ →	$\left(\dfrac{d\phi}{dt}\right)_1$	ϕ_1
1–2	First	$t_2 - t_1$	....	$(I)_{1-2}$ → Eq. 34		Eq. 35	...
2	First	t_2	from Curve → I_2		Eq. 33 → Eq. 32		third approx.
1–2	Second	$t_2 - t_1$	....	$(I)_{1-2}$ → $(\Delta E_c)_{1-2}$		$\left(\dfrac{d\phi}{dt}\right)_{1-2}$	...
2	Second	t_2	Eq. 36 → I_2		$(E_c)_2$ →	$\left(\dfrac{d\phi}{dt}\right)_2$	ϕ_2

111. Sample Calculation of the Release Characteristics of a Quick-Release Magnet

1. Problem Data. For the purpose of illustrating the method of computing developed in Art. 110, let it be required to compute the release characteristics, flux, current, condenser voltage, and holding force, against time, for the magnet and its circuit illustrated in Figs. 16a and 16b, respectively.

Figure 16a represents a conventional type of flat-faced holding or lifting magnet, designed in accordance with the methods of Art. 86.

It will be assumed that the magnet has been sufficiently slotted to make the eddy currents negligible.[6] The magnetization data are given in Fig. 17. These data are plotted in terms of the effective flux of the center pole core, which may be considered the effective flux linking with all the turns.[7] For the purpose of this problem the falling magnetization curve, taking into account the hysteresis of the iron, is necessary. This may be computed in accordance with the methods of Art. 49, Chapter IV.

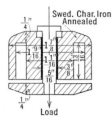

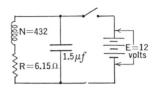

Fig. 16a. Quick-release holding magnet.

Fig. 16b. Electrical circuit for the quick-release magnet of Fig. 16a.

In calculating the force, not all the effective flux of the center pole core is effective owing to leakage. For this particular design, at the 0.005-in. air gap illustrated, the total pull between armature and pole face, properly taking into account the leakage factors, works out to be

$$\text{Force} = 0.0298\phi^2 \text{ lb.} \qquad (37)$$

where ϕ is in kilomaxwells and defined as above.

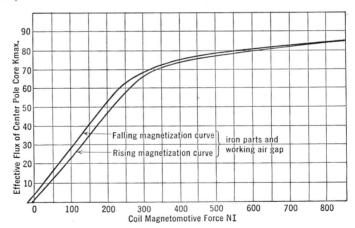

Fig. 17. Magnetization curves for the magnet of Fig. 16a, for an air-gap length of 0.005 in.

[6] This type has merely been taken as convenient for the purposes of illustration; a more practical shape would be a stack of E-type laminations for the magnet proper with a flat armature made of I strips.

[7] See the discussion on pages 366 and 367.

Art. 111] SAMPLE CALCULATION—RELEASE CHARACTERISTICS 397

Figure 16b gives all the necessary electric circuit data. The condenser of 1.5 microfarads has been designed in relation to the effective inductance of the magnet, determined from the magnetization data of Fig. 17, so as to produce an approximate free oscillation frequency of 250 cycles per second. The intention is to have the magnet release its armature at about 0.001 second.

2. Calculation of the Release Characteristics from the Time of Opening the Switch. At the instant of opening the switch the current will be $I_0 = E/R = 1.95$ amperes, and NI will be $432 \times 1.95 = 843$ ampere-turns. Referring to the magnetization curve of Fig. 17 at this point, ϕ_0 will be 84.8 kmax. By equation 30,

TABLE VIII

Point or Interval	Degree of Approx.	Time or Interval, sec.	Flux, ϕ, kmax.	Current, I, amp.	Condenser Voltage, E_c, volts	$\frac{d\phi}{dt}$, webers per sec.	Final Second Approx., ϕ, kmax.
0	Correct	0	84.8	1.95	+12	0	84.8
0-1	First	0.0001	−1.5	1.80	−120	−0.142	−1.4
1	First		83.3	1.66	−108	−0.283	83.4
0-1	Second		−1.4	1.82	−121	−0.138	−1.4
1	Second		83.4	1.68	−109	−0.276	Third Approx. 83.4
1-2	First	0.0001	....	1.48	−99	−0.388	−3.9
2	First	0.0002	80.0	1.28	−208	−0.500	79.5
2-3	First	0.0001	....	1.05	−70	−0.537	−5.4
3	First	0.0003	73.0	0.83	−278	−0.575	74.1
3-4	First	0.0001	....	0.76	−51	−0.642	−6.4
4	First	0.0004	68.0	0.69	−329	−0.709	67.7
4-6	First	0.0002	....	0.57	−76	−0.780	−15.6
6	First	0.0006	50.0	0.44	−405	−0.852	52.1
6-8	First	0.0002	....	0.37	−49	−0.923	−18.5
8	First	0.0008	34.0	0.29	−454	−0.995	33.6
8-10	First	0.0002	....	0.19	−25	−1.038	−20.8
10	First	0.0010	13.0	0.09	−479	−1.080	12.8

$$\left(\frac{d^2\phi}{dt^2}\right)_0 = -\frac{I_0}{CN} = -\frac{1.95}{1.5 \times 10^{-6} \times 432} = 3{,}010 \text{ webers per second per second}$$

Letting the first interval = 0.0001 second, the first approximation for ϕ_1 will be, by equation 31,

$$\phi_1 = 84.8 - \tfrac{1}{2} \times 3{,}010 \times 10^5 \times 10^{-8}$$
$$= 84.8 - 1.5 = 83.3 \text{ kmax.}$$

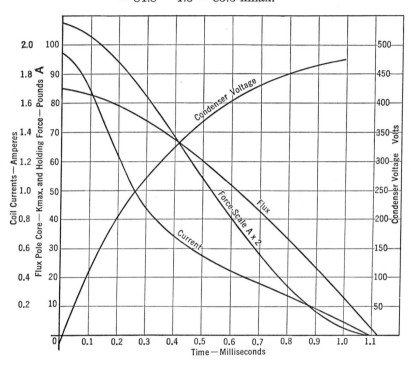

FIG. 18. Release characteristics for the magnet of Fig. 16, plotted from the tabulation of Table VIII.

Referring to the falling branch of the magnetization curve we obtain $(NI)_1 = 720$, giving $I_1 = 1.66$. The change in condenser voltage over the interval will be, by equation 34,

$$(\Delta E_c)_{0-1} = \frac{1.80 \times 0.0001}{1.5 \times 10^{-6}} = 120 \text{ volts}$$

and the condenser voltage at t_1, by equation 33, will be

$$(E_c)_1 = +12 - 120 = -108 \text{ volts}$$

Then, by equation 32,

$$\left(\frac{d\phi}{dt}\right)_1 = \frac{-108 - 1.66 \times 6.15}{432} = -0.274 \text{ weber per second}$$

and by equations 35 and 36, the second approximation for ϕ_1 will be

$$\phi_1 = 84.8 - 0.137 \times 10^5 \times 0.0001 = 83.4 \text{ kmax}.$$

Repeating this procedure a second approximation for ϕ_1 may be made. These computations are carried out in tabular form in Table VIII, following the form of Table VII. It will be noticed that the third approximation for ϕ_1 gives the same result as the second, and hence the second set of calculations may be eliminated.

The final value of ϕ_1 is now plotted and is shown on curve sheet of Fig. 18. Extrapolating this curve to $t_2 = 0.0002$, a first approximation for ϕ_2 can be found. These and all the subsequent calculations are tabulated in Table VIII.

The final results are shown plotted in Fig. 18 up to slightly beyond 0.001 second, where the current becomes zero. This is merely a convenient stopping place. The calculations could be carried into the negative lobe if desired. The force curve has been calculated by equation 37. The actual time of release can be determined from these curves if the releasing force is known.

112. Method of Rational Design of a High-Speed Magnet

1. General. Definitions. In Art. 107 and the illustration of Art. 107, Sec. 4, it was demonstrated how to predetermine the time characteristics of a magnet which has been designed. It is now proposed to outline a method whereby a design for a high-speed magnet having specified time characteristics can be obtained. This method as herein presented depends on the correlation of design constants obtained by making many sample designs. The specific data given apply only to magnets where the force increases rapidly with the stroke, that is, magnets in the range of index numbers from 800 to 30. These magnets can be either of the fully laminated type employing a square plunger, or of the slotted cylindrical plunger type.

In this discussion of high-speed magnets the following definitions and symbols will be used:

Time of Action, T_a. Total time required, from the instant of closing the switch, for the plunger to complete its stroke.

Time of Motion, T_m. Actual time the plunger is in motion in order to complete its stroke. This time will often be expressed as a percentage of the time of action, thus $T_m = K_m T_a$, where K_m is the fraction of T_a spent in actual motion.

Time Constant. The time constant of the magnet at the instant motion commences $= L/R$, where L is $N\phi/I$, $N\phi$ being the total flux linkage at the time when the action commences and I the current. R is the coil circuit resistance.

Stalled Watts, E^2/R, where E is the impressed voltage.

Mechanical Work, W_m. This is the actual mechanical work done during the stroke and is the sum of the useful work, the stored kinetic energy of motion, and the work done against friction. The unit is inch-pounds.

Mechanical Power, P_m. This is the average rate of doing mechanical work and is based on the actual mechanical work and the time of motion,

$$P_m = \frac{W_m}{T_m} \times \frac{1}{550 \times 12} \quad \text{horsepower} \tag{38}$$

Useful Work, W_u. This is the product of the average force applied to the load (excepting the acceleration force to the system and the frictional force of the plunger) by the stroke. The unit is inch-pounds.

Mechanical Efficacy, η_m. This is the ratio of the useful work, W_u, which the magnet does perform, to the work it could perform if the stroke were carried out very slowly with the current at its stall value. This is the same definition as that of Art. 36.

2. Empirical Design Data. Consideration of the operation of high-speed magnets will show that the characteristic factors just defined can be correlated in the following manner.

(1) The time constant of the magnet is related to time allowed for the current to build up to its starting value. Thus

$$\frac{L}{R} = f(T_a - T_m)$$

That such a relation should exist is fairly obvious. The less time available for the current to rise to the value where motion is incipient, the shorter the time constant must be.

In the type of magnet under discussion [8] (index number 800 to 30),

---

[8] Magnets with low index numbers like that computed in Art. 108 (index number 3.6), on the other hand, have a constant or drooping force-stroke characteristic, and the plunger will be highly saturated at the end of the stroke. In these magnets it is necessary that the starting flux be a much smaller percentage of stall flux. Otherwise the magnet may stall in the middle of its stroke. Owing to the substantially constant force of these magnets, the average acceleration will be low, and hence the

where the force rises rapidly as the stroke is completed, the flux at the start of motion should be about 90 per cent of the value it would have at that gap position if the current were equal to E/R. The reason for this is that the flux will have a tendency to stay constant or rise very slowly during motion, hence in order that the flux density in the iron during operation be at a reasonable value the flux should start high. A value higher than about 90 per cent is not feasible as then a slight decrease in supply voltage, or increase in coil temperature, might prevent operation.

The value of L/R in relation to $T_a - T_m$ depends upon the saturation of the iron at the start of the stroke with full excitation. If under these conditions the iron were completely unsaturated the time constant would have to be 43.5 per cent of the starting time in order that the flux reach 90 per cent of the final value in a time of $(T_a - T_m)$. Data giving the value of L/R in terms of the starting time for various saturations are shown in Fig. 19a. Roughly, if the working air-gap permeance curve intersects the saturation curve for the iron parts in the region of the knee, the saturation may be considered low; if it intersects at a point where the ampere-turns are about 2 to 4 times the value at the knee it may be considered medium. Intersections above this point would be considered high.

(2) The stalled power consumption will be dependent upon the average mechanical power output of the magnet. A high-speed magnet is essentially a motor. While the armature moves, a self-induced voltage of motion is acting against the current flow. The mechanical power developed is proportional to this electric power plus the rate of decrease of the stored magnetic energy of the air gap. Consequently as the action of a magnet is speeded up, its mechanical power output is increased, and the electrical input power must be increased in proportion. At constant voltage this requires an increase in current. This is obtained by using fewer turns of heavy wire. Thus the coil resistance is lowered and the stalled power input is increased. Figure 19b shows the relationship between the electric power input at stall, and the average mechanical output of the magnet.

(3) In a high-speed magnet the useful work as previously defined will always be less than the actual mechanical work. The faster the action of a given magnet the greater will the kinetic energy of the moving parts be, and the larger will the mechanical work be for a given useful work. Thus very fast magnets have a low ratio of W_u to W_m. It

percentage of motion will be high compared to that of magnets having a steeply rising characteristic. It is for this reason that the given data cannot be applied to these magnets.

would seem natural that such magnets would have a low mechanical efficacy, and that in general the mechanical efficacy should be related

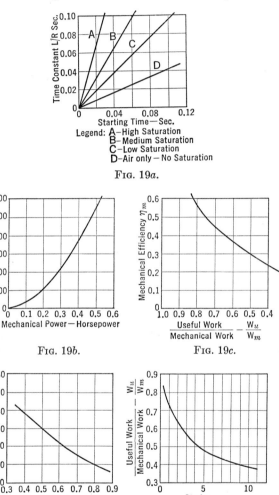

Fig. 19a.

Fig. 19b.

Fig. 19c.

Fig. 19d.

Fig. 19e.

Empirical design data for high-speed tractive magnets.

Index number = 800 to 30.

Nominal useful work = 4 to 25 in-lb.

to the ratio W_u/W_m. Such is actually the case, and the **graph of Fig. 19c** shows the empirically derived relationship.

Art. 112] RATIONAL DESIGN OF A HIGH-SPEED MAGNET 403

(4) As the ratio W_u/W_m increases, indicating lower-speed magnets, the percentage of motion, K_m, will decrease. In other words, the less the speed of a magnet, the greater will be the percentage of the time allotted to the build-up of current prior to motion. Figure 19d shows the manner of variation of these two quantities.

(5) As the ratio of the stroke to the time of action increases, more energy is diverted to the kinetic form, and the ratio W_u/W_m decreases. The manner of this variation as derived empirically is shown in Fig. 19e.

3. **Design Procedure for High-Speed Magnets.** Assume that it is required to design a magnet to move a given load over a given stroke s, in a specified time T_a, and that the static characteristic of load force against stroke has been given; further that the index number of the static load characteristic is between 800 and 30. Then we may proceed in the following manner:

(1) Estimate from curve 19e the ratio W_u/W_m.

(2) With W_u/W_m, enter curve 19d and obtain the percentage of motion K_m, and thence the estimated time of motion as $K_m T_a$.

(3) With W_u/W_m, enter curve 19c and obtain the estimated value of mechanical efficacy η_m.

(4) Using the given static load force-stroke characteristic and the estimated mechanical efficacy construct a probable stalled force-stroke curve for the desired design. This curve should have the general slope corresponding to the index number of static characteristic, and a work area which will be greater than that of the static characteristic by $1/\eta_m$.

(5) With the curve of (4) as a starting basis, make a preliminary design of a proper tractive magnet, following the directions of Chapters IX and X. This is not intended to be a complete design; it merely consists in determining the proper size of plunger or pole face, the coil ampere-turns, and the permeance of the working gap including the probable value of fringing permeance. The winding space, wire size, and coil turns cannot be determined yet.

(6) Subtract the time of motion T_m, found (2), from the time of action T_a and thence determine the time constant from Fig. 19a.

(7) From the ratio W_u/W_m, and the value of the useful work W_u, calculate W_m.

(8) From W_m calculate the average horsepower over the stroke from equation 38.

(9) Enter Fig. 19b with the average horsepower and determine the stalled power input E^2/R. Knowing E, solve for R, the coil circuit resistance.

(10) Knowing the time constant L/R, solve for L.

(11) Neglecting the reluctance of the iron, and estimating the air

path permeance P of the magnetic circuit as $0.9P_w$ to allow for the fixed gap if present, compute N, the turns of the exciting coil, as, $N = \sqrt{L/P}$.

(12) Using NI from (5), and estimating the mean length of turn of the coil, substitute into equation 17a, of Chapter VI, and solve for resistance per inch of wire:

$$R_i = \frac{E}{P_m NI} \text{ ohms per inch}$$

Referring this to Table II, Chapter VI, select the nearest size of wire.

(13) Using the turns from (11), and the wire size from (12), design the coil and coil winding space. The proportions given in Chapter X may be followed.

(14) A sketch of the magnet should be made at this point. The fixed gap should be designed to utilize 10 per cent of the coil ampere-turns. Compute the effective permeance (as regards total flux linkage) of the working gap, and if materially different than that of (5) recompute the items from (11) on. Also compute the permeance of the fixed gap.

This finishes the design of the magnet, but it is generally necessary to check its performance. This can be done by computing its dynamic characteristics by the method of Art. 107 and the illustration of Art. 108, Sec. 4, or the sample design of Art. 113.

113. Sample Design of a High-Speed Magnet

1. General. In order to illustrate the method just developed and also to prove its usefulness it will be necessary to design a magnet for a specified speed of action and then check its performance. Only in this way can the validity of the data be ascertained.

2. Statement of Problem. Design a magnet of as light a weight as is consistent with good economy, to satisfy the following data:

>Stroke = $\frac{3}{32}$ in.
>Load = 50 lb. including spring; neglect friction.
>Time of action = 0.040 sec.
>Weight of moving parts of load = 1 lb.
>Material = medium-silicon steel laminations.
>Plunger. If used, make of square cross section.
>Supply voltage = 12 volts.
>Operating temperature = 20° C.

Show a suitable mechanical arrangement of the parts, and the location of the rivets holding the laminations.

3. **Tentative Design.** The nominal index number and useful work of the required design are

$$\frac{\sqrt{50}}{0.0937} = 75.4, \text{ and } 4.7 \text{ in.-lb., respectively}$$

which come within the data of Fig. 19. The following design procedure follows that of Art. 112, Sec. 3.

(1) The ratio of stroke to time of action is $0.0937 \div 0.04 = 2.35$, which by reference to curve 19e gives $W_u/W_m = 0.61$.

(2) From curve 19d, the percentage of motion will be 21 per cent.

(3) From curve 19c, the mechanical efficacy will be 0.38.

(4) Referring to Fig. 10, Chapter IX, the proper magnet type will be the flat-faced plunger type. The proper shape of the stalled force-stroke characteristic for this type of magnet can be seen by referring to Fig. 3, Chapter IX. As the load to be overcome is 50 lb., the stalled force at $\frac{3}{32}$-in. gap should be somewhat in excess of this to allow for starting friction, small variations in voltage, and the fact that the actual flux at the start of motion is to be about 90 per cent of the stalled value (see Art. 112, Sec. 2, Item 1). With these allowances the stalled force at $\frac{3}{32}$-in. gap should be about 75 lb. The force at the end of the stroke can be estimated by taking a suitable saturation density for medium-silicon steel, say 120 kmax. per sq. in. (see Fig. 15, Chapter II), and noting from Fig. 8, Chapter X, that for an index number of 92, corresponding to the starting force of 75 lb., the air-gap flux density at the beginning of the stroke should be about 65 kmax. per sq. in. Then the force at the end of the stroke will be $75 \times (120/65)^2 = 255$ lb. With these values a tentative force-stroke curve, following the shape of Fig. 3, Chapter IX, can be constructed, as shown in Fig. 20. The area under this curve is 12.2 in-lb., which gives a mechanical efficacy of $4.7/12.2 = 0.385$. Thus this is very satisfactory. If the mechanical efficacy does not check, the force-stroke curve must be re-estimated until a check is obtained.

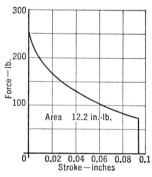

Fig. 20. Assumed static force-stroke characteristic for item 4 of the solution of the problem of Art. 113.

(5) *Preliminary Design of Magnet.* Using B_g equal to 65 from Fig. 8, Chapter X, and a force of 75 lb., the net plunger area will be, by equation 8b, Art. 72,

$$S = \frac{75 \times 72}{65^2} = 1.28 \text{ sq. in.}$$

The gross area will be 1.42, allowing a stacking factor of 0.9 for the plunger laminations. The dimensions of the square plunger will therefore be

$$\sqrt{1.42} = 1.19, \text{ call it } 1\tfrac{3}{16} \text{ in.}$$

As this magnet will only be excited a small part of the time it will be economical to have a relatively large percentage of the coil ampere-turns across the iron. Taking this as 20 per cent and allowing 10 per cent for the fixed gap, the coil ampere-turns will be, by equation 12, Chapter X,

$$NI = \frac{65 \times 0.0937}{0.70 \times 0.00319} = 2{,}730 \text{ ampere-turns}$$

If we assume the fringing flux to extend $\tfrac{1}{4}$ in. back of the center of the air gap the permeance of the working gap may be estimated as

$$P_1 = \frac{\mu S}{l} = \frac{3.19 \times 1.41}{0.09375} = \qquad 48.0$$

$$P_7 = 0.26\mu l = 0.26 \times 3.19 \times 4 \times 1\tfrac{3}{16} \text{ in.} = \qquad 4.0$$

$$P_{8b} = \frac{\mu l}{\pi} \log_\epsilon\left(1 + \frac{2t}{g}\right) = \frac{3.19 \times 4.75}{3.14} \log_\epsilon \frac{0.50}{0.0937} = \qquad 8.1$$

$$P_9 = 0.77\mu g = 4 \times 0.77\mu \times 0.0937 = \qquad 0.9$$

$$P_{10} = \frac{\mu t}{4} = \mu \times 0.22 = \qquad \underline{0.7}$$

$$P_w = 61.7 \text{ max. per ampere-turn}$$

where the subscripts refer to the permeance equations of Chapter V.

(6) The starting time of the magnet will be

$$0.04(1 - K_m) = 0.04 \times 0.79 = 0.0316 \text{ second}$$

and, by reference to the curve for low saturation[9] of Fig. 19a, the time constant of the magnet will be 0.030 second.

(7) The mechanical work will be

$$W_u \times \left(\frac{W_m}{W_u}\right) = \frac{50 \times 0.0937}{0.61} = 7.7 \text{ in-lb.}$$

---

[9] Referring back to item 4, at the start of motion the actual flux is only 90 per cent of the stalled value. As the flux does not rise rapidly with motion in this type of magnet, the motion will be carried out at low saturation.

(8) The average horsepower over the stroke will be, by equation 38,

$$P_m = \frac{7.7}{0.21 \times 0.04} \times \frac{1}{550 \times 12} = 0.139 \text{ hp.}$$

(9) From curve 19b, the stalled watts corresponding to 0.139 hp. is 120 watts. Then,

$$R = \frac{E^2}{P} = \frac{144}{120} = 1.20 \text{ ohms}$$

(10) The inductance at the beginning of the stroke will be

$$L = \left(\frac{L}{R}\right) \times R = 0.030 \times 1.20 = 0.036 \text{ henry}$$

(11) Estimating the total air-path permeance (working gap + distributed leakage + series fixed gap) of the magnetic circuit as $0.95 P_w = 58.6$, the coil turns will be

$$N = \sqrt{\frac{L}{P}} = \sqrt{\frac{0.0360}{58.6 \times 10^{-8}}} = 248 \text{ turns}$$

(12) The mean length of turn, allowing $\frac{1}{32}$-in. clearance around the plunger, $\frac{1}{16}$-in. bobbin wall, and $\frac{5}{16}$-in. net winding depth will be $4(1\frac{3}{16} + \frac{1}{16} + \frac{1}{8} + \frac{5}{16}) = 6.75$ in. The resistance per inch of wire will be, by equation 17a, Chapter VI,

$$R_i = \frac{E}{P_m N I} = \frac{12}{6.75 \times 2{,}730} = 0.652 \times 10^{-3} \text{ ohm per inch}$$

which by reference to Table II, Chapter VI, corresponds most closely to No. 19 wire.

(13) Assume the coil to be made of enameled-covered No. 19 wire (0.0379-in. diameter) wound in a phenolic bobbin $\frac{1}{16}$ in. thick. Allow $\frac{1}{32}$ in. between the coil and the moving plunger for clearance. Assume 8 layers. Then the turns per layer will be $248/8 = 31$, and the length of the coil will be

$$0.0379 \times \frac{(31 + 1)}{0.95} + \frac{1}{8} = 1.41$$

say 1.50 in. to allow some clearance for the bobbin. The height of the coil winding space will be

$$8 \times 0.0379 + \frac{1}{16} + \frac{1}{32} + \frac{1}{32} = 0.425$$

say $\frac{7}{16}$ in. The mean length of turn will be

$$4(8 \times 0.0379 + \frac{1}{8} + \frac{1}{16} + 1\frac{3}{16}) = 6.70 \text{ in.}$$

The resistance at 20° C. will be

$$6.70 \times 248 \times 0.6708 \times 10^{-3} = 1.115 \text{ ohms}$$

The current at 12 volts supplied will be

$$I(\text{stall}) = 12/1.115 = 10.76 \text{ amperes}$$

and the ampere-turns developed are

$$248 \times 10.76 = 2{,}670 \text{ ampere-turns}$$

(14) A sketch of the magnet, showing all the essential mechanical features, is shown in Fig. 21. The permeance of the main gap, whose

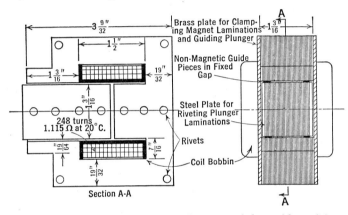

Fig. 21. Final design for the high-speed magnet of the problem of Art. 113.

flux may be considered to link with all the turns, is almost the same as that calculated in 5, and will be 62.2. The distributed permeance will be

$$P_1 = \frac{\mu S}{l} = \frac{3.19 \times 1.19 \times 0.57}{0.4375} = 4.2$$

$$P_7 = 0.26 \mu l = 0.26 \times 3.19 \times 2 \times 0.57 = 0.8$$

$$P_{8b} = \frac{\mu l}{\pi} \log_\epsilon \left(1 + \frac{2t}{g}\right) = \frac{3.19 \times 2.37}{3.14} \log_\epsilon \frac{1.69}{0.54} = 2.7$$

$$\phantom{P_{8b} = \frac{\mu l}{\pi} \log_\epsilon \left(1 + \frac{2t}{g}\right) = \frac{3.19 \times 2.37}{3.14} \log_\epsilon \frac{1.69}{0.54} =}\; 7.7$$

This distributed permeance must be multiplied by (0.96/1.50) because its flux does not link with all the turns of the coil, and then by $\frac{1}{2}$ and $\frac{2}{3}$ to get the effective permeance which may be considered in parallel with the main gap. The effective distributed permeance will be 1.64, and the total effective working gap permeance will be 62.2 + 1.6 = 63.8 max. per ampere-turn.

ART. 113] SAMPLE DESIGN OF A HIGH-SPEED MAGNET 409

The flux of the working gap may now be estimated at $0.0638 \times 2{,}670 \times 0.7 = 119$ kmax., allowing 30 per cent of NI for the iron and fixed gap. The permeance of the fixed gap, in order to use 10 per cent of the coil ampere-turns, will be

$$P_c = \tfrac{119}{267} = 0.446 \text{ kmax. per ampere-turn}$$

Letting the width of the fixed gap be $1\tfrac{3}{16}$ in. and the length along the flux lines 0.020 in., its permeance will be

$$\frac{0.00319 \times 1.19 \times 1.19 \times 2}{0.02} = 0.450$$

which with its fringing permeance of about 0.025 will make $P_c = 0.475$ kmax. per ampere-turn, which will be satisfactory.

The joint permeance of these two gaps in series is 0.0568, which is sufficiently close to the value used in item 11 so that N will not have to be recomputed. The weight of the plunger is $1.19 \times 1.19 \times 0.9 \times 2 \times 0.283 = 0.72$ lb. The total mass of the moving system will then be

$$\frac{1.0 + 0.72}{386} = 0.00445 \text{ in. mass units}$$

3. Determination of the Dynamic Characteristics of the Magnet of Fig. 21. In Fig. 22 is drawn the magnetization curve of the iron parts and the fixed air gap of the magnet. The flux plotted in this curve is that effective in producing the coil flux linkage. The net iron length and area have been taken as $6\tfrac{5}{8}$ in. and 1.27 sq. in., respectively.

The leakage coefficient of the magnet may be found from items 5 and 14, and is $(63.8/48.0 = 1.33)$. In order to find the instant at which motion commences, we must determine when the magnetic pull will overcome the load force of 50 lb. The flux density in the working gap necessary to produce this force will be, by equation 8b, Chapter VIII.

$$B_g = \sqrt{\frac{\text{Force} \times 72}{S}} = \sqrt{\frac{50 \times 72}{1.41}} = 50.3 \text{ kmax. per sq. in.}$$

The area S is the gross plunger area, as it is this area which makes up the working gap. The effective flux of the magnet will then be $50.3 \times 1.41 \times 1.33 = 94.3$ kmax., and is indicated by 0 in Fig. 22.

As the magnetization curve up to this point is practically a straight line, the time required for the field to reach this value may be calculated by the method of equation 11, Art. 95, Chapter XI.

$$I = \frac{E}{R}\left(1 - \epsilon^{-\frac{R}{L}t}\right)$$

where $R = 1.115$ and $L = N\phi/I = (248 \times 94.3 \times 10^{-5}/6.93) = 0.0337$. $I = 6.93$ amperes is obtained from the Fig. 22 for the flux of 94.3, and corresponds to NI of 1,720, shown at point F_0. Substituting into the equation 11 gives

$$6.93 = 10.76(1 - \epsilon^{-33.1t})$$

and solving we get $t = 0.0312$ second. This value is satisfactory.

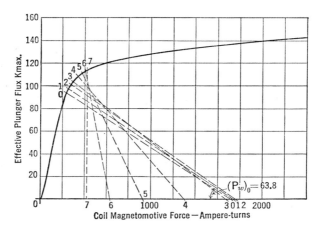

Fig. 22. Magnetization curve of the iron parts and fixed air gap of the magnet of Fig. 21.

Should this time come out very far from the desired value a new choice of turns should be made.

Letting this time be t_0,

$$\left(\frac{d\phi}{dt}\right)_0 = \frac{E - IR}{N} = \frac{12 - 6.93 \times 1.115}{248} = 0.0172 \text{ weber per second}$$

In Fig. 23 are shown curves of P_x, (dP_x/dx), and the leakage coefficient ν, all of which have been calculated in the usual manner. From these data $(P_w)_0$ and $(dP_w/dx)_0 = 63.8$ and 564, respectively. Then, following the method of Art. 107, Sec. 6, and the illustration of Art. 108, Sec. 4, we have

$$\left(\frac{d^2x}{dt^2}\right) = 8.86 \left[\frac{1,485 \times 564 \times 10^{-8} \times 0.0172}{0.00445 \times 63.8 \times 10^{-8}}\right] \times 0.0018$$

$$= 805 \text{ in. per sec. per sec.}$$

where the interval is 0.0018 second, making $t_1 = 0.033$ second.

$$\left(\frac{dx}{dt}\right) = \tfrac{1}{2} \times 805 \times 0.0018 = 0.722 \text{ in. per sec.}$$

$$x = \tfrac{1}{3} \times 0.722 \times 0.0018 = 0.000432 \text{ in.}$$

$$\phi_1 = \phi_0 + \left(\frac{d\phi}{dt}\right)_0 (t_1 - t_0) = 94.3 + 1{,}720 \times 0.0018 = 97.4 \text{ kmax.}$$

Referring this value to the magnetization curve, $(NI)_1$ will be 1,740, and I_1 will be 7.02.

$$\left(\frac{d\phi}{dt}\right)_1 = \frac{12 - 7.02 \times 1.115}{248} =$$

$$0.01685 \text{ weber per second}$$

then $(d\phi/dt)_{0-1} = 0.01702$, and

$$\phi_1 = 94.3 + 1{,}702 \times 0.0018 = 97.37 \text{ kmax.}$$

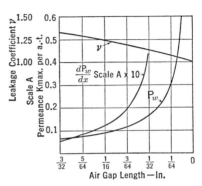

Fig. 23. Leakage coefficient, working air-gap permeance, and space rate of change of working air-gap permeance plotted as a function of air-gap length for the magnet of Fig. 21.

This completes the first step. The remainder of the computations are carried out in the same manner as the reference illustration, with the exception of the force calculation, which for this type of magnet is best done by means of equation 8b, of Art. 72, Chapter VIII, and the leakage coefficient, in the manner of Sec. 5, as follows:

$$B_g = \frac{\phi_{\text{eff.}}}{\nu S} \text{ kmax. per sq. in.}$$

where $\phi_{\text{eff.}}$ is the flux determined from Fig. 22 and $S = 1.41$ sq. in.

$$\text{Force} = \frac{B_g^2 S}{72} \text{ lb.}$$

All the calculations are set forth in Table IX, which follows the same form as Tables III and IV, with exception of dP_w/dx, F_w, which have been replaced by ν and B_g. Because of the different method of force computation the order of some columns has been changed. The operating point on the magnetization curve for each time point in the table is shown in Fig. 22 by the corresponding number. It is interesting to note

TABLE IX

Point or Interval	Time, sec.	Estimated from Curves		First Approx.		ν	B_g	Force	$\frac{d^2x}{dt^2}$	$\frac{dx}{dt}$	x	Final Values				
		I, amp.	x, in.	$\frac{d\phi}{dt}$, kmax. per sec.	ϕ, kmax.							P_w, max. per a-t.	NI, a-t.	I	$\frac{d\phi}{dt}$	ϕ
1	0.033	....	....	1,720	97.4	....	....	....	805	0.722	0.000434	64	1,740	7.02	1,685	97.37
1–2	0.002	....	....	1,790	....	1.32	54.6	58.5	1,358	2.095	0.004190	....	....	....	1,658	....
2	0.035	0.655	0.005	1,895	101.02	....	....	....	1,910	3.468	0.004624	68	1,770	7.14	1,630	100.69
2–3	0.002	....	....	1,615	....	1.27	58.1	66.0	2,755	6.223	0.012446	....	....	....	1,711	....
3	0.037	0.720	0.020	1,600	103.99	....	....	....	3,600	8.978	0.01707	75	1,680	6.78	1,792	104.11
3–4	0.002	....	....	1,989	....	1.20	64.0	80.0	5,175	14.15	0.02831	....	....	....	2,136	....
4	0.039	0.590	0.044	2,185	108.16	....	....	....	6,750	19.33	0.0454	113	1,300	5.24	2,480	108.38
4–5	0.001	....	....	2,712	....	1.12	70.6	98.0	8,770	23.72	0.0237	....	....	....	2,820	....
5	0.040	0.420	0.066	2,945	111.16	....	....	....	10,790	28.10	0.0691	204	925	3.73	3,160	111.20
5–6	0.0005	....	....	3,435	....	1.05	76.4	114.5	12,645	31.26	0.0156	....	....	....	3,430	....
6	0.0405	0.250	0.085	3,710	112.99	....	....	....	14,500	34.42	0.0847	510	630	2.54	3,700	112.92
6–7	0.0003	....	....	3,900	....	1.00	80.9	128.0	16,000	36.82	0.0110	....	....	....	....	....
7	0.0408	0.165	0.09375	4,100	114.16	....	....	....	17,500	39.22	0.0957	8	425	1.71	....	....

that during the actual operation the operating point stays very close to the knee and that the magnet operates reasonably close to optimum conditions.

The final computed dynamic characteristics are shown plotted in Fig. 24. The total time of action as read from the plunger displacement curve is 0.0408 second, which is very close to the desired value of 0.040 second.

With reference to what has been said of the effect of the static force-stroke characteristic on the dynamic characteristics it is interesting to compare Figs. 13 and 24. Figure 13 is typical of a constant or decreasing

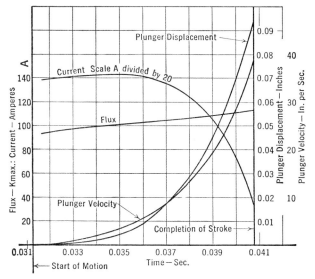

Fig. 24. Dynamic characteristics of the high-speed magnet of Fig. 21 plotted from the tabulation of Table IX.

force magnet (low index number); Fig. 24 is typical of all magnets where the static force rises rapidly with a decrease of gap length.

114. Speed Limitations of High-Speed Magnets

1. **General.** There are definite limitations to the speed that can be attained by a magnet plunger or armature. For the purpose of discussing these speed limitations it is best to classify the magnets with respect to the two fundamental actions, either direct attraction between magnetized surfaces, or the force reaction between a current-carrying conductor and an independent magnetic field (leakage flux type).

2. **Direct Attraction Type.** When the force is produced by the direct attraction between magnetized surfaces it is limited in value by the saturation of the iron. Hence, regardless of the excitation used, the maximum force that can be obtained with a solid iron plunger is about 260 lb. per sq. in., while if the plunger is of laminated medium-silicon steel this force may drop to 200 lb. These maximum forces are obtainable only near the end of the stroke when there is no leakage. Therefore, the maximum acceleration obtainable will depend only on the mass of the moving system per unit of plunger cross section.

When it is absolutely essential to develop the maximum possible accelerating force the plunger must be kept saturated. This can be done only by keeping the current near its limiting value E/R during the entire motion, and preventing a large decrease in current as shown in Fig. 24. The current may be prevented from dropping sharply by making the induced voltage small compared to the resistance drop. This voltage can be reduced only by making the coil turns less, which results in a coil of low resistance and large power consumption. Such a magnet will be inefficient and uneconomical, and may be compared to a motor whose generated voltage is low compared to the resistance drop. In general, for good economy, the direct attraction type (index number 800 to 30) should be designed to operate over the region of the saturation curve shown in Fig. 22.

An interesting consequence of the above is that for any given time of action and any given weight of plunger there will be a maximum possible stroke. This follows directly because of the limitation placed on plunger acceleration by its mass and the available accelerating force.

3. **Leakage Flux or Solenoid and Plunger Type.** In this type the force is given by equation 20c of Chapter VIII,

$$\text{Force} = HS(B_p - \mu H) = HSB_f \tag{39}$$

where B_f is the ferric flux density as defined in Art. 23, Chapter II. As soon as the ferric flux density B_f of the plunger is close to the saturation density of the plunger iron B_s, the plunger force will be directly proportional to the coil current and H. As there is no theoretical limit to the value of coil current, there is no limit to the plunger force that can be produced in this type of magnet. Practically, the instantaneous force is limited by vaporization of the copper, and the steady force by coil temperature rise. In the usual design, limited by temperature rise, the accelerations that can be obtained are considerably less than in the attraction type because the force obtainable per unit plunger cross section is less, and the plungers are relatively longer and hence heavier.

In these magnets, the induced voltage will be proportional to the product of the plunger flux and velocity,[10] thus

$$\frac{d(N\phi)}{dt} = \frac{N}{h}(\phi_p)\frac{dx}{dt} = E - IR$$

where N and h are the total turns and length of the coil, respectively. Solving this equation for the plunger velocity, we have

$$\frac{dx}{dt} = \frac{(E - IR)h}{N\phi_p} \text{ inches per second} \qquad (40)$$

As the stroke is generally long, and the force small, the motion is carried out relatively slowly. Consequently, the time for the current to build up to the starting value and the time to accelerate the plunger up to speed are generally a small part of the total time of motion. The time of action may then be quickly estimated as equal to the quotient of the stroke and the average plunger velocity.

If the magnet starts its stroke with the plunger reasonably well saturated, which is usual, ϕ_p will be fairly well determined by the saturation properties of the plunger material, and will be approximately equal to $B_p S$, where B_p is the plunger flux density produced by a magnetic intensity of NI/h ampere-turns per inch. The current I will then be directly proportional to the load force. When the stroke is reasonably long, and the total load force including friction is constant, the plunger will quickly accelerate to constant velocity. The current will then stay constant at a value just sufficient to overcome the load, and may be calculated by equation 39. Denoting this current by I_L, the time of action will be

$$T_a = \frac{\text{Stroke} \times N\phi_p}{(E - I_L R)h} \text{ seconds} \qquad (41)$$

PROBLEMS

1. Compute the effective depth of penetration in medium-silicon steel of a 60-c.p.s. sinusoidal wave of flux for peak flux densities of 64.5, 100, and 115 kmax. per sq. in. Assume $\rho = 16$ microhm-inch. Is the statement of page 365, that greater depths of penetration can be secured by increasing the iron saturation, true? *Note:* The average effective permeability of a piece of steel, as regards depth of penetration, can be determined by plotting the instantaneous permeability as a function of time and then finding the average value over the cycle. This permeability-time curve will depend on the wave form of the flux.

---

[10] Except when the plunger is close to the open end of the coil.

To illustrate the solution, the permeability-time curves of Fig. 25 have been drawn. Figure 25a represents the hysteresis loop for medium-silicon steel for $B_m = 64.5$; Fig. 25b shows the manner in which the differential permeability, or the slope

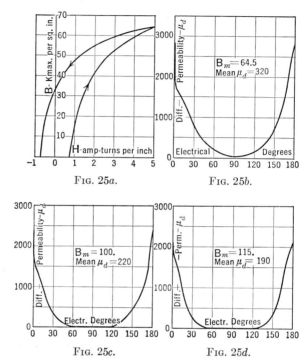

Fig. 25a. Fig. 25b.

Fig. 25c. Fig. 25d.

Differential permeability of iron plotted as a function of time for various **maximum** values of sinusoidal flux density waves.

of the hysteresis loop, varies as a function of time for a sinusoidal flux wave. Figures 25c and d show the differential permeability derived in similar manner for maximum loop densities of 100 and 115, respectively.

The final results are tabulated below:

B_m	Avg. Diff. μ	Depth of Penetration, in.
64.5	320	0.065
100	220	0.078
115	190	0.084

2. Figure 26 represents the field structure of a small magnetic motor and Fig. 27 the core structure of a small power transformer. Comment on the position of the rivets, as regards their effect on the production of circulating currents in the core structure, for each case. Sketch a better arrangement.

3. Compute for the flat-faced plunger magnet designed in Art. 87, Chapter X, the time required for the flux to build up to the value required to lift the rated load of 123 lb. Use the method of Art. 104 with ϕ as the variable. Assume all the data the same as tabulated for the final design, except that the winding is to be assumed at 20° C. Use the magnetization curve of Fig. 10 of the article, and add to it any points required. *Note*: This magnetization curve is plotted in terms of the useful flux of the working gap, whereas, for the purpose of the flux-time characteristic curve com-

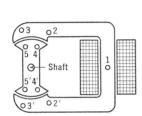

Fig. 26. Magnetic circuit of a magnetic motor.

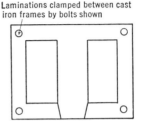

Fig. 27. Magnetic circuit of a shell-type transformer.

putation of this problem, the effective flux linking with all the turns is required. This effective flux will be, for all practical purposes, that of the plunger as listed in Table III of Sec. 6, Art. 87, Chapter X.

4. Repeat Problem 3, using the method of Art. 105 with I as the variable. In this case, in order to get the rated force of 123 lb., the current must rise to 1.227 amperes.

5. Check the answers of Problems 3 and 4 by the method of Art. 106.

6. Compute for the magnet of Problem 3 the time required for the armature to complete its full stroke of 0.125 in. Assume that the load on the armature consists of the following:

Weight to be accelerated (entire mechanical system including armature) = 5.0 lb.
Friction force to be overcome (assume constant).................... = 23.0 lb.
Initial spring force... = 100.0 lb.

The spring constant is 400 lb. per in. *Note*: The time required for incipient motion to occur will be the same as that of Problems 3, 4, and 5, and hence need not be recomputed. For computing the dynamic characteristic following this interval, it will be necessary to compute curves of the working air-gap permeance (including associated leakage paths) and its derivative as a function of gap length, besides the complete magnetization curve. These curves will be similar to those of Figs. 6 and 7.

7. Compute the release characteristics for the magnet of Fig. 16, assuming all the data to remain the same as stated, except that the air gap is to be taken as of a nominal zero length (see footnote 2, Chapter IV) instead of 5 mils. *Note*: Magnetization curve of Fig. 17 must be recomputed as the air gap has been reduced.

8. Redesign the coil (wire size and turns) of the sample design of Art. 113, so that the time of action will be approximately 0.030 sec. Do not make any changes in mechanical design or dimensions.

9. Make a complete preliminary design for a high-speed magnet that will meet the following specifications:

Stroke = $\frac{1}{4}$ in.
Load = 60 lb., including spring; neglect friction.
Time of action = 0.02 sec.
Weight of moving parts of load = 5 lb.
Material = medium-silicon steel, laminations.
Plunger. If used, make of square cross section (pole face does not necessarily have to be flat).
Supply voltage = 120 volts.
Operating temperature = 20° C.

CHAPTER XIII

ALTERNATING-CURRENT MAGNETS

115. General

The alternating-current magnet,[1] in its instantaneous magnetic relationships, behaves exactly the same as the direct-current magnets discussed in the previous chapters; thus the instantaneous force and exciting ampere-turns are related to the flux by the same equations and graphical constructions as have been used for direct current. However, the electric-circuit relationships will be different, as the impressed voltage will now be equal to the sum of the voltage induced in the exciting coil by the continuous alternation of the flux and the resistance voltage.[2] Two cases may arise: (1) operation from a constant-voltage, constant-frequency supply; and (2) operation from a constant-current, constant-frequency supply. Considering the former, which is the more common, the voltage equation will be

$$e = \frac{d(N\phi)}{dt} + iR \tag{1}$$

where the lower-case letters are used to indicate the instantaneous values of the alternating quantities. This equation is identical with equation 1 of Chapters XI and XII. In Chapter XI we were interested in the transient response due to a suddenly applied voltage of constant magnitude; in Chapter XII the restrictions regarding saturation and armature motion of Chapter XI were removed. Now it is proposed to investigate the steady-state response due to a steadily applied alternating voltage. To simplify the discussion we shall, at present, make two assumptions or stipulations:

[1] It will be assumed throughout this chapter that the alternating-current electromagnets under discussion are thoroughly laminated, and that eddy currents in the iron parts have been eliminated. See Art. 103 for a discussion of eddy currents and methods of laminating.

[2] This statement assumes that the plunger is stationary, or that the plunger motion is carried out slowly enough so that the induced voltage due to motion is negligible. In general, the alternating-current magnet is a relatively fast-acting magnet, because the voltage induced by reason of plunger motion is always small compared to the impressed voltage.

1. That the iron and air gaps of the magnetic circuit are so designed in relation to the maximum cyclic flux density that the flux is substantially proportional to the exciting current.

2. That operation is from a constant-voltage, constant-frequency supply of sinusoidal wave form.

In accordance with these assumptions the exciting current and flux will be sinusoidal, and we may write for the instantaneous value of flux

$$\phi = \phi_m \sin \omega t \qquad (2)$$

where ϕ_m is the maximum value of cyclic flux density, and ω is the angular frequency of the supply in radians per second. As current and flux are a simultaneous cause and effect, they will be in phase, and

$$i = I_m \sin \omega t \qquad (3)$$

These instantaneous relationships between flux and current, and their method of derivation from the magnetization curve, are shown in Fig. 1.

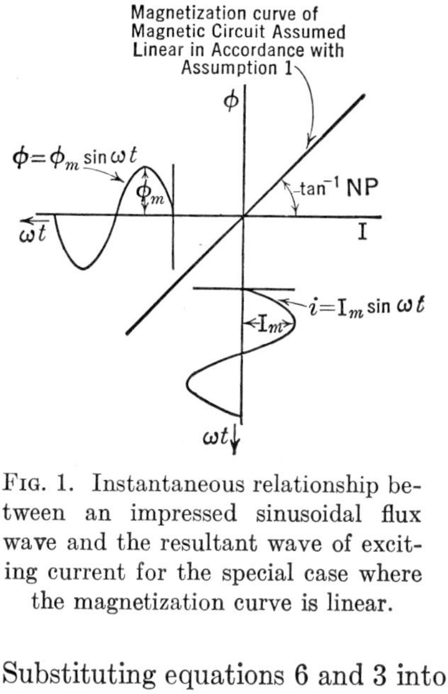

Differentiating (2), we have

$$\frac{d\phi}{dt} = \omega \phi_m \cos \omega t \qquad (4)$$

In accordance with assumption 1,

$$\phi_m = F_m P = N I_m P \qquad (5)$$

which, when substituted into (4), will give

$$\frac{d\phi}{dt} = \omega N P I_m \cos \omega t \qquad (6)$$

Fig. 1. Instantaneous relationship between an impressed sinusoidal flux wave and the resultant wave of exciting current for the special case where the magnetization curve is linear.

Substituting equations 6 and 3 into 1, we have

$$e = \omega(N^2 P) I_m \cos \omega t + I_m R \sin \omega t \qquad (7)$$

where $N^2 P$ will be recognized as the inductance of the circuit, L. Equation 7 may then be written as

$$e = \omega L I_m \cos \omega t + I_m R \sin \omega t \qquad (8)$$

In Fig. 2a the reactance voltage $\omega L I_m \cos \omega t$, and the resistance voltage $I_m R \sin \omega t$, together with their sum, the supply voltage e, are shown drawn in their correct relative phase positions. An easy way of combining the two component voltages is shown in Fig. 2b. Here the vectors oa and ob, with a magnitude equal to the maximum amplitude of their respective sinusoids, are drawn in their proper phase positions for $t = 0$. Projecting them on the vertical axis OA of Fig. 2a, by the dashed lines shown, will give the initial values of the sinusoids. If these vectors are allowed to rotate counterclockwise at an angular frequency of the supply, ω radians per second, as shown by the arrow, their projections on line OA will give the instantaneous magnitude of the sinusoids. These vectors, as they rotate, will keep their 90° relationship. The resultant of the vectors is oc, which is drawn by completing the

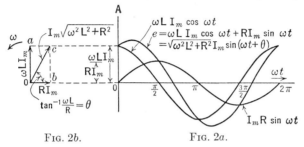

Fig. 2b. Fig. 2a.

Instantaneous sinusoidal relationships between current and voltage for a simple R, L, circuit, and the corresponding vector relations.

parallelogram. Vector oc is the maximum cyclic value of the impressed voltage, and its magnitude, by the geometry of the figure, is

$$E_m = I_m \sqrt{\omega^2 L^2 + R^2} \qquad (9)$$

Its projection on line OA, as it rotates, will give the instantaneous amplitude of the impressed voltage, which when plotted on the time scale gives the sinusoid shown. This resultant sinusoid e may also be obtained by adding the component ones together graphically. Dividing equation 9 through by $\sqrt{2}$, the relation will be obtained in terms of the root-mean-square voltage and current, designated by E and I, respectively,

$$E = I\sqrt{\omega^2 L^2 + R^2} \qquad (10)$$

The quotient of the voltage by the current will be the magnitude of the impedance of the circuit; it equals

$$|Z| = \frac{E}{I} = \sqrt{\omega^2 L^2 + R^2} \qquad (11)$$

As impedance is a vector quantity, its complete expression must include the phase shift between the current and voltage; it is

$$Z = \sqrt{\omega^2 L^2 + R^2} \underline{/\theta} \qquad (12)$$

where θ, the phase angle, is $\omega L/R$ of Fig. 2.

Equation 1, as regards its steady state alternating-current solution, can now be written as

$$E = IZ \qquad (13)$$

or Z may be broken into its vector components,

$$Z = jX + R$$

where $X = \omega L$ is called the reactance of the circuit, and j is the usual symbol of complex algebra to denote that the component X has a phase advance of 90° over component R. Then (13) may be written

$$E = jIX + IR \qquad (13a)$$

In Fig. 3 the vector diagram for equation 13a is shown.

The reactance voltage may be expressed in terms of the maximum cyclic flux ϕ_m, as follows:

$$IX = \omega LI = \frac{\omega N \phi_m}{I_m} \frac{I_m}{\sqrt{2}} = \frac{\omega N \phi_m}{\sqrt{2}}$$

Substituting $2\pi f$ for ω, we have

$$\text{Reactance voltage} = E_r = \frac{2\pi f N \phi_m}{\sqrt{2}} = 4.44 f N \phi_m \text{ r.m.s. volts} \qquad (14)$$

By equation 13a, the magnitude of the reactance voltage may also be expressed in terms of E,

$$E_r = |IX| = \sqrt{E^2 - I^2 R^2} \qquad (15)$$

In a well-designed alternating-current magnet the reactance voltage will be large compared to the resistance voltage. Consequently, if the supply voltage E is constant, the reactance voltage will also be substantially constant, regardless of the current magnitude. Thus, if IR is 10 per cent of E, $IX = E\sqrt{1 - 0.1^2} = 0.995E$; if it is 20 per cent of E, $IX = 0.98E$. It therefore follows, by reference to equation 14, that when a magnet operates on a constant voltage supply, the flux linkage $N\phi_m$ will be constant regardless of the plunger position. If the flux may be considered to link with all the coil turns N, the flux ϕ_m will be constant. This fact is the basis of the peculiar characteristics of the constant-voltage alternating-current magnet which stand out in comparison to those of the constant-voltage direct-current magnet. In particular, these characteristics are the following:

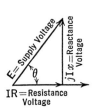

Fig. 3. Vector diagram for equation 13a.

1. The exciting current starts at a high value in the open-gap position and drops to a relatively low value in the closed-gap position.

ART. 116]   FORCE OF A SINGLE-PHASE MAGNET

2. The force-stroke curve, for long-stroke magnets, starts high in the open-gap position, drops rapidly to a minimum, and then rises to a high value at the closed-gap position; for short-stroke magnets it is similar to that of the direct-current magnet, but does not rise to such high closed-gap values.

3. There is a definite minimum relationship between the volt-ampere input to the magnet in the open-gap position and the work the magnet can do. This is known as the volt-ampere limitation.

These characteristics will be discussed in detail in later articles of this chapter.

When the magnet operates on a constant-current, constant-frequency supply of sinusoidal wave form, the mathematical relationships derived still apply, provided that the linear relationship between flux and current is maintained valid. The volt-ampere limitation noted for constant-voltage operation will still apply, but the force-stroke characteristic will be exactly similar to constant-voltage, direct-current operation. The exciting voltage will start at a low value in the open-gap position and rise to a relatively high value at the closed-gap position.

116. Nature of the Force of an Alternating-Current Magnet

1. **Single-Phase Operation—Direct-Attraction Type.** In Fig. 4 is shown a diagrammatic sketch of a single-phase, direct-attraction type magnet. Assuming a sinusoidal supply voltage and the reactance voltage large compared to the resistance voltage, the flux wave will be

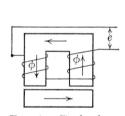

Fig. 4. Single-phase direct-attraction-type magnet.

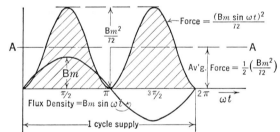

Fig. 5. Instantaneous and average force relationships for the single-phase magnet of Fig. 4.

sinusoidal. The flux density in the working gap will then be sinusoidal, and its instantaneous value will be $B_m \sin \omega t$. The instantaneous force between the armature and pole faces per square inch of surface will then be, by equation 8b, of Chapter VIII,

$$\text{Instantaneous force} = \frac{(B_m \sin \omega t)^2}{72} \text{ lb. per sq. in.}$$

In Fig. 5 the wave of sinusoidal flux density is shown along with the instantaneous-force curve, which is the square of the flux wave. The force pulsates from zero to a maximum twice each cycle. Mathematically, it can be resolved into a constant force plus a sinusoidally alternating force having twice the frequency of the supply.

$$\text{Instantaneous force} = \frac{B_m^2 \sin^2 \omega t}{72} = \frac{B_m^2}{72}\left[\frac{1 - \cos 2\omega t}{2}\right]$$

$$= \frac{1}{2}\left(\frac{B_m^2}{72}\right) - \frac{1}{2}\left(\frac{B_m^2}{72}\right) \cos 2\omega t \text{ lb. per sq. in.} \quad (16)$$

The first term is the average force over the cycle, and represents what is usually the useful force:

$$\text{Useful force} = \text{Average force} = \frac{B_m^2}{144} \text{ lb. per sq. in.} \quad (17)$$

This force may also be expressed in terms of the r.m.s. flux density, $B = B_m/\sqrt{2}$; then

$$\text{Average force} = \frac{B^2}{72} \text{ lb. per sq. in.} \quad (18)$$

which is exactly the same force formula as for direct-current operation. This average value is shown by the dashed line A–A of Fig. 5, and is exactly midway between the zero and peak value of the instantaneous-force curve.

The second term of equation 16 is the alternating component of the force. It has a frequency double that of the supply, and accounts for the characteristic hum or chatter heard from alternating-current magnets and transformers. This component of the force is generally not useful, and various methods, the most common of which is the shading coil, are used to suppress it.

When designing an alternating-current magnet, all calculations and comparisons should be made in terms of the maximum flux density, B_m. It is this value which is definitely limited by the saturation of the iron. Thus, if a comparison is to be made between the holding power of an alternating-current and direct-current magnet, both will have the same limiting flux density and hence the same peak force, but the average or useful force of the alternating-current one will be only one-half that of the direct-current one. Consequently, it follows that, as regards average force produced, the alternating-current magnet utilizes the iron only one-half as effectively as does the direct-current magnet.

2. Single-Phase Operation—Leakage-Flux Type. For this magnet the force is given by equation 20a of Chapter VIII,

$$\text{Force} = H\phi$$

As both H and ϕ are sinusoidal and in phase, the average force produced will be one-half the maximum; all the other comments made in connection with the direct-attraction type will apply.

3. **Polyphase Operation—Two-Phase.** Suppose that the pole faces of a magnet are split into two equal sections, as shown in Fig. 6a, and that each section carries a sinusoidally varying flux of the same magnitude, but in quadrature phase relation as shown in Fig. 6b. The instantaneous-force curve for each section will be a pulsating double-frequency sinusoid, but owing to their phase displacement, the total instantaneous force of the pole face will be constant. The two component force curves in their proper phase relation, together with their constant resultant, are shown in Fig. 7.

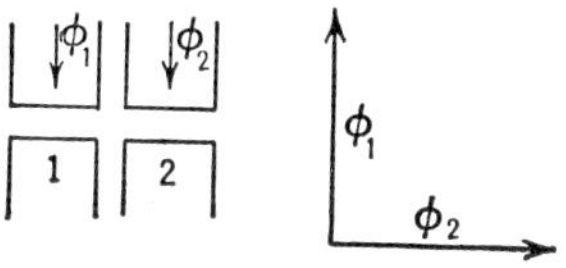

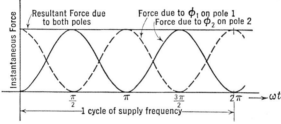

FIG. 6a.   FIG. 6b.

Two-phase magnet system and vector diagram of fluxes.

$$\text{Force due to } \phi_1 = \frac{1}{2} \times \frac{(\phi_1)_m^2}{72 S_1} [1 + \cos 2\omega t]$$

$$\text{Force due to } \phi_2 = \frac{1}{2} \times \frac{(\phi_2)_m^2}{72 S_2} [1 - \cos 2\omega t]$$

FIG. 7. Instantaneous and average force relationships for the two-phase magnet of Fig. 6.

Adding these equations, the resultant force effective over the entire pole face is

$$\text{Total force} = \frac{1}{2} \frac{(\phi_1)_m^2}{72 S_1} + \frac{1}{2} \frac{(\phi_2)_m^2}{72 S_2}$$

As $(\phi_1)_m = (\phi_2)_m$, and as $S_1 = S_2$,

$$\text{Total force} = \frac{(\phi_1)_m^2}{72 S_1} = \frac{(\phi_2)_m^2}{72 S_2}$$

which shows that the total force is constant and equal to that produced by half the pole face carrying a constant flux equal to the peak sinusoidal value. While the total force is constant with this arrangement, the iron is still used at one-half its maximum effectiveness.

The arrangement of Fig. 6a, even though it will produce a constant force over the entire pole face, is not ideal, because it will not keep the point of application of the force at the same place. This will shift from the center of section 1 to the center of section 2, at a frequency double that of the supply. This often causes chattering if the pole or armature are not absolutely flat. It may be avoided by splitting section 2 into two equal parts and placing half on each side of section 1, as shown in Fig. 8. This will keep the center of action of the force stationary.

Practically, to actually attain the quadrature polyphase arrangement of Fig. 6a, one would have to build a magnet like that shown in Fig. 9, where coil circuits 1 and 2 receive their excitation from different phases of a two-phase supply. The arrangement shown is obviously nothing more than two independent single-phase magnets whose cores are fastened together. Though the total pull on the armature is constant, its center of application will shift from a to b at double the supply frequency. A better arrangement, which would eliminate the space shift of the point of application of the force, would be to wind phase 1 on poles 1 and 4, and phase 2 on poles 2 and 3. This would necessitate a yoke of larger cross section between poles 2 and 3.

Fig. 8. Two-phase magnet pole face arrangement giving a constant force at a fixed point of application.

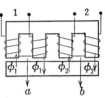

Fig. 9. A two-phase magnet which gives a constant force but a varying point of application. If phase 1 is wound on pole cores 1 and 4, and phase 2 on pole cores 2 and 3, an ideal arrangement can be obtained.

A more practical arrangement of the two-phase magnet of Fig. 9 is shown in Fig. 10a. Here the two center cores are combined to form one core, common to both magnetic circuits, as is frequently done when building two-phase transformers. The center core cannot be wound with coils of either phase, as its flux is the vector resultant of the phase fluxes as shown in Fig. 10b; its area need only be $\sqrt{2}$ times greater than that

Fig. 10a. Fig. 10b.

A two-phase magnet arrangement similar to a two-phase transformer, and the vector diagram of the fluxes.

of an outer core. The force produced by the center core will be that due to the resultant flux, and will pulsate between zero and its maximum, just like that of the outer cores. Its phase position will be midway between that of the outer poles, and when combined with them will produce a total force on the armature which is constant, but the point of application of this force will alternate between a and b at twice the supply frequency.

In order to produce the ideal arrangement of Fig. 8, a two-phase magnet must be arranged as shown in Fig. 11, or in an equivalent man-

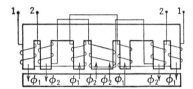

FIG. 11. A two-phase magnet equivalent to that of Fig. 8.

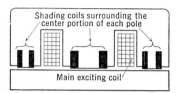

FIG. 12. A shaded pole single-phase magnet arrangement which can be made to give almost constant force for a particular gap length.

ner. This will give not only a constant resultant force but also a constant point of application of the resultant force. A properly designed, shaded-pole, single-phase magnet can be made to approach the ideal of Fig. 8 very closely for a particular gap length. Such a magnet is illustrated in Fig. 12. The shading coil causes the flux passing through it to be out of phase with that passing through the unshaded portion of the pole. The theory and design of shading coils will be taken up in Art. 123.

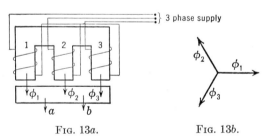

FIG. 13a. FIG. 13b.

Three-phase magnet system and vector diagram of fluxes.

4. **Polyphase Operation—Three-Phase.** In Fig. 13a is illustrated a three-phase magnet, having an exciting coil on each leg connected to a different phase. The flux waves of the three legs will be 120° of out phase, as shown in Fig. 13b; their respective instantaneous forces are shown in Fig. 14.

As can be seen from this figure, the resultant force of all poles, which is the sum of the instantaneous forces of the individual poles, is constant. The point of application of the resultant will shift from (a) midway between poles 1 and 2 to (b) midway between poles 2 and 3 at twice the supply frequency. Therefore, as regards constancy of force and the shifting of the point of application of the force, it is identical with the two-phase magnet. Likewise, the total force developed by all poles is 1.5 times the peak force of one pole, and as there are 3 poles, the effective use of the iron as compared to direct-current use is only one-half.

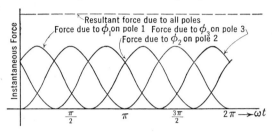

Fig. 14. Instantaneous and average force relationships for the magnet of Fig. 13.

Another arrangement of a three-phase magnet is shown in Fig. 15. Here, three single-phase U-shaped magnets with an exciting coil on one leg of each U, are placed with the spare legs toward the center, forming a Y. Each coil is connected to a separate phase. The force due to each magnet is the same as illustrated in Fig. 14. If the phase rotation is 1–2–3 as shown in Fig. 14, when magnet 1 pulls with maximum strength, 2 and 3 pull with one-quarter of maximum strength; one-sixth of a cycle later 3 will pull with maximum strength and 1 and 2 with one-quarter strength, etc. Thus the point of application of the force will shift around a circle at twice the frequency of the supply, but in a direction opposite to the phase rotation of the supply.

5. **Summary of Polyphase Characteristics.** Summarizing, we may state that:

1. Balanced polyphase magnets produce a total force which is constant.

2. The point of application of the resultant force will shift at a frequency twice that of the supply.

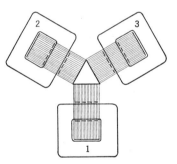

Fig. 15. A three-phase magnet consisting of three single-phase magnets arranged in the form of a Y.

3. The effective use of the iron on a polyphase magnet is only one-half that which would be obtained with the same amount of iron on direct current.

4. The use of shading coils on polyphase magnets is desirable if chattering is to be reduced to a minimum.

117. Volt-Ampere Limitation of Alternating-Current Magnets

Consider the case of the constant-voltage magnet in which the flux and hence the flux linkage is assumed to be proportional to current, as represented by the line 01 of Fig. 16. Let I_1 be the current required to produce the flux linkage ($N\phi$) at the beginning of the stroke. Then as the plunger completes its stroke, moving from 1 to 2, the exciting current will fall from I_1 to I_2 owing to the increase of self-inductance as the gap closes. If the resistance of the coil is small, the reactance voltage will be substantially equal to the supply voltage, and the flux linkage will remain constant during the motion as explained in Art. 115. The work done during this motion is equal to the change in stored energy of the magnet, which is equal to the area 0–1–2–0 of the flux linkage-current loop. Expressing this area in terms of the dimensions of the figure, we have

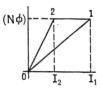

FIG. 16. Flux-linkage current loop for a constant-voltage magnet.

$$\text{Average } \Delta W = \tfrac{1}{2} N\phi (I_1 - I_2) \quad \text{joules} \tag{19}$$

where ϕ and I are r.m.s. values. The reactance voltage of the coil will be, by equation 14,

$$\text{Reactance voltage} = E_r = \frac{2\pi f N \phi_m}{\sqrt{2}} = 2\pi f N \phi \tag{14}$$

Substituting this into (19), we have

$$\text{Average } \Delta W = \frac{1}{4\pi f} E_r (I_1 - I_2) \quad \text{joules} \tag{20}$$

Changing from joules to inch-pounds,

$$\text{Average } \Delta W = \frac{0.706}{f} E_r (I_1 - I_2) \quad \text{in-lb.} \tag{21}$$

The product $E_r(I_1 - I_2)$ is the change in reactive or quadrature volt-amperes taken by the magnet due to plunger motion. Calling this ΔQ, we have

$$\text{Average } \Delta W = \frac{0.706}{f} \Delta Q \quad \text{in-lb.} \tag{22}$$

Thus, at a given frequency a definite change in the quadrature volt-ampere excitation of the magnet must occur to produce a specified mechanical work. For instance, to produce 1 in-lb. of mechanical work from a magnet operating on a 60-c.p.s. power supply will require

$$\Delta Q = \frac{(\Delta W)f}{0.706} = \frac{1.0 \times 60}{0.706} = 85 \text{ volt-amperes change}$$

and on a 25-c.p.s. supply,

$$\Delta Q = \frac{1.0 \times 25}{0.706} = 35.4 \text{ volt-amperes change}$$

This is only the change in reactive power supplied to the magnet and does not represent the reactive power input necessary to operate the magnet. The total reactive power input to the magnet at the beginning of the stroke Q_1 is equal to

$$Q_1 = \Delta Q + Q_2$$

where Q_2 is the reactive power necessary to excite the magnet at the end of the stroke in the closed-gap position. Q_2 cannot be reduced to zero because of volt-amperes required to produce the constant flux ϕ through the reluctance of the iron and any air gaps which may be left in the magnetic circuit. The ratio $\Delta Q/Q_1$ is a measure of the volt-ampere efficiency of the magnet. If the iron is not highly saturated, and if there are no air gaps at the end of the stroke, it can be made to approach quite closely to unity. When there is an air gap at the end of the stroke, as is usually necessary in plunger magnets, or relays, it will vary between about 0.75 and 0.25.

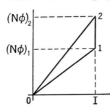

Fig. 17. Flux-linkage current loop for a constant-current magnet.

If the magnet operates on a constant-current supply, the flux linkage-current loop will be changed from that of Fig. 16 to that shown in Fig. 17. In this case the flux linkage of the coil will be $(N\phi)_1$ at the beginning of the stroke and will rise to $(N\phi)_2$ at the end of the stroke. The work done during this motion will again be equal to the change in stored energy of the magnet, which is equal to the area 0–1–2–0 of the flux linkage-current loop,

$$\text{Average } \Delta W = \tfrac{1}{2}I[(N\phi)_2 - (N\phi)_1] \text{ joules} \qquad (23)$$

Substituting for $(N\phi)$ in terms of the reactance voltage from equation 14, we have

$$\text{Average } \Delta W = \frac{1}{4\pi f} I[(E_r)_2 - (E_r)_1] \text{ joules} \qquad (24)$$

Changing from joules to inch-pounds,

$$\text{Average } \Delta W = \frac{0.706}{f} I[(E_r)_2 - (E_r)_1] \quad \text{in.-lb.} \tag{25}$$

The product $I[(E_r)_2 - (E_r)_1]$ is the change in reactive volt-amperes taken by the magnet due to plunger motion.

This result is seen to be exactly the same as for the constant voltage case, and hence we can conclude that the volt-ampere limitation of equation 22 applies to any alternating-current magnet.

118. The Alternating-Current Magnet vs. the Direct-Current Magnet

In the last three articles we have derived the fundamental relationships applying to alternating-current magnets. It appears wise to pause at this point and summarize their limitations in comparison with those of direct-current magnets, so that an intelligent decision can be made as to whether to use an alternating-current magnet or a direct-current magnet in combination with a rectifier. Modern dry disk rectifiers, of the copper oxide and selenium types, and electronic rectifiers offer such a convenient and efficient way of converting from alternating to direct current that, unless there are special circumstances, it is hard to justify the use of an alternating-current magnet.

The limitations of alternating-current magnets as compared with direct-current magnets, operating with the same maximum flux density, are summarized as follows:[3]

1. **Force Limitation.** For a given pole face area, the average force on alternating current is only one-half that obtained with direct current. This applies to both single and polyphase operation.

2. **Weight Limitation.** For a given force and stroke the alternating-current magnet will be much heavier than the direct-current one because (a) at least twice as much iron must be used to develop the same force and (b) generally more copper is required to carry the large reactive power for alternating-current operation.

3. **Volt-Ampere Limitation.** For alternating-current operation from 60- and 25-c.p.s. power supplies, a minimum of 85 and 35.4 volt-amperes, respectively, per inch-pound of work is required under ideal conditions. In actual operation this minimum will usually be considerably higher.

[3] This comparison is based on operation within the range of substantially linear proportionality between flux and current. While it is economical to saturate direct-current magnets, it is decidedly poor practice to saturate an alternating-current one, as then the exciting volt-amperes become disproportionately high, causing the volt-ampere efficiency $\Delta Q/Q_1$ to be very low. Bearing this in mind, the comparison made is somewhat optimistic about alternating-current operation.

Direct-current magnets have no volt-ampere limitation unless speed of action is a consideration.

4. Speed of Action. Alternating-current magnets are inherently fast acting, because (1) the time-constant is of the same order as the period of one cycle, and (2) the self-induced voltage due to motion is generally small compared to the impressed voltage. In a direct-current magnet the speed of action can be altered at will by changing the ratio of the self-induced voltage due to motion and the impressed voltage. In general, a direct-current magnet can be designed to accomplish the same work in the same time with a smaller peak volt-ampere input than on alternating current. Thus the high-speed magnet designed in Art. 113 requires a peak of 14 amperes $\times$ 12 volts = 168 volt-amperes, to do 4.7 in-lb. of work in 0.04 second. On a supply frequency of 25 c.p.s., with a volt-ampere efficiency of 50 per cent, an alternating-current magnet would require

$$4.7 \times 35.4 \times 2 = 333 \text{ volt-amperes}$$

while on a 60-c.p.s. supply it would require 800 volt-amperes. It is quite likely, however, that when operating on the 60-cycle supply the time of action would be shorter than the 0.04 second required on direct current.

5. Eddy-Current Limitation. All alternating-current magnets must be laminated or slotted, while only high-speed direct-current magnets must be so treated. A poorly laminated alternating-current magnet will have excessive eddy currents. This not only will decrease the effectiveness of the iron below the limitations noted in 1 and 2, but also will result in greatly increased heating in the iron parts and an excessively high volt-ampere consumption.

6. Nature of Service. Where the magnet is to be operated from a large power supply system on intermittent duty, many of the foregoing limitations noted do not necessarily offer serious objection to the use of an alternating-current magnet. The large volt-ampere input at the beginning of the stroke will not materially affect the regulation of an adequate power supply. If the magnet is designed to have no air gaps at the end of the stroke, the volt-ampere input will drop to a very low value at the completion of the stroke, and operation of the magnet will merely act like a momentary surge on the system. Heating in the closed-gap position will be comparable to that of an unloaded transformer of the same size, if the iron parts are well laminated. The contactors which handle the magnet, however, will be called upon to handle the initial volt-ampere input.

119. Characteristics of the Alternating-Current Solenoid and Plunger or Leakage-Flux Type of Magnet

1. Description of Modes of Construction. In Figs. 18, 19, and 20 are shown various ways of building alternating-current solenoid and plunger-type magnets.

The magnet of Fig. 18 is of the fully laminated type having a square plunger. The laminations of the outer return magnetic circuit are clamped between ribbed cast-iron frames which also serve to guide the plunger. The bolts which clamp these frames should avoid passing through the laminations, if possible. The plunger is also made of laminated iron and is riveted between heavy side plates as shown. For best results these side plates should be silicon steel. The coil is self-supporting, and it is so held as to stay out of contact with the plunger. The stationary stop A is arranged to protrude slightly from the main body of the laminations, so that its face, which comes in contact with the plunger, may be accurately machined after the laminations have been assembled and clamped. This prevents undue humming and vibration at the end of the stroke and also gives maximum possible permeance, thereby reducing the required volt-ampere excitation at the end of the stroke to a minimum. Sometimes this stop is made in the form of a plunger arranged to be clamped between the frames. This allows the stroke to be adjustable and the working gap to be placed in the most advantageous position with respect to the coil.

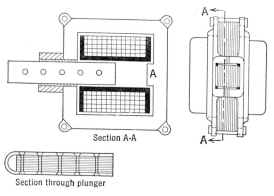

Fig. 18. Alternating-current solenoid and plunger-type magnet built in the form of a shell-type transformer.

A section through the axis of a slotted cylindrical type of magnet is shown in Fig. 19a, and in Fig. 19b are shown sections of the plunger and shell to illustrate the method of slotting. The method of providing the return magnetic circuit through the laminated end washers independent of the bearings allows the clearance between these washers and the plunger to be reduced to a few thousandths of an inch necessary for mechanical clearance. This increases the permeance of the fixed cylindrical gap and so reduces the volt-amperes consumed in this gap by the plunger flux. The clearance between the other washer and the plunger should be designed to give the desired force characteristic. The concentricity of the plunger and washers is maintained by fitting the bearings and washers tightly in the cast-iron end plates. The rim of this plate, in contact with the washers, should be slotted like the shell. The washers should have one radial slot as shown in Fig. 19c to eliminate the concentric flow of eddy currents due to that portion of the plunger

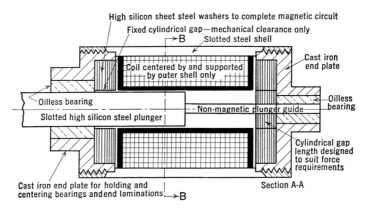

FIG. 19a. Cylindrical type of alternating-current solenoid and plunger magnet.

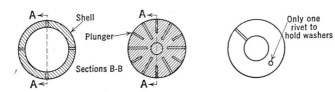

FIG. 19b. Sectionthrough shell and plunger of magnet of Fig. 19a.

FIG. 19c. Section through stack of high-silicon steel washers.

flux which passes axially through the washers. The coil is wound on a phenolic bobbin the flanges of which fit tightly in the shell and so keep the bobbin tube out of contact with the plunger. This construction is unique, in that it eliminates the conventional brass tube used to guide the plunger. A brass tube is undesirable because of the difficulty of slotting it to minimize eddy currents, and at the same time keeping it mechanically rigid. The brass tube, because of its thickness, also causes the reluctance of the fixed cylindrical gap to be high, and so greatly increases the volt-ampere consumption. A phenolic tube is impractical because of this consideration alone; its thickness would be

ART. 119] CHARACTERISTICS—SOLENOID AND PLUNGER TYPE 435

greater than that of the brass tube and would therefore further decrease the volt-ampere efficiency.

Where volt-ampere efficiency is of no great consequence the method of construction illustrated in Fig. 20 is cheap. The plunger is of square cross section and consists of high-silicon steel strips riveted between thicker steel plates as shown. These outer plates should also be made of silicon steel, if obtainable. The plunger is guided by a square brass tube which is slotted on each of its four sides to the extent shown in Fig. 20b. The flux is returned from the plunger to the shell by means of the silicon-steel washers shown. These washers should have one radial slot and should be held together by only one rivet as shown in

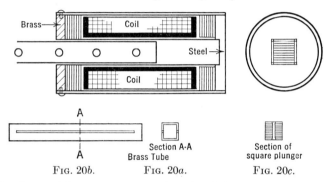

FIG. 20b.   FIG. 20a.   FIG. 20c.

A cylindrical type of alternating-current solenoid and plunger magnet having a square section laminated plunger, and a slotted square brass guide tube for the plunger.

Fig. 19c. The outer shell should have at least one slot which need extend only between the outside brass end plates.

2. **Constant-Voltage Characteristics.** In Fig. 21a is shown a diagrammatic sketch of a square plunger constant-voltage alternating-current solenoid and plunger magnet built like that illustrated in Fig. 18. In Fig. 21b the force-stroke and current-stroke characteristics are shown. The plunger displacements of Fig. 21b have been made to register with the position of the plunger in the coil of Fig. 21a. These curves represent actual data taken on a magnet having the dimensions shown.

Starting with plunger fully withdrawn beyond the edge of the coil, position a, it is seen that the current is very high. This high value of current is necessary to produce the required flux linkage through the low-permeance magnetic path. Movement of the plunger from a to b produces only a very slight improvement in the permeance and hence the exciting current is reduced very little. As the plunger moves from b to the stop at f, the permeance increases linearly and the current falls to a low value at f.

In this particular magnet the low current at the end of the stroke is made possible by the high permeance of the fixed cylindrical gap pro-

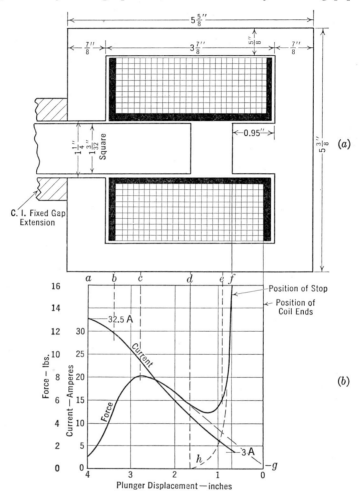

Fig. 21a. Section through an actual solenoid and plunger type of alternating-current magnet built in the form of a shell-type transformer.

Fig. 21b. Experimental force-stroke, and current-stroke curves for the magnet of Fig. 21a, when operating from a 60 c.p.s. constant-voltage circuit of 57 volts.

duced by the long cast-iron extension. The volt-ampere efficiency is therefore very high, equaling

$$\text{Volt-ampere efficiency} = \frac{\Delta Q}{Q_1} = \frac{I_1 - I_2}{I_1} = \frac{32.5 - 3.0}{32.5} = 0.908$$

The force-stroke data given show the actual electromagnetic pull, the plunger friction having been eliminated by taking the average of readings with the gap opening and closing. The force on the plunger is due to two effects: leakage flux force from a to d, and direct attraction between e and f. The transition between these two types of force action occurs between d and e. The leakage flux force is equal to

$$\text{Force} = \phi_L H$$

where H, the magnetic intensity of the coil, will vary directly with the exciting current. In the region from a to c the leakage permeance between the plunger and shell increases proportionally to the distance of the plunger insertion, and the leakage flux ϕ_L would increase almost as the square, if the exciting current remained constant. However, the exciting current drops, and hence the force does not increase as rapidly as the square. At c the leakage flux has reached its limiting value determined by the supply voltage, and so the force, from this point on, drops almost proportionally to the decrease in exciting current. If the stop were moved out of influence of the plunger flux, the force would continue dropping as shown by the dashed-line extension to g. At point d an appreciable amount of the plunger flux passes directly to the stop and so produces a direct attraction force. This force will increase about inversely as the square of the separation between the plunger and stop and produces the dashed force curve hf. The diversion of flux from the leakage path to the stop further augments the decrease of the leakage force toward the end of the stroke. Adding the two dashed curves, between d and f, we obtain the balance of the force-stroke characteristic.

The data of Fig. 21b can be used to check the theoretical deduction made in Art. 117, that at 60 c.p.s. a change in excitation of 85 volt-amperes is required for each inch-pound of mechanical work. The area under the force-stroke curve as determined by a planimeter measurement is 19.9 in-lb., which theoretically corresponds to a ΔQ of $19.9 \times 85 = 1,690$ volt-amperes. The actual change in volt-amperes input, according to the current curve, is $(32.5 - 3.0)57 = 1,680$ volt-amperes, which checks reasonably close with the theoretical value.

3. **Constant-Current Characteristics.** When the current is constant the force-stroke characteristic will be identical in shape with the constant-voltage direct-current characteristic. If the r.m.s. value of the alternating exciting current equals the direct exciting current, the average force on alternating current will equal the constant force on direct current, provided that the iron is not saturated. Constant-current operation is seldom obtained except on series lighting circuits or the series magnet of an arc light.

120. Design of the Constant-Voltage Alternating-Current Solenoid and Plunger, or Leakage-Flux Type of Magnet

1. General. The design of a constant-voltage alternating-current solenoid and plunger magnet is somewhat complicated by the nature of its force-stroke characteristic. Consider a cylindrical magnet constructed as shown in Fig. 19 and shown schematically in Fig. 22a. In Fig. 22b, and registering with the plunger positions of Fig. 22a, is shown the general shape of the force-stroke characteristic in the operating range of the plunger. This curve is of the same nature as that of Fig. 21b. For design purposes this curve may be analyzed on two bases:

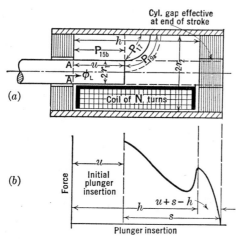

FIG. 22a. Schematic sketch of a cylindrical-type solenoid and plunger magnet showing design symbols.

FIG. 22b. Typical force-stroke curve for the magnet of Fig. 22b.

1. That the plunger will pull its load slowly through the stroke. In this case the inertia of the plunger system will not assist when the magnetic force is low, and hence the minimum magnetic force must be slightly larger than the load force plus friction.

2. That the plunger will act very rapidly. In this case the inertia of the plunger system will assist when the magnetic force is low, and the area under the force-stroke curve must equal the sum of the useful work, the work done against friction, and the work to accelerate the mass of the plunger and load to its final velocity.

On basis 1, the plunger and coil system should be designed to produce a specified minimum force at a plunger insertion slightly less than h, and should be checked against excessive saturation at the beginning of the stroke when the plunger insertion is u.

On basis 2, the plunger and coil system should be designed to produce a minimum specified area under the force-stroke curve, or what is equivalent to the same thing, a minimum specified change in reactive power when the plunger is moved over its useful stroke.

Considering these two design bases, then, the designer should be

ART. 120] DESIGN OF THE SOLENOID AND PLUNGER TYPE   439

able to compute the following items in order to design an alternating-current solenoid and plunger magnet.

(a) The minimum force throughout the stroke.
(b) The saturation of the iron, force, and volt-ampere consumption at the beginning of the stroke.
(c) The volt-ampere consumption at the end of the stroke.

The method of computing these items will be developed in the following sections.[4]

2. **Minimum Force throughout the Stroke.** This will occur when the plunger is separated from the stop or cylindrical gap a distance equal to about $r_2 - r_1$, as illustrated in Fig. 23. At this plunger position the flux will leak away radially to the outside shell from the cylindrical surface of the plunger through the path P_{15b}, and from the ends of the plunger through the paths P_{17} and P_{19a}. Convenient analytical relations, suitable for design purposes, can be derived if the permeances P_{17} and P_{19a} can be added in with P_{15b}. This can be done with sufficient accuracy, at least for preliminary designs, by considering the plunger to be inserted the full coil length h, and neglecting all permeances other than that of path P_{15b}. Then, the total distributed leakage permeance between the hypothetical plunger of length h and the shell will be, by equation 15b, Chapter V,

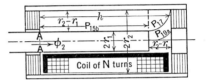

FIG. 23. Magnet of Fig. 22a with plunger at position of minimum force.

$$P_{15b} = \frac{2\pi\mu h}{\log_\epsilon r_2/r_1} \qquad (26)$$

Following the reasoning of Sec. 3, Art. 90, Chapter X, for the magnetic circuit equation, and letting K equal the part of the coil magnetomotive force which is used in the fixed cylindrical gap, and neglecting the magnetomotive force used in the iron,[5] the average magnetomotive force effective across the leakage path will be

$$F_{\text{avg.}} = \frac{(1 - 2K)NI}{2}$$

---

[4] Although the method is developed for the cylindrical type of Fig. 19, it will apply equally as well to the shell type of Fig. 18, provided that the proper permeance formulas are used.

[5] This is permissible because for efficient alternating-current operation the iron cannot be allowed to saturate and hence the percentage of the coil magnetomotive force consumed in the iron will be small.

The total leakage flux will then be [6]

$$\phi_L = \frac{\pi\mu h}{\log_\epsilon r_2/r_1}(1 - 2K)NI \qquad (27)$$

All this flux passes through section A–A of the plunger, producing the maximum plunger flux density at A–A, and gradually leaks away radially to the shell. The flux linkage produced by this flux can be estimated by equation [7] 3, of Art. 46, Chapter IV,

$$(N\phi) = \frac{2}{3}\frac{\pi\mu h}{\log_\epsilon r_2/r_1}(1 - 2K)N^2 I \qquad (28)$$

If the resistance drop of the coil is small,[8] the reactance voltage may be taken equal to the supply voltage. Then the actual r.m.s. value of the flux linkage can be found by solving equation 14 for $(N\phi)$

$$(N\phi) = \frac{(N\phi_m)}{\sqrt{2}} = \frac{E}{2\pi f} \qquad (29)$$

Equating 29 and 28, we have

$$I = \frac{E}{2\pi f} \times \frac{3}{2}\frac{\log_\epsilon r_2/r_1}{\pi\mu h N^2(1 - 2K)} \qquad (30)$$

Substituting this value of I into equation 27, the total leakage flux ϕ_L may be found.[9]

$$\Phi_L = \frac{3}{2} \times \frac{E}{2\pi f N} \qquad (27a)$$

The average force on the plunger will then be, by equation 20b, Chapter VIII

$$\text{Force} = 8.86 \times 10^{-5} \frac{NI}{h} \phi_L \quad \text{lb.} \qquad (31)$$

This, on the basis of the assumption made, is a fairly close approximation to the minimum force during the stroke.

[6] All values of flux, current, voltage, etc., used in this article are r.m.s., unless otherwise designated.

[7] This neglects the slight change in distribution produced by the fixed cylindrical air gap.

[8] This is the usual case, especially when the plunger is near the end of the stroke.

[9] Obviously, to solve equation 30, K must be known. The method used is to solve 27a for Φ_L and 30 for I, neglecting K. With this value of Φ_L and the approximate value of I, and the known permeance of the fixed cylindrical gap, K may be determined and equation 30 resolved using this value of K. As K will be small, one approximation should give sufficient accuracy.

3. The Saturation of the Iron, Force, and Volt-Ampere Consumption at the Beginning of the Stroke.

Referring to Fig. 22a, the plunger is shown with its initial insertion u at the beginning of the stroke. The flux of the plunger will leak to the shell through the leakage paths P_{15b}, P_{17}, and P_{19a}. The total flux linkage produced by this flux will be given by equation 29. The manner of computing the actual flux ϕ_L passing through section A–A of the plunger and hence the maximum plunger flux density which will occur at this section, the force, and the volt-ampere consumption can be carried out as follows:[10]

The flux passing through path P_{15b} is distributed, and will be, in the manner of equation 27,

$$\phi_{15b} = P_{15b} \frac{(1 - 2K)}{2} \left(\frac{Nu}{h}\right) I_1 \qquad (32)$$

where K represents the portion of the total ampere-turns developed in the length u of the coil, which is used up across the fixed cylindrical gap, and I is the exciting current when the plunger is at the beginning of its stroke.[11]

Estimating the average ampere turns effective across paths P_{17} and P_{19a} as $(1 - K)(u + r_2 - r_1)(N/h)I_1$, the flux through these paths will be

$$\phi_{17} + \phi_{19a} = (P_{17} + P_{19a})(1 - K)(u + r_2 - r_1)\frac{N}{h} I \qquad (33)$$

The total flux passing through the fixed cylindrical gap will then be the sum of (32) and (33),

$$\phi_L = \left[\left(\frac{1 - 2K}{2}\right) P_{15b} + (P_{17} + P_{19a})(1 - K)\left(\frac{u + r_2 - r_1}{u}\right)\right]\frac{Nu}{h} I \qquad (34)$$

The expression in the brackets is an equivalent permeance, concentrated at a distance u from the beginning of the coil, which, when multiplied by the total ampere-turns developed in the coil length u, gives the total plunger flux ϕ_L. The equivalent magnetic circuit is sketched in Fig. 24. Designating this permeance as A, we have

$$\phi_L = A \frac{Nu}{h} I_1 \qquad (35)$$

---

[10] This method assumes that the wave form of current remains substantially sinusoidal; i.e., the peak flux density occurring in the iron must not be so high that the peak magnetomotive force used in the iron as compared to that in the air paths will determine the current wave form.

[11] When designing a magnet, K can be chosen arbitrarily, the fixed cylindrical gap being designed later to require the assumed ampere-turns equal to $K\frac{Nu}{h}I$.

The peak flux density occurring at section A–A of the plunger will be

$$(B_m)_{A-A} = \frac{\sqrt{2}\phi_L}{\pi r_1^2} \qquad (36)$$

The flux linkage produced by ϕ_L is obtained by multiplying the flux of each path by the average turns with which it links, and adding the component linkages, thus:

$$(N\phi) = \left[\left(\frac{1-2K}{3}\right)P_{15b} + (P_{17} + P_{19a}) \right.$$
$$\left. (1-K)\left(\frac{u + r_2 - r_1}{u}\right)^2\right]\left(\frac{Nu}{h}\right)^2 I_1 \qquad (37)$$

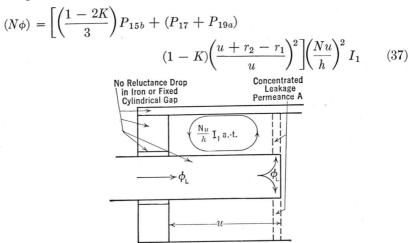

Fig. 24. Equivalent magnetic circuit for the magnet of Fig. 22a at the beginning of the stroke.

Designating the expression in the brackets by D,

$$(N\phi) = D\left(\frac{Nu}{h}\right)^2 I_1 \qquad (38)$$

Equating 38 and 29, and solving for the exciting current, we have

$$I_1 = \frac{E_r}{2\pi f D}\left(\frac{h}{Nu}\right)^2 \qquad (39)$$

where E_r is the reactance voltage of the coil.[12]

The average force due to the leakage flux will then be given by equation 31,

$$\text{Force} = 8.86 \times 10^{-5} \frac{NI}{h}\phi_L \text{ lb.} \qquad (31)$$

---

[12] If the resistance drop is small, E_r may be taken equal to E, the supply voltage; if not, (39) must be solved by successive approximation; first use the supply voltage for E_r and solve for I, then resolve, letting $E_r = \sqrt{E^2 - (IR)^2}$.

The exciting volt-amperes will equal the product of the r.m.s. exciting current I_1 of equation 39 and the r.m.s. supply voltage E.

$$Q_1 = EI_1 \tag{40}$$

4. The Volt-Ampere Consumption at the End of the Stroke. In a well-designed alternating-current magnet the air gaps left in the magnetic circuit at the end of the stroke should be small, and hence the magnetomotive force used in the iron parts will be a large part of the total excitation. Hence, in computing the volt-ampere excitation at the end of the stroke, that used by the iron must be taken into account.

Consider a magnet constructed like that of Fig. 19 and shown schematically at the end of the stroke in Fig. 25.[13] Because of the relatively high permeance of the cylindrical gaps at the plunger ends, the plunger flux may be considered to pass through the entire length of the plunger without leaking across to the shell or return path.[14] The maximum flux of the plunger will be, by equation 14,

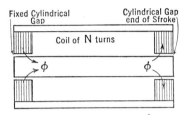

Fig. 25. Magnet of Fig. 22a with the plunger at end of stroke.

$$\phi_m = \frac{E}{4.44fN} \tag{41}{}^{15}$$

As the supply voltage has been assumed sinusoidal, the flux wave will be sinusoidal. In order to determine the wave of exciting current, the rising and falling magnetization curves for the entire magnetic circuit, between the cyclic limits of $+\phi_m$ and $-\phi_m$, must be plotted. When computing these curves, the magnetic intensity in the iron for the various values of flux density must be determined from a normal hysteresis loop for the particular iron. The extremities of the loop must pass through $+B_m$ and $-B_m$, where $B_m = \phi_m/S$, and S is the area of the particular iron part in question. Figure 26 shows an assumed rising and falling magnetization curve for the entire magnetic circuit, and the method of projecting the flux wave on the magnetization loop in order to obtain the wave of exciting magnetomotive force. Points on the flux wave are projected on the magnetization loop in the order a–b–c–d–e–f–g–h, and

---

[13] This discussion applies equally well to a magnet built like that of Fig. 18 or Fig. 27.

[14] This makes the magnetic circuit similar to that of a transformer at no load.

[15] At the end of the stroke the exciting current will be at its lowest value and hence the IR drop may be neglected.

produce progressively the wave of coil magnetomotive force [16] shown. The portion of the wave between π and 2π radians is not shown, as it is merely a repetition of that from 0 to π with the ordinates reversed.

The r.m.s. value of the wave can be determined by squaring each ordinate, and taking the square root of the mean height of the square curve obtained, or by any other equivalent method. Designating this r.m.s. value by F_2, the exciting current at the end of the stroke I_2 is

$$I_2 = \frac{F_2}{N} \qquad (42)$$

and the exciting volt-amperes at the end of the stroke will be

$$Q_2 = EI_2 \qquad (43)$$

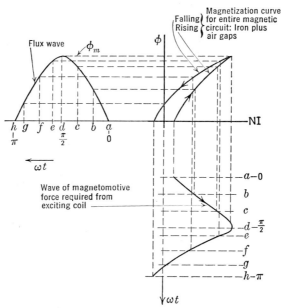

Fig. 26. Graphical construction for determining the wave of exciting current of a magnetic circuit.

5. Design Procedure. Owing to the restriction of constant flux linkage which is imposed by reason of the constant supply voltage, the

---

[16] If the eddy-current loss is appreciable and known, the magnetomotive force can be corrected to include this component, as follows. The maximum value of the eddy-current component of exciting magnetomotive force will be $\sqrt{2}P_eN/E$, where P_e is the eddy-current loss in watts. This component will be sinusoidal, and should be added to magnetomotive force wave of Fig. 26 by plotting it as a + cosine function, and adding the ordinates.

design of the alternating-current solenoid and plunger magnet is best carried out by trial. In the following paragraphs, suggestions as to possible procedure are outlined.

Design Limited by Minimum Force. With this limitation the design should be so chosen that the calculations, as outlined in Sec. 2, will give the proper minimum force. However, to avoid making too many trials, care must be exercised in selecting the values of NI/h and ϕ_L to be substituted into equation 31 to obtain the required minimum force. As the plunger is moved from its minimum-force position to its position at the beginning of the stroke, the permeance of the leakage path for the plunger flux will decrease. This will cause the exciting current to rise in order to maintain the flux. The average turns with which the flux links will also decrease, and hence the flux must be increased in order to maintain constant flux linkage. This calls for a further increase in exciting current.

These effects should be roughly estimated on the basis of the values of u and h assumed for the tentative design. Then the values of NI/h and ϕ_L chosen for the minimum force position can be selected so as to avoid excessive coil currents and plunger flux densities at the start of the stroke. Just what is to be considered as an excessive coil current will vary greatly with the application. If the magnet operates only infrequently, and if the current at the end of the stroke is low enough to cause no heating problem, high starting currents are economical. Peak plunger flux densities at the beginning of the stroke should generally be high compared to those used in transformer design, provided the iron is well laminated.

Design Limited by Area under Force-Stroke Curve. This is easier to handle than the last case. Here, the total work developed by the plunger is known, and hence the change in volt-ampere consumption ΔQ can be calculated by equation 22. Depending on the size and number of air gaps which will be present at the end of the stroke, a value for the volt-ampere efficiency should be assumed. Then Q_1, the required volt-ampere excitation at the beginning of the stroke, can be computed by dividing ΔQ by the volt-ampere efficiency. A magnet can then be designed to consume this number of volt-amperes at the beginning of the stroke. The maximum current density and plunger flux density which are permissible will depend as before on the duty cycle of the application and how well the iron is laminated. The volt-ampere consumption at the end of the stroke should be computed for the proposed design so that ΔQ can be checked.

121. Characteristics of the Direct-Attraction Type of Alternating-Current Magnet

1. Description of Construction. Figure 27a shows a direct-attraction-type, short-stroke magnet. The stationary iron circuit is constructed of laminations which are clamped between ribbed cast-iron frames. The plunger is of square cross section, made of a stack of laminations riveted between heavier steel plates and ground to size, as

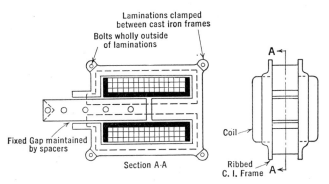

FIG. 27a. Alternating-current direct-attraction type of magnet built in the form of a shell-type transformer.

shown in Fig. 27b. The stack of laminations forming the stationary plunger should be faced, in order that surface contact can be obtained between the plungers at the end of the stroke. The placement of rivets and clamping bolts should be given careful consideration to avoid spurious eddy-current paths. The arrangement shown, with the clamping bolts outside of the laminations and the plunger rivets in a row along the center, is the best possible. The laminations of the stationary plunger can be held together by driving wedges between the plunger and the coil.

When the index number is too low for the economical use of the flat faced plungers of Fig. 27a, the coned plungers of Fig. 27c may be used.

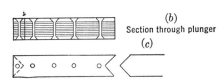

FIG. 27b. Section through the plunger of the magnet of Fig. 27a.

FIG. 27c. Alternate type of plunger pole face for the magnet of Fig. 27a when the index number is low.

An alternative construction for a short-stroke magnet can be arranged like that of Fig. 19, using round solid silicon-steel plungers with slots. These can have either flat or coned faces.

Art. 121] CHARACTERISTICS OF DIRECT-ATTRACTION TYPE 447

Figure 28 shows a short-stroke magnet of the clapper type. The stationary part of the magnetic circuit is made up of U-shaped laminations which can be riveted together between heavier steel plates. These rivets should be placed along a line of flux, so that no flux passes between any two rivets. This will prevent the formation of an eddy-current path by the two rivets and the connecting outer steel plates. Likewise the clapper armature should have one rivet at each end, as shown, or some in the center, as shown by the dotted circles. If both center and end rivets are used, they must be moved to the extreme left edge to avoid producing paths for eddy currents. The clapper should be hinged to the stationary member in such a way that it comes in good surface contact with both poles at the end of its stroke.

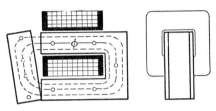

FIG. 28. Short-stroke clapper-type magnet.

The face of the clapper and the abutting pole face can be coned or otherwise shaped to increase the useful stroke for low index numbers.

For very short strokes or magnets having high index numbers, an arrangement like that of Fig. 4, or a single-phase adaptation of Fig. 10a with the exciting coil on the center leg, can be used. The latter is equivalent to flat-faced armature type of Chapters IX and X.

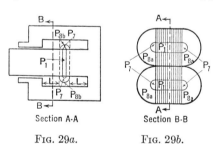

Section A-A

FIG. 29a.

Section B-B

FIG. 29b.

Leakage and fringing flux paths occurring in a direct-attraction type of magnet like that of Fig. 27a.

All these types of short-stroke magnets may have shading coils embedded in their pole faces to reduce the pulsation of pull and hence reduce chattering. The theory and design of shading coils for this application are taken up in Art. 123.

2. **Constant-Voltage Characteristics.** Consider a magnet built like that of Fig. 27a and shown schematically in Fig. 29 to illustrate the various leakage paths. Referring to Fig. 29a, almost all the plunger flux will pass through the main gap permeance P_1 and its fringing permeances P_7 and P_{8b}. The balance of the flux, which is the distributed leakage flux, will pass through the two ends of the coil over the length L, and can be assumed to follow the paths P_1, P_7, and P_{8a}, shown in Fig. 29b. The total flux linkage produced in all these paths can be expressed in terms of an

equivalent permeance P_e, all of whose flux links with all the turns. The flux of this equivalent permeance will be constant as the plunger moves, and can be found by solving equation 14,

$$\phi = \frac{E_r}{2\pi f N} \qquad (44)$$

where ϕ is the r.m.s. flux, N the total turns of the coil, and E_r the reactance voltage which can be taken equal to the supply voltage if the IR drop is small.

The flux which passes through the working gap of permeance P_1 will be ϕ/ν, where $\nu = P_e/P_1$ is the leakage coefficient based on flux linkage. The force developed in the working gap will be, by equation 18,

$$\text{Average force} = \left(\frac{\phi}{\nu}\right)^2 \frac{1}{72S} \text{ lb.} \qquad (45)$$

As ϕ is constant, the force will vary inversely with the square of the leakage coefficient. This force will not rise as rapidly as on the corresponding direct-current magnet, because the flux of the direct-current one is limited only by the saturation of the magnetic circuit and the available magnetomotive force, while on constant-voltage alternating-current magnets the flux remains constant.

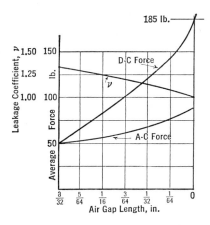

FIG. 30. Curves showing the relative force-stroke characteristics that will be obtained from the same magnet (magnet of Fig. 21, Chapter XII) when operating from a diect-current supply and a constant-voltage, alternating-current supply.

For an example, consider the magnet of Art. 113, Sec. 3, illustrated in Fig. 21, Chapter XII, which is of the type we are discussing. The leakage coefficient,[17] as a function of air-gap length, is shown in Fig. 30, reproduced from that of Fig. 23, Chapter XII. Suppose that the average force at the beginning of the stroke is to be 50 lb. Then the flux which may be considered as linking with all the turns is, by equation 45,

$$\phi = \sqrt{\text{Force} \times \nu^2 \times 72S} = \sqrt{50 \times 1.33^2 \times 72 \times 1.41} = 94.7$$
r.m.s. kmax.

[17] This leakage coefficient, though not based on flux linkage, is sufficiently close to it for comparative purposes.

where S is the gross area of the stack of laminations = 1.41 sq. in. The maximum flux density [18] in the plunger at the end of the stroke will be

$$B_m = \frac{\sqrt{2}\phi}{K_s S} = \frac{\sqrt{2} \times 94.7}{1.27} = 105.3 \text{ kmax. per sq. in.}$$

where K_s is the stacking factor of the laminations taken as 0.9. The average force at the end of the stroke will be $50 \times 1.33^2 = 88.5$ lb. On direct current, with the magnetomotive force constant at the value required to give the same starting force, it can be determined from Fig. 22, Chapter XII, that the force at the end of the stroke will be 185 lb. In Fig. 30 these comparative force-stroke characteristics for alternating- and direct-current operation are shown plotted. The relative values of force shown in these curves are typical for short-stroke magnets.

3. **Constant-Current Characteristic.** The constant-current characteristic will be of a shape similar to that obtained on direct current. For the same peak magnetomotive force, the force on direct current will be twice that obtained with sinusoidal alternating current, provided that the iron is unsaturated. As the iron saturates, the alternating-current force tends to approach that obtained on direct current.

122. Design of the Constant-Voltage Direct-Attraction Type of Alternating-Current Magnet.

1. **General.** With a constant-voltage supply the wave form of flux and flux density will be sinusoidal. The peak flux density allowed in the plunger at the end of the stroke is the limiting factor of the design. This may be selected on the basis of an economical design, and the force at the end of the stroke per unit plunger area calculated. If the leakage coefficient at the beginning of the stroke can be estimated, it is then possible to compute the unit force at the beginning of the stroke. This will allow the plunger area to be determined from the required force. Thence the plunger flux at the end of the stroke can be found and the required turns on the coil computed.

At this point a tentative estimate of the closed-gap volt-ampere consumption can be made by selecting dimensions for the fixed air gap and computing its volt-ampere consumption. To this should be added an estimated volt-ampere consumption for the iron. This may be deter-

---

[18] A peak flux density of 105.3 kmax. per sq. in. is high, because for high-silicon steel this will require a peak H of 175 ampere-turns per inch, and as the length of the iron is $6\frac{5}{8}$ in., the iron alone will consume a peak magnetomotive force at the end of the stroke of 1,160 ampere-turns. This will result in a low volt-ampere efficiency and a high core loss.

mined by estimating the volume of iron required and the volt-ampere excitation [19] required per unit volume of iron.

The total volt-ampere consumption divided by the supply voltage will give the exciting current. The coil space and wire size may now be designed to carry this current. A complete sketch of the preliminary design may be now made, and final calculations to check its performance should be made.

As the design of the short-stroke magnet is fairly definite, it appears best to derive the specific design equations and to illustrate their use by carrying through a specific design as was done in Chapter X.

2. Design Data. For the purpose of illustration, let it be required to design a constant-voltage alternating magnet of the type of Fig. 27, having given the following data:

Stroke = 0.125 in. = g.
Force = 50 lb.
Voltage = 115 volts.
Frequency = 60 c.p.s.

Excitation = continuous with plunger at end of stroke.
Material = No. 26 U.S.S. gauge medium-silicon steel.

Temperature rise.[20] The average temperature rise of coil (by change of resistance), above surrounding ambient temperature of 20° C., shall not exceed 70° C. from a cold start, if the excitation is applied continuously for 2 minutes with the plunger in the open-gap position; nor shall a 55° C. rise be exceeded if the excitation is maintained continuously in the closed-gap position.

3. Choice of Pole Face Type and Estimation of Leakage Coefficient. Some of the data of Chapters IX and X for the design of direct-current magnets can be made applicable to the short-stroke magnet if the latter is converted over to an equivalently designed direct-current magnet. For the same area of plunger, and the same peak flux density, the direct-current magnet will develop twice the force. Now the leakage coefficient of a magnet, and hence the magnet type, depend on the ratio of the plunger area to stroke. Therefore the index-number data of Chapter IX can be applied to the alternating-current magnet, if the equivalent index number of the latter is determined by doubling the required alternating-current force.

For the required design the equivalent index number will be $\sqrt{50 \times 2} \div 0.125 = 80$, which by reference to Fig. 10, Chapter IX,

---

[19] This is often available in manufacturers' data, or it may be estimated from the data of Chapter II.

[20] Temperature limitations on a magnet of this type are difficult owing to great disparity between heating in the open- and closed-gap positions. It is probably best to design the coil to withstand the maximum excitation for a limited period, and then check the final temperature rise in the closed-gap position.

indicates the flat-faced plunger type. Referring to Fig. 8, Chapter X, it is found that for direct-current flat-faced plunger magnets a suitable working gap flux density, for an index number of 80, would be 63 kmax. per sq. in.; while at a very high index number, say 2,000, a suitable flux density would be 89 kmax. per sq. in. Inasmuch as the leakage coefficient at the very high index number is practically unity, the ratio of these two flux densities may be taken as an approximate value of the leakage coefficient for the required design. Thus,

$$\nu = \tfrac{89}{63} = 1.41$$

4. **Design Equations.** The fundamental design equations applying to the short-stroke alternating-current magnet are as follows:

Force Equation. Equation 45 may be used directly.

$$\text{Average force} = \frac{1}{72S}\left(\frac{\phi}{\nu}\right)^2 \text{ lb.} \qquad (46)$$

where S is the gross plunger area, and ν is the leakage coefficient, as defined in Art. 121, Sec. 2.

Flux Equation. ϕ of equation 46 is the r.m.s. plunger flux in the closed-gap position, as defined in Art. 121, Sec. 2, and is, by equation 44,

$$\phi = \frac{E_r \times 10^5}{2\pi fN} \text{ kmax.} \qquad (47)$$

where N is the total turns of the coil.

Maximum Plunger Flux Density.

$$B_m = \frac{\sqrt{2}\phi}{K_s S} \text{ kmax. per sq. in.} \qquad (48)$$

where K_s is the stacking factor of the laminations. B_m should lie between limits of 80 and 100 kmax. per sq. in. Judgment must be used in selecting this peak flux density. If the time of excitation is short compared to the thermal time constant, heating is not a limitation and the choice should depend on the desired volt-ampere efficiency or on considerations of size. A low value of B_m will reduce the volt-ampere consumption and heating at the end of the stroke, while a high value of B_m will give minimum size. When the time of excitation is long or continuous, heating becomes the design limitation.

Heating Equation. This may be taken directly from Art. 66, Chapter VII, as

$$\theta_f = \frac{qP}{2khP_m} \qquad (49)$$

where all the symbols are as defined in the reference, and P is the total power loss of the magnet, equal to coil copper loss plus the iron loss, $P = I^2R +$ Core loss.

Voltage Equation. The supply voltage will equal the vector sum of the reactance voltage and the resistance drop, as derived in equation 15,

$$E = \sqrt{E_r^2 + (IR)^2} \qquad (50)$$

Volt-Amperes Excitation Consumed by an Air Gap. The easiest way to develop an expression for this is in terms of a fictitious inductance made up of the air-gap permeance P_c and all the turns of the exciting coil, thus

$$\text{Volt-amperes} = E_r I = \omega L I \times I$$

But $L = N^2 P_c$; therefore

$$\text{Volt-amperes} = \omega N^2 P_c I^2 = \frac{\omega (NIP_c)^2}{P_c}$$

$$\text{Volt-amperes} = \frac{2\pi f \phi^2}{P_c}$$

If ϕ is in (r.m.s.) kilomaxwells and P_c in kilomaxwells per ampere-turn, then

$$\text{Excitation for air gap} = \frac{2\pi f \phi^2 \times 10^{-5}}{P_c} \quad \text{volt-amperes} \qquad (51)$$

Another convenient form of this equation is

$$\text{Excitation for air gap} = 2\pi f \phi NI \times 10^{-5} \qquad (51a)$$

where NI is the r.m.s. magnetomotive force.

5. Preliminary Design.

Plunger Area. Letting $B_m = 90$, $K_s = 0.9$, and $\nu = 1.41$, equation 46 may be solved for S, the gross plunger area as follows:

$$\phi = \frac{B_m K_s S}{\sqrt{2}}$$

Substituting this into equation 46, we have

$$\text{Average force} = \frac{1}{72S}\left(\frac{B_m K_s S}{\sqrt{2\nu}}\right)^2$$

which when solved for S gives

$$S = \frac{\text{Avg. force} \times 144\nu^2}{B_m^2 K_s^2} \qquad (52)$$

ART. 122]   DESIGN OF THE DIRECT-ATTRACTION TYPE   453

Substituting the known and assumed values into 52,

$$S = \frac{50 \times 144 \times 1.41^2}{90^2 \times 0.9^2} = 2.18 \text{ sq. in.}$$

Equivalent Plunger Flux. The flux of the plunger ϕ will be

$$\phi = \frac{B_m}{\sqrt{2}} K_s S = \frac{90}{\sqrt{2}} \times 0.9 \times 2.18 = 125 \text{ kmax.}$$

Coil Turns. The necessary turns on the exciting coil can then be found by solving equation 47 for N, and substituting,

$$N = \frac{E_r \times 10^5}{2\pi f \phi} = \frac{115 \times 10^5}{2\pi \times 60 \times 125} = 244 \text{ turns}$$

Tentative Dimensions. At this point a sketch of the tentative design should be drawn.

Assume the cross section of the plunger to be a square of G inches per side. Then $G = \sqrt{S} = \sqrt{2.18} = 1.475$ in. Let the length of the fixed air gap between the plunger and the stationary laminations be 0.02 in. and the width of the gap measured along the plunger be 1.5 in. The return iron circuit should have at least the same area as the plunger, hence its width can be made $G/2$. For a first trial the width of the coil space may be made $G/3$ and its length $1\frac{1}{2}G$. A sketch of this tentative design is shown in Fig. 31.

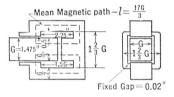

Fig. 31. Sketch showing proportions assumed for the tentative design of the magnet of the problem of Art. 122.

Exciting Current, Closed Gap. The volt-ampere consumption is most conveniently estimated by breaking the magnetic circuit into its component parts and computing their excitation requirements separately. These parts will be: the fixed air gap between the plunger and the stationary laminations, the residual gap between the plunger ends in the closed-gap position, and the iron circuit.

The fixed gap will consist of two parts in parallel, each 0.02 in. in length and having an area of 1.5×1.475 sq. in. Their joint permeance will be

$$P_c = \frac{\mu S}{l} = \frac{3.19 \times 10^{-3} \times 2.0 \times 1.475 \times 1.5}{0.02} = 0.706 \text{ kmax. per ampere-turn}$$

The volt-amperes consumed in the fixed gap will then be, by equation 51,

$$\text{Excitation, fixed gap} = \frac{2\pi f \phi^2 \times 10^{-5}}{P_c}$$

$$= \frac{2\pi \times 60 \times 125^2 \times 10^{-5}}{0.706} = 83 \text{ volt-amperes}$$

Assume that at the end of the stroke there is an effective residual gap of 0.002 in.; then the permeance of the gap will be

$$P_c = \frac{3.19 \times 10^{-3} \times 2.18}{0.002} = 3.48 \text{ kmax. per ampere-turn}$$

and the volt-ampere excitation will be

$$\text{Excitation, residual gap} = \frac{2\pi f \phi^2 \times 10^{-5}}{P_c}$$

$$= \frac{2\pi \times 60 \times 125^2 \times 10^{-5}}{3.48} = 17.0 \text{ volt-amperes}$$

For $B_m = 90$ kmax. per sq. in. the peak magnetic intensity in the iron will be, by reference to Fig. 15, Chapter II, for medium-silicon steel, 27 ampere-turns per inch. Assuming for the purposes of estimation that the magnetomotive force wave form is sinusoidal,[21] the r.m.s. magnetic intensity will be $27/\sqrt{2} = 19.1$. The net cross section of the iron circuit is $K_s S = 0.9 \times 2.18 = 1.96$ sq. in., and from Fig. 31 its average length is $17/3G = 8.35$ in. The r.m.s. magnetomotive force for the iron may then be taken as $8.35 \times 19.1 = 160$. Using equation 51a, the volt-amperes consumed in iron excitation will be

$$\text{Excitation, iron} = 2\pi f \phi NI \times 10^{-5}$$

$$= 2\pi \times 60 \times 125 \times 160 \times 10^{-5} = 75 \text{ volt-amperes}$$

The total volt-ampere consumption in the closed-gap position will then be $83 + 17 + 75 = 175$ volt-amperes. The exciting current in the closed-gap position will therefore be

$$I_2 = \frac{\text{v-a.}}{E} = \frac{175}{115} = 1.52 \text{ amperes}$$

[21] This assumption will make the volt-ampere consumption higher than it actually is, and hence is on the safe side. It is useful when the iron saturation does not exceed the knee of the saturation curve; for saturations above this it gives values which are much too high, owing to wave-shape distortion.

Exciting Current, Open Gap. In the open-gap position, the equivalent permeance of the working gap will be

$$P_c = \nu \frac{\mu S}{l} = \frac{1.41 \times 0.00319 \times 2.18}{0.125} = 0.0785 \text{ kmax. per ampere-turn}$$

and the required volt-ampere excitation of the gap will be

$$\text{Excitation, main gap} = \frac{2\pi \times 60 \times 125^2 \times 10^{-5}}{0.0785} = 751 \text{ volt-amperes}$$

The total volt-ampere excitation in the open-gap position will be $83 + 751 + 75 = 909$ volt-amperes, and the current will be

$$I_1 = \frac{909}{115} = 7.91 \text{ amperes}$$

Coil Design. The gross winding depth of the coil is $G/3$, say 0.5 in. Allowing $\frac{1}{16}$ in. clearance between the coil and the plunger, 30 mils for tape on each side of the coil, and $\frac{1}{32}$ in. for insulation between the coil and the outside laminations, the net winding depth will be $0.5 - 0.155 = 0.345$ in. The gross winding length of the coil is $1.5G$, say 2.25 in. Allowing $\frac{1}{16}$ in. for clearance and insulation and 30 mils for tape at each end, the net winding length will be $2.25 - 0.185 = 2.065$ in. The net winding area will be $0.345 \times 2.065 = 0.712$ sq. in. The mean length of a turn will be $4(G + 2 \times \frac{1}{16} + 2 \times 0.03 + 0.345) = 8.02$ in.

A coil for this type of magnet can be easily form wound, using double-cotton-covered wire, and varnishing each layer as wound. It should then be baked and taped. The space available for each wire will be $0.712/244 = 0.00292$ sq. in. This allows an over-all diameter of the wire of 0.054 in., corresponding to No. 17 with double cotton covering. Let us therefore assume a coil of 244 turns of No. 17 d.c.c. wire.

The coil resistance will be $8.02 \times 244 \times 0.422 \times 10^{-3} = 0.825$ ohm at 20° C., and 90° C. it will be $0.825 \times 1.275 = 1.05$ ohms, where 0.422×10^{-3} and 1.275 are the specific wire resistance and temperature factors from Tables II and III, respectively, of Chapter VI.

Temperature Rise During the 2-Minute Interval of Open-Gap Excitation. The weight of the wire of the coil will be $(8.02 \times 244)/(12 \times 161.3) = 1.01$ lb. The thermal capacity of the copper will be $180 \times 1.01 = 182$ joules per degree Centigrade, where 180 is the specific thermal capacity of copper from Table I, Chapter VII. Assuming an average coil resistance of 0.94 ohm during the 2-minute period of excitation on open gap, the heat energy released in the coil will be $I^2R\Delta t = 7.91^2 \times$

0.94 × 120 = 7,040 joules. The temperature rise during the interval [22] will be 7,040 ÷ 182 = 38.7° C. This is satisfactory.

Temperature Rise with Continuous Excitation in the Closed-Gap Position. The total inside and outside surface area of the coil of Fig. 31 is $(4G + 6\tfrac{2}{3}G)1.5G = 16G^2 = 34.8$ sq. in. The heat-dissipation coefficient for this coil, as it is self-supporting and out of contact with the metal parts, may be found from curve 2, Fig. 7b, Chapter VII; it is 0.00635 watt per sq. in. per °C. temperature difference, at a temperature rise of 55° C. The total heat-dissipating ability of the coil is 0.00635 × 34.8 = 0.221 watt per °C. temperature difference.

The heat to be radiated by the coil is the copper loss plus the core loss.[23] The copper loss will be $I_2^2 R = 1.52^2 \times 1.002 = 2.3$ watts. The volume of iron is 1.96 × 8.35 = 16.3 cu. in., and its weight will be 16.3 × 0.274 = 4.48 lb. From manufacturers' data the core loss for No. 26 gauge medium-silicon steel at $B_m = 90$ will be 2.2 watts per lb., at 60 c.p.s. Therefore the core loss will be 4.48 × 2.2 = 9.85 watts. The total heat to be dissipated by the coil will be 2.3 + 9.85 = 12.15 watts. The temperature rise of the coil will therefore be 12.15 ÷ 0.221 = 55.0° C. This is satisfactory.

6. Check of Preliminary Design Using Stock Sizes.

Final Sizes. The dimensions calculated for the preliminary design may be rounded off, by making $G = 1.5$ in.

Coil Design. For the coil let us assume 8 layers; then the turns per layer will be 244 ÷ 8 = 30.5 (say 31), and the net winding length will be [(31 + 1)/0.95]0.0548 = 1.85 in. Making the allowances required at each end, which we may take at about $\tfrac{1}{8}$ in., the gross length of the winding space can be made $2\tfrac{1}{16}$ in. The net winding depth will be 8 × 0.0548 × 0.95 = 0.416 in., and adding to this the allowances suggested in the preliminary design, the gross depth of the winding space may be taken as $\tfrac{9}{16}$ in. The mean length of turn will now be 4(1.5 + 2 × 0.092 + 0.416) = 8.4 in. The resistance at 75° C. will be

$$R = 8.4 \times 248 \times 0.422 \times 10^{-3} \times 1.2162 = 1.07 \text{ ohms}$$

Reactance Voltage. Taking the preliminary value of exciting current of 7.91 amperes, the reactance voltage in the open gap position will be

$$E_r = \sqrt{115^2 - (7.91 \times 1.07)^2} = 114.7 \text{ volts}$$

---

[22] This method assumes that the interval is sufficiently short that no appreciable heat flows from the copper.

[23] Because the heat-dissipation coefficients of Chapter VII have been derived on the basis that the coil is in free air, or surrounded by iron which has no heat generated in it, the iron loss should be added to that which the coil must dissipate. This is a convenient approximation and is on the safe side.

Equivalent Flux. The equivalent flux of the plunger will be, by equation 47,

$$\phi = \frac{E_r \times 10^5}{2\pi f N} = \frac{114.7 \times 10^5}{2\pi \times 60 \times 248} = 122.3 \text{ kmax.}$$

Leakage Coefficient. A section along the axis of the plunger showing the iron parts and coil of the magnet drawn to scale with complete dimensions is shown in Fig. 32. In the lower half of the sketch the various bands of leakage and fringing fluxes are shown. The dimensions indicate the extent of the bands. These paths are the same as shown in Fig. 29. This figure should be referred to in order to see the shape of the leakage-flux path and the formulas for its computation in the plane normal to the axis. Designating the various leakage paths by their formulas, as derived in Chapter V, the leakage coefficient based on flux linkage, as described in Art. 121, Sec. 2, is computed as follows:

The main gap permeance is

$$P_1 = \frac{\mu S}{l} = \frac{3.19 \times 1.5^2}{0.125} = 57.5$$

The fringing permeances are

$$P_7 = 0.26\mu l = 0.26 \times 3.19 \times 1.5 \times 4 = 4.97$$

$$P_{8b} = \frac{\mu}{\pi} \log_\epsilon \frac{r_2}{r_1} = \frac{3.19 \times 1.5 \times 4}{3.14} \log_\epsilon \frac{0.716}{0.125} = 10.62$$

$$P_9 = 0.077 \mu g = 4 \times 0.077 \times 3.19 \times 0.125 = 0.12$$

$$P_{10} = \frac{\mu t}{4} = \frac{4 \times 3.19 \times 0.295}{4} = 0.94$$

$$\text{Total} = 16.65$$

The leakage permeances on one side of the main gap are

$$P_1 = \frac{\mu S}{l} = \frac{2 \times 3.19 \times 0.57}{0.5625} = 6.49$$

$$P_7 = 0.26\mu l = 4 \times 0.26 \times 3.19 \times 0.57 = 1.88$$

$$P_{8b} = \frac{\mu l}{\pi} \log_\epsilon \frac{r_2}{r_1} = \frac{4 \times 3.19 \times 0.57}{3.14} \log_\epsilon \frac{33}{9} = 3.02$$

$$\text{Total} = 11.39$$

As regards flux linkage; the flux which passes through the main gap links with all the turns; the fringing flux may be considered to link with all the turns because, if the path of its mean flux line is examined, it will be found to include all but about 2 per cent of the turns; the leakage flux, considering all the turns of the coil, is only half effective as there are two paths in series, and it is $(57 \div 92.5)^2$ effective because the path

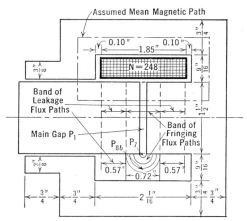

Fig. 32. Final design of the magnet of the problem of Art. 122. Lower half of the figure shows the assumed distribution of fringing and leakage fluxes.

covers only 0.57 in. out of a possible 0.925, and is $\frac{1}{2} \times \frac{2}{3}$ effective, owing to its distribution.

Therefore, the equivalent permeance as regards flux linkage will be

$$P_e = 57.5 + 16.65 + 11.39 \frac{1}{2} \times \left(\frac{57}{92.5}\right)^2 \times \frac{1}{2} \times \frac{2}{3}$$

$$= 57.5 + 16.65 + 0.72 = 74.9$$

The leakage coefficient, as regards flux linkage, is

$$\nu = \frac{74.9}{57.5} = 1.305$$

This is less than estimated in the preliminary design, which is partly due to the relatively short length of the coil.

Plunger Force. The plunger force in the open-gap position will be, by equation 46,

$$\text{Average force} = \frac{1}{72 \times 1.5^2} \left(\frac{122.3}{1.305}\right)^2 = 54.2 \text{ lb.}$$

and in the closed-gap position will be

$$\text{Average force} = \frac{1}{72 \times 1.5^2} (122.3)^2 = 92.5 \text{ lb.}$$

These forces are satisfactory.

Maximum Plunger Flux Density. The maximum value of the plunger flux density occurring at the end of the stroke will be, by equation 48,

$$B_m = \frac{\sqrt{2}\phi}{K_s S} = \frac{\sqrt{2} \times 122.3}{0.90 \times 1.5^2} = 85.5 \text{ kmax. per sq. in.}$$

Core Loss. The net area of the magnetic circuit is $1.5^2 \times 0.9 = 2.03$ sq. in., and its mean length is 8.25 in., as shown on Fig. 32. For the purpose of computing the core loss and the excitation for the iron, the small extension used to produce the fixed gap will be neglected. The volume of the iron path is $2.03 \times 8.25 = 16.7$ cu. in., and its weight will be $16.7 \times 0.274 = 4.57$ lb. Manufacturers' data give the core loss as 2.0 watts per lb. at $B_m = 84.5$, $f = 60$ c.p.s., for No. 26 U. S. gauge

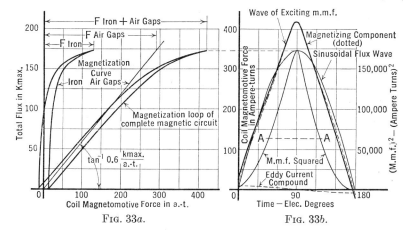

Fig. 33a.   Fig. 33b.

Magnetic data and graphical construction necessary to obtain the r.m.s. value of the exciting current of the magnet of Fig. 32.

medium-silicon steel. The core loss will therefore be $4.57 \times 2.0 = 9.2$ watts.

Calculation of Closed-Gap Exciting Current. This is done by following the procedure outlined in Art. 120, Sec. 4. In Fig. 33a are shown the magnetization curves, for both rising and falling fields, for the iron and the residual air gaps of the magnetic circuit. The magnetization curve for the iron has been computed from a normal hysteresis loop of iron for $B_m = 85.5$, using an area of 2.03 sq. in. and a mean length of 8.25 in. for the iron. The air-gap curve has been constructed by drawing a straight line through the origin, having a slope equal to the joint permeance of the fixed air gap and the residual gap between the plungers. By adding the magnetomotive-force components of these curves for the same value of flux, the heavy loop which is the magnetization curve for the entire circuit is obtained.

By projecting the sinusoidal flux wave of Fig. 33b on the heavy loop in the manner illustrated in Fig. 26, the ordinates of the dashed curve, entitled "magnetizing component," are obtained as the abscissas of the loop. This is the true wave of magnetomotive force required to establish the sinusoidal flux in the complete magnetic circuit. However, this does not exactly correspond to the exciting current, because the latter must have in addition a component to neutralize the eddy currents produced in the iron parts. This is a power component whose phase position will lead the flux curve by 90°. If the flux is sinusoidal it will be sinusoidal. Its maximum value will be

$$(NI_e)_m = \left(\frac{\sqrt{2}P_e}{E}\right) N \text{ ampere-turns}$$

where P_e is the total power loss due to eddy currents in the laminations or solid iron structure of the magnet. Inasmuch as we have assumed the laminations to be perfectly insulated, and the bolts and rivets carefully placed so as to avoid eddy currents, we need consider only the eddy-current loss in the laminations. Estimating this as 25 per cent of the total core loss, $P_e = 9.2 \times 0.25 = 2.3$ watts, and

$$(NI_e)_m = \frac{\sqrt{2} \times 2.3}{115} \times 248 = 7.0 \text{ ampere-turns}$$

Adding together the two dashed-line components, the heavy full-line wave of exciting magnetomotive force of Fig. 33b is obtained. The r.m.s. value of this wave is obtained by squaring each ordinate and plotting them as indicated. The mean height of the curve can be obtained by a planimeter and is shown as the line A–A at 63,800 (ampere-turns)^2. The exciting current will then be

$$I_2 = \frac{\sqrt{63,800}}{248} = 1.02 \text{ amperes}$$

This does not appear to check too well with the preliminary value of 1.52 amperes. However, owing to the change of dimensions and turns, the preliminary value should be recomputed if a comparison is to be made. Repeating the preliminary estimate:

Excitation, all residual air gaps $= \dfrac{2\pi \times 60 \times 122.3^2 \times 10^{-5}}{0.600} = 94$ volt-amperes

ART. 122]   DESIGN OF THE DIRECT-ATTRACTION TYPE   461

Excitation, iron $= 2\pi \times 60 \times 122.3 \times \dfrac{130}{\sqrt{2}} \times 10^{-5} = 42$ volt-amperes

This gives a total excitation of 136 volt-amperes corresponding to an exciting current of $136/115 = 1.18$ amperes. This checks better, but as explained before,[21] is high because it is based on sinusoidal exciting wave for the iron of the same peak value as the iron actually takes. Examination of Figs. 33a and 33b shows that the magnetomotive force required to excite the iron does not become appreciable until $\phi = 125$ kmax. and, from that point on, causes the wave form to shift from sinusoidal to triangular. Where the transition will occur and how much the wave will peak depend entirely on the relative values of the iron and air-gap magnetization curves.

Calculation of Open-Gap Exciting Current. The iron excitation will be practically the same as in the closed-gap position and may be obtained by subtracting the air-gap excitation from the total computed excitation of the last paragraph, thus:

Excitation, iron only $= 115 \times 1.02 - 94 = 23$ volt-amperes

The equivalent permeance of the main gap including all leakage and fringing permeance will be $\nu\mu S/l = 1.305 \times 0.00319 \times 1.5^2/0.125 = 0.075$ kmax. per ampere-turn. Adding this in series with the corrected permeance of the fixed gap, which is 0.72 kmax. per ampere-turn, we have for their joint permeance 0.0679 kmax. per ampere-turn. Then the excitation for the air gaps will be:

$$\text{Excitation, air gaps} = \dfrac{2\pi \times 60 \times 122.3^2 \times 10^{-5}}{0.0679} = 831 \text{ volt-amperes}$$

The open-gap exciting current will therefore be

$$I_1 = \dfrac{831 + 23}{115} = 7.42 \text{ amperes}$$

Volt-Ampere Efficiency. This will be

$$\text{Volt-ampere efficiency} = \dfrac{I_1 - I_2}{I_1} = \dfrac{7.42 - 1.02}{7.42} = 0.863$$

Mechanical Work Check. The work as calculated from the area under the force-stroke curve should equal that calculated from the change in reactive power.

Mechanical work, computed by the change in reactive power at $f = 60$ c.p.s., equals

$$\dfrac{\Delta Q}{85} = \dfrac{(7.42 - 1.02)}{85} \times 115 = 8.66 \text{ in-lb.}$$

while computing by the area under force-stroke curve, it may be estimated from the starting force of 54.2 lb., and the closed-gap force of 92.5 lb., as

$$\tfrac{1}{8}[54.2 + \tfrac{1}{3}(92.5 - 54.2)] = 8.39 \text{ in-lb.}$$

This estimate is on the low side because the curve is slightly straighter than parabolic, as assumed in averaging. It is therefore seen that the two methods check very closely.

Temperature Rise—Closed Gap. The temperature rise in the closed-gap position is due to the iron loss plus the copper loss. The iron loss has been computed as 9.2 watts, and the copper loss will be $1.02^2 \times 1.07 = 1.1$ watts, making a total loss of 10.3 watts. The total inside and outside surface area of the coil will be $(6.74 + 10.3)1.91 = 32.6$ sq. in. The final temperature rise will be

$$\theta_f = \frac{10.3}{32.6 \times 0.00635} = 49.7° \text{ C.}$$

This is satisfactory.

Maximum Allowable Time of Excitation Open Gap. The weight of copper in the final coil is 1.04 lb., and its thermal capacity is 187 joules per degree Centigrade rise. The resistance of the coil at its average temperature of $(20 + 90) \div 2 = 55°$ C. will be $1.07 \times 289 \div 309 = 1.0$ ohm. The average rate of heat dissipation in the coil will be $I^2R = 7.42^2 \times 1.0 = 55$ watts. Then the time of continuous excitation for a maximum temperature rise of 70° C. will be

$$\text{Time} = \frac{187 \times 70}{55} = 238 \text{ sec., say 4 min.}$$

Power Factor. The power factor will equal the losses divided by the exciting volt-amperes: The losses will equal the iron loss of 9.2 watts plus the copper loss which is 55 watts in the open-gap position and 1.1 watts in the closed-gap position.

$$\text{Power factor, closed gap} = \frac{9.2 + 1.1}{117} = 0.088$$

$$\text{Power factor, open gap} = \frac{9.2 + 55}{834} = 0.075$$

Magnet Weight. The weight of the iron of the magnetic circuit and the copper is 4.77 and 1.04 lb., respectively, making the total weight of these parts 5.81 lb. This does not include the weight of the cast-iron clamping frame and insulation or bearing guides.

Final Results.

Supply voltage = 115 volts, 60 c.p.s.
Stroke = $\frac{1}{8}$ in.
Final temperature rise, continuous excitation, closed gap = 50° C.
Maximum allowable time of excitation on open gap without exceeding a coil temperature of 70° C. = 4.0 min.
Coil turns = 248.
Wire = No. 17 double cotton covered.
Coil resistance at 75° C. = 1.07 ohm.
Volt-ampere efficiency = 0.863.

	Average Force, lb.	Exciting		Power Input, watts	Supply Power Factor
		Current r.m.s., amp.	Volt-Amp.		
Open gap	54.2	7.42	854	64.2	0.075
Closed gap	92.5	1.02	117	10.3	0.088

123. Theory of Shading Coils for Single-Phase Magnets

1. General. The fundamental purpose of a shading coil on a single-phase magnet is to mitigate the pulsations in force between the armature and pole face. This is of importance only when the armature is in contact with the pole face, because only then do the force pulsations allow the armature to hammer on the pole face and produce the undesirable effect known as "chattering." In a well-designed shaded pole magnet the least force should never drop below the load force when the armature is against its stop.

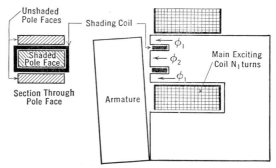

Fig. 34. Section through a clapper-type single-phase alternating-current magnet with a shading coil.

Figure 34 shows a section through a clapper type of alternating-current magnet having a shading coil embedded in the pole face. The shading coil itself may consist of a heavy slug of copper or brass, or of a coil of several turns of copper wire either short-circuited on itself or connected to an external resistor. One convenient way of making these coils, which gives the highest space factor and lowest resistance, is to cut off pieces of a square copper tube of the correct size and thickness. Whether the coil is of one heavy turn or of several lighter turns is of no importance. Either will have the same shading effect if the quotient of its resistance by the square of the turns remains constant.

As was explained in Art. 116, Sec. 3, the instantaneous force of a split-pole magnet will be constant, provided that the maximum force of each part is the same, and their fluxes 90° out of phase. The shading coil splits the pole into two parts, and causes the fluxes of these parts to be out of phase. However, it is impossible to attain a 90° phase relationship between these fluxes, and hence ideal shading cannot be attained.

Referring to Fig. 34, let ϕ_1 and ϕ_2 be the actual [24] fluxes existing in the unshaded and shaded poles, respectively. As will be shown later, the phase angle α between ϕ_2 and ϕ_1 depends on the ratio of the reactance of the shading coil to its resistance. For α to be 90° this ratio must be infinite, which means that the shading coil resistance must be made zero. Assuming that it is possible to have a shading coil with zero resistance, it will still be impossible to attain ideal shading, because the flux of the shaded pole approaches zero as the shading-coil resistance approaches zero. This is necessarily so because the ampere-turns of the shading coil must remain finite, and hence in the limit the induced voltage of the coil must be zero in order that the current be finite.

However, for any given set of conditions, there is an optimum design of shading coil which will produce the least pulsation in force. It is the purpose of this article to develop a sound theory of shading coils and to set forth the requirements of an optimum design.

2. Fundamental Theory of the Shaded-Pole Single-Phase Magnet.

For the purpose of analyzing the operation of shading coils it is only necessary to determine the flux relations in the region of the pole faces. Figure 35 shows the magnetic circuit in the region of the pole face of a shaded-pole magnet. Let ϕ_1 and ϕ_2 be the actual [25] fluxes passing through the unshaded and shaded portions of the pole, respectively, including all the effective fringing flux.[26] Then ϕ_2, the actual flux

[24] This means the actual fluxes existing in the presence of the shading coil, including the effect of the shading-coil ampere-turns. These fluxes would be measured by winding an exploring coil directly over the iron of each of the pole sections.

[25] By "actual" is meant the real flux issuing forth from the respective poles, in the presence of the shading coil and including the effect of the shading-coil ampere-turns. This is the flux that would be measured by an exploring coil wrapped around the base of each pole.

[26] Fringing flux in the immediate vicinity of the pole face merely alters the reluctance from that of the simple path shown in Fig. 35. It does not in any way invalidate the theory to be developed so long as it is considered in determining the effective reluctance of the path.

through the shading coil, will lag ϕ_1, as shown in Fig. 36. The voltage e, induced in the shading coil by ϕ_2, will lag ϕ_2 by 90°, but the current I of the shading coil will be in phase with e, because ϕ_2 is the actual flux linking with the shading coil while I is flowing. Hence I flows against the resistance of the shading coil only. α is the angle by which ϕ_2 lags ϕ_1.

Consider now the local magnetic circuit of Fig. 35. It is possible to go from point A to B by two paths, as shown. The magnetic potential between these points will be the same, no matter what path is taken. Path 1 goes direct from A to B through the reluctance [27] R_1/ν_1 of the

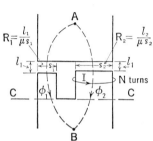
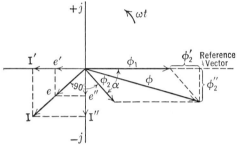

Fig. 35. Magnetic circuit of a shaded-pole magnet in the region of the shaded pole face.

Fig. 36. Vector diagram of the fluxes and the shading coil induced voltage and current, for a shaded pole face.

air gap under the unshaded pole, while path 2 goes through R_2/ν_2, the reluctance [27] of the air gap under the shaded pole, and thence through the shading coil. The flow of current in the shading coil will always be in such a direction as to tend to prevent any change in flux through it, with the result that the effective reluctance of the path is increased.

Equating the magnetomotive force between A and B via the two paths, we have the general vector equation [28]

$$\phi_1 \frac{R_1}{\nu_1} = \phi_2 \frac{R_2}{\nu_2} - NI \qquad (53)$$

where the minus sign in front of NI indicates that the magnetomotive

---

[27] R is the reluctance of the rectangular or working gap only, and R/ν is the total reluctance of the flux path, including the fringing and leakage paths, and is equal to the joint reluctance of the working gap and the fringing and leakage paths. ν is therefore the leakage coefficient defined in the usual way, and will be the ratio of the total flux entering the pole at section C-C, Fig. 35, to that passing through the working gap of area S.

[28] This equation neglects the reluctance drop of the iron as it is assumed that the iron is unsaturated, and hence its reluctance drop will be small compared to that of the air gap.

force of the shading coil is inherently negative. The vector diagram of Fig. 37 will make this clearer. Here the magnetomotive-force vector $\phi_1(R_1/\nu_1)$ is taken as a datum vector in the positive direction. Relative to this vector it can be seen that NI will always be in the third quadrant and hence the resultant of $\phi_2(R_2/\nu_2)$ and NI must be obtained as a vector difference, in order that the algebraic sum of the magnetomotive forces around the closed loop AB add to zero.

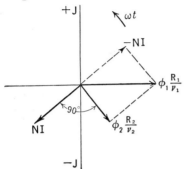 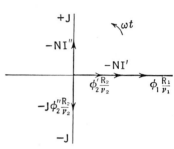

Fig. 37. Vector diagram of the reluctance magnetomotive force drops and the mangnetomotive force of the shading coil for a shaded pole face.

Fig. 38. Resolution of the vectors of Fig. 37 into their real and imaginary components.

Resolving the vectors of Fig. 37 into their real and imaginary parts, we have the components as shown in Fig. 38. Equating the real components of this figure, we have

$$\phi_1 \frac{R_1}{\nu_1} = \phi_2' \frac{R_2}{\nu_2} - NI' \tag{54}$$

The in-phase component of shading-coil current can be determined from the quadrature component of shading-coil flux, as shown in Fig. 36.

$$I' = \frac{e'}{r} = \frac{-\omega N \phi_2''}{r} \tag{55}$$

Substituting equation 55 into 54, we have

$$\phi_1 \frac{R_1}{\nu_1} = \phi_2' \frac{R_2}{\nu_2} + \frac{\omega N^2 \phi_2''}{r} \tag{56}$$

This and subsequent equations can be greatly simplified if the symbol K is introduced for the ratio of reactance to resistance of the shading coil, thus

$$K = \frac{\nu_2 \omega N^2}{R_2 r} = \frac{\nu_2 \, \omega N^2 \mu S_2}{l_2 r} \tag{57}$$

This is a logical substitution, as K is a tangible design factor of the shading coil which is independent of the size of the magnet. In considering the reactance of the shading coil equal to $\nu_2 \omega N^2 / R_2$, the reluctance of iron path through the main exciting coil has been neglected, as it is small compared to that of the air gap. The working air-gap reluctance, $R_2 = l_2/\mu S_2$, automatically takes into account all possible variations in area or length of the air gap under the shaded pole.

Substituting equation 57 into 56, we have

$$\phi_1 \frac{R_1 \nu_2}{R_2 \nu_1} = \phi_2' + K\phi_2'' \tag{58}$$

Equating the imaginary components of Fig. 38, we have

$$-NI'' = -\left(-j\phi_2'' \frac{R_2}{\nu_2}\right) \tag{59}$$

The quadrature component of the shading coil current, I'', may be found from the in-phase component of the shading coil flux,

$$I'' = \frac{e''}{r} = \frac{-j\omega N \phi_2'}{r} \tag{60}$$

Substituting equations 60 and 57 into 59, we have

$$K\phi_2' = \phi_2'' \tag{61}$$

By substituting 61 into 58, we can solve for the components ϕ_2' and ϕ_2'' of the shaded-pole flux in terms of that of the unshaded pole. Thus

$$\phi_2' = \phi_1 \frac{1}{1+K^2} \frac{R_1 \nu_2}{R_2 \nu_1} \tag{62}$$

$$\phi_2'' = \phi_1 \frac{K}{1+K^2} \frac{R_1 \nu_2}{R_2 \nu_1} \tag{63}$$

As

$$\phi_2 = \sqrt{(\phi_2')^2 + (\phi_2'')^2} \tag{64}$$

we may combine 62 and 63 and obtain

$$\frac{\phi_2}{\phi_1} = \frac{R_1 \nu_2}{R_2 \nu_1} \frac{1}{\sqrt{K^2+1}} \tag{65}$$

The tangent of the phase angle α is, from 61,

$$\tan \alpha = \frac{\phi_2''}{\phi_2'} = K \tag{66}$$

The total flux of the magnetic circuit may be found from Fig. 36 by applying the law of cosines; it is

$$\phi = \sqrt{\phi_1^2 + \phi_2^2 + 2\phi_1\phi_2 \cos \alpha} \qquad (67)$$

It may also be expressed in terms of the flux of the unshaded pole as follows:

$$\phi = \sqrt{(\phi_2'')^2 + (\phi_1 + \phi_2')^2}$$

which, when simplified, yields

$$\phi = \phi_1 \frac{R_1\nu_2}{R_2\nu_1} \frac{1}{1+K^2} \sqrt{K^2 + \left[\frac{1+K^2}{\frac{R_1\nu_2}{R_2\nu_1}} + 1\right]^2} \qquad (68)$$

3. Instantaneous Force Relations in a Shaded-Pole, Single-Phase Magnet. The instantaneous force produced under a pole is, by reference to Art. 116,

$$\text{Instantaneous force} = \frac{(B_m \sin \omega t)^2 S}{72} \text{ lb. per sq. in.}$$

where B_m is the peak flux density in the working gap. The force produced under the unshaded pole will be

$$\frac{(\phi_1)_m^2 \sin^2 \omega t}{\nu_1^2 72 S_1} \text{ lb.} \qquad (69)$$

and that under the shaded pole will be

$$\frac{(\phi_2)_m^2 \sin^2 (\omega t - \alpha)}{\nu_2^2 72 S_2} \text{ lb.} \qquad (70)$$

The total instantaneous force of both poles will be

$$\text{Instantaneous force} = \frac{\phi_1^2}{36 S_1 \nu_1^2} \sin^2 \omega t + \frac{\phi_2^2}{36 S_2 \nu_2^2} \sin^2 (\omega t + \alpha) \qquad (71)$$

These actual force components, together with the flux waves that produce them, are shown in Fig. 40a.

Expanding $\sin^2 \omega t$, and collecting terms, we have

$$\text{Instantaneous force} = \frac{\phi_1^2}{72 S_1 \nu_1^2} + \frac{\phi_2^2}{72 S_2 \nu_2^2} - \frac{\phi_1^2}{72 S_1 \nu_1^2} \cos 2\omega t$$
$$- \frac{\phi_2^2}{72 S_2 \nu_2^2} \cos (2\omega t - 2\alpha)$$

The first two terms are the average force of the magnet, and the last two terms are the double-frequency alternating components. The alternating components may be represented vectorially as shown in Fig. 39. Combining these components graphically, as illustrated, the total alternating component of the force is obtained. This vector will become smaller as α approaches 90°, and its magnitude, by the law of cosines, will be

$$\sqrt{\left(\frac{\phi_1^2}{72S_1\nu_1^2}\right)^2 + \left(\frac{\phi_2^2}{72S_2\nu_2^2}\right)^2 + 2\left(\frac{\phi_1^2}{72S_1\nu_1^2}\right)\left(\frac{\phi_2^2}{72S_2\nu_2^2}\right)\cos 2\alpha}$$

The instantaneous force of the shaded-pole magnet may therefore be written in terms of its average and alternating components, as follows:

$$\text{Average force component} = \frac{1}{72}\left(\frac{\phi_1^2}{S_1\nu_1^2} + \frac{\phi_2^2}{S_2\nu_2^2}\right) \text{ lb.} \quad (72)$$

Alternating force component

$$= -\sqrt{\left(\frac{\phi_1^2}{72S_1\nu_1^2}\right)^2 + \left(\frac{\phi_2^2}{72S_2\nu_2^2}\right)^2 + 2\left(\frac{\phi_1^2}{72S_1\nu_1^2}\right)\left(\frac{\phi_2^2}{72S_2\nu_2^2}\right)\cos 2\alpha}$$

$$\left[\cos(2\omega t - \theta)\right] \quad (73)$$

The total instantaneous force and its components are shown in Fig. 40b, and are drawn to correspond correctly with actual forces of Fig. 40a.

By examining equation 73 in conjunction with Fig. 39 it can be seen that, if the average force of the shaded pole equals that of the unshaded pole, the alternating force of the magnet will become zero when α is 90°. The criterion for ideal shading is, therefore, that the average force of the shaded pole equal that of the unshaded pole and that the flux through the shading coil lag that of the unshaded pole by 90°.

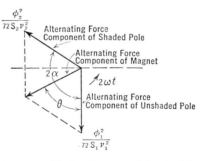

Fig. 39. Vector diagram of the double-frequency alternating components of force in a shaded pole face.

4. **Optimum Conditions for Minimum Force Variation.** As mentioned before, the conditions necessary for ideal shading are impossible to attain. However, it is possible to attain optimum conditions which will give the least pulsation in force. When deriving an optimum a normal set of conditions

must be imposed. For our purpose it is a reasonable design condition to specify that the average force of the magnet must remain constant, or

$$p_1\phi_1^2 + p_2\phi_2^2 = m = \text{constant} \tag{74}$$

where

$$p_1 = \frac{1}{72S_1\nu_1^2} \quad \text{and} \quad p_2 = \frac{1}{72S_2\nu_2^2}$$

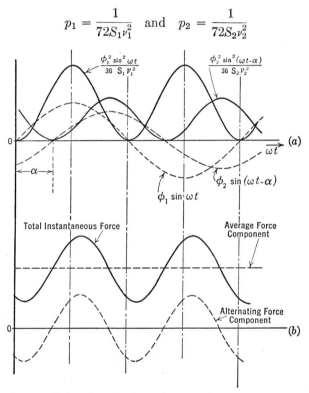

Fig. 40a. Waves of flux through the poles of a shaded pole face and the force waves produced by these fluxes shown in their proper relative phase positions
Fig. 40b. Total instantaneous force wave produced by both pole faces of Fig. 40a and its resolution into an average and alternating force component.

Solving equation 74 for ϕ_1^2 and substituting into the magnitude of 73, we have

Alternating force component

$$= -\sqrt{(m - p_2\phi_2^2)^2 + p_2^2\phi_2^4 + 2p_2(m - p_2\phi_2^2)\phi_2^2 \cos 2\alpha} \tag{75}$$

If this expression is differentiated with respect to ϕ_2, letting α be constant, the value of ϕ_2 producing the least alternating force component can be found.

[Art. 123] OPTIMUM CONDITIONS FOR SHADED-POLE MAGNETS 471

$$\frac{d(\sqrt{\ })^2}{d\phi_2} = 2(m - p_2\phi_2^2)(-2p_2\phi_2) + 4p_2^2\phi_2^3$$
$$+ 4p_2 m\phi_2 \cos 2\alpha - 8p_2^2\phi_2^3 \cos 2\alpha = 0$$
$$-4mp_2\phi_2 + 4p_2^2\phi_2^3 + 4p_2^2\phi_2^3 + 4p_2 m\phi_2 \cos 2\alpha - 8p_2^2\phi_2^3 \cos 2\alpha = 0$$
$$4mp_2\phi_2(\cos 2\alpha - 1) - 8p_2^2\phi_2^3(\cos 2\alpha - 1) = 0$$
$$(4mp_2\phi_2 - 8p_2^2\phi_2^3)(\cos 2\alpha - 1) = 0$$

As α has been assumed constant, and not necessarily equal to 90°, the second term cannot equal zero. Therefore,

$$4mp_2\phi_2 - 8p_2^2\phi_2^3 = 0$$

or

$$m = 2p_2\phi_2^2$$

Substituting this value into 74 we find that

$$p_1\phi_1^2 = p_2\phi_2^2$$

or

$$\frac{\phi_1^2}{72S_1\nu_1^2} = \frac{\phi_2^2}{72S_2\nu_2^2} \tag{76}$$

as the optimum condition. Thus for any given value of α the average force of the shaded pole should be equal to that of the unshaded pole. It has already been shown that under any conditions α should be as near 90° as is possible. Therefore it may be stated that an optimum design, giving the least pulsation in force, is obtained when the average force under each pole is made equal and α is made as near 90° as possible.[29]

5. **Design Factors for Optimum Shading.** Letting

$$\frac{\phi_1^2}{72S_1\nu_1^2} = \frac{\phi_2^2}{72S_2\nu_2^2} \tag{76}$$

as the optimum condition, we have

$$\left(\frac{\phi_2}{\phi_1}\right)^2 = \frac{S_2\nu_2^2}{S_1\nu_1^2} \tag{77}$$

From equation 65 we have

$$\frac{\phi_2}{\phi_1} = \frac{R_1\nu_2}{R_2\nu_1}\frac{1}{\sqrt{K^2+1}} = \frac{l_1 S_2 \nu_2}{l_2 S_1 \nu_1}\frac{1}{\sqrt{K^2+1}} \tag{65}$$

[29] This is for the condition that the total average force of both poles remain constant. Other conditions, however, which might be equally logical might be imposed. Thus the total flux of both poles ϕ might be held constant.

Squaring 65 and equating to 77, we have

$$\frac{S_2 \nu_2^2}{S_1 \nu_1^2} = \frac{l_1^2 S_2^2 \nu_2^2}{l_2^2 S_1^2 \nu_1^2} \left(\frac{1}{K^2 + 1} \right)$$

which, when solved for K, gives

$$K = \sqrt{\frac{l_1^2 S_2}{l_2^2 S_1} - 1} \tag{78}$$

Now, as $\tan \alpha = K$, K should be made as large as possible in order that α be made to approach 90°. Therefore S_1 and l_2 should be made as small as possible, and l_1 and S_2 as large as possible.

The practical limitations in the choice of these four factors are discussed below:

S_1 is limited by the maximum saturation density in the iron and the requirement that the average force of the unshaded pole must be one-half the required average force of the magnet.

l_2 for a holding magnet may be reduced to the point where the pole faces touch. If the pole faces are carefully finished so as to make good contact, l_2 may be considered to be as small as 0.001 in. In general, however, l_2 is determined arbitrarily by the conditions of the problem. It should be remembered that optimum shading can be produced at only one gap length.

S_2 is limited by the maximum size allowed for the entire pole face. A reasonable value is between S_1 and $2S_1$.

l_1 is limited by the allowable volt-ampere input to the main exciting coil, or the power to be dissipated in the shading coil, and the leakage. The longer l_1 is made, the more magnetomotive force will be required for the unshaded gap, and the higher the leakage of this pole face will be. As l_1 is increased and the magnetomotive force across the unshaded gap increases, the magnetomotive force of the shaded coil must increase in proportion. This will correspondingly increase the current of the shading coil and the power loss. As a practical consideration, the power loss of the shading coil is the most logical limitation. In general, the force variation will be decreased as this power loss is increased.

124. Sample Shading Coil Computation

1. General. For the purpose of illustrating the actual procedure necessary to make an optimum design for a shaded pole face, and also to give some idea of the order of magnitudes involved, a complete design has been worked out in the following sections.

As the design of the main iron circuit and exciting coil do not depend particularly on the shading coil, we shall concern ourselves only with the design of the shading coil and the pole faces. The change in excitation of the main coil due to shading will also be investigated. Likewise the copper loss in the shading coil will be computed.

2. **Design Data.** A 115-volt, single-phase, 60-c.p.s. magnet is to develop an average pull of about 50 lb. in the closed-gap position with a minimum of chattering. Design a clapper type of magnet suitable for the purpose, using the following data:

> Lamination material = No. 26 gauge medium-silicon steel.
> Stacking factor, $K_s = 0.9$.
> Minimum air gap under shaded pole = 0.002 in.
> Maximum air gap under unshaded pole = such that the power loss in the shading coil circuit will not exceed 30 watts.

3. **Preliminary Design.** We shall first make a preliminary design for optimum conditions, following the suggestions of Art. 123, Sec. 5.

Maximum Flux Density of Unshaded Pole $(B_m)_1$. Choose a value of 110 kmax. per sq. in. This seems high, but it must be remembered that the leakage coefficient of the unshaded pole will be high and hence the average peak density throughout the unshaded pole will be much lower.

Leakage Coefficient for Unshaded Pole ν_1. Assume 1.15 for a preliminary value.

Area of Unshaded Pole Face S_1. By equation 76 the average force of the unshaded pole should equal one-half of the force to be developed by the magnet for optimum shading. By equation 17

$$\text{Average force (unshaded pole)} = \left[\frac{(B_m)_1 K_s}{\nu_1}\right]^2 \frac{S_1}{144}$$

or

$$S_1 = \frac{144 \times \text{A.F.}}{\left[\dfrac{(B_m)_1 K_s}{\nu_1}\right]^2} = \frac{144 \times 25}{\left(\dfrac{110 \times 0.9}{1.15}\right)^2} = 0.487 \text{ sq. in.}$$

Minimum Gap Length under Shaded Pole l_2. The minimum value of this gap has been already set at 0.002 in. in the problem data.

Maximum Area under Shaded Pole S_2. The value taken for this depends merely on how large the magnet is to be made. Let us, for the purpose of this design, assume a main pole core section 1 in. by $1\frac{1}{2}$ in., with a slot for the shading coil 0.104 in. wide. The pole-face dimensions will then be as shown in Fig. 41, with the shaded pole face having 1.87 times the area of the unshaded face. This allows the shading coil to be almost square in cross section, which is the optimum shape for maximum K.

Flux of the Shaded Pole ϕ_2. The average force of this gap is to be made the same as that of the unshaded pole. Solving the shaded pole portion of equation 72, we have,

$$\phi_2 = \sqrt{72 \text{ A.F.} \times S_2 v_2^2}$$

$$= \sqrt{72 \times 25 \times 0.909 \times 1.03^2}$$

$$= 41.5 \text{ kmax. (r.m.s.)}$$

where v_2 is assumed as 1.03.

Voltage Induced in Shading Coil E_s.

$$E_s = 2\pi f N \phi_2 \times 10^{-5}$$

$$= 2\pi \times 60 \times 1 \times 41.5 \times 10^{-5} = 0.156 \text{ volt}$$

Resistance of Shading Coil r. As the power loss in the shading coil circuit is to be limited to 30 watts, the resistance of the shading coil circuit will be

$$r = \frac{E^2}{P} = \frac{0.156^2}{30} = 0.000815 \text{ ohm}$$

Current of Shading Coil I_s.

$$I_s = \frac{E_s}{r} = \frac{0.156}{0.000815} = 191 \text{ amperes}$$

Dimensions of the Shading Coil. The dimensions of the shading coil must be such that the proper resistance is obtained and also such that the temperature rise will be satisfactory. The above computations have been carried out on the basis of a one-turn coil. For continuous operation it is obvious that the temperature rise for 30 watts dissipation will be excessive. If the time of excitation is sufficiently short, so that heating is not a prime consideration, the coil could be made of a slug of copper, or if more thermal capacity is desired, it could be made of a slug of high-resistivity metal such as brass or german silver. The higher the resistivity of the metal the larger the cross section must be, and hence, the larger the thermal capacity and heat-dissipating surfaces will be. For this particular design it is best to make the coil of several turns of copper so designed that resistance of the coil itself will be only a fraction of the desired

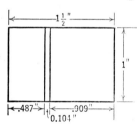

Fig. 41. Tentative proportions for shaded pole face for preliminary design of magnet of problem of Art. 124.

resistance. The required resistance can then be obtained by means of an external resistor. This will allow a portion of the 30 watts generated, in proportion to the external resistance, to be dissipated away from the coil proper and so make possible a lower temperature rise.

Let us assume the coil to be made of 5 turns of No. 12 chromoxide-covered wire (84 mils diameter), insulated from the laminations by mica. Allowing 10 mils for mica, the mean length of a turn will be $2(0.084 + 0.909 + 0.020) + 2(0.084 + 1.0 + 0.020) = 4.24$ in. Assuming the coil temperature to be 90° C., the resistance of the shading coil proper will be

$$r = 5 \times 4.24 \times 0.1323 \times 10^{-3} \times 1.275 = 0.00358 \text{ ohm}$$

Because the coil is to have 5 turns, its current will be only $\frac{1}{5}$ of that of a one-turn coil; it is $\frac{191}{5} = 38.2$ amperes. The power dissipated in the coil proper will be [30]

$$P = I^2 r = 38.2^2 \times 0.00358 = 5.2 \text{ watts}$$

As the resistance of the shading coil circuit varies as N^2, its resistance will be $0.000815 \times 25 = 0.0204$ ohm, and the value of the external resistor will be $0.0204 - 0.00358 = 0.01682$ ohm. As a check, the total power dissipation will be

$$P = I^2 r = 38.2^2 \times 0.0204 = 29.8 \text{ watts}$$

The depth of the shading coil slot can be made $5 \times 0.084 + 0.02 + 0.06 = 0.5$ in., where the last two terms are for mica and clearance, respectively, as shown in Fig. 42.

Factor K for the Shading Coil.

$$K = \frac{\nu_2 \omega N^2}{R_2 r} = \frac{\nu_2 \omega N^2 \mu S_2}{l_2 r}$$

$$= \frac{1.03 \times 2\pi \times 60 \times 5^2 \times 3.19 \times 10^{-8} \times 0.909}{0.002 \times 0.0204} = 6.9$$

Length of the Unshaded Gap l_1. For the force variation to be a minimum, we have, from equation 78,

$$K = \sqrt{\frac{l_1^2 S_2}{l_2^2 S_1} - 1}$$

---

[30] No attempt has been made to compute the temperature rise of the coil. The assumed 90° C. rise has been taken as a convenient value. In use the shading-coil circuit when hot would have to be adjusted to a resistance of 0.0204 ohm.

solving this for l_1, we get

$$l_1 = l_2 \sqrt{(K^2 + 1) \frac{S_1}{S_2}}$$

$$l_1 = 0.002 \sqrt{(6.9^2 + 1) \frac{0.487}{0.909}}$$

$$l_1 = 0.0104 \text{ in.}$$

This completes the preliminary design.

4. Final Design—Check. In Fig. 42 is shown a sketch of the proposed shaded-pole magnet of the preliminary design.

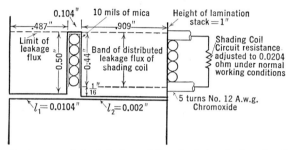

FIG. 42. Preliminary and final design of the shaded pole face for the magnet of the problem of Art. 124.

Leakage Coefficients. The first step in making the final design will be to compute the leakage coefficients. Consider first the shaded pole. It will have two bands of leakage flux, that from the pole edge to the beginning of the shading coil which links with all the turns of the shading coil, and that in the region of the shading coil which is only partially linked. We then have:

Complete linkage.

$$P_1 \text{ (main gap)} = \frac{\mu S}{l} = \frac{3.19 \times 0.909}{0.002} = 1{,}450$$

$$P_7 = 0.26\mu l = 0.26 \times 3.19 \times 2.818 = \qquad 2.3$$

$$P_{8b} = \frac{\mu l}{\pi} \log_\epsilon \frac{r_2}{r_1} = \frac{3.19 \times 2.818}{\pi} \log_\epsilon \frac{0.0635}{0.001} = 11.9$$

$$P_{11} = 0.52\mu l = 0.52 \times 3.19 \times 1 = \qquad 1.7$$

$$P_{12b} = \frac{2\mu l}{\pi} \log_\epsilon \frac{r_2}{r_1} = \frac{2 \times 3.19 \times 1}{\pi} \log_\epsilon \frac{0.033}{0.002} = 5.7$$

$$\overline{\qquad 21.6}$$

ART. 124]   SAMPLE SHADING COIL COMPUTATION   477

The distributed linkages are:

$$P_{8b} = \frac{\mu l}{\pi} \log_e \frac{r_2}{r_1} = \frac{3.19 \times 2.818}{\pi} \log_e \frac{0.5104}{0.0635} = 5.9$$

$$P_1 \text{ (coil space)} \frac{\mu S}{l} = \frac{3.19 \times 0.477}{0.104} = 14.6$$

$$P_7 \text{ (edges of coil space)} = 0.26 \mu l = 0.26 \times 3.19 \times 2 \times 0.44 = 0.8$$

$$\overline{21.3}$$

For the purpose of determining the effective flux linkage and hence the reactance of the shading coil, these permeances must be multiplied by $\frac{1}{3}$ because of the distribution. The leakage coefficient to be used in determining K of the coil will then be

$$\nu_2 \text{ (for } K\text{)} = \frac{1{,}450 + 21.6 + \frac{1}{3} \times 21.3}{1{,}450} = 1.0198$$

For determining the force the leakage coefficient will be

$$\nu_2 \text{ (force)} = \frac{1{,}450 + 21.6 + \frac{1}{2} \times 21.3}{1{,}450} = 1.022$$

For the unshaded pole we have:

$$P_1 = \frac{\mu S}{l} = \frac{3.19 \times 0.487}{0.104} = 149$$

$$P_7 = 0.26 \mu l = 0.26 \times 3.19 \times 1.974 = 1.6$$

$$P_{8b} = \frac{\mu l}{\pi} \log_e \frac{r_2}{r_1} = \frac{3.19 \times 1.974}{\pi} \log_e \frac{0.505}{0.0052} = 9.2$$

$$P_{11} = 0.52 \mu l = 0.52 \times 3.19 \times 1 = 1.7$$

$$P_{12b} = \frac{2 \mu l}{\pi} \log_e \frac{r_2}{r_1} = \frac{2 \times 3.19 \times 1}{\pi} \log_e \frac{0.033}{0.0104} = 2.4$$

$$\overline{14.9}$$

and the leakage coefficient for this pole will be

$$\nu_1 = \frac{149 + 14.9}{149} = 1.100$$

Total Flux of Unshaded, Shaded, and Main Pole. The flux of the unshaded pole will remain the same as in the preliminary design, ϕ_1.

$$\phi_1 = \frac{(B_1)_m S_1 K_s}{\sqrt{2}} = \frac{110 \times 0.487 \times 0.9}{\sqrt{2}} = 34.1 \text{ kmax.}$$

The ratio of the fluxes of the shaded and unshaded poles will be, by equation 77,

$$\left(\frac{\phi_2}{\phi_1}\right)^2 = \frac{S_2 \nu_2^2}{S_1 \nu_1^2} = \frac{0.909}{0.487} \times \left(\frac{1.022}{1.100}\right)^2 = 1.61$$

$$\frac{\phi_2}{\phi_1} = 1.27$$

and

$$\phi_2 = 34.1 \times 1.27 = 43.3 \text{ kmax.}$$

From equation 57, we have

$$K = \frac{\nu_2 \omega N^2 \mu S_2}{l_2 r} = \frac{1.0198 \times 2\pi \times 60 \times 5^2 \times 3.19 \times 10^{-8} \times 0.909}{0.002 \times 0.0204} = 6.81$$

The phase angle between ϕ_2 and ϕ_1 will be

$$\tan \alpha = K = 6.81$$

$$\alpha = 81.65°$$

By reference to equation 67 the flux of the main pole ϕ will be

$$\phi = \sqrt{34.1^2 + 43.3^2 + 2 \times 34.1 \times 43.3 \times \cos 81.65°}$$
$$= \sqrt{1{,}170 + 1{,}850 + 427}$$
$$= 58.9 \text{ kmax.}$$

Current and Power Loss of Shading Coil. The induced voltage of the shading coil will be

$$E_s = 2\pi f N \phi_2 = 2\pi \times 60 \times 5 \times 43.3 \times 10^{-5} = 0.818 \text{ volt}$$

The current of the shading coil will be

$$I = \frac{E}{r} = \frac{0.818}{0.0204} = 40.0 \text{ amperes}$$

The total power dissipated in the shading coil circuit will be

$$P = I^2 r = 40.0^2 \times 0.0204 = 32.6 \text{ watts}$$

of which $40.0^2 \times 0.00358 = 5.73$ watts will be dissipated in the coil directly, and the balance in the external resistor.

Average Force. By equation 72, the average force of the magnet will be:

SAMPLE SHADING COIL COMPUTATION

$$\text{Average force} = \left(\frac{\phi_1}{\nu_1}\right)^2 \frac{1}{72S_1} + \left(\frac{\phi_2}{\nu_2}\right)^2 \frac{1}{72S_2}$$

$$= \left(\frac{34.1}{1.100}\right)^2 \frac{1}{72 \times 0.487} + \left(\frac{43.3}{1.022}\right)^2 \frac{1}{72 \times 0.909}$$

$$= 27.4 + 27.5$$

$$= 54.9 \text{ lb.}$$

Force Pulsation. By equation 73 the magnitude of the force pulsation will be

Alternating force component

$$= \sqrt{27.4^2 + 27.5^2 + 2 \times 27.4 \times 27.5 \cos(2 \times 81.65)}$$

$$= \sqrt{751 + 757 - 1{,}440}$$

$$= 8.25 \text{ lb.}$$

On a percentage basis the force pulsation will be

$$\text{Percentage force pulsation} = \frac{8.25}{54.9} = 15 \text{ per cent}$$

Maximum and Minimum Forces. The maximum force will be

$$\text{Maximum force} = 54.9 + 8.25 = 63.2 \text{ lb.}$$

and the minimum force will be

$$\text{Minimum force} = 54.9 - 8.25 = 46.6 \text{ lb.}$$

Design of the Main Exciting Coil of Magnet. The design of the magnet, exclusive of the pole face, and the main exciting coil can be carried out in the same manner as in Art. 122.

Neglecting the leakage flux [31] shunting the shaded section of the pole face directly from pole arm to pole arm, the required turns on the main coil will be, by equation 14,

$$N = \frac{E_r}{2\pi f \phi} = \frac{115}{2\pi \times 60 \times 58.9 \times 10^{-5}} = 518$$

The magnetomotive force across the main gap can be most easily determined from the unshaded pole:

$$(B_m)_1 = \frac{\sqrt{2}\phi_1}{\nu_1 S_1} = \frac{\sqrt{2} \times 34.1}{1.100 \times 0.487} = 90.0 \text{ kmax. per sq. in.}$$

---

[31] Owing to the high magnetomotive force across the main gap due to the shading-coil ampere-turns, this may be appreciable.

The peak magnetomotive force across this gap

$$(F_m)_1 = (H_m)_1 l_1 = \frac{(B_m)_1}{\mu} l_1 = \frac{90.0}{0.00319} \times 0.0104 = 293 \text{ ampere-turns}$$

The required excitation for the main coil can be determined by properly [32] combining this with the magnetomotive forces required to excite the iron and overcome the reluctance of the fixed gap, as illustrated in Art. 122.

The total power input to the main coil will be the sum of the iron losses and the copper losses of the shading coil and of the main exciting coil.

Final Results.

Length of gap under shaded pole	= 0.002 in.
Length of gap under unshaded pole	= 0.0104 in.
Average pull of shaded pole	= 27.4 lb.
Average pull of unshaded pole	= 27.5 lb.
Total average pull	= 54.9 lb.
Maximum total pull	= 63.2 lb.
Minimum total pull	= 46.6 lb.
Percentage force pulsation	= 15 per cent
Shading coil power loss	= 32.6 watts, of which 5.73 watts are dissipated in the coil itself and the balance in an external resistor.

Phase angle lag of flux of shaded pole = 81.6°

PROBLEMS

1. An alternating-current magnet has a winding of 450 turns. The effective permeance which may be considered to produce flux linking with all the turns is 14.1 and 100 max. per ampere-turn at the start and finish of the stroke, respectively. Assuming this permeance to be independent of the flux, calculate the r.m.s. exciting current at the start and finish of the stroke if it is operated from a 120-volt, 60-c.p.s., constant-voltage supply.

2. Calculate the reactance voltages of the magnet of Problem 1 if it is operated from a 2.0-ampere, 60-c.p.s., constant-current supply.

3. Calculate for the magnets of Problems 1 and 2 their mechanical work and volt-ampere efficiency.

4. If the resistance of the coil of the magnet of Problem 1 is 1 ohm, calculate the reactance voltage of the coil at the start of the stroke for Problem 1 and the coil terminal voltage for Problem 2.

---

[32] The magnetomotive force across the main gap will be in phase with ϕ_1, while the magnetomotive forces of the other parts of the circuit will be substantially in phase with ϕ.

5. Calculate the average and maximum force of a flat-face direct-attraction type of magnet which has a 25-c.p.s. sinusoidal flux of 200 kmax. maximum value issuing forth from a pole face 1.5 in. by 1.5 in.

6. A 60-c.p.s. solenoid and plunger magnet has a square plunger $\frac{1}{2}$ in. on a side and is surrounded by a coil of 200 turns per inch of winding length. The plunger is far removed from the open end of the solenoid. Assuming the coil current to be 5 amperes r.m.s., and the flux density at the entering end of the plunger to be 60 kmax. per sq. in. r.m.s., compute the average force on the plunger.

7. If the magnet of Fig. 10a has 500 turns per coil and is to be operated from a 220-volt, two-phase, three-wire, 60-c.p.s. power supply, compute the dimensions of the parts so that the maximum flux density will be 100 kmax. per sq. in. Draw a sketch to illustrate.

8. Same as Problem 7 except for the magnet of Fig. 13a operated from a 220-volt, three-phase, three-wire, 60-c.p.s. power supply.

9. Assuming that a magnet can be built having a volt-ampere efficiency of 80 per cent compute the highest frequency at which it will be possible to design a magnet having a rating of 20 in-lb. and 1.0 kv-a.

10. A magnet built as illustrated in Fig. 22a has the following data:

Plunger diameter $(2r_1) = \frac{1}{2}$ in.
Inside diameter of shell $(2r_2) = 1\frac{1}{2}$ in.
Length of coil space $(h) = 2$ in.
Axial length of washers at each end $= \frac{1}{2}$ in.
Diameter of hole for plunger in both end washers $= \frac{9}{16}$ in.
Plunger insertion $= 1$ in.
Supply power 120 volts, 60 c.p.s. sinusoidal wave form.

Assuming a maximum plunger flux density of 100 kmax. per sq. in. calculate: (a) the required coil turns; (b) the coil current; (c) the force; (d) the exciting volt-amperes.

11. A magnet like that of Art. 122 is to be designed for a force of 10 lb. through a stroke of $\frac{1}{4}$ in. Assume a 60 c.p.s., single-phase, 120 volt supply voltage. Excitation to be continuous with the plunger at the end of its stroke. Compute the preliminary value of: (a) the equivalent index number; (b) the best type of pole face; (c) dimensions of plunger section; (d) coil turns; (e) exciting current open gap.

12. Lay out a suitable final design for the magnet of Problem 11 and compute the exciting current in the closed gap position.

13. Plot a curve of the relative average force and percentage force variation as a function of α for a shaded pole magnet having $S_1 = S_2$ and $l_1 = l_2$. Assume the total flux ϕ of both poles to be constant.

14. Repeat Problem 13 for the conditions $2S_1 = S_2$, and $l_1 = l_2$.

15. Design a one turn brass shading coil to take the place of the copper shading coil shown in Fig. 42. The relative conductivity of the usual yellow brass may be taken as 14 per cent, and its temperature coefficient the same as that of copper. Let the length of the new shading coil remain the same as that of the old one.

16. Assuming the magnet of Art. 124 to have the brass shading coil designed in Problem 15 how will the phase angle α vary with the temperature of the shading coil? Calculate α for the shading coil at 20, 90, and 150° C.

17. Determine for the same conditions as in Problem 16 the manner in which the percentage force variation will change with the temperature of the shading coil. Assume the flux of the main coil to remain constant at 58.9 kmax.

18. What would α have to be for the magnet designed in Art. 124 if the percentage force pulsation is to be increased to 30 per cent? The total flux ϕ is to remain at 58.9 kmax.

19. If the new value of α for Problem 18 is to be obtained by making l_2 larger, leaving other things the same, what will the new value of l_2 be and what will be the power loss in the shading coil? Assume v_2 to remain constant.

20. If the new value of α for Problem 18 is to be obtained by increasing the resistance of the shading coil, leaving other things the same, what will the new value of this resistance be and what will be the power loss in the shading coil?

CHAPTER XIV

RELAYS

125. General

It is not the purpose of this chapter to describe in any detail the multifarious types of relays available, or the various purposes for which they are used, but rather simply to classify them briefly, and to discuss in detail their electromagnetic characteristics and the method of predetermining those characteristics which are common to all types.

Fundamentally, a relay is an electromagnetic switch, designed to open or close a circuit when the current through its exciting coils is either caused to flow or interrupted, and in some applications merely varied in magnitude. From the point of view of the definition, relays can be further subdivided into two distinctly different types: (1) electromagnetic contactors, capable of handling heavy currents at power voltages, used for the control of large motors, etc.; and (2) electromagnetic relays, designed to handle only small currents at moderate voltages, intended for light circuit switching operations in communication, signaling, automatic regulators, and remote-control systems in general.

Type 1, used for heavy industrial control, generally require magnets having large strokes and forces. Because they are operated from commercial power supply systems, power economy is of no great advantage, but size and cost are. They are therefore designed with temperature rise as a limitation. They properly fit into the class of magnets considered in Chapters X, XI, XII, and XIII, and will not be considered in this chapter.

Type 2, used for light circuit switching operations, generally require magnets having very short strokes with relatively small forces. Their mode of operation generally dictates power economy or sensitivity as a limitation rather than heating. Thus they are often required to operate from long lines where the available power is small, from expensive sources of energy such as dry-cell batteries or the plate circuits of vacuum tubes, and in differential applications where a small change or reversal in current must produce reliable operation. The magnetic characteristics of this type of relay, their optimum design, and the predetermination of their cycle of operation are discussed in detail in this chapter.

126. Classification of Relays

The light-duty circuit-switching relays discussed in the last article may be classified as follows:

Light-duty circuit-switching operations in

1. Signaling systems.
    a. Railroad.
    b. Traffic.
    c. Miscellaneous—annunciator, call, stock-market quotation, fire alarm, burglar alarm, etc.
2. Communication systems.
    a. Telephone.
    b. Telegraph.
    c. Radio.
    d. Teletype.
    e. Tickers.
    f. News printers.
3. Regulators and control systems.
    a. Voltage regulators.
    b. Current regulators.
    c. Frequency regulators.
    d. Speed regulators.
    e. Temperature regulators.
    f. Humidity regulators.
    g. Illumination regulators.
    h. Pressure, etc., regulators.

Relay operation is secured in response to changes in voltage, current, frequency, speed, phase, temperature, humidity, sound, light, color, pressure, length, or practically any physical condition or property.

127. Contacting Problem in Relays [1]

General. The fundamental purpose of a relay is to make contacts, and unless this is done reliably the relay as a whole is a failure. The reliability of a contact depends upon many factors, the more important of which are contact material, contact pressure, contact deflection and follow-through, contact size in relation to current, and arrangements for arc suppression.

---

[1] Many of the data presented in this section and in particular the data of Table I and Fig. 1 have been taken from a booklet published by P. R. Mallory & Co., Indianapolis, Indiana, entitled "Electrical Contacts, Engineering Data."

1. Contact Materials.

The principal contact materials used are tungsten, silver, and platinum. In addition to these, gold, molybdenum, and special alloys such as iridium platinum, copper tungsten, silver tungsten, silver tungsten-carbide, silver molybdenum, silver molybdenum-carbide, and other alloys having silver, gold, palladium, or platinum as a base.

In choosing a contact material the following factors are important:

   a. Type of load, resistance or inductive.
   b. Type of break, slow or quick.
   c. Frequency of use, number of contacts per minute.

The circuit supplying an inductive load is more difficult to interrupt than that supplying a resistive load, for the reason that the stored energy of the inductance is generally dissipated in the arc at the breaking contact. If an alloy with a low melting point is used, the contacts will often fuse together. A slow break is also hard on the contacts, as it will permit the normal circuit voltage to build up a sustained arc across the contacts. The use of a condenser of proper size across the contacts will help greatly to prevent this. Contacts which are used with great frequency must be capable of dissipating heat rapidly and hence should have a high heat conductivity, or they should be capable of withstanding high temperatures. The more important physical data of the various contact materials are given in Table I.

TABLE I

Some Physical Properties of Contact Materials

	Tungsten	Molybdenum	Hard Platinum	Silver
Density	19.3	10.2	21.4	10.5
Tensile strength, lb. per sq. in.	490,000	260,000	54,000	40,000
Brinell hardness	290	147	90	59
Melting point, °C	3,400	2,620	1,755	960
Boiling point, °C	5,830	3,620	3,910	1,955
Thermal conductivity, cal. per cm. cube per °C. at 18°C.	0.476	0.346	0.166	1.01
Electrical resistivity, microhm cm. 20°C.	5.51	5.7	10.0	1.64

Silver. Pure silver and its alloys are extensively used as contact materials where the service is not severe. Silver has many desirable properties—high electrical conductivity, high thermal conductivity, and low contact resistance. The oxide and sulphide of silver are also relatively good conductors. Likewise, silver is relatively inexpensive. Its chief disadvantages are that its melting and boiling points are low and it is soft. A low melting point produces a tendency for the contacts to weld or fuse together under severe overload conditions, while the low boiling point makes it easy to sustain an arc when the contacts separate. Softness causes the material to wear away rapidly. Silver is extensively used for relays and other applications where the load is light, or where cost is a paramount consideration.

Tungsten. Tungsten contacts were developed primarily for automobile battery ignition to replace the more costly platinum contacts, which required frequent adjustment and replacement on account of the rapid transfer of contact material by the current, and rapid wear caused by the hydrocarbon atmosphere. The high melting and boiling points of tungsten make it particularly useful in applications where arcing conditions are severe. Its chief disadvantages are that it has a relatively high contact resistance and requires a high contact pressure. The more important characteristics of properly made tungsten contacts are:

1. High density.
2. High melting point.
3. High electrical conductivity.
4. High heat conductivity.
5. Low vapor pressure.
6. Suitable degree of hardness.
7. Toughness to withstand hammering action.
8. Slightly oxidizable so as to cause the arc formed to spread over the entire area, resulting in even wear.
9. Uniformity as to grain size, and structure.

Besides their application in the ignition field, tungsten contacts are used in all the more common commercial applications where high contact pressures are available and low contact drops are not essential. Some of these applications are: adding, vending, business, teletype, violet-ray, and miscellaneous other types of machines; vibrators for supplying high-voltage direct current from low-voltage batteries; etc.

Tungsten contacts are always welded to a base-metal backing or support, usually in the form of a rivet or screw.

Molybdenum. Molybdenum, which is somewhat similar to tungsten in its inherent characteristics, has been found very satisfactory for a few special applications where the contacts must resist mechanical wear, be substantially free from fusing characteristics, and operate under light pressure, comparatively high voltage, low current, and at a medium frequency.

Molybdenum is more ductile than tungsten and can be cold worked within certain limits. Since molybdenum is used chiefly for low currents, the contact is often made by force-fitting a small piece of molybdenum wire into a base-metal backing.

Platinum. Platinum, for electrical contacts, is usually alloyed with iridium, or other members of the platinum family. This increases its hardness and wear-resistant qualities. These alloys have a very high melting point, are not attacked by acids, and do not oxidize. They are generally used on sensitive relays and other devices which operate at low contact pressures and where a low contact resistance and freedom from the possibility of fusion are imperative. Fusion becomes a difficult problem where a low-pressure, slow-breaking contact supplies an inductive load.

2. **Contact Pressure.** Contact pressure is a very important factor in determining contact reliability. In ordinary service it is inevitable that a layer of dust or dirt will be formed on the contact. Unless the contact pressure is sufficient, this layer may prevent the contacts from coming together. K. W. Graybill reports in the *Telephone Engineer*[2] the results of tests conducted on horizontal-type relays to determine the effect of dust and dirt. The test consisted in operating a hundred pairs of contacts for each condition of pressure, follow, etc., for many thousands of times, the total number of contact failures in each group being recorded. The relays under test were mounted on the roof of a factory building and covered in such a manner as to shut out rain, but permit smoke and atmospheric dust to circulate freely around them. He states that with regard to contact pressure only "Three groups, each consisting of one hundred pairs of make contacts, were operated with contact pressures of 5, 15, and 30 grams, respectively. During the first half-million operations the first group had 4,500 failures per pair of contacts; the second group had 213 failures per pair; and the third group had only 2 failures per pair. The test conditions in this case were, of course, much more severe than any operating conditions ever encountered. However, the advantage of contact pressure is strikingly proved."

---

[2] See "Characteristics of Strowger Relays," "a series of five articles which discuss the important points in the design of relays, with an analysis of their action and various reference tables, governing their adjustment," by K. W. Graybill, Engineering Department, Automatic Electric Company. This is a series of five articles reprinted from the *Telephone Engineer* into Sales Circular 1545 by the Automatic Electric Company.

Another factor which determines contact reliability is the frequency of use. Quite obviously a contact which is used frequently is less affected by the small accumulations of dirt which may prevent the contacts from coming into actual physical contact.

In general, for the ordinary non-power switching application of a relay with low contact resistance contacts, a contact pressure of about 1 oz. is adequate. Where relays are exposed to dirt and grease, and cannot be adequately protected by covers, the contacts should be cleaned periodically with carbon tetrachloride.

When the relay is used to control light power circuits of moderate voltage, where high contact resistance is not objectionable, tungsten contacts with higher contact pressures are used. A moderate contact pressure for tungsten is about 4 oz., a fairly high pressure is about 8 oz., and a high contact pressure is about 12 oz.

3. **Contact Deflection and Follow-Through.** The contact deflection required to operate a contact is made up of two parts: (1) the actual physical separation of the contacts, and (2) the follow-through. The required physical separation of the contacts is determined primarily by the circuit voltage, or the surge voltage produced on opening the contacts. The surge voltage in turn depends on the characteristics of the circuit and on the methods taken to suppress arc formation. In general, for the same supply voltage, an inductive circuit will require a greater contact separation than a resistive circuit.

Arc suppression is obtained in a combination of ways: by means of condensers or resistances across the contacts, high speed of contact separation, the use of a contact material having a low arcing characteristic, and in cases of severe load, use of blow-out coils. The proper size of condenser can be best determined experimentally. It should be large enough to prevent arc hangover and transfer of contact metal due to the momentarily high voltage that occurs during point separation. Too large a condenser is detrimental, as it produces a bad spark on closing the contact and also makes the circuit respond more slowly. A resistance across the contact is also beneficial but does not allow the circuit to be completely opened.

Speed of contact separation is very important. If the contacts are allowed to open slowly, an arc is drawn during the time of separation. If the circuit is inductive this arc may be sustained for a long time. Where a clean break is very important special methods should be used to insure fast contact separation.

Contact materials vary in their arcing characteristics. This is determined to a great extent by the vaporization temperature of the metal. In this respect tungsten is superior to other metals, its vaporiza-

ART. 127]   CONTACT DEFLECTION AND FOLLOW-THROUGH   489

tion temperature being 5830° C. as compared to 3910° C. for platinum, which is next highest.

In many light-duty relays the necessary contact separation need be only a few thousandths of an inch, while for heavy duties separations as large as $\frac{1}{8}$ in. may be necessary. No general rules can be given because of the large number of variables involved.

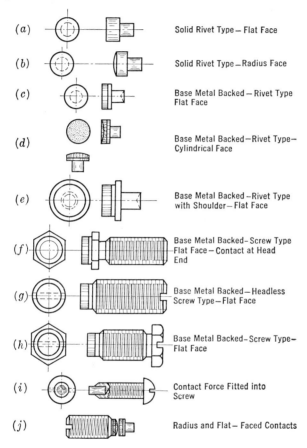

FIG. 1. Various standard contact shapes. (Courtesy of P. R. Mallory & Co., Inc., Indianapolis, Indiana.)

Follow-through is the term applied to that part of the contact motion which occurs after the contact has been made. In contacts mounted on leaf springs, it is necessary so that contact pressure can be built up by deflecting the springs. The follow-through of the contacts after the contact has been made generally results in one contact sliding over the other and hence produces a wiping action which tends to keep the contact clean.

4. Contact Size and Shape. The size of a contact is determined by the current to be carried, the frequency of operation, the permissible temperature rise, and the desired contact life. Standard contacts can be had in a variety of sizes and shapes. Rivet-type contacts are available in face diameters varying from $\frac{1}{16}$ to $\frac{5}{8}$ in. In Fig. 1 are shown various standard contact shapes. The ideal shape of the contact is a flat face producing the largest possible contact area. However, it is generally very difficult to keep two flat-faced contacts in perfect alignment, and therefore it is quite common to use a flat-faced contact working against a radius-faced contact as shown in Fig. 1j. This makes a practically self-aligning combination.

The current-carrying capacities for various sized contacts for both direct and alternating current and for the common contact materials are about as follows:

Current Range	Contact Diameter
1 amp. max.	$\frac{1}{16}$ to $\frac{1}{8}$ in.
1 to 5 amp.	$\frac{1}{8}$ to $\frac{3}{16}$ in.
5 to 15 amp.	$\frac{3}{16}$ to $\frac{3}{8}$ in.
15 to 30 amp.	$\frac{3}{8}$ to $\frac{5}{8}$ in.

128. Relay Sensitivity as Affected by Design Limitations

General. The smallness of the power input to the relay necessary to make it perform its assigned task is a measure of its sensitivity. The sensitivity which can be attained in a relay design will depend on the values assigned to the factors tabulated below:

1. Contact pressure.
2. Contact deflection and tolerance.
3. Differential current sensitivity.
4. Minimum air gap.
5. Size or weight.
6. Speed of action.
7. Delayed action.
8. Single- or double-coil design.

In the following sections the manner in which these factors affect sensitivity is discussed.

1. Contact Pressure. The required contact pressure of the relay determines the operating force for which the magnet must be designed. As stated before, for the ordinary light-duty circuit control relay, **the contact pressure should not be less than about 1 oz.** Multiple contact relays will require larger forces as determined by the arrangements of the contacts. When computing the required magnetic force, the required contact force must be multiplied by the ratio of the length of the lever arm from the contacts to the armature hinge to that between the point of application of the magnetic force and the hinge. To this must be added the force required to deflect any springs which are used to return the contacts. Quite obviously the magnitude of this force will determine, to a great extent, the necessary size of the pole core and the number of ampere-turns.

2. **Contact Deflection and Tolerance.** The necessary contact deflection, multiplied by the proper lever arm ratio, will determine the minimum stroke of the magnet at its working air gap. To this must be added the maximum variation in stroke required for manufacturing tolerance, in order to obtain the maximum required stroke. Tolerance in the stroke adjustment of the relay is a vital necessity if reliability is to be insured. This tolerance should be based upon the manufacturing requirements and the type of service expected from the relay. To the maximum stroke must be added the minimum allowable air gap under the pole core face to obtain the maximum air-gap length under the pole core. This will be the initial length of the air gap at the beginning of the stroke and together with the required force, as discussed in Sec. 1, will determine the maximum work the relay is to perform. As the total air-gap length at the beginning of the stroke is increased the relay sensitivity will decrease.

3. **Differential Current Sensitivity.** In many applications the relay must release on a current other than zero. If the ratios of the release to actuation force, and the release to actuation current, are close to unity the minimum air-gap length will be a large percentage of the initial gap length, and the relay will be relatively insensitive. Neglecting the effect of hysteresis, this may be analyzed as follows:

Let the subscript 1 designate quantities at the beginning of the stroke, and subscript 2, quantities at the end of the stroke. Then, referring to Art. 72, Sec. 1, we may write for the actuation and release forces

$$(\text{Force})_1 = \frac{\mu S}{2} \left[\frac{(F_a)_1}{l_1} \right]^2$$

$$(\text{Force})_2 = \frac{\mu S}{2} \left[\frac{(F_a)_2}{l_2} \right]^2$$

where S is the area of the working gap assumed to remain constant, and F_a and l are the ampere-turns and length of the working gap, respectively. Letting K equal the ratio of the release force to the actuating force, we may write

$$\sqrt{K}\left[\frac{(F_a)_1}{l_1}\right] = \left[\frac{(F_a)_2}{l_2}\right]$$

and

$$\frac{l_2}{l_1} = \frac{1}{\sqrt{K}}\frac{(F_a)_2}{(F_a)_1}$$

If we assume that the iron has no residual effect and is very permeable, its magnetomotive force may be neglected; then $(F_a)_1 \propto I_1$, and $(F_a)_2 \propto I_2$. Substituting these we have

$$\frac{l_2}{l_1} = \frac{1}{\sqrt{K}}\left(\frac{I_2}{I_1}\right) \quad (1)$$

If we designate the required stroke of the relay armature by s, the initial and minimum values of the gap length may be found as follows:

$$l_1 - l_2 = s$$

or

$$l_1 = \frac{s}{(1 - l_2/l_1)} \quad (2)$$

and

$$l_2 = l_1 \times \left(\frac{l_2}{l_1}\right) \quad (3)$$

To illustrate the effect of these ratios let us calculate the required air gaps for a relay having a required stroke of 0.030 in. for two cases: (1) $K = 1.25$, and $I_2/I_1 = 0.9$; (2) $K = 2.0$, and $I_2/I_1 = 0.5$. Substituting into equations 1, 2, and 3, the results tabulated below have been calculated. Assuming the required force to be the same in each

	s	K	I_2/I_1	l_2/l_1	l_2	l_1
Case 1.....	0.03	1.25	0.9	0.805	0.124	0.154
Case 2.....	0.03	2.00	0.5	0.355	0.017	0.047

case, it is obvious that the relay of case 2 will be much more sensitive than that of case 1. If we neglect magnetic leakage and hysteresis, and assume the exciting coils of the same size, the power required to operate the relay of case 2 will be approximately $(47 \div 154)^2 = 0.0909$ times as great as that required for the relay of case 1.

The above type of analysis can be made to include the effect of hysteresis, in an approximate way, by adding to the required release current I_2, a current component which will simulate the effective coercive magnetomotive force. This current component may be computed as $H_c l_i / N$ amperes, where H_c is the average coercive magnetic intensity of the iron, l_i the length of the iron circuit, and N the turns on the exciting coil. This will have the effect of increasing the ratio of I_2/I_1, and hence will cause an increase in l_2.

4. **Minimum Air Gap.** The allowable least length of the air gap between the armature and pole face at the end of the stroke is determined, fundamentally, by the residual magnetism of the relay. If this residual effect were zero no air gap would be necessary at the end of the stroke. However, as there always is some coercive magnetomotive force available from the iron circuit of the relay, an air gap must be provided to prevent the residual flux from being large enough to hold the armature when the coil is de-energized. Inasmuch as this air gap causes the total gap at the beginning of the stroke to be larger, it is undesirable from the standpoint of relay sensitivity. The use of magnetic materials having low coercive magnetic intensities is of great importance, if this effect is to be minimized. To this end the softest and purest grades of iron, or special materials such as Hipernik or Permalloy, should be used, depending upon the application. If the full potentialities of any of these materials is to be realized they must be carefully annealed after fabrication.

The effect of residual magnetism can be reduced or eliminated by using strong return springs. This will allow the minimum air gap to be made small, but will not necessarily enhance the sensitivity of the relay as the initial required force will be increased. If relay sensitivity is to be made high, it is absolutely essential that a magnetic circuit material having a low coercive intensity be used.

The required minimum length of gap necessary to effect release of the armature on breaking the exciting circuit can be determined in the manner of the last section. The effective current due to hysteresis, when the exciting circuit is broken, will be

$$I_2 = \frac{H_c l_i}{N} \qquad (4)$$

where the symbols are the same as defined in the last section. Using this value of I_2, the ratio I_2/I_1 may be computed, which, together with the value of K and the required stroke, will allow l_2, the minimum air gap, to be computed by means of equations 1, 2, and 3.

5. **Size or Weight.** If there is no limitation to the size or weight of the relay there will be no limitation to the sensitivity that can be obtained. Thus the cross section of the magnetic circuit could be increased indefinitely, and the ampere-turns for a given power input could be kept constant by making the cross section of the copper larger in the ratio of the increased mean length of turn. However, as this process goes on, the flux of the relay would increase proportionally to the magnetic circuit section, but the coil turns and resistance would remain constant. The inductance of the relay would then be proportional to the magnetic circuit section, and as the coil resistance remains constant, the time constant of the relay would also vary in proportion to the iron section. The net result is that, even if the size could be made unreasonably large, the relay would become impractical owing to its high time constant and consequent slow speed of action. The heavy weight of the moving parts would also contribute to producing a sluggish action.

Practically it is desirable to keep a relay small. Where the power available for operation is too small to permit this, it is often advisable to use a vacuum-tube amplifier or other power-amplifying device preceding the relay.

6. **Speed of Action.** High speed of action is generally obtained at a sacrifice of sensitivity. As pointed out in Chapter XII, high-speed magnets are essentially motors, and the speed of action determines the mechanical power developed. For a given relay, the greater the power input, the faster it will operate.

The above statements are generalizations and should be qualified. In the first place, relays are generally small pieces of apparatus with small moving parts and so are inherently fast. The average quick-acting relay of the telephone type operates in a time of about 10 to 20 milliseconds, and much shorter times are possible with the proper design. Relays of the polar type, which are inherently sensitive, have been made to follow 400-c.p.s. pulses. When the relay operation is carried out in a time of about 5 milliseconds or less, it may be considered high speed.

Two essential prerequisites for high-speed relays are light moving parts, and lamination of the magnetic circuit. The method of constructing the magnetic circuit of flat strip stock automatically increases the resistance of the eddy-current paths so that further laminating is generally not necessary. The use of high-silicon relay steels, in the form of strip stock or bars, further assists in decreasing time delay due to eddy currents. Where a round pole core of any appreciable [3] size, or a polar

[3] When used in this sense the word appreciable must be qualified; it depends not only on dimensions but also upon the speed of action required. In other words, the size is appreciable when it becomes difficult to obtain a satisfactory depth of penetration of the flux.

enlargement, is used, satisfactory laminating can be effected by means of radial slots.

7. **Delayed Action.** Delayed action in relays is obtained by placing copper lag coils, in the form of sleeves or slugs, on the main pole core. Their use and function do not particularly affect the sensitivity of the relay other than by reason of the loss of winding space occupied by the lag coil copper.

8. **Single- or Double-Coil Design.** The double-coil design offers two very important advantages over single-coil design when relay sensitivity is considered: (1) Two coils will permit about twice the coil copper to be used with the same over-all coil length, and with a reduction in interpolar leakage; (2) two working pole faces eliminate the necessity

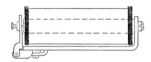

Fig. 2. Single-coil relay with hinged type of armature.

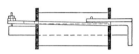

Fig. 3. Single-coil relay of U type with special high-permeability pin hinge joint.

for the fixed air gap between the armature and yoke and so allows all the ampere-turns, other than those in the iron, to be made available in the working gap.

Figures 2, 3, and 4 show various magnetic circuit structures used commercially for single-coil relays. Figure 2 shows a hinged type of

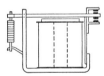

Fig. 4. Common commercial form of general-purpose single-coil relay.

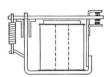

Fig. 5. Modified form of the relay of Fig. 4 designed to give a high-permeability joint at the armature hinge.

armature so arranged as to give a high permeance to the fixed gap between the armature and heel piece or yoke. Figure 3 shows a U-type relay with a special pin hinge joint giving a fixed gap of high permeability. Figure 4 shows a common commercial form of general-purpose relay, and Fig. 5 is a modification of Fig. 4 designed to increase the permeance of the fixed gap.

In all these relays the magnetomotive force for sending the flux across the fixed gap represents a waste in the sense that it produces no useful force. In the relay of the type of Fig. 4, the magnetomotive force across the fixed gap is often of the same magnitude as that across the working gap. In the two-coil design, illustrated in Fig. 6, all the magnetomotive force in the air gaps is useful in producing force. Thus, for the same developed coil ampere-turns, a two-coil design can be made to produce more force or mechanical work than a single-coil design. For the same total winding length of pole core, the two-coil relay will be only one-half as long as the single-coil design. This not only results in a shorter magnetic circuit, but also gives a more efficient magnetic circuit from the standpoint of leakage. For the same length of pole core, the two-coil design will permit the use of twice as much coil copper and hence give the same ampere-turns with only one-half the power input.

FIG. 6. Two-coil relay.

129. Relay Coils

The usual coil, on the better-made relay, is of the paper-section or cotton-interwoven type. For data on the design of these coils see Chapter VI. Owing to the fine sizes of wire which are often necessary, special problems are encountered which need not be considered when dealing with heavy wire coils. In particular, moisture condensation, arising from sudden changes in temperature or high-humidity climatic conditions, causes considerable trouble with electrolysis and corrosion.

Where climatic conditions are severe it is wise to use a coil which is completely impregnated so as to exclude moisture from the windings. The best coil for this purpose is the cotton-interwoven coil. The wick action of the cotton fiber allows the impregnating compound or varnish to permeate the whole coil. If a paper-section coil is used the ends should be carefully sealed with an impregnating compound.

Where electrolysis conditions are severe it is always desirable to connect the negative terminal of the power supply directly to the coil. If any leakage of current then occurs, the coil will be less affected, as destructive corrosion always takes place at the positive pole of the battery.

130. High-Speed Relays

The time of action of a relay is determined in exactly the same manner as that of a tractive magnet. For exact details, reference should be made to Chapter XII. The usual relay is of the direct attraction type and its dynamic characteristics will be of the same general form as illustrated in Fig. 24, Chapter XII. In general the average relay is relatively fast acting compared to a tractive magnet. This is so for two reasons: (1) the armature is of relatively light weight and the required stroke is small, thus making the kinetic energy of the moving parts a small portion of the total work done; and (2) its physical size is small, its time constant is small, and hence the time required for the current to build up to its operate value will be short.

As with any other type of magnet, the more sensitive the relay is on a power basis, the slower acting it will be. As an example the high-silicon steel relay designed in Art. 135 and illustrated in Fig. 16 is a sensitive relay. An approximate idea of its time of action can be found by determining its time constant. Referring to the rising magnetization curve of Fig. 20, Art. 135, Sec. 10, the permeance of the entire magnetic circuit in the open-gap position will be

$$P = \frac{\nu_y \phi_u \times 10^{-5}}{F} = \frac{1.113 \times 2.86 \times 10^{-5}}{41.7} = 0.0746 \times 10^{-5} \text{ weber per ampere-turn}$$

From Sec. 17 the coil turns and resistance are 2,250 and 65 ohms, respectively. The coil inductance will therefore be

$$L = N^2 P = 2{,}250^2 \times 0.0746 \times 10^{-5} = 3.77 \text{ henries}$$

and its time constant will be

$$\frac{L}{R} = \frac{3.77}{65} = 0.058 \text{ second}$$

The ratio of the operate ampere-turns to the normal ampere-turns is $41.7 \div 52 = 0.8$. The time required for the current to rise to the operate value may be determined by substituting into equation 11, Chapter XI. This gives, on simplification,

$$\epsilon^{-\frac{R}{L}t} = \epsilon^{-17.25 t} = 0.2$$

$$t = 0.0933 \text{ second}$$

Thus this relay will take 0.0933 second after the switch is closed before the armature commences to move.[4] This is relatively slow. The average quick-acting relay of the telephone type will operate in the order of 0.015 second or less.

[4] As the magnetization curve is practically a straight line from zero to the operate flux ϕ_u, the permeance as computed will be constant, and hence this time calculation is quite precise.

High-speed operation in a relay is obtained in exactly the same manner as in a tractive magnet. Care must be taken to reduce eddy currents in the magnetic core material to a minimum. Eddy currents can be greatly mitigated by: (1) the use of flat strip when practical; (2) slotting of round pole cores as illustrated in Fig. 19b, Chapter XIII; and (3) the use of a special relay steel, such as high-silicon steel or ferronickel, which has a high resistivity. In addition to these precautions regarding eddy currents, the moving parts should be made as light as possible, and the stroke reduced to a minimum. After these factors have been made as favorable as convenient, the only thing which can be done to improve the speed of operation is to decrease the time required for the current to reach its operate value. This can be done in two ways: (1) decrease the time constant of the coil circuit; (2) increase the operating power input leaving the coil circuit time constant unchanged.

The time constant can be reduced in two ways: (a) improve the magnetic design of the relay by decreasing leakage flux, etc., so that the flux linkage, and hence the coil inductance, will be reduced; (b) increase the coil circuit resistance by using either a series resistance or a finer wire size. Method b will necessitate the use of a higher operating voltage. Method 2, calling for an increase in power input, requires either an increase in operating voltage or a decrease in turns. The limitation of this method is heating due to the excessively large value of stall or steady-state current equal to E/R.

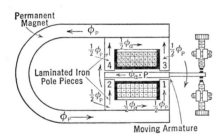

Fig. 7. High-speed polarized type of relay.

Where extremely high relay speeds are required, a polarized form of relay illustrated in Fig. 7 is very successful. In this form of relay the armature is pivoted at point P between the laminated pole faces of a permanent magnet arranged as shown. The control spring torque acting on the armature can be arranged to give two effects: (1) sufficiently weak to give instability, resulting in the armature contact resting on one or the other of the outer contacts, depending on the direction of the last current pulse through the armature coil; (2) sufficiently strong to (a) hold the armature centered, giving an armature motion proportional to the armature current for small deflections, or (b) bias the armature against one set of pole faces, making it responsive to only one polarity of current. Type (1) is usually employed for relay work. Type (2), because of the stiff control spring, is less sensitive than type (1), but also has a higher natural frequency. With the armature centered, it has been widely employed as a loud speaker driver.

The operation of type (2a) is explained as follows: With no current through the armature coil the armature flux ϕ_a will be zero, and the flux of the permanent magnet ϕ_p will divide equally between the pole faces as illustrated. As the flux of each gap is $\frac{1}{2}\phi_p$, the force between the armature and pole face of each gap will be the same, and the net torque available to rotate the armature will be zero. If armature current is now allowed to flow, an armature flux ϕ_a will be produced. This flux will divide equally between the two poles faces as shown, when the armature is in its center position and also for small deflections off the center position. Thus, the total flux in gaps 1 and 4 will be $(\frac{1}{2}\phi_p + \frac{1}{2}\phi_a)$, and that in gaps 2 and 3 will be $(\frac{1}{2}\phi_p - \frac{1}{2}\phi_a)$. The corresponding forces will be proportional to these fluxes squared, and the net torque available to rotate the armature will be proportional to the difference of the squares of the fluxes or $(\frac{1}{2}\phi_p + \frac{1}{2}\phi_a)^2 - (\frac{1}{2}\phi_p - \frac{1}{2}\phi_a)^2 = \phi_p\phi_a$. As ϕ_p is constant, the torque will be directly proportional to ϕ_a. If ϕ_a is reversed, by reversing the armature current, the direction of the torque will be reversed.[5]

131. Time Delayed Relays

Short time delays, of the order of 0.1 or 0.2 second, can be readily produced by means of a lag coil. The theory of operation of the lag coil and the method of computing its characteristics are fully explained in Arts. 97, 98, and 99 of Chapter XI. In applying the lag coil to a

---

[5] Theoretically, as soon as the armature moves from its central position, the balanced bridge arrangement effective for the armature flux is disturbed, and a portion of the armature flux will be shunted through the yoke of the permanent magnet. The reluctance of the path for armature flux (magnetic circuit assumed unsaturated) will decrease as the armature is moved off center owing to the shunt path through the yoke, and an additional force term will be introduced that varies as the square of the current. For a relay this additional term is desirable as it makes the action of the armature rapid and positive. In a device like a loud-speaker unit, where an armature motion proportional to current is desired, this additional force term causes distortion. For such applications the armature flux must be prevented from going through the yoke of the permanent magnet so that the linear force relation derived in the text will be valid. The armature flux that passes through the permanent magnet may be minimized by using a permanent-magnet material of low incremental permeability or by the use of a saturated core in series with the yoke.

A further consequence of armature motion is that the distribution of the flux of the permanent magnet will be altered and a portion will be shunted through the armature. If the armature is unsaturated, an additional torque proportional to the angular deflection will result from this cause. This torque is of the same nature as the torque produced by the control spring and hence can be balanced out by a stronger spring.

relay two forms are commonly used, as illustrated in Figs. 8 and 9. Figure 8 shows the lag coil in the form of a copper sleeve extending for the full length of the pole core, while Fig. 9 shows the copper concentrated into the form of a slug at the armature end of the pole core.

The lag coil in the form of the sleeve will produce more time delay in operation and release than an equal amount of copper in the form of a slug. The reason for this is that the sleeve will link with more flux, and has a lower resistance due to its shorter mean length of turn.

When the copper slug is placed on the yoke or heel end of the relay, as illustrated in Fig. 10, the relay will be fast on operation and slow on release. On energizing the coil, the flux can build up rapidly through the leakage path between the pole core under the coil and the side

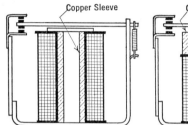
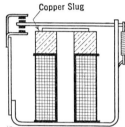
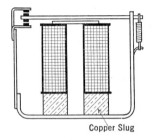

FIG. 8. Time-delay relay having the lag coil in the form of a copper sleeve.

FIG. 9. Time-delay relay having the lag coil in the form of a slug around the pole-face end of the pole core.

FIG. 10. Relay with copper slug on yoke or heel end of pole core, giving fast operation and slow release.

bracket, giving fast operation. After operation the flux will be slowly diverted from the leakage path to the copper slug. On de-energizing the relay the self-induced currents in the slug will help sustain the field and produce slow release.

The use of a lag coil is only suitable for producing short time delays, up to about 0.15 second for operation and 0.25 second on release. For time delays longer than this, special methods are required.

A method of producing a long time delay is by means of two relays, one of which has a weighted spring armature, and the other of which is of a quick-actuation, slow-release type, as illustrated in Fig. 10. The circuit through the coils of the slow-release relay is controlled by a contact on the weighted spring armature relay, while the circuit to be controlled is actuated by the contacts on the slow-release relay. When it is desired to break the controlled circuit, the circuit to the weighted armature relay is broken. This releases the weighted spring armature, which oscillates back and forth, making and breaking the circuit to the slow-release relay. As this relay is slow to release but quick to actuate, it will hold on during the periods of open circuit and quickly replenish its field strength during the periods of closed circuit. When the weighted spring armature finally stops oscillating it will leave the coil circuit to the slow release relay open, allowing it to break its circuit. Time delays of the order of 15 seconds can be obtained in this way.

A long time delay on actuation can be produced by the reverse of the above scheme, having one relay whose armature strikes a weighted spring and another which is slow to operate. When the former is actuated, the armature closes, striking the weighted spring and causing it to oscillate. This spring carries contacts which actuate the coil of the slow-to-operate relay. The slow-to-operate relay will not be completely energized until the amplitude of the oscillation of the weighted spring is very small. Final operation of this relay closes the circuit to be controlled. Inasmuch as the slow-to-operate relay is also slow to release, the successive contacts of the oscillating spring will tend to accumulate. Hence the operate time delay that can be produced is much shorter than with the weighted armature scheme described for slow release.

Another method, which is equally applicable for either delayed operation or release, is described in connection with Fig. 3, Chapter XI. This method, which is electronic, can be accurately controlled and is capable of producing time delays as high as 30 seconds with reliability.

132. Alternating-Current Relays

Relays for direct operation on alternating current can be successfully built provided that a shading coil is used to prevent undue chattering. In relays for relatively small forces the iron parts can be made solid. Figure 11 shows a construction identical to that of Fig. 4 with the armature slotted to receive the shading coil, while Fig. 12 shows a similar construction with a slot to provide room for a shading coil in the pole face. In this type of construction, high-silicon steel is very advantageous. Its high resistivity permits the use of larger pole cores before slotting, to minimize the effect of eddy currents, becomes imperative.

Where the best type of construction is desired, the core should be laminated as illustrated in Fig. 13.

The design of an alternating-current magnet for a relay follows the same principle as those designed for tractive magnets. Chapter XIII gives all the necessary information for the design of short-stroke alternating-current magnets and the design of shading coils to prevent chattering.

In general it is best to arrange the contacts of an alternating-current relay on springs, so that a slight chattering of the armature will not open the contact.

Regarding sensitivity, the same comparison can be made between the alternating-current relay and the direct-current relay as between the

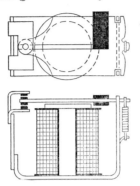

FIG. 11. Alternating-current relay with a shading coil on the armature.

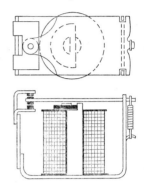

FIG. 12. Alternating current relay with a shading coil in the face of the pole core.

respective tractive magnets. The alternating-current relay has the same volt-ampere limitation as the tractive magnet and hence its sensitivity,

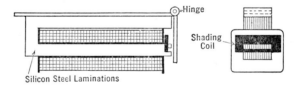

FIG. 13. Alternating-current relay with a laminated pole core.

on a volt-ampere basis, is of necessity much lower than for the same relay excited with direct current.

133. The Cycle of a Simple Relay

In Fig. 14 is shown a simple type of relay with a hinged armature provided with a spring to hold it open. The open and closed positions of the armature are designated by subscripts 1 and 2, respectively, as shown. In Fig. 15 are shown the magnetization loops of the relay. Loop a–b–e–a represents the magnetic cycle obtained, if the magnetomotive force is varied from zero to its maximum value and then back to zero continuously, with the armature open or in the non-operated position. Loop f–c–d–f represents the same magnetic cycle with the armature in its closed or operated position. Points e and c are determined by the maximum coil magnetomotive force $= NI_m = NE/R$.

In actual operation we start with the armature open, and the current at zero, as shown by point a. As the current is gradually increased the flux will rise along curve a–e, the rising branch of the open-gap hysteresis loop, until some point b is reached, where the flux is just sufficient to operate the armature. This point must be below e; otherwise the relay would be inoperative on the normal supply voltage E. From b the armature moves to its closed-gap position and the flux rises to point c. The exact path that the flux follows from b to c depends on the speed with which the armature moves. With the armature in the closed-gap position the current is gradually lowered and the flux falls along c–f, the falling branch of the closed-gap hysteresis loop, to point d, where the flux has fallen to a value just sufficient to release the armature. From d the armature moves to its open-gap position and the flux falls to a, completing the cycle. As long as the armature is allowed to move in this sequence while the current is varied between zero and I_m, the flux will follow the cycle.

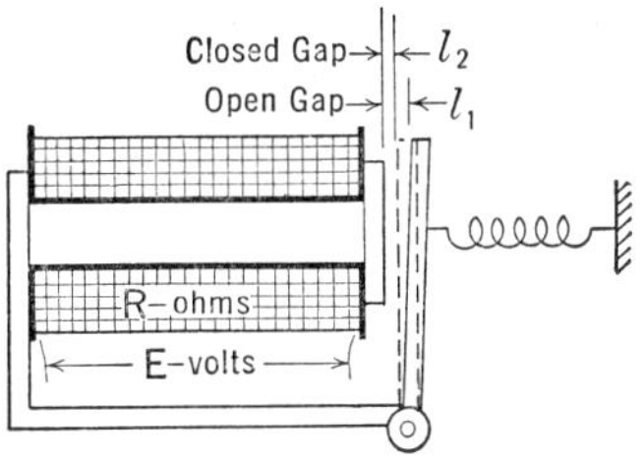

FIG. 14. Simple hinged relay showing open and closed positions of armature.

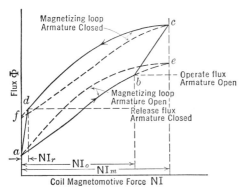

FIG. 15. Magnetization loops of the simple relay of Fig. 14 showing the magnetic cycle obtained when the relay is put through a normal sequence of operations.

The important points of the cycle are points b and d, which determine the operate and release currents, respectively. The operate magnetomotive force NI_0 is determined by the length of the open air gap, the spring force, and the rising magnetization curve of the iron. The release magnetomotive force, NI_r, is determined by the length of the closed air gap, the spring force, and the falling magnetization curve of the iron.

When it is desired to make the release current very close to the operate current, then, as already explained in Art. 128, Sec. 3, the closed-gap length must be a large percentage of the open-gap length, and the coercive magnetomotive force of the iron must be reduced to a minimum.

134. Optimum Design Considerations of Relays

The actual design of the magnetic circuit of a relay must be treated on the same basis as for tractive magnets. Thus, for maximum work from a given available coil magnetomotive force, the optimum conditions as derived in Art. 85 must be adhered to. Those conditions state that the differential permeance of the iron circuit including the fixed air gaps must be made equal to the permeance of the working gap. This is true regardless of the shape of the pole faces used or the effects of hysteresis.

The usual relay is of one of the types illustrated in Figs. 2, 3, 4, and 5, which represent modifications of the fundamental horseshoe-type magnet discussed in Art. 91. The optimum conditions as set forth in Fig. 25, Art. 91, however, do not necessarily apply to relays, even if they are of the exact horseshoe type. The reason for this is that the data of Art. 91 have been computed to give optimum designs with temperature rise as a limitation, and also because these data have been evolved for magnets doing considerably more work than the usual relay.

In general, because of the low values of magnetomotive force available for relay operation, the iron and air-gap flux densities which are economical will be much lower than for tractive magnets. Likewise, the polar enlargements that should be used to obtain optimum conditions are generally larger.

Another condition which will affect optimum proportions in a relay as compared to a tractive magnet is the possibility of advantageously employing special alloys such as Permalloy, 50 per cent ferronickel, or high-silicon steel.

The coils of a relay should be so proportioned as to give the greatest number of ampere-turns for a given power input.

135. Sample Design of a Simple Relay: Comparison of Results Using Three Types of Magnetic Material

For the purpose of illustrating the method of computing the actual operate and release currents of a relay, it is proposed to design and calculate the performance of a relay meeting the following specifications:

Type. Single-coil type with inexpensive and simple magnetic circuit like that of Fig. 5. Double-throw switch with upper contact closed when coil is de-energized.

Size. Overall coil length 1 in.; diameter $\frac{15}{16}$ in.

Weight. Minimum possible, consistent with other requirements.

SAMPLE DESIGN OF A SIMPLE RELAY

Voltage. Relay is to operate on 1.5 volts.
Coil resistance. Maximum possible and still meet force requirements.
Stroke. 0.030 in. at contacts.
Contact force. On either up or down contact the force is not to be less than 1 oz.
Operate current. A current equal to 0.8 the normal current must be just sufficient to operate the relay.
Release current. Release is to be at zero coil current, but for the purpose of obtaining a factor of safety, the release must occur at not less than 25 per cent of the maximum current.
Material. Make designs using three types of materials:
 1. Swedish charcoal iron.
 2. Relay steel, high-silicon steel, sample 9
 3. 47 per cent ferronickel, sample 11.

Assume that all parts have been properly annealed after manufacture.

Preliminary Design

1. **Tentative Proportions.** In Fig. 16 is shown a suitable relay construction similar in magnetic-circuit design to that of Fig. 5. The coil and contact separation given satisfy those of the specifications; the other proportions necessary to start the preliminary design have been chosen to keep the over-all size as small as possible and prevent undue leakage. In Fig. 16, part 1 is the armature; 2 the armature fins used to produce

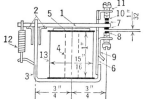

Fig. 16. Tentative design for the relay of the problem of Art. 135.

a low-reluctance fixed gap; 3 the bracket or yoke supporting the armature, pole core, and contacts; 4 the pole core; 5 the polar enlargement; 6 a piece of phenolic insulation supporting the contact and terminal pieces 9 and 11; 7 the contacts on the armature tongue; 8 and 10 screw-type contacts; 12 the spring holding the armature against the upper contact; and 13 the coil. The armature is shown with the upper contact closed under the influence of the spring. The coil leads can be brought out to eyelets on the phenolic piece, and the moving contact can be brought out from the armature by means of a pigtail.

2. **Force-Current Requirements.** As the force on the upper contact produced by the spring is to be 1 oz., the magnetic pull required at the center of the pole core to just break this contact will be $1 \times 1.5 \div 0.75 =$

2 oz., where 1.5 in. and 0.75 in. are the lever arms from the contact and pole center to the hinge, respectively. In order to close the lower contact against the spring and develop a contact force of 1 oz., a total magnetic pull of 4 oz. will be necessary. However, to release the armature, the magnetic pull need drop to only 2 oz., which is the normal spring pull at the center of the armature. It has been assumed that the change of spring force due to the armature motion is negligible. The required armature motion at the center of the pole face will be $0.03 \times 0.75 \div 1.5 = 0.015$ in. In the table below are shown the force-current requirements at the two extreme armature positions. I is the normal coil current equal to E/R.

Armature Position	Electromagnetic Force, oz.	Coil Current	
Open gap	2	$0.8I$	Actuation
Closed gap	4	$0.8I$	Actuation
Closed gap	2	$0.25I$	Release

3. **Open- and Closed-Gap Length.** Referring to Art. 128, Sec. 3,[6] the open- and closed-gap lengths, designated as l_1 and l_2, respectively, may be found as follows. Substituting into equation 1, we have

$$\frac{l_2}{l_1} = \frac{1}{\sqrt{K}} \left(\frac{I_2}{I_1}\right) = \frac{0.25I}{0.8I} = 0.313$$

where K, the ratio of the release to actuating force, equals 1. Substituting into equation 2, we have

$$l_1 = \frac{s}{(1 - l_2/l_1)} = \frac{0.015}{(1 - 0.313)} = 0.022 \text{ in.}$$

and

$$l_2 = l_1 - s = 0.007 \text{ in.}$$

4. **Size of Polar Enlargement.** This depends on the required force, the available magnetomotive force, and the open-gap length. If it is desired to keep the required exciting ampere-turns as small as possible,

---

[6] This leaves out of consideration the residual effect of the iron. The magnitude of this could be estimated at this time, but because three different materials are being used, it is not convenient. The exact effect of the residual flux density will be apparent when the finished designs are compared.

the polar enlargement should be made as large as is convenient, and the iron circuit should be so proportioned that its differential permeance equals that of the working air gap produced by the selected polar enlargement. This is the same criterion for optimum economy as derived in Art. 85, Chapter X. If, on the other hand, the relay power input has been specified, it will be possible to estimate, from the size of the space allotted to the coil, an approximate value of the coil ampere-turns. Assuming a suitable proportion of these ampere-turns across the open-gap length, the flux density of the working gap, and hence the area necessary to produce the required force, can be estimated. The area of the iron circuit and the permeance of the fixed gap must then be proportioned to use their allotted ampere-turns. The final result, to be optimum, must satisfy the criterion of Art. 85.

The former method of procedure gives the most sensitive design for a given space limitation; the latter method gives the smallest design for a given ampere-turn input.

This design is of the first kind, as the specifications call for the highest possible coil resistance with 1.5 volts applied. If the bracket and armature are made as wide as the coil, or slightly wider, they serve as a protection to the coil and give a design of reasonable proportions. Let us then assume a circular polar enlargement 1 in. in diameter as the largest convenient size. When the design has been finally completed, it will be possible to check how close to the optimum this choice is (see Sec. 18).

5. Flux Density of Working Gap. The actual required force of 2 oz. at open gap is to be developed at 0.8 normal excitation. As a sensitive relay is seldom saturated, we may assume that the force will be $2 \div 16 \times (1 \div 0.8)^2 = 0.195$ lb. with normal excitation. The average working gap flux density will then be

$$B = \sqrt{\frac{72 \text{ Force}}{S}} = \sqrt{\frac{72 \times 0.195}{\pi \times 0.5^2}} = 4.23 \text{ kmax. per sq. in.}$$

6. Size of Pole Core. If we tentatively estimate the leakage coefficient as 1.25, the flux at the base of the pole core and in the yoke will be

$$\phi_y = \nu_y BS = 1.25 \times 4.23 \times 0.785 = 4.15 \text{ kmax.}$$

The proper flux density for the pole core is hard to predetermine. If the relay is to be made as sensitive as possible, the iron should be operated in the region of its maximum permeability. A flux density a little higher than that corresponding to maximum permeability is preferable, as the slight reduction in permeability is more than compensated for by the decrease in pole core section. A small pole core section is desirable as it permits a shorter mean length of turn on the exciting coil and hence will allow more ampere-turns for a given power input, and coil size. However, in the final analysis, the criterion of Art. 85 must be applied to determine the optimum flux density.

In Fig. 17 are shown relative permeability curves for the three most important relay steels, Swedish charcoal or American ingot iron, high-

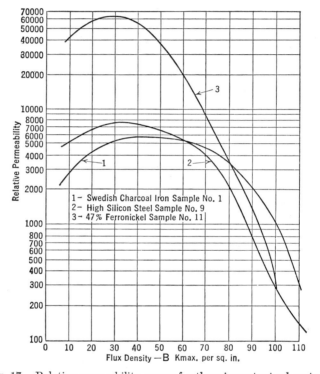

Fig. 17. Relative permeability curves for three important relay steels.

silicon steel (sample 9), and 47 per cent ferronickel (sample 11). In special cases where extreme sensitivity is desired Permalloy may be found useful.

For the purpose of this design let us try a value of pole core flux density equal to 70 kmax. per sq. in. for all three materials. Then the pole core area will be

$$S \text{ pole core} = \frac{\phi_y}{B_y} = \frac{4.15}{70} = 0.0593 \text{ sq. in.}$$

corresponding to a pole core diameter of 0.275 in., which can be rounded off to $\frac{9}{32}$ in.

7. Yoke or Bracket. This should be made of strip stock 1 in. wide, so as to protect the coil and cover the polar enlargement. Its area should be the same as the pole core, and hence a thickness of $\frac{1}{16}$ in. will be suitable.

8. Leakage Coefficient. Owing to use of only one coil the leakage field in this type of magnet is inclined to be dissymmetrical in comparison with that of a bipolar magnet. However, if the permeance of the fixed hinge joint is large compared to that of the working gap, the fringing and leakage field may be considered to follow the conventional paths as designated by the permeance formulas. If the permeance of this joint is low, there will be a relatively large magnetic potential between the armature and the yoke, with the result that an appreciable amount of flux will leak directly from the polar enlargement to the yoke. Because of these considerations the point of maximum flux in this type of magnetic circuit will not be in the yoke but in the pole core near the yoke.

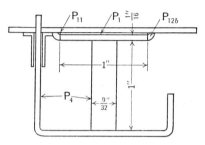

FIG. 18. Iron circuit of the relay of Fig. 16 showing the leakage paths of the flux of the main pole.

For our purpose let us use the leakage paths as sketched in Fig. 18. Then:

$$P_1 = \frac{\mu S}{l} = \frac{3.19 \times 0.785}{0.022} = 114 \text{ max. per ampere-turn}$$

where $0.022 = l_1$ is the average length of the air gap in the open armature position.

$$P_{11} = 0.26 \mu l = 0.26 \times 3.19 \times \pi \times 1 = 2.6 \text{ max. per ampere-turn}$$

$$P_{12b} = \frac{2\mu l}{\pi} \log_\epsilon \left(1 + \frac{t}{g}\right) = \frac{2 \times 3.19 \times \pi \times 1}{\pi} \log_\epsilon \frac{84}{22} = 7.2 \text{ max. per ampere-turn.}$$

In the last two fringing permeance calculations, the armature has been assumed to be wider than the polar enlargements to simplify the leakage paths. This is permissible, as the portion so affected is small, and the result obtained is slightly on the high side. The leakage coefficient referred to the armature flux will then be,

$$\nu_a = \frac{\phi_a}{\phi_u} = \frac{114 + 9.8}{114} = 1.086$$

In calculating the distributed permeance between the bracket and the pole core, the vertical part of the bracket supporting the armature has been taken as a plane of infinite extent. This is on the safe side and will tend to compensate for some of the extraneous leakage paths present due to the dissymmetry. Then

$$P_4 = \frac{2\,\mu\pi l}{\log_\epsilon \dfrac{d}{r_1}} = \frac{2 \times 3.19 \times \pi \times 1}{\log_\epsilon \dfrac{1.5}{0.141}} = 8.5 \text{ max. per ampere-turn}$$

If we consider the flux through the entire pole core and yoke to be equal to the armature flux plus two-thirds of the pole core to bracket leakage flux,[7] the leakage coefficient as referred to the pole core and yoke will be

$$\nu_y = \frac{\phi_a + \tfrac{2}{3}\phi_L}{\phi_u} = \frac{123.8 + 8.5 \times \tfrac{1}{2} \times \tfrac{2}{3}}{114} = 1.113$$

9. Permeance of Fixed Hinged Joint. In order that the design be efficient, the permeance of the hinged joint and its associated leakage paths should be high compared to that of the working gap. As this relay is intended to be as sensitive as possible, special care has been taken to increase the permeance of this gap by means of armature fins. Figure 19 shows the contemplated construction of the gap. In order to get a high permeance, the armature fins must be brought close to the yoke bracket. This clearance is shown as 0.01 in. in Fig. 19 and is just enough to give a $\tfrac{1}{32}$-in. contact motion in the worst possible armature position, taking into account the 0.002-in. clearance of armature around the bracket pins.

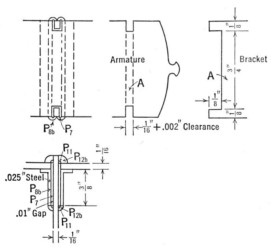

Fig. 19. Details of the hinged joint of the relay of Fig. 16, showing the paths of fringing flux.

---

[7] See Art. 46. This is a good approximation in this case, because the iron is to be operated near its range of constant permeability.

Where the armature rests in contact with the bracket on surface A, the air gap cannot be taken as zero because, unless the relay is a handmade job, this will be a punched surface and slightly uneven. It is therefore on the safe side to consider a small air gap to exist. Taking this at 0.002 in., the permeance of the gap will be

$$P_1 \text{ armature to bracket} = \frac{\mu S}{l} = \frac{3.19 \times \frac{1}{16} \times 1\frac{3}{8}}{0.002} = 137 \text{ max. per ampere-turn}$$

The direct permeance from the armature fins to the bracket will be

$$P_1 \text{ fins to bracket} = \frac{\mu S}{l} = \frac{3.19 \times 1 \times \frac{3}{8} \times 2}{0.01} = 239 \text{ max. per ampere-turn}$$

Armature to bracket fringing permeances:

$$P_{11} = 0.52 \mu l = 0.52 \times 3.19 \times \tfrac{5}{16} \times 2 \qquad\qquad = 1.04$$

$$P_{12b} = \frac{2\mu l}{\pi} \log_\epsilon \left(1 + \frac{t}{g}\right) = \frac{2 \times 3.19 \times \tfrac{5}{16} \times 2}{\pi} \log_\epsilon \frac{0.062}{0.002} = 4.30$$

$$P_7 = 0.26 \mu l = 0.26 \times 3.19 \times \tfrac{1}{4} \qquad\qquad = 0.21$$

$$P_{8b} = \frac{\mu l}{\pi} \log_\epsilon \left(1 + \frac{2t}{g}\right) = \frac{3.19 \times \tfrac{1}{4}}{\pi} \log_\epsilon \frac{0.062}{0.002} = 0.86$$

$$\underline{\phantom{6.4 \text{ max. per a-t.}}}$$
$$6.4 \text{ max. per a-t.}$$

Fin to bracket fringing permeances:

$$P_{11} = 0.52 \mu l = 0.52 \times 3.19 \times 2 \qquad\qquad = 3.32$$

$$P_{12b} = \frac{2\mu l}{\pi} \log_\epsilon \left(1 + \frac{t}{g}\right) = \frac{2 \times 3.19 \times 2}{\pi} \log_\epsilon \frac{0.025}{0.002} = 10.1$$

$$P_7 = 0.26 \mu l = 0.26 \times 3.19 \times 1.5 \qquad\qquad = 1.24$$

$$P_{8b} = \frac{\mu l}{\pi} \log_\epsilon \left(1 + \frac{2t}{g}\right) = \frac{3.19 \times 1.5}{\pi} \log_\epsilon \frac{0.057}{0.002} = 5.03$$

$$\underline{\phantom{19.7 \text{ max. per a-t.}}}$$
$$19.7 \text{ max. per a-t.}$$

The total permeance of the hinge joint will be $137 + 6.4 + 239 + 19.7 = 402$ max. per ampere-turn.

10. Rising Magnetization Curve, Open Gap. Before this can be calculated, the armature thickness must be decided on. Inasmuch as the armature and yoke leakage coefficients are so nearly equal, their fluxes will be about the same. Therefore it is convenient, and on the safe side, to make the armature $\frac{1}{16}$ in. thick.

All the iron calculations should be carried out using an unsymmetrical hysteresis loop for the iron, drawn between the proper flux density limits, as discussed in Art. 28. As pointed out in Art. 20, the rising branch of such a loop for the higher and more important densities is well simulated by the normal magnetization curve. We shall therefore use the normal magnetization curve in this computation. The computations for the three materials, for one sample point on each, are shown below in tabular form.

Part	Effective		Flux	Flux, kmax.	B, kmax. per sq. in.	H, a-t. per in.	F, a-t.
	Length, in.	Area, sq. in.					
				Swedish Charcoal or American Ingot Iron Annealed Sample 1 $\nu_a = 1.086$, $\nu_y = 1.113$			
Armature.........	0.75	0.0625	$\nu_a \phi_u$	4.26	67	4.4	3.3
Pole core and yoke..	2.88	0.0625	$\nu_y \phi_u$	4.37	70	4.9	14.1
Fixed air gap......	$P = 402$ max. per a-t.		$\nu_a \phi_u$	4.26	..	...	10.6
Useful gap........	$P = 114$ max. per a-t.		ϕ_u	3.92	..	...	34.4
						Total =	62.4

Part	ϕ	High-Silicon Steel Sample 9			47% Ferronickel Sample 11	
		B, kmax. per sq. in.	H, a-t. per in.	F, a-t.	H, a-t.	F, a-t.
Armature.............	4.26	67	4.6	3.5	1.85	1.4
Pole core and yoke.......	4.37	70	5.2	15.0	2.5	7.2
Fixed air gap...........	4.26	..	...	10.6	....	10.6
Useful gap.............	3.92	..	...	34.4	....	34.4
				Total = 63.5	Total =	53.6

The final results are shown plotted in Fig. 20.

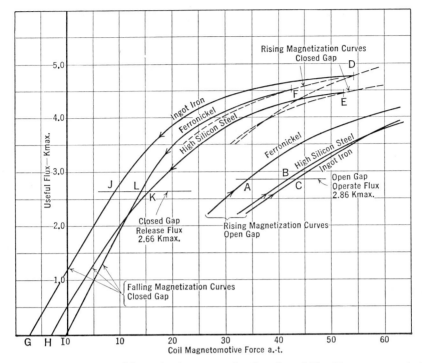

Fig. 20. Magnetization curves for the relay of Fig. 16.

11. Operate Ampere-Turns. The magnetic force required to operate the armature in the open-gap position is 2 oz. The flux density in the working gap necessary to produce this force is

$$B = \sqrt{\frac{72 \text{ Force}}{S}} = \sqrt{\frac{72 \times 0.125}{\pi \times 0.5^2}} = 3.65 \text{ kmax. per sq. in.}$$

The useful flux will then be

$$\phi_u = BS = 3.65 \times 0.785 = 2.86 \text{ kmax.}$$

Referring this to Fig. 20 on the rising magnetization curves, the operate ampere-turns for the various steels will be:

Point C Swedish or American ingot iron = 43.4 a-t.
Point B high-silicon steel   = 41.8 a-t.
Point A 47% ferronickel   = 34.2 a-t.

The normal ampere-turns of the coil, according to the specification, will be 1.25 times the operate value.

12. Leakage Coefficient, Closed Gap.

$$P_1 = \frac{\mu S}{l} = \frac{3.19 \times 0.785}{0.007} = 368 \text{ max. per ampere-turn}$$

$P_{11} =$ same as for open gap $= 2.6$

$$P_{12b} = \frac{2\mu l}{\pi} \log_\epsilon \left(1 + \frac{t}{g}\right) = \frac{2 \times 3.19 \times \pi \times 1}{\pi} \log_\epsilon \frac{84}{7} = 15.6$$

$$\nu_a = \frac{368 + 18.2}{368} = \frac{386}{368} = 1.050$$

The interpolar leakage permeance will be the same as for the open gap, 8.5 and

$$\nu_y = \frac{368 + 18.2 + \tfrac{2}{3} \times \tfrac{1}{2} \times 8.5}{368} = \frac{389}{368} = 1.058$$

These leakage coefficients are so nearly equal that the armature, pole core, and yoke may be considered as one piece when the rising and falling magnetization curves at closed gap are computed.

13. Rising Magnetization Curve, Closed Gap. A portion of this curve must be computed in order to find out the flux and force in this position. A sample point is computed below in tabular form. The three curves are shown plotted in Fig. 20.

	Effective		Flux	Flux, kmax.	B, kmax. per sq. in.	Ingot Iron		High-Silicon Steel		47% Ferronickel	
$\nu_a = 1.050$ $\nu_y = 1.058$	Length, in.	Area, sq. in.				H, a-t. per in.	F, a-t.	H, a-t. per in.	F, a-t.	H, a-t. per in.	F, a-t.
Armature pole core and yoke	3.63	0.0625	$\nu_y \phi_u$	5.0	80	7.6	27.6	11.3	41.0	7.0	25.4
Fixed air gap..	$P = 402$ max. per a-t.		$\nu_y \phi_u$	5.0	....	...	12.4	...	12.4	...	12.4
Useful air gap..	$P = 368$ max. per a-t.		ϕ_u	4.73	....	...	12.9	...	12.9	...	12.9
							52.9		66.3		50.7

14. Flux and Force in the Closed-Gap Position, Operate Ampere-Turns. With the ampere-turns held at the operate value, the relay magnet must develop a force of 4 oz. in the closed-gap position. The useful flux in this position is determined by the intersection of the vertical lines drawn from points C, B, and A, with the rising magnetization curves in the closed-gap position, for the ingot iron, silicon steel, and ferronickel, respectively. The force will then be $(\phi_u)^2/72S$ lb.

These forces are given in the tabulation below. They are all above the minimum requirement of 4.0 oz.

Closed-Gap Force with Operate Ampere-Turns

Material	Useful Flux	Force, oz.
Ingot iron...............	4.31	5.3
Silicon steel..............	4.15	4.9
Ferronickel..............	4.24	5.1

15. Flux and Force in the Closed-Gap Position, Normal Ampere-Turns. If the operate ampere-turns are multiplied by 1.25, the normal coil ampere-turns will be obtained. The values of these, together with the useful flux and force produced, are given in the tabulation below.

Closed Gap Force with Normal Ampere-Turns

Material	Normal Ampere-Turns	Point on Curves	Useful Flux	Force, oz.
Ingot iron.........	54.2	D	4.77	6.4
Silicon steel........	52.3	E	4.45	5.6
Ferronickel........	42.7	F	4.53	5.8

16. Falling Magnetization Curve, Closed Gap. It is this demagnetization curve which will determine the release current of the relay. As there is a family of demagnetization curves, one for each value of magnetic intensity to which the iron has been magnetized before breaking the current, it is necessary to compute H_m for the iron in the closed-gap position. The useful flux may be determined from points D, E, and F of the closed-gap rising magnetization curve. Multiplying this by the leakage coefficient, the iron flux will be obtained, and dividing this flux by the iron area, the maximum loop flux density, B_m, will be obtained. Reference to the normal magnetization curves for these materials at B_m will give H_m. These data are tabulated below.

Material	Point on Curve	Useful Flux	Flux of Iron	B_m Iron	H_m Iron
Ingot iron............	D	4.77	5.05	80.7	7.8
Silicon steel..........	E	4.45	4.70	75.0	7.6
Ferronickel..........	F	4.53	4.80	76.7	5.0

The demagnetization curves, corresponding to the values of H_m listed above, are most easily constructed from the falling branches of the hysteresis loops for the different materials. These demagnetization curves may be constructed from the loop data of Figs. 12a, 12f, and 12g, and the residual flux and coercive intensity data of Figs. 13a and 13b of Chapter II. In order to change the loop coordinates from kilomaxwells per square inch and ampere-turns per inch to total flux and magnetomotive force for the entire iron circuit of the relay, the ordinates must be multiplied by the area of the iron circuit [8] 0.0625 sq. in., and the abscissas by 3.63 in., which is the length of the total iron circuit. The resulting demagnetization curves for the iron only are shown in Fig. 21.

The demagnetization curves of the entire magnetic circuit of the relay in the closed-gap position may now be computed by adding to the iron magnetomotive force of Fig. 21, for each value of useful flux, the magnetomotive force for the effective permeance of the working air gap (closed position) and the fixed-hinge air gap. This permeance will be from Secs. 9 and 12, $P_0 = 1 \div (1/389 + 1/402) = 197.5$ max. per ampere-turn. The computations, shown below in tabular form, are self-explanatory.

			Magnetomotive Force					
			Ingot Iron		High-Silicon Steel		Ferronickel	
Iron Flux	Useful Flux	Air Gaps	Iron	Total Coil	Iron	Total Coil	Iron	Total Coil
4.23	4.00	21.2	+0.6	21.8	+11.8	33.0	+6.0	27.2
3.70	3.50	18.6	−2.7	15.9	+ 6.5	25.1	+1.5	20.1
3.17	3.00	15.9	−4.3	11.6	+ 3.1	19.0	+0.2	16.1
2.64	2.50	13.3	−5.3	8.0	+ 0.6	13.9	−0.15	13.15
2.12	2.00	10.6	−6.0	4.6	− 0.9	9.5	−0.3	10.3
1.59	1.50	8.0	−6.5	1.5	− 1.9	6.1	−0.35	7.65
1.06	1.00	5.3	−6.8	−1.3	− 2.5	2.8	−0.4	4.9
0.53	0.50	2.7	−7.1	−4.4	− 2.9	−0.2	−0.4	2.3
0.0	0.0	0.0	−7.3	−7.3	− 3.3	−3.3	−0.4	−0.4

The results of the tabular computation, columns 1 and 5, 1 and 7,

---

[8] This simple treatment which considers all the iron parts as one is possible only because their various areas and leakage coefficients are such that the flux density is substantially the same in all parts. When this is not the case the method of Problem 11, Art. 49 must be followed.

ART. 135] SAMPLE DESIGN OF A SIMPLE RELAY 517

and 1 and 9, are shown plotted in Fig. 20 as curves D–G, E–H, and F–I, for the ingot iron, high-silicon steel, and ferronickel, respectively.

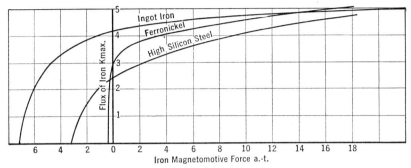

FIG. 21. Demagnetization curves for the iron only of the relay of Fig. 16.

17. Release Ampere-Turns. The release in the closed-gap position occurs with a magnetic force of 2 oz. as set forth in Sec. 2. The useful flux required to give this force in the closed-gap position is

$$\phi_u = \sqrt{72 S \text{ Force}} = \sqrt{72 \times 0.785 \times 0.125} = 2.66 \text{ kmax}.$$

18. Operate and Release Ampere-Turns. These may be read directly from Fig. 20 and are tabulated below.

	Operate		Release		Normal Excitation	
	Point	a-t.	Point	a-t.	Point	a-t.
Ingot iron.............	C	43.4	J	9.1	D	54.2
High-silicon steel........	B	41.8	K	15.7	E	52.3
Ferronickel.............	A	34.2	L	14.1	F	42.8

19. Coil Design. Let us use a paper-section coil designed in accordance with the data of Art. 59. Allowing 0.04 in. for the core tube thickness and 0.005 in. for the cover, the net winding depth of coil will be:

```
    0.938 specified outside coil diameter
   −0.282 pole core
   −0.040 core tube
   −0.003 clearance over pole core
   −0.005 coil cover
   ─────
 2)0.608
    0.304 net winding depth
```

The mean length of a turn will then be

$$\pi(0.282 + 2 \times 0.04 + 2 \times 0.003 + 0.304) = \pi \times 0.672 = 2.11 \text{ in.}$$

Coil for Ingot Iron. The required wire resistance at 20° C., ohms per inch, will be

$$R_i = \frac{E}{P_m NI} = \frac{1.5}{2.11 \times 54.2} = 0.0131 \text{ ohm per inch}$$

Referring to Table II, Chapter VI, the nearest size of wire is No. 32 (diameter = 0.009 in. over enamel), which has 0.01368 ohm per inch. From Table IV of Chapter VI, the minimum paper margin, interlayer paper thickness, and turns per inch are 3/32 in., 0.0013 in., and 103, respectively. Then

$$\text{Coil layers} = \frac{0.304}{0.009 + 0.0013} = 29.5, \text{ say } 30$$

$$\text{Turns per layer} = (0.938 - 2 \times 3/32 - 0.02)103 = 75$$

where 0.01 in. has been allowed at each end of the coil for a washer of insulating material to protect the coil leads.

$$\text{Coil turns} = 30 \times 75 = 2{,}250$$
$$\text{Length of wire} = 2{,}250 \times 2.11 = 4{,}750 \text{ in.}$$
$$\text{Resistance at } 20° \text{ C.} = 4{,}750 \times 0.01368 = 65 \text{ ohms}$$
$$\text{Current at } 20° \text{ C.} = 1.5/65 = 0.0231 \text{ ampere}$$
$$\text{Ampere-turns at } 20° \text{ C.} = 2{,}250 \times 0.0231 = 52$$

This will be satisfactory.

Coil for High-Silicon Steel. The required ampere-turns for this design are 52.3. Therefore the coil design for the ingot iron core will be satisfactory.

Coil for Ferronickel. Designing this coil in the same manner as that for the ingot iron, the following results are obtained: Wire size No. 33; layers = 33; turns per layer = 83; turns = 2,740; length of wire = 5,780 in., resistance at 20° C. = 99.5 ohms; current at 20° C. = 0.0151 ampere; and the ampere-turns = 41.4.

20. Check of Optimum Size of Polar Enlargement. In order to check how close the final design approaches optimum conditions, it is necessary to plot the magnetization curve of the iron parts and fixed air gap in the open-gap position. The data for this may be obtained directly from Sec. 10 by omitting the ampere-turns across the useful gap. These curves are shown plotted in Fig. 22.

Following the method of Art. 85, the line a–b is the air-gap permeance line for the high-silicon steel and ingot-iron relays. Its intersections at points c and d with the magnetization curves of the two relays, respectively, give the operating points in the open-gap position with normal ampere-turns. At operating point d, the magnitude of the slope of the magnetization curve is exactly equal to that of the line a–b, and hence the ingot iron relay is at its optimum operating point with the 1 in.-diameter polar enlargement. For the high-silicon steel relay, however,

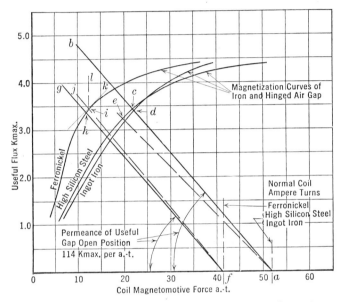

Fig. 22. Graphical construction for determining how close to the optimum size the polar enlargements of the relay of Fig. 16 are.

the optimum operating point is at e instead of c. In order to operate at this point, the permeance of the useful gap must be that of line a–e, equal to 100 max. per ampere-turn. This corresponds to a polar enlargement of 0.936-in. diameter.

For the ferronickel relay the air-gap permeance line is f–g, which gives the actual operating point at h. The optimum operating point, however, is at i, where the tangent to the magnetization curve i–k makes the same angle with the vertical i–l as does the air-gap permeance line f–j. The slope of f–j corresponds to a useful air-gap permeance of 119 max. per ampere-turn, which will be obtained by a polar enlargement of 1.02 in. in diameter.

In all cases the actual operating point is so close to the optimum that no great gain can be made by changing the size of the polar enlargement.

In the table below a comparison is made of the actual polar enlargement and the optimum in regard to the available work.

	Actual			Optimum		
	Diameter of Polar Enlargement, in.	Operating Point	Useful Work, in-lb.	Diameter of Polar Enlargement, in.	Operating Point	Useful Work, in-lb.
Ingot iron.............	1.00	d	0.00446	1.00	d	0.00446
High-silicon steel.......	1.00	c	0.00457	0.936	e	0.00461
Ferronickel............	1.00	h	0.00439	1.020	i	0.00439

21. Comparison of Design Results for the Three Materials. In the tabulation below is a summary of the final design results.

1.5 Volts Applied at 20° C.	Ingot Iron		High-Silicon Steel		Ferronickel	
		Percentage of Normal		Percentage of Normal		Percentage of Normal
Normal current, ma......	23.1	100	23.1	100	15.1	100
Operate current, ma. Force 2 oz. at open gap	19.3	83.5	18.6	80.5	12.5	82.5
Force at open gap with normal current, oz.	3.25	162*	3.35	167	3.23	161
Force at closed gap with operate current, oz.	5.3	132†	4.9	122	5.1	127
Force at closed gap with normal current, oz.	6.25	156†	5.61	140	5.72	143
Release current, ma. Force 2 oz. at closed gap	4.05	17.5	6.98	30.2	5.15	34.1

* The normal or specified force at open gap is 2 oz.
† The normal or specified force at closed gap is 4 oz.

22. Discussion of Results. The table of the last section shows that none of the designs meet the specifications exactly.

The high-silicon steel and the ferronickel designs almost meet the specifications exactly, except that their minimum operate currents are 80.5 and 82.5 per cent of normal, respectively, instead of 80 per cent. The ferronickel design is much more sensitive than the high-silicon steel design, using only 65.5 per cent as much current. Both these designs release at a current higher than required, and hence it would be possible to redesign them with a shorter closed-gap length. This would improve the sensitivity of both.

The ingot-iron design falls down particularly because its release current is too low. This is because the length of the air gap in the closed position is too short. When calculating the length of this gap in Sec. 3, the coercive magnetomotive force of the iron was neglected, resulting in the error. This effect could have been taken into account by estimating this magnetomotive force as a percentage of the normal excitation and applying a correction to the release current percentage in the manner suggested in Art. 128, Sec. 3. Thus, in the ingot-iron design, the release current should have been estimated as 32.5 per cent instead of the specified 25 per cent, when the length of the air gap was computed. Had this been done, the ingot-iron design would have shown even less current sensitivity.

Generally speaking, the use of high-silicon steel or ferronickel instead of ingot iron in a relay design is justified only when the requirements are so exacting that the increased current sensitivity or the reduction of weight that may be obtained makes the difference between success and failure. It should be particularly pointed out that the materials considered in the above designs were assumed to be carefully annealed after fabrication. In most relay work the decrease of residual force that can be made by annealing more than justifies the expense. Materials such as sample 4, S.A.E. 10–20, $\frac{1}{2}$ hard strip stock are very inferior for relay work on account of their high coercive intensities.

Another advantage of the ferronickel and the high-silicon steel that should not be overlooked is their high resistivity. This is of material benefit when the relay is to be extremely fast acting.

136. Design of a Bipolar Sensitive Relay for Operation from a Vacuum Tube

In order to illustrate the method of designing a sensitive relay for operation from a vacuum tube, the following example is presented. The particular relay chosen is not the ordinary contacting relay, but rather an electromechanical device for controlling the operation of a machine. It has been assumed that it is desired to operate this relay device in response to a train of fast impulses received by either radio or telegraph. In order to be effective, the relay must develop a relatively large holding force and be capable of releasing its load practically instantaneously. The method followed in this example would be equally applicable to any relay.

The particular requirements of the design are set forth below:

1. *Normal Load:* The relay is normally actuated and must hold an actual load on its armature of 5 lb.
2. *Factor of Safety:* To provide a factor of safety, it must develop a holding force of not less than 10 lb. with normal excitation.
3. *Release Current:* While holding the load of 5 lb. it must be capable of releasing on not less than 0.1 normal excitation, following the application of normal excitation.
4. *Time of Release:* Its time of release, following normal excitation, must not be greater than 0.0001 second.
5. *Open-Gap Force:* It should be capable of producing a minimum force of 1 lb. through a working stroke of 0.01 in., with normal excitation.
6. *Signal Voltage:* There is no restriction on the magnitude of the signal voltage required to operate the relay, other than that it must not be necessary for the grid of the relay tube to go positive to produce operation.
7. *Other Restrictions:* It is desired to keep the weight of the over-all unit, including the power supply for the relay tube, at a minimum.
8. *Circuit Arrangement:* The type of circuit arrangement desired is shown in Fig. 23 below.

1. Analysis of Problem. *a.* As the over-all weight is to be kept at a minimum, there must be a compromise between relay weight and sensitivity. High sensitivity requires a heavy relay but only a small power input. A small input will allow a lighter weight power supply. On the other hand, a lower relay sensitivity will make the relay lighter and require a larger power input. A reasonable way to handle this situation is to design the relay entirely on a magnetic basis to meet the force requirements. The coil can then be designed to give the required ampere-turns and maximum over-all weight economy in conjunction with a selected tube and the necessary power supply.

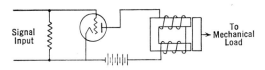

Fig. 23. Diagrammatic circuit arrangement of a bipolar relay operating from the plate circuit of a high-vacuum tube.

b. The release time stipulation of 0.0001 second requires careful lamination of the core and a very short time constant on release. The short time constant can be achieved in two ways: (1) by making the closed length of the air gap as short as possible and thereby avoiding a large stored energy in the relay, and (2) the use of a vacuum tube having a high effective plate resistance. A short air gap will necessitate the use of a material, such as ferronickel, having a very low coercive intensity.

2. Preliminary Design. In accordance with the above analysis, let us design the pole-face structure of a relay magnet meeting the following specifications.

   a. 1-lb. force at 0.01-in. stroke.
   b. 10-lb. force at closed gap.
   c. Release current not less than 10 per cent rated.
   d. Material = laminated ferronickel, No. 29, U.S.S. gauge.

(a) *Open- and Closed-Gap Lengths.* Referring to equation 1 of Art. 128, Sec. 3, we can calculate the ratio of the closed-gap length to the open-gap length as

$$\frac{l_2}{l_1} = \sqrt{\frac{1}{K} \frac{I_2}{I_1}}$$

$$\frac{l_2}{l_1} = \frac{1}{10} \times \sqrt{\frac{1}{10}} = 0.0317$$

then the closed-gap length will be, by equations 2 and 3,[9]

$$l_2 = \frac{s \dfrac{l_2}{l_1}}{\left(1 - \dfrac{l_2}{l_1}\right)} = \frac{0.01 \times 0.0317}{1 - 0.0317} = 0.00033 \text{ in.}$$

This is so small that for practical purposes we shall make the gap 0.001 in. Such a gap can be obtained by plating the carefully finished surfaces with 0.0005 in. of chromium.

The open gap length will then be

$$l_1 = s + l_2 = 0.01 + 0.001 = 0.011 \text{ in.}$$

(b) *Necessary Flux Ratio, Closed to Open Gap.* As the force on the

---

[9] This computation neglects the effect of the coercive magnetic intensity of the iron, but, as explained in Art. 128, Sec. 3, this effect will be very small with ferronickel.

armature equals $\phi^2/72S\nu^2$ the ratio of the forces in the closed- and open-gap positions will be

$$\frac{\text{Force}_2}{\text{Force}_1} = \left(\frac{\phi_2\nu_1}{\phi_1\nu_2}\right)^2$$

and

$$\frac{\phi_2}{\phi_1} = \frac{\nu_2}{\nu_1}\sqrt{\frac{\text{Force}_2}{\text{Force}_1}}$$

Assuming that $\nu_1 = 1.25$ and $\nu_2 = 1.05$, we have

$$\frac{\phi_2}{\phi_1} = \frac{1.05}{1.25}\sqrt{10} = 2.66$$

Because the ratio of the open- to closed-gap lengths is 11 to 1, the only way that the flux ratio can be held to 2.66 is by saturation of the magnetic circuit at closed gap.

(c) *Design Procedure for Determining the Approximate Size of Pole Core, Ampere-Turns, and Exciting Current.* No set procedure can be outlined at this point owing to many possible ways of handling the problem. Because of the saturation requirement noted in (b), it seems logical to start with the pole core.

If there were no polar enlargement,[10] the required size of pole core, assuming a saturation density of 80 kmax. per sq. in. in the air gap, would be

$$S = \frac{72F}{B^2} = \frac{72 \times 5.0}{80^2} = 0.056 \text{ sq. in.}$$

where 5.0 lb. is taken as the force of one pole of the bipolar magnet. This would represent a size of about 0.236 in. square. Therefore let us try a stack of laminations $\frac{1}{4}$ in. by $\frac{1}{4}$ in. in the form of a U. Allowing a stacking factor, K_s, equal to 0.9, the net area will be 0.0562 sq. in.

The required ampere-turns can be determined from the open-gap requirements. The flux density in the working gap necessary to produce a force of 0.5 lb. per pole is

$$B = \sqrt{\frac{72F}{S}} = \sqrt{\frac{72 \times 0.5}{0.25 \times 0.25}} = 24 \text{ kmax. per sq. in.}$$

[10] A polar enlargement is not desirable in this design because laminating is necessary. If used it will necessitate two joints in the magnetic circuit which would probably more than offset the possible gain. Because of the relatively large force and short gap length the optimum can be approached quite closely without a polar enlargement.

The required ampere-turns per coil will then be

$$(F) \text{ one coil} = \frac{Bl_1}{\mu} = \frac{24 \times 0.011}{0.00319} = 83 \text{ ampere-turns}$$

In order to determine the current required to excite the coil, it will be necessary to make a tentative coil design. Let us assume a paper-section,[11] enameled-wire, square-section coil using No. 40 wire. A reasonable proportion for the coil will be about $\frac{3}{4}$ in. square, and $1\frac{1}{2}$ in. long.

If we allow 0.04 in. for the core tube and 0.005 in. for a cover, the net winding depth of the coil will be $(0.375 - 0.045 - 0.125) = 0.205$ in. Referring to Table IV, Chapter VI, the thickness of a layer of wire and paper will be 0.0043 in., the paper margin will be $\frac{1}{16}$ in., and the turns per inch 251. Then the number of layers will be $0.205 \div 0.0043 = 48$, and the turns per layer $(1.5 - 0.125)251 = 345$. The turns per coil will be $345 \times 48 = 16{,}580$. The mean length of turn will be $4(\frac{1}{4} + 2 \times .04 + 0.205) = 2.14$ in. The resistance of each coil at 20° C. will be

$$R = 16{,}580 \times 2.14 \times 0.08742 = 3{,}090 \text{ ohms}$$

For 83 ampere-turns, the current will be $83 \div 16{,}580 = 0.005$ ampere. Thus with the size of coil tentatively chosen, and a minimum wire size of No. 40, the smallest operating current is 5 milliamperes. This merely represents a lower current limit of design.

The approximate value of inductance of the magnet will be

$$L = \frac{N\phi}{I} = \frac{NBSK_s}{I} = \frac{16{,}580 \times 2 \times 100 \times 10^{-5} \times 0.25 \times 0.25 \times 0.9}{0.005}$$

$L = 37.3$ henries, and its time constant will be

$$\frac{L}{R} = \frac{37.3}{3{,}090 \times 2} = 0.00604 \text{ second}$$

(d) *Choice of Tube—Modification of Coil to Suit Tube.* In choosing a tube the following should be kept in mind: In order to attain the release time desired, the time constant must be made approximately 0.0001 sec. This is $\frac{1}{60}$ of the normal time constant of the relay. Therefore a tube

---

[11] A paper-section coil is used in order to have sufficient insulation for the high voltage found in plate circuits. The choice of No. 40 wire will give the highest coil resistance that is practical. When working out of a vacuum tube it is generally desirable to make the coil impedance equal to the tube impedance in order to get optimum power conditions. As this design is to use a pentode tube, which has a very high internal resistance, the highest resistance coil will be best. The use of a wire finer than No. 40 is not recommended unless very special precautions are taken to avoid corrosion and breakage.

having an effective internal resistance of at least 360,000 ohms should be used. As the resistance of the relay will then be small compared to that of the tube, maximum power output will be obtained with the highest possible relay resistance. For any selected tube the lowest possible plate voltage should be used.

The 6J7 pentode, operating with the screen at 100 volts, is about the best suited. It will have a plate current of about 7 milliamperes with a plate voltage of 100 volts, and zero grid bias. Using 7 milliamperes plate current instead of 5, the wound length of the coil can be reduced from $1\frac{3}{8}$ in., as in the tentative design, to $1\frac{3}{8} \times \frac{5}{7} = 0.98$. Letting this be 1 in., the resistance of both coils will be $3{,}090 \times 2 \times 1/1.375 = 4{,}500$ ohms.

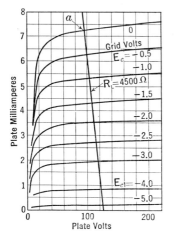

FIG. 24. Family of plate current-plate voltage characteristics for a 6J7 pentode with the screen at 100 volts, showing the operating line for a load of 4500 ohms and a battery supply voltage of 125 volts.

In Fig. 24 are shown the average plate characteristic curves for a 6J7 tube. The load line for 4,500 ohms is drawn from a battery supply voltage of 125 volts. This voltage is chosen because it is the lowest value which will put the operating point at zero grid bias, point (a) well beyond the knee of the curve. In order to reduce the current to 0.1 normal for release, the grid bias must be raised to -4.2 volts. During the grid swing, while the plate current is changing from 7 to 0.7 milliampere, the effective plate resistance will be over 1,500,000 ohms. The time constant will therefore be of the magnitude of about

$$\frac{37.3 \times \left(\frac{1.00}{1.375}\right)^2}{1{,}500{,}000} = 13 \times 10^{-6} \text{ second}$$

This is so short that the release time will probably be limited by eddy currents in the laminations.[12]

The number of turns on each coil will be $16{,}580 \times \dfrac{1}{1.375} = 12{,}020$

(e) *Final Design.* In Fig. 25 is shown the final circuit arrangement, and in Fig. 26, the final relay design.

---

[12] A 14-mil iron has been selected because it is the thinnest commercial sheet stock available. A 6-mil iron is to be preferred.

3. Check of Preliminary Design.

(a) *Check of Open-Gap Force.* By reference to Fig. 24, it will be seen that the normal relay current at operating point (a) is 7.2 milliamperes. The total ampere-turns of both coils will then be

$$12{,}020 \times 2 \times 0.0072 = 173 \text{ ampere-turns}$$

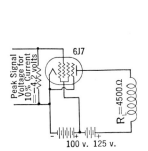

FIG. 25. Final circuit arrangement for the relay of the problem of Art. 136 operating from a 6J7 pentode.

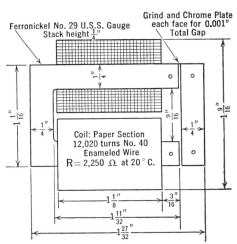

FIG. 26. Final design of the relay for the problem of Art. 136.

The air-gap flux density in the open-gap (0.011 in.) position, if all the magnetomotive force is across the gaps, will be

$$B = \mu H = 0.00319 \times \frac{173}{2 \times 0.011} = 25.1 \text{ kmax. per sq. in.}$$

At flux densities in this region the magnetic intensity in the ferronickel will be so low that the reluctance drop in the iron and the effect of leakage flux on the force may be neglected. The total pull will therefore be

$$F = \frac{B^2 S}{72} = \frac{25.1^2 \times 0.0625 \times 2}{72} = 1.09 \text{ lb.}$$

(b) *Check of Closed-Gap Force.* In order to find the closed-gap force, it will be necessary to plot the magnetization curve for the iron parts and also to calculate the leakage permeances.

The permeance of the main gap will be

$$P_1 = \frac{\mu S}{l} = \frac{3.19 \times 0.0625}{0.002} = 99.6 \text{ max. per ampere-turn}$$

The fringing permeances from the pole cores to the armature will be

$$P_7 = 0.26\mu l = \frac{0.26 \times 3.19 \times 0.75}{2} = 0.31$$

$$P_{8b} = \frac{\mu l}{\pi} \log_e \left(1 + \frac{2t}{g}\right) = \frac{3.19 \times 0.75}{2\pi} \log_e \left(1 + \frac{0.5}{0.001}\right) = 2.37$$

$$P_{11} = 0.52\mu l = \frac{0.52 \times 3.19 \times 0.25}{2} = 0.21$$

$$P_{12b} = \frac{2\mu l}{\pi} \log_e \left(1 + \frac{t}{g}\right) = \frac{2 \times 3.19 \times 0.25}{2\pi} \log_e \left(1 + \frac{0.18}{0.001}\right) = 1.32$$

$$\text{Total} = 4.21$$

The distributed permeances are:

$$P_1 = \frac{\mu S}{l} = \frac{3.19 \times 1.16 \times 0.25}{0.5625} = 1.64$$

$$P_7 = 0.26\mu l = 0.26 \times 3.19 \times 1.16 \times 2 = 1.96$$

$$P_{8b} = \frac{\mu l}{\pi} \log_e \left(1 + \frac{2t}{g}\right) = \frac{3.19 \times 1.16 \times 2}{\pi} \log_e \left(1 + \frac{0.5}{0.56}\right) = 1.48$$

$$\text{Total} = 5.08$$

As the distributed permeance is only one-half effective as regards the leakage flux produced, the effective distributed permeance is 2.54.

The leakage coefficients will then be:

$$\nu_a = \frac{99.6 + 4.21}{99.6} = 1.042$$

$$\nu_y = \frac{99.6 + 4.21 + 2.54}{99.6} = 1.068$$

$$\nu_{pc} = 1.042 + \tfrac{2}{3}(1.068 - 1.042) = 1.059$$

The length, areas, and relative flux densities of the various parts of the iron circuit are tabulated below. A stacking factor, $K_s = 0.9$, has been used in computing the area of the laminated iron parts.

Part	Length, in.	Area, sq. in.	Relative Flux	Relative Flux Densities
Armature	1.06	0.0562	$\nu_a \phi_u$	$0.985\ B_{pc}$
Pole cores	2.68	0.0562	$\nu_{pc} \phi_u$	B_{pc}
Yoke	1.06	0.0562	$\nu_y \phi_u$	$1.008\ B_{pc}$
Air gap			ϕ_u	

Because the flux densities in the various iron parts are so nearly alike, they have been assumed equal in plotting the magnetization curve. Magnetic data for ferronickel were taken from curve 11 of Fig. 11b. Chapter II. The magnetization curve is shown in Fig. 27. The operating point is shown at (a), where the flux of the pole cores is 5.2 kmax. The force will be

$$F = \frac{2}{72S}\left(\frac{\phi_{pc}}{\nu_{pc}}\right)^2 = \frac{2}{72 \times 0.0625}\left(\frac{5.2}{1.059}\right)^2 = 10.8 \text{ lb.}$$

(c) *Check of Release Current.* The actual load which the armature must hold in the closed-gap position is 5 lb. This load is to be released at 0.1 normal current which corresponds to 17.3 ampere-turns.

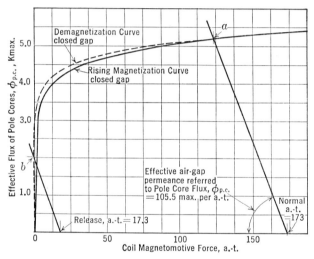

Fig. 27. Graphical construction for determining the force of the relay of Fig. 26 at its operating and release ampere-turns.

In order to calculate the flux in the closed-gap position, for small currents, the demagnetization curve of the iron parts must be drawn. This is shown as a dashed line, and has been computed from data of Figs. 13a and 13b, and Fig. 12g of Chapter II. The coercive intensity of the ferronickel is so small that the descending branch of the loop as drawn almost coincides with the axis of ordinates.

To locate the operating point with 17.3 ampere-turns applied, the air-gap permeance line having a slope of -105.5 max. per ampere-turn is drawn as shown. The intersection of this line with the demagnetization curve at (b) is the desired operating point. The pole core flux at this point is 2.01 kmax. The force, with 17.3 ampere-turns applied, will then be:

$$F = \frac{2}{72S}\left(\frac{\phi_{pc}}{\nu_{pc}}\right)^2 = \frac{2}{72 \times 0.0625}\left(\frac{2.1}{1.059}\right)^2 = 1.75 \text{ lb.}$$

As this is well below the load force of 5.0 lb., there will be no difficulty in releasing with 0.1 normal current.

Final Results

Tube = 6J7.
Plate supply voltage = 125 volts.
Screen voltage = 100 volts.
Relay coils = 12,020 turns No. 40 enameled wire, per coil.
Relay resistance = 4,500 ohms at 20° C.
Operation data = 7.2 milliamperes at zero grid bias; 10.8 lb. closed and 1.09 lb. open gap.
Release data = 0.72 milliampere at −4.1 volts grid bias; 1.75 lb. at closed gap.

APPENDIX

UNITS — DEFINITIONS — FUNDAMENTAL PHYSICAL CONCEPTS

The Joule-Inch System. The system of units employed for physical quantities in this book is based upon the *joule* [1] as a unit of energy, the *inch* as the unit of length, and the *second* as the unit of time. It is primarily an electrical system and need not have any particular length unit associated with it. The inch as the length unit is chosen merely for convenience. The above three dimensions are sufficient for dealing with concepts of mechanics; however, when dealing with other physical sciences it is necessary to introduce other dimensions such as *temperature* in heat, and *current* or *electric quantity* in electricity. The most important units of the system are defined below:

The *joule* is the unit of *energy* and is equal to 10^7 ergs.

The *coulomb* is the unit of *electric quantity* and is equal to the quantity of electricity which will deposit 0.001118 gram of silver from a silver solution.

The *second* is the unit of *time* and is equal to 1/86,400 part of the mean solar day.

The *inch* (in.) is the unit of *length* and is equal to 1/39.37 of the length of the international meter.

The *degree Centigrade* (deg. cent. or ° C.) is the unit of temperature and is one-hundredth part of the temperature difference between the freezing and boiling points of water at normal atmospheric pressure.

The remaining mechanical, electrical, and thermal units are defined by and derived from the above five fundamental units in the usual manner.

Table I gives a comparison of the units of the commonly used systems and the factors necessary to convert from these systems to the joule-inch-second system. In particular, it should be noticed that the only difference between the joule-inch-second system of this book and the meter-kilogram-second system which has now become standard [2] is the unit of length. The inch is to be preferred for a design book for use in this country as the inch is used almost exclusively for dimensioning working drawings of machine parts.

[1] Energy, or the change in energy, though not the easiest physical concept, is the basis or cause of most of the familiar physical phenomena and hence should be considered fundamental. An electric current, mechanical motion, or the flow of a fluid are a means of transmitting energy from one place to another, and phenomena which occur in an electric, mechanical, or hydraulic system are thus due to changes of energy in such a system.

The storage of energy gives rise to certain very common physical concepts; thus, a magnetic field or magnetic flux is considered a manifestation of energy stored in a particular way by particular agents, while an electrostatic field, a gravitational field, the velocity of a body, or the temperature of a body are manifestations of energy stored in other ways and by other agents.

A change of energy, stored by any one of the above agents, gives rise to transient phenomena, which as a group constitute by far the largest number of ordinary occurrences and certainly the most difficult from the standpoint of physical analysis.

[2] The International Committee of Weights and Measures at a meeting in October, 1935, decided that the actual substitution of the meter-kilogram-second absolute system of electrical units for the international system was to take place on January 1, 1940.

TABLE I

UNITS OF THE COMMON SYSTEMS AND THEIR CONVERSION FACTORS TO THE JOULE-INCH-SECOND SYSTEM

Name of Quantity	Symbol	Defining Equation	English System	Meter-Kilogram-Second System	Irrational C.G.S. Electromagnetic System	Joule-Inch-Second System Unit
Fundamental Units						
Energy	W	....	1 in-lb. $\rightleftharpoons$ 0.1128	1 joule $\rightleftharpoons$ 1	1 erg $\rightleftharpoons 10^{-7}$	joule
Length	x or l or s	....	1 in. $\rightleftharpoons$ 1	1 meter $\rightleftharpoons$ 39.37	1 cm. $\rightleftharpoons$ 0.3937	inch
Time	t	....	1 sec. $\rightleftharpoons$ 1	1 sec. $\rightleftharpoons$ 1	1 sec. $\rightleftharpoons$ 1	second
Electric quantity	Q	....	1 coulomb $\rightleftharpoons$ 1	1 coulomb $\rightleftharpoons$ 1	1 abcoulomb $\rightleftharpoons$ 10	coulomb
Temperature	θ	....	1 deg. Fahr. $\rightleftharpoons$ 5/9	1 deg. cent. $\rightleftharpoons$ 1	1 deg. cent. $\rightleftharpoons$ 1	degree centigrade
Derived Energy Units						
Energy density	w	$w = dW/dV$	1 in-lb./cu. in. $\rightleftharpoons$ 0.1128	1 joule/cu. m. $\rightleftharpoons 2.54^3 \times 10^{-6}$	1 erg/cu. cm. $\rightleftharpoons 0.39372 \times 10^{-7}$	joule/cubic inch
Power	P	$P = dW/dt$	1 horse power $\rightleftharpoons$ 746	1 watt $\rightleftharpoons$ 1	1 erg/sec. $\rightleftharpoons 10^{-7}$	watt
Power density	p	$p = dP/dV$	....	1 watt/cu. m. $\rightleftharpoons 2.54^3 \times 10^{-6}$	1 (erg/sec.)/cu. cm. $\rightleftharpoons 2.54^3 \times 10^{-7}$	watt/cubic inch
Derived Mechanical Units						
Area	S	$S = l^2$	1 sq. in. $\rightleftharpoons$ 1	1 sq. m. $\rightleftharpoons 39.37^2$	1 sq. cm. $\rightleftharpoons 0.3937^2$	square inch
Volume	V	$V = l^3$	1 cu. in. $\rightleftharpoons$ 1	1 cu. m. $\rightleftharpoons 39.37^3$	1 cu. cm. $\rightleftharpoons 0.3937^3$	cubic inch
Velocity	v	$v = dx/dt$	1 in./sec. $\rightleftharpoons$ 1	1 m./sec. $\rightleftharpoons$ 39.37	1 cm./sec. $\rightleftharpoons$ 0.3937	inch/second
Acceleration	a†	$a = d^2x/dt^2$	1 in./sec.^{2} $\rightleftharpoons$ 1	1 m./sec.^{2} $\rightleftharpoons$ 39.37	1 cm./sec.^{2} $\rightleftharpoons$ 0.3937	inch/second^2
Force	$\mathcal{F}$	$\mathcal{F} = dW/ds$	1 lb. $\rightleftharpoons$ 0.1128	1 newton * $\rightleftharpoons$ 1/39.37	1 dyne $\rightleftharpoons 2.54 \times 10^{-7}$	joule/inch
Mass	M	$M = \mathcal{F}/a$	1 lb./(in./sec.^{2}) $\rightleftharpoons$ 0.1128	1 kg. $\rightleftharpoons$ 6.44 $\times 10^{-4}$	1 gram $\rightleftharpoons 6.44 \times 10^{-7}$	(joule/inch)/(inch/second^2)
Angle	θ	....	1 radian $\rightleftharpoons$ 1	1 radian $\rightleftharpoons$ 1	1 radian $\rightleftharpoons$ 1	radian
Torque	D	$D = dW/d\theta$	1 in-lb./rad. $\rightleftharpoons$ 0.1128	1 joule/rad. $\rightleftharpoons$ 1	1 erg/rad. $\rightleftharpoons 10^{-7}$	joule/radian
Angular velocity	ω	$\omega = d\theta/dt$	1 rad./sec. $\rightleftharpoons$ 1	1 rad./sec. $\rightleftharpoons$ 1	1 rad./sec. $\rightleftharpoons$ 1	radian/second
Derived Electric Units						
Current	I	$I = dQ/dt$	....	1 ampere $\rightleftharpoons$ 1	1 abamp. $\rightleftharpoons$ 10	ampere
Voltage	E	$E = dW/dQ$	....	1 volt $\rightleftharpoons$ 1	1 abvolt $\rightleftharpoons 10^{-8}$	volt
Resistance	R	$R = E/I$	....	1 ohm $\rightleftharpoons$ 1	1 abohm $\rightleftharpoons 10^{-9}$	ohm
Current density	i	$i = dI/ds$	....	1 amp./sq. m. $\rightleftharpoons 1/39.37^2$	1 abamp./sq. cm. $\rightleftharpoons 2.54^2 \times 10$	ampere/square inch
Electric intensity	e	$e = dE/dl$	....	1 volt/m. $\rightleftharpoons$ 1/39.37	1 abvolt/cm. $\rightleftharpoons 2.54 \times 10^{-8}$	volt/inch
Resistivity	ρ	$\rho = e/i$	....	1 ohm/m. $\rightleftharpoons$ 39.37	1 abohm/cm. $\rightleftharpoons 0.3937 \times 10^{-9}$	ohm/inch
Capacitance	C	$C = dQ/dE$	....	1 farad $\rightleftharpoons$ 1	1 abfarad $\rightleftharpoons 10^9$	farad
Derived Magnetic Units						
Magnetic flux	ϕ	$\Delta\phi = \int E\,dt/N$	....	1 weber $\rightleftharpoons$ 1	1 maxwell $\rightleftharpoons 10^{-8}$	weber
Magnetomotive force	$\mathcal{F}$	$\mathcal{F} = dW/d\phi$	....	1 amp.-turn $\rightleftharpoons$ 1	1 gilbert $\rightleftharpoons 10/4\pi$	ampere-turn
Permeance	P	$P = \phi/\mathcal{F}$	....	1 weber/amp.-turn $\rightleftharpoons$ 1	1 maxwell/gilbert $\rightleftharpoons 4\pi \times 10^{-9}$	weber/ampere-turn
Magnetic flux density	B	$B = d\phi/dS$	....	1 weber/sq. m. $\rightleftharpoons 1/39.37^2$	1 gauss $\rightleftharpoons 2.54^2 \times 10^{-8}$	weber/square inch
Magnetic intensity	H	$H = dF/dl$	....	1 amp.-turn/m. $\rightleftharpoons$ 1/39.37	1 oersted $\rightleftharpoons$ 2.02	ampere-turn/inch
Permeability	μ ‡	$\mu = B/H$ §	....	1 weber/amp.-turn m. $\rightleftharpoons$ 1/39.37	1 gauss/oersted $\rightleftharpoons 3.19 \times 10^{-8}$	weber/ampere-turn inch
Inductance	L	$L = d(N\phi)/dI$	....	1 henry $\rightleftharpoons$ 1	1 abhenry $\rightleftharpoons 10^{-9}$	henry

NOTES: * 1 newton = 1 joule per meter.
† Acceleration of gravity is 386 in./sec.^{2} at sea level and 45° latitude.
‡ The permeability of a vacuum is $\mu = 3.192 \times 10^{-8}$ weber per amp.-turn in.
§ This is the normal permeability. Several other permeabilities are defined. See Arts. 21 and 22

INDEX

A

Acceleration
  of gravity, 532
  unit, 532
Air gap
  effect on residual force, 9-10, 106-109
  equivalent length of faced joint, 86
  energy stored, 33, 78, 251
  graphical method of determination of a-t. for actual magnet, 222-225
  optimum length for choke, 102-105
  use in polarized apparatus, 103
  volt-ampere excitation of, 452
Allegheny Electric Metal, 42; *also see* Ferronickel
Alloys with low Curie point, 43
Alnico, 7, 13
  demagnetization curve, 67
  external energy curve, 67
  general, 46-47
  general data, 70
  hysteresis loop, 54
  magnetization curve, 68
  normal permeability curve, 68
Alternating-current magnets, 419-482
  average force, 423-424
  change in reactive power, 429-431
  characteristics of
    constant-current magnets, 423
    constant-voltage magnets, 422-423
    comparison with d-c. magnets, 431-432
  constant-current operation, 430-431, 437
  constant-voltage operation, 429-430
  direct-attraction type, 446-463
    characteristics of, 446-449
      comparison of a-c. and d-c. force-stroke curves, 448-449
      coned laminated face, 446
      constant-current characteristic, 449
      construction, 446-447

Alternating-current magnets, direct attraction type—*Continued*
      effect of leakage coefficient on force, 448
      force-stroke characteristic, 448
    direct-attraction-type design, 449-463
      calculation of
        exciting current closed gap, 459-460
        leakage coefficient, 457-458
      check of
        mechanical work, 461
        volt-ampere limitation, 461-462
      choice of pole-face type, 450
      coil design, 455
      core loss, 456, 459
      estimation of leakage coefficient, 451
      exciting current
        closed gap, 453-454
        open gap, 455, 461
      fundamental design equations, 451-452
      general procedure, 449-450
      power factor, 462
      preliminary design, 452-456
      temperature rise, 462
        of coil, 455-456
      volt-ampere efficiency, 461
      volt-ampere excitation for air gap, 452
      wave form of exciting current, 459-461
  effect of resistance drop, 422
  exciting current, 443-444, 453-455, 459-460
  force, nature of, 423-428
  force limitation, 431
  fundamental circuit theory, 419-423
  general, 5-6, 419
  instantaneous force, 423-428
  leakage flux type, *see* solenoid and plunger type

533

INDEX

Alternating-current magnets — *Continued*
  phase relation between flux and current, 420
  polyphase magnet arrangement, 425-428
  pulsation of force, 423-428
    single-phase, 423-424
    three-phase, 427-428
    two-phase, 425-427
    with shading coils, 427, 469, 479
  reactance voltage, 420-422
  reactive power, total, 430
  shading coils, *see* Shading coils; Shaded-pole magnets
  single-phase operation, 423-425
  solenoid and plunger type, 433-445
    current at minimum force position, 440
    current-stroke characteristics, 435-437
    design, 438-445
    design procedure, 444-445
    force at the beginning of the stroke, 442
    force-stroke characteristic, 435-437
    general design basis, 438-439
    laminated construction, 433, 435
    minimum force throughout stroke, 439-440
    minimum iron flux density at beginning of stroke, 441-442
    slotted construction, 433-434
    volt-amperes
      at beginning of stroke, 442-443
      at end of stroke, 443-444
  speed of action, 432
  steel for, 41
  stipulations regarding mathematical derivations, 420
  three-phase operation, 427-428
  two-phase operation, 425-427
  use of rectifiers, 431
  volt-ampere efficiency, 430, 436, 461
  volt-ampere limitation,
    check, 437, 461-462
    constant-current operation, 430-431
    constant-voltage operation, 429-430
    experimental, 437
    general, 423, 431

Alternating-current magnets — *Continued*
  wave form of exciting current, 443-444, 459-460
  weight limitation, 431
Alternating-current relays, 6, 501-502
Alternating-current *vs.* direct-current magnets
  effective use of iron, 424-428
    single-phase, 424
    three-phase, 428
    two-phase, 425-426
Aluminum
  density, 188
  resistivity, 155
  temperature coefficient of resistance, 156
  thermal capacity, 188
  wire resistance, 158
American ingot iron
  general, 39
  general data, 69
  magnetization curves, 48
American wire gauge, 156, 157
Anhysteretic magnetization curves, 22, 62, 63
  use of, in polarized apparatus, 104
Angle, unit, 532
Angular velocity, unit, 532
Annealing
  importance in relays, 521
  to remove machining strains, 38, 40, 42
Arc suppression, relays, 488
Area, unit, 532
Asbestos insulation, 160-161, 162
Atkinson, F. W., 161
Attraction between magnetized faces, 196-200
Audio-frequency transformers, 102
  hysteresis cycle, 35
Average force, 423-424
  single-phase, 423-424
  three-phase, 427-428
  two-phase, 425

B

Bakelite, 162
Bensin, Igor, 21
Bipolar magnet
  characteristics, 230-231

INDEX

Bipolar magnet—*Continued*
  check of preliminary design, 326-333
  coil design, 326-328
  design, 321-333
  design equation, 322-324
  deviation from inverse square law, 230-231
  effect of magnetic leakage, 230-231
  effect of saturation, 230-231
  empirical design data, 323
  flux density in pole cores, 325
  force
    from energy stored in working air gap, 332
    from saturation curve, 332
  force formula, derivation, 201-202
  force-stroke characteristic, 230
  general, 321-322
  leakage coefficient, 142-144
    experimental check, 147-149
    modification of formula when polar enlargements are close, 328-329
  leakage field, 122-125
  magnetic circuit calculation, 328-331
  magnetization curve of iron parts, 330-331
  magnetomotive force used in iron, 323
  method of determining optimum, 330-331
  polar enlargements, design, 324
  preliminary design, 324-326
  ratio of coil length to build, 325
  temperature rise, 327-328
  use of polar enlargements, 230-231, 242
  useful work-stroke characteristic, 230
  variation in flux density between pole cores, 124, 126
  weight, 332
  weight economy, 231, 332
    vs. index number, 242
Bobbin-wound coils, *see* Coils
Boiling point of contacting materials, 485
Boyle, Joseph C., 363
Brass
  density, 188
  thermal capacity, 188

Brass tubing, sizes commercially available, 249
Brinell hardness, contacting materials, 485
Bush, V., 339

C

Capacitance, unit, 532
Carbon steel, 45
  coercive intensity data, 55
  demagnetization curve, 67
  external energy curve, 67
  general data, 70
  residual flux density data, 55
Cast iron
  general, 40, 45
  general data, 69
  magnetization curves, 48
Casting permanent magnet steels, 47
C.G.S. system, conversion factors to joule-inch-second system, 532
Choke coil
  saturated
    current-time characteristic
      general derivation, 369-372
      illustrative example, 382-384
    flux-time characteristic
      general derivation, 366-369
      illustrative example, 382-384
    time for flux to rise to given value
      general derivation, 372-373
      illustrative problem, 384-385
  steel for, 41
Chokes, optimum air gap length, 102-105
Chromium steel
  demagnetization curve, 67
  external energy curve, 67
  general, 45
  general data, 70
  magnetization curve, 68
  normal permeability curve, 68
Chromoxide insulation, 160-161
Circuits
  for time delay, *see* Time-delay circuits
  magnetic, *see* Magnetic circuit calculations
Cobalt
  alloys, 42-43
  general data, 69, 70

Cobalt—*Continued*
  critical temperature, 43
  general data, 69
Cobaltchrome steel
  demagnetization curve, 67
  external energy curve, 67
  general, 45
  general data, 70
Cobalt-iron alloys, *see* Ferrocobalt
Cobalt steel
  coercive intensity data, 26, 55
  demagnetization curve, 67
  external energy curve, 67
  general, 45-46
  general data, 70
  magnetization curve, 68
  normal permeability curve, 68
  residual flux density-data, 26, 55
Coercive intensity data, *see also hysteresis loops for particular material*
  definition, 15
  hard magnetic materials, 55, 70
  method of extrapolating for saturation value of, 25-26
  soft magnetic materials, 55, 69
  variation with maximum magnetizing intensity, data, 26, 55
Coil calculations
  ampere-turns
    method, 171, 173
    sample problems, 175
  resistance
    method, 171
    sample problem, 175-177
  space factor, *see* Space factor of coils
    method, 167
    sample problem, 175
  turns
    method, 171
    sample problems, 174
  voltage required, 173
  wire diameter
    method, 173, 174
    sample problem, 175-177
Coil design
  a-c. magnet, 455
  high-speed magnet, 407
  illustrative problems,
    bipolar magnet, 326-328

Coil design—*Continued*
  illustrative problems—*Continued*
    flat-faced armature magnet, 257-259
    flat-faced plunger magnet, 271-273
    full conical plunger magnet, 284-286
    leakage flux magnet, 314-315
    tapered plunger magnet, 297-299
  splitting the wire size, 297
  two wire sizes, 286
Coil insulating materials, 162
Coil redesign to give required temperature rise, 298, 327-328
Coil shape for tractive magnets, 248, 255, 269, 282, 311, 323
Coils, 151-177
  bobbin wound
    bobbin of brass tube with iron end flanges, 271-273, 297-299, 314-315
    coil insulation, 168, 169
    fabricated brass bobbin, illustrative design, 257-259
    general, 153, 155
    molded spool type, 154
    resistance density, 170-172
    space factor, 170, 172
    turn density, 170, 172
    wire insulation, 168
  changing dimensions to secure required temperature rise, 298
  classifications, 151
  cotton-interwoven, 153
  effect of change of one wire size, 249
  form wound
    coil insulation, 168-170
    electrolysis, 496
    general, 155
    resistance density, 170, 172
    space factor, 170, 172
    turn density, 170, 172
    wire insulation, 168
  general, 3, 151
  heat-dissipation capacity
    definition, 178
    dependence on final temperature, 182
  heat-dissipation coefficient
    data, 185
    dependence on final temperature, 184-186
  moisture resisting, 153

INDEX

Coils—*Continued*
  molded spool, 154
  paper section
    advantages, 152
    allowance in coil wall thickness, 164, 166
    coil covers, 164
    cover, 152
    crushing the paper margin, 164
    data for economical manufacture, 163-169
    excessive paper overlap, 153
    general, 151-153
    illustrative problems
      bipolar magnet, 326, 328
      relay, 517-518
    impregnating materials, 164
    impregnation, 152
    interlayer material, 163
    leads, 152
    paper margin, 164, 165
    resistance density, 168, 169
    space factor, 153, 167, 169
    specifications, 166
    turn density, 168, 169
    turns per inch, 163, 165
  random wound, 154
  resistance density, definition, 162
  ribbon wound, 155
  space factor, definition, 162
  strap-wound, 155
  temperature rise
    experimental method of measurement, 181, 182
    exponential heating law, 181
    final *vs.* power input, 185
    formula for final values, 186-187
    ideal cooling curve, 180, 181
    ideal heating curve, 178, 180
    test data, 183
    time curves, 190-192
    use of semi-log paper, 181-183
    verification of ideal law, 181-184
  thermal capacity, 187-190
    definition, 178
    dependence on surrounding iron, 182, 184
  thermal time constant, 180
  turn density, definition, 162
  use of two wire sizes, 176
    illustrative example, 286

Cold-rolled steel
  coercive intensity data, 55
  energy returned to electric circuit, data, 57
  general, 39-40
  general data, 69
  hysteresis energy loss, 57
  hysteresis loops data, 51
  magnetization curves, 48
  residual flux density data, 55
Comparison of a-c. and d-c. magnets, 431-432
Compensating windings, effect of, 217
Condenser and resistance for time delay, 337-338, 353-359
Condensers
  arc suppression in relays, 488
  for quick release, 391
Cones
  coaxial full, axial force, 206-209
  coaxial truncated, axial force, 205-206
Conical plunger magnet, *see* Full conical plunger magnet
Constant-current, a-c. magnets, *see* Alternating-current magnets
Constant-permeability alloys, 7, 43
Constant-voltage, a-c. magnets, *see* Alternating-current magnets
Contacts for relays, *see* Relays
Conversion from conical to equivalent flat-faced plunger, 279
  illustrative problem, 280
Cooling curve of ideal coil, 180-181
Copper
  density, 188
  inferred absolute zero, 156, 159
  resistivity, 155
  temperature coefficient of resistance, 156
  temperature resistance table, 160
  thermal capacity, 188
  wire resistance, 158
Copper loss in d-c. electromagnet, 81, 82
Copper wire, splaying, 159
Core loss
  in a-c. field, 30-32
  of a-c. magnet, 456, 459
  of silicon steels, 59, 60

Cotton insulation, 159-160
  breakdown temperature, 162
  density, 188
  thermal capacity, 188
Cotton-interwoven coils, see Coils
Coupled circuits, 345-353, 356-359, 361-362
Critical temperature, definition, 43
Curie point, definition, 43
Current, unit, 532
Current-carrying conductor, force due to independent field, 216-218
Current density, unit, 532
Current-stroke characteristics, a-c. solenoid and plunger, 435-437
Current-time characteristic
  iron core with saturation
    general derivation, 369-372
    illustrative example, 382-384
  iron core without saturation, see Time-delay circuits
  magnet with motion, see High-speed magnets
Curvilinear squares, 127-128
"Cut and try" method, magnetic circuit calculation, 96
"Cut and try" solutions
  flux distribution in long plunger, 316-320
  tapered plunger magnet, 293-295
Cycle of d-c. electromagnet, 73-77
Cyclic state, definition, 16
Cylinder and cone, coaxial, axial force, 209-210
Cylinders, coaxial axial force, 203
Cylindrical-faced plunger magnet
  characteristics, 233-234
  effect of
    fringing flux, 203
    magnetic stop, 234
    radial leakage flux from plunger, 204
    saturation, 234
  force formula, derivation, 203-204
  force-stroke characteristics, 233-234
  plunger leakage flux pull, 234
  replacing by tapered plunger, 242
  useful work-stroke characteristic, 234
  weight economy, 235

Cylindrical-rotary armature magnet, torque formula, derivation, 204-205
Cylindrical-shaped gap and plug, axial force, 212

D

Demagnetization curves, see also Hysteresis loops
  discussion of properties, 44
  graphical calculations, 107-111
  hard magnetic materials, 67
  of relay, 517
  soft magnetic alloys, 9
  soft steel magnet, 107, 108
Demagnetizing iron, process, 14
Density
  coil materials, 188
  contacting materials, 485
  hard magnetic materials, 70
  metals, 188
  soft magnetic materials, 69
Depth of penetration, 364-365
  effect of
    iron saturation, 415-416
    wave form, 415
Design data, empirical
  bipolar magnet, 323
  flat-faced armature magnet, 255
  flat-faced plunger magnet, 269
  full conical plunger magnet, 282
Design factors
  d-c. magnets, 82
  relays, 83
Design of
  a-c. magnets, see Alternating-current magnets
  d-c. magnets, see *particular type*
  high-speed magnets, see High-speed magnets
  relays, see Relays
  time-delay magnets, see Time-delay magnets
  tractive magnets, see Tractive magnets
Dielectric strength
  chromoxide insulation, 161
  enamel insulation, 159
  glass insulation, 161

INDEX

Dielectric strength—*Continued*
  glassine paper, 163
  Kraft paper, 163
Differential permeability, definition, 23
Direct conversion of electric energy into mechanical work, 217
Direct-current magnets
  available mechanical work, 77
  build-up current, not saturated, 339-342
  comparison with a-c. magnet, 431-432
  copper loss, 81
  current-time characteristic
    general derivation, 369-372
    illustrative example, 382-384
  design, **4**, **245**; *see also particular magnet type*
  design factors, 82
  effective inductance, saturation present, 370
  effect of
    coercive magnetomotive force, 75
    eddy currents, 80, 364
    grain direction, 38
    index number on dynamic characteristics, 413
  energy
    changes during cycle of operation, 73-77
    returned to electric circuit, 75
  equivalent circuit for temperature-rise time calculations, 192-194
  factors entering into efficient design, 2
  flux current loop, 73-75
  flux-time characteristic, saturated
    general derivation, 366-369
    illustrative example, 382-384
  general, **4**, **245**
  heating, 81
  high-speed, *see* High-speed magnets
  hysteresis loss during cycle, 33, 76
  importance of iron used, 78
  loss due to
    fixed air gaps, 78
    initial plunger position, 78, 79
    stored energy of iron, 77
  loss in work ability due to initial armature position, 150
  magnetic efficacy, 77, 79
  mechanical efficacy, 80

Direct-current magnets—*Continued*
  mechanical work to overcome residual effect, 75
  potential work ability, 77
  quick-release characteristics
    general, 391-393
    general derivation, 393-395
    illustrative problem, 395-399
  residual flux, definition, 74
  residual flux linkage, definition, 74
  residual force, definition, 74
  shaping of pole faces, 79
  space-time characteristic
    evaluation of initial force, 378-379
    illustrative problems, 385-386, 410-411
    experimental check, 390
    force at end of stroke, estimation of, 390, 405
    general derivation, 373-380
    illustrative problem, 385-391
  speed limitations, 413-415
  temperature rise as design limitation, 82
  temperature-rise time calculation, 190-192
  temperature-rise time curves, experimental results, 191, 194
  theory of operation, 73-83
  thermal capacity, 187-190
  time delay, *see also* Time-delay magnets
    general, 81
    normally present, 339
  time for flux to rise to a given value
    not saturated
      illustrative problem, 409-410
    saturated
      general derivation, 372-373
      illustrative problem, 384-385
  time for incipient motion
    not saturated, 409-410
    saturated, 382-385
  weight economy *vs.* index number, 242
Displacement factor, definition, 35
Displacing electromagnets by hard magnetic materials, 13
Distributed leakage flux
  around a bipolar magnet, 123-126

Distributed leakage flux—*Continued*
  effective magnetomotive force, 222
  general deviation, 97-99
  illustrative problems
    for bipolar magnet, 144
    for flat-faced armature magnets, 101, 146
    for flat-faced plunger magnets, 141
    for plunger magnets, 222-224, 303
  method of accurate solution for leakage flux magnet, 316-320
  method of approximate solution for leakage flux magnet, 309-310
Dust in relay contacts, 487-488

E

Eccentric cylindrical surfaces, radial side pull, 212-213
Eddy-current loss, 30-32
  effect of grain size, 31
Eddy currents
  effect on
    d-c. magnets, 80
    exciting current of a-c. magnets or transformers, 444
    high-speed magnets, 364-366
    relay operation, 494, 498
  limitation in a-c. magnets, 432
  method of minimizing
    a-c. magnets, 433-435
    high-speed magnets, 364-366
    quick-release magnets, 391-392
  mitigation of, 41
Edgar, R. F., 36-37
Electric intensity, unit, 532
Electric quantity, unit, 532
Electrical bar steel, 41
Electrical sheet steel, 41
Electrodynamic problems, *see* Space-time characteristics or High-speed magnets
Electrolysis, effect on coils, 496
Electromagnetic hammers, 6
Electromagnets, *see* Alternating-current magnets or Direct-current magnets
Electonic means of time delay, 337-338
Elmen, G. W., 43
Empire cloth, 162

Enamel insulation, 159
Energy
  basis for system of units, 531
  changes during a normal magnetic cycle, 27-29
  changes during unsymmetrical hysteresis cycles, 32-35
  conversions in an electromagnet, 73-77
  density, unit, 532
  in air gap, 33, 78, 251
    optimum condition, 252
  losses during normal hysteresis cycles, 28
  losses during unsymmetrical hysteresis cycles, 35
  required to demagnetize iron, 28, 33-34, 56-57
  returned to electric circuit from iron, 33, 56, 57
  stored by iron, 33
  stored in air gap
    by hard magnetic material, 70
    of electromagnet, 251
  unit, 532
Equation for normal hysteresis energy loss, 29
Equipotential lines, 118, 121, 124, 125, 127
Evaluation of areas of unsymmetrical hysteresis loops for d-c. apparatus, 34
Evershed, S., 47
Exciting current of a-c. magnet or transformer, 443-444, 453-455, 459-460
External energy curves, data, 67
Ewing, J. A., 86

F

Ferric flux density, 23
Ferric permeability, 24
Ferric reluctivity, 24
Ferrocobalt
  demagnetization curve, 9
  general, 7, 42-43
  general data, 69
  hysteresis loops, 53
  magnetization curves, 49

Ferrocobalt—*Continued*
  residual force effects, 9, 10
  tractive effort, 8
Ferromagnetic materials, recent advances, 6-13
Ferronickel
  coercive intensity data, 55
  demagnetization curves, 9
  general, 7, 42
  general data, 69
  hysteresis loops, 53
  magnetization curves, 49
  permeability-flux density, curve for, 508
  relay, sample design, 504-521
  residual flux density data, 55
  residual force effects, 9, 10
  tractive effort, 8
  use in relays, 7
Fiber board, 162
Fibering, 11
Field mapping, 116, 127
Filter choke, 102-105
Fixed cylindrical gap design
  flat-faced plunger magnet, 273
  full conical plunger magnet, 280, 286-287
  general, 248, 267, 268
  high-permeance, 280-281
  leakage flux magnet, illustrative problem, 315
  tapered plunger magnet, 299
Flat-faced armature magnet (flat-faced lifting magnet)
  ampere-turns in iron, 254
  characteristics, 228-230
  check of preliminary design, 257-266
  coil shape optimum, 255
  coils, 257-259
  design, 253-266
  design data, empirical, 255
  design equations, 254
  deviation from inverse square law, 229
  force formula, derivation, 201-202
  force from energy stored in working gap, 265
  force from saturation curve, 265
  force-stroke characteristic, 228-230
  flux density in air gap, 255
  general, 253

Flat-faced armature magnet—*Continued*
  leakage coefficient
    derivation, 144-147
    simplified formula, 147
    special proportions, 146
    experimental check, 148-149
    illustrative problem, 101, 261-263
  magnetic circuit calculation, 99-102, 260-264
  magnetic leakage, effect of, 229-230
  method of determining optimum, 263-265
  operated from dry cell, 102, 112
  preliminary design procedure, 254-256
  saturation, effect of, 229-230
  temperature rise, 259
  useful work-stroke characteristic, 229
  weight, computation, 266
  weight economy, 230, 266
  weight economy *vs.* index number, 242, 255
Flat-faced lifting magnet, *see* Flat-faced armature magnet
Flat-faced plunger magnet
  characteristics, 231-232
  check of preliminary design, 271-279
  coil design, 271-273
  design, 266-279
  design equations, 267-268
  deviation from inverse square law, 232
  effect
    of magnetic leakage, 232
    of saturation, 232
  empirical design data, 269
  fixed cylindrical gap design, 273
  force
    due to fringing flux, 214
    formula, derivation, 201-202
    from energy stored in working gap, 276
    from saturation curve, 276
  force-stroke characteristics, 232
  general, 266
  leakage coefficient
    derivation, 139-142
    example, 273
    experimental verification, 147
  magnetic circuit calculations, 274-275
  method of determining optimum, 274

Flat-faced plunger magnet—*Continued*
preliminary design procedure, 268-270
temperature rise, 273
time-temperature rise calculations, 277-278
useful work-stroke characteristic, 232
weight, computation, 276-277
weight economy, 232, 277
weight economy *vs.* index number, 242, 269
Flux as function of time, saturation present, 366-369
general equation and discussion, 338-339
Flux-current loop of d-c. magnet, 73-77
Flux density, effect on depth of penetration, 365, 416
Flux distribution in plunger of leakage flux magnet, 317, 320
Flux produced by a given magnetomotive force, *see* Magnetic circuit calculations
Flux-time characteristic
iron core saturated
general derivation, 366-369
illustrative example, 382-384
time for definite flux change, 372-373
illustrative example, 384-385
magnet shunted by condenser, saturated
general derivation, 393-395
illustrative example, 395-399
Force
a-c. magnets, 423-428
at end of stroke, estimation of, 390, 405
effect of air gap on residual, 106-109
error produced by neglecting fringing and leakage, 102
from energy stored in air gap
bipolar magnet, 332
flat-faced armature magnet, 265
flat-faced plunger magnet, 276
full conical plunger magnet, 289
graphical calculation
bipolar magnet, 330, 332
flat-faced armature magnet, 265
flat-faced plunger magnet, 276
full conical plunger magnet, 288-289

Force—*Continued*
graphical calculation—*Continued*
general, illustrative problem, 221-226
tapered plunger magnet, 300-306
graphical determination from flux-current loop, 79
high momentary values, 308, 414
leakage flux magnet, 316, 319
magnetic
circular parallel plane surfaces, 202
coaxial cylindrical and conical surfaces, 209-210
coaxial cylindrical surfaces, 203
coaxial full conical surfaces, 206-209
coaxial truncated conical surfaces, 205-206
current-carrying conductor in independent magnetic field, 216-218
cylindrical-shaped gap and plug, 212
eccentric cylindrical surfaces, 212-213
effect
of fringing flux, 213-214
of residual flux, 196, 199
flux constant during motion, 196-199
general, 196
general case, flux proportional to current, 214-215
general magnetic force formula, 196-200
graphical evaluation of, 221-226
in terms of
air-gap quantities, 196-198
inductance, 214-215
total magnet quantities, 198, 200
leakage flux from plunger, 218-225
loss due to magnetizing plunger, 200-201, 219-220
magnetomotive force constant during motion, 199-200
non-coaxial cylindrical surfaces, 212-213
parallel plane circular faces, 202
parallel plane surfaces, 201
rectangular-shaped gap and plug, 212
solenoid and plunger, 218-220
wedge-shaped gap and plug, 210-212
wire in magnetic field, 216-218

INDEX 543

Force—*Continued*
of relay at release a-t., 529-530
residual of soft steel magnet, 106-109
unit, 532
value for optimum magnetic conditions, 251-252
Force characteristics of a-c. magnets, 423-428
Force-distance curve, experimental verification, 224
Force formulas, general, 4, 196
Force limitation in a-c. magnets, 431
Force relation in shaded pole magnets, 468-469
Force-stroke characteristic
a-c. magnet, 448
a-c. solenoid and plunger, 435-437
method of computation, 221-226, 300-306
shaping to suit requirements, 240-241
Form-wound coils, *see* Coils
Friction in plunger magnets, 228, 237, 242, 334
Fringing flux, force produced by, 213-214
Fringing flux paths
bipolar magnet with polar enlargements, 142-143
flat-faced cylindrical plunger magnet, 139-140
flat-faced lifting magnet, 144-145
stepped-cylindrical-faced plunger, 221
tapered-plunger magnet, 302
Fringing permeances, *see special formulas and* Permeance calculations
Frölich, 23
Full conical plunger magnet
calculation of leakage coefficient, 286-287
characteristics, 232-233
check of preliminary design, 284-291
coil design, 284-286
design, 279-291
design equations, 280-281
deviation from inverse square law, 233
effect of fringing flux in conical gap, 208-209
empirical design data, 282

Full conical plunger magnet—*Continued*
equivalent flat-faced plunger, 279-280
estimation of side pull in conical gap, 334
fixed cylindrical gap design, 280, 286-287
force formula
derivation, 206-209
including fringing effect, 289
force from energy stored in working gap, 289
force from saturation curve, 288-289
force-stroke characteristics, 232-233
general, 279-280
magnetic circuit calculation, 287-288, 290
method of determining optimum, 289
preliminary design procedure, 281-283
temperature rise, 285-286
temperature rise calculation, 285-286
useful work-stroke characteristic, 233
weight, computation, 290-291
weight economy, 233, 291
weight economy *vs.* index number, 242, 282

G

General Cable Co., 164
Glass insulation, 161
Glassine paper, 163
Goss, Norman P., 38
Grain orientation of magnetic materials, 11-12
Graphical differentiation, 222
Graphical integration
general derivation, 372-373
illustrative example, 384-385
Graphical magnetic circuit calculations
ampere-turns for a series of flux values, 367
and plunger positions, 375, 409-410
ampere-turns for air gap of actual magnet, 222-225
ampere-turns given, 96-97
check of optimum polar enlargement size for relay, 519-520
demagnetization curves, 107-111
evaluation of magnetic force, 221-226
flux for series of air-gap lengths, 223-224

544 INDEX

Graphical magnetic circuit calculations
—*Continued*
  flux given, 96
  force at end of stroke, 390
  force for any plunger position and flux, 376
  force of relay
    at operate a-t., 529
    at release a-t., 529-530
  force-stroke characteristic, 221-226
  force-stroke curve of tapered plunger magnet, 300-306
  incremental inductance, 103
  method of determining optimum magnet design
    bipolar magnet, 330-331
    flat-faced armature magnet, 263, 265
    flat-faced plunger magnet, 274
    full conical plunger magnet, 288-289
    generalized scheme, 252-253
    tapered plunger magnet, 306
  operate current of a relay, 513
  optimum gap length for polarized apparatus, 105
  parallel unsymmetrical circuit, 90-94
  release current of a relay, 517
  residual flux of soft steel magnet, 107-109
  series circuit
    iron only, 88-89
    iron plus air, 96-97
    variable air gap, 104
Graybill, K. W., 487
Guillemin, E., 339
Gumlich, E., 45

H

Hard magnetic materials
  Alnico, 46
  carbon steel, 45
  cast iron, 45
  chrome steel, 45
  cobalt-chrome steel, 45
  cobalt steel, 45
  demagnetization curves, 67
  dispersion-hardened alloys, 46
  displacing electromagnets, 13

Hard magnetic materials—*Continued*
  effect of temperature, 12
  effect of vibration, 12
  external energy curves, 67
  general, 12
  general data, 70
  hysteresis loops, 54
  magnetization curves, 68
  normal permeability curves, 68
  Nipermag, 46-47
  tungsten steel, 45
Hard rubber, 162
Hazeltine, L. A., 102, 240, 241, 309
Heat-dissipation
  capacity, definition, 178
  coefficient, 184-186
Heating
  equation applied to tractive magnets, 247-248
  equation for coils, 186-187
  general, 82
  large magnets, 253
  magnet coils, *see also* Temperature rise
    exact, 192-194
    experimental, 181-184
    final value, 186
    ideal, 178-181
High-speed magnets
  alternating current, 6, 432
  current-time characteristic, no motion, derivation, 369-372
  eddy currents, 364-366
  effect of index number on dynamic characteristic, 413
  flux-time characteristic—no motion
    general derivation, 366-369
    illustrative example, 382-384
  general, 5, 363-364
  placement of rivets, 364, 366
  quick release, *see* Quick-release magnets
  rational design—short stroke
    check of illustrative design, 409-413
    design procedure, 403-404
    effect of
      change of speed, 401-403
      change of stroke, 403
      power output, 401
      saturation, 401-402

High-speed magnets—*Continued*
  rational design—short stroke—*Continued*
    empirical design data, 400-403
    general, 399
    illustrative design, 404-409
    stalled power consumption, 401
    time constant, 400-401
  space-time characteristics
    current required to produce incipient motion, 382
    evaluation of initial motion, 378-379
    illustrative problem, 385-386, 410-411
    experimental check, 390
    general derivation, 373-380
    illustrative problem
      long stroke, 385-391
      short stroke, 409-413
    time to produce incipient motion, 372-373
      illustrative problem, 384-385
    speed limitations, 413-415
    time for flux to rise to given value—no motion
      general derivation, 372-373
      illustrative problem, 384-385
    time for incipient motion
      no saturation, 409-410
      saturation present, 382-385
High-speed relays, 496-499
Hipernik, 11, 42, 493
Horseshoe magnet, *see* Bipolar magnet
Hydrogen annealed iron, 39
Hydrogen annealing
  ferrocobalt, 11
  ferronickel, 11
  pure iron, 11
Hydrogenized iron, 69
Hysteresis
  displacement factor, definition, 35
  effect in relays, 14, 493
  energy changes
    during symmetrical cycle, 27-29
    during unsymmetrical cycle, 32-35
  energy loss
    in d-c. magnets, 32-35, 56, 57, 76, 78
    of normal loop, 27
  general, 14

Hysteresis—*Continued*
  normal cycle in soft iron, 14
  normal loss, 29-30
  Steinmetz, equation and coefficient, 29-30
  unsymmetrical cycles in a-c. apparatus, 35-38
Hysteresis loops
  Alnico, 54
  cold-rolled steel
    S.A.E. 10-10, 15, 51
    S.A.E. 10-20, 51
  effect of air gap, 33
  evaluation of energy of, 27-29
  ferrocobalt, 53
  ferronickel, 53
  minor, 18, 36, 37, 111
  normal, definition, 16
  silicon steel
    high, 52
    medium, 52
  Swedish charcoal iron
    annealed, 50
    unannealed, 50
  unsymmetrical loops
    for a-c. polarized apparatus, 36
    for d-c. magnets, 32-35
Hysteresis loss
  displacement factors for unsymmetrical loops, 61
  evaluation in polarized apparatus, 35-38
  in d-c. magnets, 32-35
  normal
    for silicon steels, 61
    for soft steels, 56-57, 69
  separation from eddy-current loss, 30-32

I

Ideal work, computed for different experimental magnets, 239
Impedance, 421-422
Importance of impurities in soft magnetic materials, 9, 12
Impregnated cloth, 162
Impregnating compound
  density, 188
  thermal capacity, 188
Incremental inductance, 103

Incremental permeability
  data, 64, 65, 66
  definition, 17, 20
  discussion, 17-22
  importance of wave shape, 19
  iron-cobalt alloys, 42
  permanent magnets, 111
Index number
  computed for different experimental magnets, 239
  definition, 241
  dependence of air gap flux density on, 247-248
  pole-face shape as function of, 243
  various types of tractive magnets, 243
  weight economy as function of, 242
Inductance
  effective, saturation present, 370
  of coil, 103
  unit, 532
Inductive circuit
  build-down of current, 342-344
  build-up of current, 339-342
  time constant, 342
Initial permeability, 16
  of soft magnetic materials, 69
Instantaneous force of a-c. magnets
  single-phase, 423-424
  three-phase, 427-428
  two-phase, 425
Insulation, coils, 162
Insulation, wires, *see* Wire insulation
Intermittent duty magnets
  choice of temperature rise, 244
  illustrative examples, 193, 266
Intermittent excitation, maximum permissible duration, 277-278
Interpolar leakage, 97-99, 141, 144, 146
Iron
  American ingot, 39
  cast, 40, 45
  critical temperature, 43
  density, 188
  grain size, 11
  magnetization curves, 48
  malleable cast, 40
  Swedish charcoal, 40
  thermal capacity, 188
    effective in a magnet, 187-190
Iron-cobalt alloys, 42

Iron loss
  effect of, on performance of electromagnet, 78
  total, 30-32
Iron magnetomotive force in tractive magnets, 248
Iron-nickel alloys, 42
Iron-nickel-chromium-silicon alloys, 43
Iron-nickel-cobalt alloys, 43
Ironclad solenoid, *see* Leakage flux magnet

J

Jackson, L. R., 43
Joule-inch system of units, 531

K

Kennelly, method of extrapolating for saturation density, 23, 25
Kirchhoff's second law, 338
Kraft paper, 163, 164

L

Lag coils
  for relays, 500
  for time delay, 337
    approximate solution, 351-353
    build-down of m.m.f., 348-351
    build-up of m.m.f., 345-348
Lamellar field, 127
Laminated mica, 162
Laminating in a-c. magnets, 433-435
Leakage, plunger
  effect in stepped cylindrical plunger magnet, 223-225, 234
  force effect in all plunger magnets, 234-237
  tapered plunger magnet, 205, 293
Leakage coefficients
  definition, 100
  derivation
    bipolar magnet with polar enlargements, 142-144
    flat-faced armature magnet, 144-147
    flat-faced cylindrical plunger magnet, 139-142
    full conical plunger magnet, 207-208, 279-280
    tapered plunger magnet, 302-303

INDEX 547

Leakage coefficients—*Continued*
  illustrative problems
    bipolar magnet, 228-230
    flat-faced armature magnet, 261-263
    flat-faced cylindrical plunger magnet, 273
    flat-faced square plunger magnet, 457-458
    full conical plunger magnet, 286-287
    relay, 509-510
    shaded-pole magnet, 476-477
Leakage flux
  derivation of formulas for special magnets, 139-147
  distributed between pole cores, 97-99
  effect in
    bipolar magnet, 230-231
    cylindrical-faced plunger magnet, 234
    flat-faced armature magnet, 229-230
    flat-faced plunger magnet, 232
  equivalent of distributed path, 140-141
  extent of appreciable field, 123-126
  flux linkage of distributed flux, 99
  force due to, 218-225
  formulas
    derivation for special magnets, 139-147
    experimental verification
      bipolar magnet with square polar enlargements, 147-149
      flat-faced cylindrical plunger, 147
      flat-faced lifting magnet, 148-149
  reluctance drop due to distributed flux, 98-99
  ring, uniformly wound, 85
  transition from fringing to leakage flux in cylindrical plunger magnet, 140
Leakage flux magnet
  a-c. operation, *see* Alternating-current magnets
  characteristics, 238
  check of force, 316, 319
  check of preliminary design, 314, 321
  coil design, 314-315
  design, 308-321
  design equations, 309-310

Leakage flux magnet—*Continued*
  distribution of flux in plunger, 316-319
  effect of saturation, 238
  empirical design data, 311
  fixed cylindrical gap, 315
  force formula derivation, 218-220
  force on alternating current, 424-425
  force-stroke characteristics, 238
  general, 308
  loss in force due to magnetizing plunger, 200-201, 219-220, 309
  loss of force due to flux passing out of open end, 220
  magnetic circuit derivations and calculations
    accurate solution, 316-320
    approximate solution, 309-310
  plunger flux density, approximate solution, 309-310
  position of maximum plunger flux density, 319
  preliminary design procedure, 311-313
  required coil length, 308, 312
  speed limitation, 414-415
  temperature rise, 315
  time of action, 415
  use of tapered end, 238
  velocity of plunger, 415
  weight, 320-321
  weight economy, 321
  weight economy *vs.* index number, 242
Leakage flux paths
  bipolar magnet with polar enlargements, 122-125, 143
  flat-faced armature magnet, 99, 144, 146, 261-262
  flat-faced cylindrical plunger magnet, 140
  leakage flux magnet, 217
  stepped-cylindrical plunger magnet, 221-223
  tapered plunger magnet, 302-303
Legg, V. E., 6, 51, 69
Length, unit, 532
Lifting magnet, *see* Flat-faced armature magnet
Linearity in time-delay circuits, 339
Loss of work due to initial plunger position, 79

M

Machining strains
  annealing to remove, 38, 40
  importance of removing, 39, 42
Magnet for short-time excitation, 193, 266
Magnet types, *see particular type*
  a-c. magnets
  bipolar
  cylindrical-faced plunger
  cylindrical rotary armature
  flat-faced armature
  flat-faced plunger
  full conical plunger
  horseshoe
  ironclad solenoid
  leakage flux
  lifting
  shaded pole
  solenoid and plunger
  stepped-cylindrical-faced plunger
  tapered plunger
  truncated conical plunger
Magnet with shunt condenser—saturated
  flux-time characteristic, 393-399
  release characteristic, 393-399
Magnetic circuit calculations
  bipolar magnet, 328-331
  current given, 86, 88, 96, 221-224, 300-305
  cyclic variation in permanent magnets, 111
  equivalent length of butt joint, 86
  exciting current of air gap
    derivation, 452
    illustrative examples, 454, 455, 459-460
  exciting current of a-c. magnet or transformer
    illustrative example, 453-455, 459-460
    method, 443-444
  falling magnetization curve of
    magnet, 106-109
    relay, 515-517
  flat-faced armature magnet, 90-102, 260-264
  flat-faced plunger magnet, 274-275

Magnetic circuit calculations—*Continued*
  flux distribution in a long plunger, step-by-step method, 316-320
  flux given, 85, 87, 96, 99-102
  flux limited by saturation, 90-94
  flux-linkage of an a-c. solenoid and plunger, 441-442
  full conical plunger magnet, 287-288, 290
  general, 2, 84
  graphical, *see* Graphical magnetic circuit calculations
  importance of computing sufficient points on magnetization curve, 88-89
  incremental permeability, 102-105
  interpolation of magnetization curve, 89
  iron of
    different cross sections, 86
    uniform cross sections, 85
  leakage flux magnet, 309-310, 316-320
  leakage flux of a-c. solenoid and plunger magnet, 439-440
  magnetization loop for magnet, 459
  magneto, 109-111
  parallel circuit
    symmetrical, 59
    unsymmetrical, 90-94
  permanent magnets, 106-111
  polarized filter choke, 102-105
  residual flux of soft steel magnet, 107-109
  rising magnetization curve of relay, 511-514
  series circuit
    no leakage, 94-97
    with distributed leakage, 99-102
  stepped-cylindrical plunger magnet, 221-224
  tabular methods, *see* Tabular magnetic circuit calculations
  tapered plunger magnet, 300-306
  variable cross section, 89
Magnetic cycle of relay, 502-504
Magnetic data
  coercive intensity *vs.* magnetizing intensity, 55
  demagnetization curves, 67

Magnetic data—*Continued*
  dependence of total core loss
    on frequency, 60
    on maximum flux density, 59
  displacement factors, 61
  external energy curves, 54, 67
  for iron and iron alloys, 47-70
  general
    hard magnetic materials, 70
    soft magnetic materials, 69
  incremental permeability, 64, 65, 66
  magnetization curves, 48, 49, 58, 68
  normal hysteresis
    loops, 14, 15, 50-54
    loss, 56, 57, 61
  normal permeability curves, 15, 58, 68, 508
  polarized magnetization curves, 62, 63
  residual flux density *vs.* magnetizing intensity, 55
  unsymmetrical hysteresis loss in d-c. magnets, 56, 57
Magnetic efficacy, 77, 79
  tapered plunger magnet, 306
  use in estimating required coil ampere turns, 293, 295
  various types of tractive magnets, 239, 240
Magnetic energy stored by iron, 28
Magnetic flux, unit, 532
Magnetic flux density, unit, 532
Magnetic force, *see* Force, magnetic
Magnetic hardness of magnetic materials, 25
Magnetic intensity, unit, 532
Magnetic materials
  for electromagnets, 39-44; *see also* Soft magnetic materials
  for permanent magnets, 44-47; *see also* Hard magnetic materials
  grain orientation, 11-12
Magnetic properties of iron, 2, 14-70
Magnetic type loud-speaker unit, 498-499
Magnetization curves
  Allegheny Electric Metal, 49
  Alnico, 68
  American ingot iron, 48
  cast iron, 48
  cast steel, 48

Magnetization curves—*Continued*
  chromium steel, 68
  cobalt steel, 68
  definition, 14
  ferrocobalt, 49
  ferronickel, 49
  for magnets, *see* Magnetic circuit calculations
  Magtiz, 48
  malleable cast iron, 48
  method of extrapolating, 23-24
  mild cold-rolled steel, 15, 48
  normal, definition, 16
  Permalloy, 49
  Permendur, 49
  silicon steel, 23, 58
  Swedish charcoal iron, 48
  tungsten steel, 68
  use of, illustration, 85-86
Magnetomotive force
  required for a given flux, *see* Magnetic circuit calculations
  unit, 532
Magnets for high momentary force, 308, 414
Magtiz
  coercive intensity data, 55
  general, 40, 69
  magnetization curve, 48
  residual flux density, 55
Malleable cast iron, 40
  general data, 69
Mass, unit, 532
Mathematical flux fields
  non-parallel plane surfaces, 117
  parallel concentric cylinders, 117, 120
  parallel cylinder and plane, 120
  parallel non-concentric cylinders, 118, 119
  parallel plane surfaces, 116
  sphere and plane, 121
  spheres of same radii, 120
Maximum permeability
  definition, 16
  of soft magnetic materials, 69
Mechanical efficacy, 80, 400, 405
  tapered-plunger magnet, 306
  various types of tractive magnets, 239-240
Mechanical methods of time delay, 337

Mechanical power output of a magnet, 400, 401, 407
Mechanical work, 400, 406
  available, computed for different experimental magnets, 239
  of a-c. magnet by change of reactive power, 461
  of electromagnet, 76-77
Melting point of contacting materials, 485
Micarta, 162
M.K.S. system, conversion factors to joule-inch-second system, 532
Moisture, effect on coils, 496
Moisture-resisting coils, 153, 496
Moisture-resisting wire insulation, 161
Molybdenum, physical properties, 485
Moore, A. D., 127, 194, 253

N

Nicaloi, 42
Nickel
  critical temperature, 43
  general data, 69
Nickel alloys, general data, 69
Nickel-cobalt alloys, general data, 69
Nickel-iron alloys, magnetization curves, 49
Nipermag, 13, 47
  demagnetization curve, 67
  external energy curve, 67
  general data, 70
Niwa and Asami, 37
Non-magnetic steels, 44

O

Optimum magnetic conditions
  in tractive magnets, 251-252
  in tapered plunger magnet, 292, 304, 306
Orthogonal fields, 127

P

Paper
  density, 188
  thermal capacity, 188
Paper-margin, coils, 152, 164, 165

Paper-section coils, see Coils
Paraffin
  density, 188
  thermal capacity, 188
Permalloy, 7, 42
  general data, 69
  magnetization curves, 49
Permanent magnet materials, see Hard magnetic materials
Permanent magnets
  demagnetization curve calculations, 106-111
  effect of periodic cycle, 110-111
  energy available in air gap, 44, 67, 70
  general, 106
  incremental permeability calculations, 110-111
  materials for, 44-47
  optimum proportions, 110
  steels for
    criterion of usefulness, 44
    demagnetization curves, 67
    general data, 70
Permeability
  differential, definition, 23
  effective in polarized cores, 105
  ferric, 24
  incremental
    definition, 17
    general, 17-23
  initial, definition, 16
  maximum, definition, 16
  normal
    definition, 16
    effect of superposed a-c. fields, 22
  of vacuum, 532
  reversible, definition, 18
  unit, 532
Permeance, unit, 532
Permeance calculations, 116-149
  bipolar relay, 527-528
  estimation of permeance, 126-130
  field mapping, 127
  fixed-cylindrical gap, 222, 273, 286, 299, 315
  fixed-hinged joint of relay, 510-511
  general, 3, 116
  methods employing functions of a complex variable, 129, 130
  methods of estimating, 121-130

INDEX 551

Permeance calculations—*Continued*
  special formulas for mathematical fields
    non-parallel plane surfaces, 117
    parallel concentric cylinders, 117-118, 120
    parallel cylinder and plane, 120
    parallel non-concentric cylinders, 118-119
    parallel plane surfaces, 116
    sphere and plane, 121
    spheres of same radii, 120
  special formulas for method of estimation of permeance of probable flux paths, 130-139
    coaxial cylindrical and conical surfaces, 139, 209-210
    coaxial cylindrical surfaces, 134-139
    coaxial full conical surfaces, 139, 207
    coaxial truncated conical surfaces, 139, 205
    corner to corner in same line, 132
    corner to perpendicular plane, 133-134
    cylinder to surrounding cylinder, 134-135
    cylinder to surrounding edge, 134-135
    cylinder to surrounding perpendicular face, 134-137
    cylindrical edge to surrounding cylindrical edge
      in different planes, 137-138
      in same plane, 138
    cylindrical face to surrounding annulus in same plane, 138-139
    edge to
      edge in same line, 133
      parallel edge, 131
      parallel plane, 133-134
      perpendicular plane, 133-134
      surrounding cylinder, 134-136
    face to face in same plane, 131-132
    face to perpendicular plane, 133-134
    perpendicular face to surrounding cylinder, 134-137
  working gap of full conical plunger magnet, 289

Permendur, 7, 43
  general data, 69
  magnetization curves, 49
Perminvar, 7, 43
  general data, 69
Phenolic material, density, 188
Platinum, physical properties, 485
Platinum-cobalt alloy, 70
Plunger magnets
  clearance around plunger, 249, 271
  cylindrical-faced plunger magnet, *see type*
  design of fixed cylindrical gap, *see* Fixed cylindrical gap design
  flat-faced, *see type*
  flux distribution in a long plunger, 316-320
  force due to plunger leakage flux, 225, 305
  friction, 228, 242, 296
  full conical, *see type*
  index number, 241-242
  leakage flux, *see type and* Leakage flux
  loss due to magnetizing plunger, 200-201, 219-220, 223-225, 305
  method of
    force-stroke computations, 221-226
    handling distributed leakage flux, 222-224
  solenoid and plunger, *see* Leakage flux
  stepped-cylindrical-faced plunger magnet, *see type*
  tapered plunger magnet, *see type*
  truncated conical plunger magnet, 237, 242
Polar enlargements
  check for optimum in relay, 518-520
  method of design, 324, 506-507
Polarized alternating-current apparatus
  hysteresis loops, 35-38
  magnetic circuit calculations, 102-105
  optimum length air gap in choke, 102-105
Polarized magnetic cores, 19
Polarized magnetization curves, 62, 63
Polarized relays, 82, 498-499
Polarized transformers, 102
  steel for, 41

Polarizing flux density, 19-20
  in a-c. apparatus, 37
  method of determination, 22
Pole-face shape
  effect on force-stroke characteristic, 4, 231-243
  optimum, 242-243
Polyphase magnet arrangements, 425-428
Potential work ability of an electromagnet, 77
Power, unit, 532
Power density, unit, 532
Power factor of a-c. magnet, 462
Pure iron, 10

Q

Quick-acting magnets, see High-speed magnets
Quick-release magnets
  calculations of release time for magnet shunted by condenser, 393-399
    general derivation, 393-395
    illustrative problem, 395-399
  effect of stored energy, 392
  elimination of eddy-current effects, 391-392
  general, 391-392
  method of demagnetizing, 391-392
  methods of breaking circuit, 391-392
Quick-release relay, 521-530

R

Rapid-acting magnets, 80; see also High-speed magnets
Rate of change of permeance, method of determination, 222, 224
Reactance voltage, 420-422
Reactive power
  change of, in a-c. magnets, 429-431
  total, in a-c. magnets, 430
Reactor, variable, 102, 113, 114
Rectangular-shaped gap and plug, axial force, 212
Rectifiers
  for a-c. magnets, 6, 431
  half-wave, 102

Relay steels, 41
  comparison of three types, 520
  permeability-flux density curves for, 508
  use of, 498, 505
Relays, 483-530
  a-c., 501-502
  annealing relay steel, 521
  arc suppression, 488
  classification of, 484
  coils, 496
  condenser for arc suppression, 488
  contact deflection, 488-489, 491
  contact follow-through, 488-490
  contact materials, 485-487
    molybdenum, 485, 486
    platinum, 485, 487
    silver, 485, 486
    tungsten, 485, 486
  contact pressure, 487, 490-491
  contact size and shape, 489-490
  delayed action, 495, 499-501
  design
    air-gap length, 491-493, 506, 523
    coil design, 517-518
    comparison of results for three relay steels, 520-521
    flux density of working gap, 507
    force, closed gap, 514-515
    leakage coefficient calculation, 509-510, 514
    magnetization curve
      falling, closed gap, 515-517
      rising
        closed gap, 514
        open gap, 511-513
    operate ampere-turns, 513, 517
    optimum air-gap considerations, 507, 518-520
    permeance of fixed-hinged joint, 510-511
    polar enlargements, 506-507
      check for optimum, 518-520
    pole core size, 507-509, 524
    release ampere-turns, 517
  design factors, 83
  design for operation from vacuum tubes, 521-530
    choice of tube, 525-526
    coil design, 525

INDEX 553

Relays—*Continued*
 design for operation from vacuum tube—*Continued*
  force
   closed gap, 529
   open gap, 527
   preliminary design, 523-526
  design for quick release, 521-530
  design of simple relay, 504-521
  differential current sensitivity, 491-493
  double coil design, 495-496
  eddy current, mitigation of, 494, 498
  effect of
   dust on contacts, 487-488
   hysteresis, 493, 503
   moisture, 496
  electrolysis, 496
  general, 6, 483
  high-speed, 494, 496-499
  inductive circuits, 485, 488
  magnetic cycle of, 502-504
  minimum length gap
   as determined by hysteresis, 493
   as determined by release current, 491-493
  polar enlargements
   check for optimum size, 518-520
   size of, 506-507
  polarized, 82, 498-499
  power consumption as a design limitation, 82
  required stroke, 491
  sensitive moving coil, 112
  sensitivity as affected by design limitations, 490-496
   contact deflection, 491
   contact pressure, 490-491
   delayed action, 495
   release current, 491-493
   residual magnetism, 493-494
   single or double coil design, 495
   size or weight, 494
   speed of action, 494
  single coil design, 495
  size as affecting sensitivity, 494
  speed of contact separation, 488
  time for current to reach operate value, 497

Release time, magnet with shunt condenser
 general derivation, 393-395
 illustrative problem, 395-399
Reluctance, *see* Permeance
Reluctance calculations, *see* Permeance calculations
Reluctivity, *see* Permeability
Residual effects in commercial soft magnetic materials, 8
Residual flux, method of removal, 392
Residual flux density
 definition, 15
 hard magnetic materials, 70
 method of extrapolating, for saturation value of, 25
 soft magnetic materials, 69
 variation with maximum magnetizing intensity, 26
 data, 55, 69, 70
Residual force
 effect of air gap, 9-10, 107-109
 in soft steel magnets, 106-109
 of relays, 6
Residual magnetism in relays, 493
Resistance
 aluminum wire, 158
 change with temperature, 156, 160
 coil, method of computing, 171, 173
  illustrative problem, 175-177
   bipolar magnet, 327
   flat-faced armature magnet, 259
   flat-faced plunger magnet, 272
   full conical plunger magnet, 285
   leakage flux magnet, 315
   tapered plunger magnet, 298
 copper wire, 158
 density of coils, 162, 168, 169, 170, 172
 temperature correction table, 160
 temperature equation, 156, 159
 unit, 532
 voltage, 420-422
Resistivity
 aluminum, 155
 copper, 155
 effect on depth of penetration, 364-365
 effect of vanadium, 8
 hard magnetic materials, 70
 molybdenum, 485
 platinum, 485

Resistivity—*Continued*
 silver, 485
 soft magnetic materials, 69
 tungsten, 485
 unit, 532
Reversible permeability
 definition, 18
 hard magnetic materials, 70
Ribbon-wound coils, *see* Coils
Ring sample, 85
Rising magnetization curve, 15
Rivets
 in high-speed magnets, 364-366
 placement of, 433, 446, 447
Root-mean-square exciting current, method of determination, 444, 459-460
Rouault, Charles, 20
Rules for wire gauge use, 156
Russell, H. W., 43

S

Sandford and Cheney, 25
Saturation, 23
 effect in
  bipolar magnet, 230-231
  cylindrical-faced-plunger magnet, 234
  flat-faced armature magnet, 229-230
  flat-faced plunger magnet, 232
  leakage flux magnet, 238
  stepped-cylindrical-faced plunger magnet, 235
  tapered plunger magnet, 236
  truncated conical plunger magnet, 237
 effect on
  depth of penetration, 415-416
  speed in plunger type magnets, 414
Saturation curve for magnets, *see* Magnetic circuit calculations
Saturation density
 iron-cobalt alloys, 42
 soft magnetic materials, 69
Saturation flux density, 23
Sensitive relays
 annealing, 521
 effect of grain direction, 38
 50% nickel iron, use of, 42
 Permalloy, use of, 42
 sample design, 504-521

Separation of eddy-current and hysteresis loss, 31
Series magnetic circuit, 87
Shaded-pole magnets, 463-480
 analysis of magnetic circuit, 464-468
 average force, 469, 479
 criterion for ideal shading, 464, 469
 criterion for optimum shading, 471
 flux of shaded pole, 467, 478
 flux of unshaded pole, 477
 general, 6, 427, 463
 instantaneous force relation, 468-469
 leakage coefficient calculations, 476-477
 length of gap under unshaded pole, 472, 475-476
 magnetomotive force across main gap, 479-480
 mathematical analysis of flux relation, 464-468
 optimum conditions for minimum force variation, 469-471
 optimum design, 464, 471-472
 optimum design conditions, 471-472
 phase-angle lag of shaded-pole flux, 466, 467, 478
 power loss in shading coil, 472, 478
 preliminary design procedure, 473-476
 pulsation of force, 469, 479
 total flux, 468, 478
Shading coils
 for a-c. magnets, 6, 427, 463
 for a-c. relays, 501-502
 for polyphase magnets, 428
 for single-phase magnets
  design, 472-480
  theory, 463-472
Shaping of force-stroke characteristic, 240-241
Shaping of pole faces, 79; *see also* Pole-face shape
Short-time excitation
 design of magnet for, 266
 determination of temperature rise, 277-278
 general, 187
Side pull in conical plunger magnet, 334
Side pull in plunger magnets, 212-213
Silicon, addition to iron, 10, 11

INDEX

555

Silicon steel
  anhysteretic magnetization curves, 62, 63
  bar stock, 41
  core loss as a function of
    flux density, 59
    frequency, 60
  demagnetization curves, 9
  depth of penetration of flux, 365, 416
  determination of eddy current loss, 31
  differential permeability, 416
  displacement factor data, 61
  general data, 69
  high permeability and low loss by special rolling methods, 38
  hysteresis energy loss
    data, 61
    constants, 30
  hysteresis loops, 52
  incremental permeability, 64, 65, 66
  magnetization curves, 58
  normal permeability curves, 58, 62
    flux density curve for, 58, 508
  polarized magnetization curves, 62, 63
  residual force effects, 9-10
  sheet steel, 41
  tractive effort curves, 8
Silk insulation, 159-160
Silver, physical properties, 485
Simm, L. G. A., 20
Simple inductive circuit
  build-down of current, 342-344
  build-up of current, 339-342
  time constant, 342
Sintering of hard magnetic materials, 13
Slotting in a-c. magnets, 433-434
Slow-acting magnets, 80; *see also* Time-delay magnets
Soft magnetic materials, *see also particular material*
  Allegheny Electric Metal, 42
  American ingot iron, 39
  cast iron, 40
  cold-rolled steel, 39
  constant permeability alloys, 43
  electrical bar steel, 41
  electrical sheet steel, 41
  general, 7
  general data, 69
  hydrogen annealed iron, 39

Soft magnetic materials—*Continued*
  iron, malleable cast, 40
  iron-cobalt alloys, 42
  iron-nickel alloys, 42
  iron-nickel-chromium-silicon alloys, 43
  iron-nickel-cobalt alloys, 43
  Nicaloi, 42
  Permalloy, 7, 42
  Permendur, 7, 43
  Perminvar, 7, 43
  silicon steel, 41
  steel
    cast, 40
    S.A.E. 11-12, 40
    S.A.E. 10-10, 40
    S.A.E. 10-20, 40
  Swedish charcoal iron, 40
Solenoid and plunger axial force, 218-220
Solenoid and plunger magnet, *see* Leakage flux magnet
Solenoid magnet, *see particular type*
Solenoidal field, 127
Space factor of coils, 162, 167, 169, 170, 172
  flat-faced armature magnet, 257-259
  flat-faced plunger magnet, 272
  full conical plunger magnet, 285, 286
  illustrative problems, 175
  leakage flux magnet, 315
  paper-section coil, 169, 327
  tapered plunger magnet, 298
Space-time characteristics of magnet
  experimental check, 390
  general derivation, 373-380
  method of evaluating initial motion
    general derivation, 378-379
    illustrative problem, 385-386
Spark energy of electromagnet, 75
Specific gravity
  hard magnetic materials, 70
  soft magnetic materials, 69
Speed limitations of d-c. magnets, 413-414
Speed of action, a-c. magnets, 432
Splaying, copper wire, 159
Splitting wire size on coils, 285-286
Spooner, Thomas, 21, 22, 31, 35, 36, 37, 38, 111

556 INDEX

Spooner's formula of incremental permeability, 21
Square plunger magnet, calculation of fringing and leakage permeances, 406-408
Steel
  carbon, 45
  cast, 40
  chrome, 45
  cobalt, 45
  cobaltchrome, 45
  cold-rolled, 39.
  density, 188
  electrical sheet, 41
  ground bars, sizes commercially available, 249
  S.A.E. 10-10, 39
  S.A.E. 10-20, 40
  S.A.E. 11-12, 40
  S.A.E. 18-8, 44
  silicon, 41
  thermal capacity, 188
  tungsten, 45
Steinmetz, C. P., 23, 364
Steinmetz's equation for normal hysteresis loss, 29
Step-by-step calculations
  current-time characteristics—no motion
    general derivation, 369-372
    illustrative example, 382-384
  flux distribution in long plunger, 316-320
  flux-time characteristic—no motion
    general derivation, 366-369
    illustrative example, 382-384
  quick-release magnet with condenser
    general derivation, 393-395
    illustrative example, 395-399
  space-time characteristics
    experimental check, 390
    general derivation, 373-380
    illustrative example, 385-391
    short-stroke tractive magnet, 409-413
Stepped-cylindrical-faced plunger magnet
  characteristics, 235-236
  effect of magnetic stop, 235-236
  effect of saturation, 235

Stepped-cylindrical-faced plunger magnet—*Continued*
  force-stroke characteristic, 224, 235, 334
  graphical evaluation of force, 221-226
  loss due to magnetizing plunger, 223-225
  method of force-stroke calculation, 221-226
  plunger leakage flux pull, 223-225, 234
  replacing by tapered plunger, 242
  useful work-stroke characteristics, 235
  weight economy, 236
Stock dimensions in tractive magnet design, 248-249
Stop, magnetic
  use in cylindrical-faced plunger magnet, 234
  use in stepped-cylindrical-faced plunger magnet, 235
  use in tapered plunger magnet, 236, 307
Stored energy of magnet, effect on quick release, 392
Strap-wound coils, *see* Coils
Stroke, value for optimum magnetic conditions, 251-252
Structure of magnetic materials, 11
Swedish charcoal iron
  coercive intensity, 55
  demagnetization curve, 9
  energy returned to electric circuit, 56
  energy to demagnetize, 56
  general, 40
  general data, 69
  hysteresis energy loss, 56
  hysteresis loops data, 50
  magnetization curve, 48
  permeability-flux density curve, 508
  residual flux density data, 55
  residual force effects, 9
  tractive effort curve, 8

T

Tabular computation, *see also* Step-by-step calculations
  force-stroke curve of tapered plunger magnet, 305
  force-stroke curve of **stepped-cylindrical-faced plunger magnet**, 225

INDEX

Tabular magnetic circuit calculations
 bipolar magnet, 330-331
 falling magnetization curve for relay, 516
 flat-faced armature magnet, 264
 flat-faced plunger magnet, 275
 flux distribution in plunger of leakage flux magnet, 316-320
 full conical plunger magnet, 290
 parallel unsymmetrical circuit, 92-94
 residual flux of soft steel magnet, 107-109
 rising magnetization curve for relay, 511-514
 series circuit, iron only, 87, 88
 series with fringing and distributed leakage, 102
 tapered plunger magnet, 300-301
Tapered plunger magnet
 characteristics, 236
 check of preliminary design, 297-307
 coil design, 297-299
 constant force possible, 236
 "cut-and-try" design solution, 294-296
 design, 291-307
 design equation, 293-294
 fixed cylindrical gap design, 299
 force due to plunger leakage, 234, 305
 force formula, derivation, 209-210
 force-stroke characteristics, 236, 291-292
 general, 236, 291-292
 graphical evaluation of force-stroke characteristic, 300-306
 loss in force due to magnetizing plunger, 305
 magnetic circuit calculations, 300-306
 magnetic efficacy, 306
 magnetic stop
  effect of, 239
  use of, 307
 mechanical efficacy, 306
 method of determining optimum, 306
 plunger, friction, 296
 preliminary design procedure, 294-296
 saturation, effect of, 236, 292
 temperature rise, 299
 useful work-stroke characteristic, 236

Tapered plunger magnet—*Continued*
 weight, 306-307
 weight economy, 236, 307
 weight economy *vs.* index number, 242
 working-gap permeance, 302-303
Temperature-resistance
 correction table, 160
 equation, 156, 157
Temperature rise
 as design limitation of d-c. magnets, 82
 as affected by intermittent duty, 244
 during short period of excitation, 277-278, 455-456
 effect of surrounding iron, 182-184
 effect on
  speed of leakage flux magnet, 414
  weight economy, 243-244
 exact solution, 192-194
 experimental heating law for coils, 181-184
 experimental method of measurement, 181-182
 general, 3-4
 ideal coil, 178-180
 illustrative problems
  bipolar magnet, 327-328
  direct attraction type a-c. magnet, 456
  flat-faced armature magnet, 259
  flat-faced plunger magnet, 273
  full conical plunger magnet, 285-286
  leakage flux magnet, 315
  method of changing dimensions to give required rise, 298, 327-328
  tapered plunger magnet, 299
 intermittent excitation, 187, 277-278, 462
 magnets and coils, 178-195
 time
  calculation, 190-192
  curves, experimental results, 191-194
  variable power input, 195
Temperature-sensitive magnetic alloys, 43
Temperature unit, 532
Tensile strength, contacting materials, 485

Thermal capacity, 178
  data for various magnetic materials, 188
  of coils and iron parts, 187-190
Thermal conductivity of contacting materials, 485
Thermal time constant, 180
  illustrative problem, method of computation, 277-278
Time
  for flux to rise to a given value, 372-373
    illustrative example, 384-385
  unit, 532
Time constant
  magnet with lag coils, approximate, 352-353
  simple inductive circuit, 342
  simple magnet, approximate, 360
Time-delay action, 81
Time-delay magnets, 336-362
  condenser and resistance, 337-338, 353-356
  electronic, 337-338
  experimental check
    magnet with lag coil, build-down, 349-350
    magnet with shunt condenser, build-down, 355-356
    magnet with shunt condenser and lag coils, build-down, 350-359
    simple magnet, build-down, 343-344
  general, 5, 336
  lag coil, general, 336
  linearity, 339
  mechanical devices, 337-338
  method of computing solution, general scheme, 338-339
  selector magnet, 361-362
  series inductance, general, 337
  solution of differential equations
    magnet with lag coil
      approximate solution, 351-353
      build-down, 348-351
      build-up, 345-348
    magnet with series inductance, 342
    magnet with shunt condenser, build-down, 353-356
    magnet with shunt condenser and lag coils, build-down, 356-359

Time-delay magnets—*Continued*
  solution of differential equations—*Continued*
    simple magnet
      build-down, 342-344
      build-up, 339-342
Time of action and motion
  computation of, *see* Space-time characteristics of magnet
  defined, 399
  leakage flux magnet, 415
Time of excitation, method of determining actual time for intermittent excitation, 277-278
Time-temperature rise computation, illustrative problem, 190-192, 277-278
Torque
  magnetic
    coaxial cylindrical surfaces, 204
    general case, flux proportional to current, 216
    general magnetic torque formula, 198, 199
    in terms of
      air-gap quantities, 198
      inductance, 216
      total magnet quantities, 199
  unit, 532
Tractive effects in commercial soft magnetic materials, 8
Tractive magnets, *see also* Direct-current magnets
  a-c., *see* Alternating-current magnets
  air-gap magnetomotive force as function of stroke, 221-225, 300-305
  bipolar, *see* Bipolar magnets
  coil magnetomotive force in terms of magnetic efficacy and work, 293-295
  comparison of different types, 239, 242
  cylindrical plunger magnet, *see* Cylindrical-faced plunger magnet
  design, *see particular magnet type for details*
    brass tubes, sizes commercially available, 239
    check of
      final design, 249-250
      optimum, 251-253

INDEX 559

Tractive magnets—*Continued*
  design—*Continued*
    clearance between plunger and tube, 249
    coil shape, 248
    coil space factor, 246
    design symbols, 246
    effect of
      change of one wire size, 249
      index number on working-gap flux density, 247-248
      size on working-gap flux density, 248
    ferrocobalt, 7
    force, optimum, 251-253
    fundamental design equations, 246-247
    general considerations, 245
    generalized scheme for design, 247-250
    heating equation, solution of, 248
    magnetic circuit equation, solution of, 248
    magnetomotive force used in fixed air gap, 248
    maximum useful work, 251
    optimum conditions for fixed excitation, 251-253
    plunger clearance, 249
    steel, round bars, sizes commercially available, 249
    stroke, optimum, 251-252
    use of stock dimensions, 248-249
    wire size, 247-248
  d-c. magnets, *see* Direct-current magnets
  effect of different pole-face types, 239-242
  flat-faced armature magnet, *see* Flat-faced armature magnets
  flat-faced plunger magnet, *see* Flat-faced plunger magnets
  flux current loop, 73-77
  force from flux current loop, 78-79
  force-stroke calculations, method, 221-226
  full conical plunger, *see* Full conical plunger magnets
  general, 228
  heating, 81-82

Tractive magnets—*Continued*
  high-speed magnets, *see* High-speed magnets
  horseshoe magnets, *see* Bipolar magnets
  index number, 241-243
  leakage flux, *see* Leakage flux magnets
  loss of work due to
    hysteresis, 78
    initial plunger position, 78-79
    stored energy, 77-78
  magnetic efficacy, 77-79
  mechanical efficacy, 80
  potential work ability of, 77
  selecting best pole-face type, 241-243
  shaping of force-stroke characteristics, 240-241
  solenoid and plunger, *see* Leakage flux magnets
  stepped-cylindrical-faced plunger magnet, *see* Stepped-cylindrical-faced plunger magnets
  tapered plunger magnet, *see* Tapered plunger magnets
  time delay, *see* Time-delay magnet
  time-delay action, 81
  truncated conical plunger magnet, *see* Truncated conical plunger magnets
  useful work, 80, 228
Transformers
  audio-frequency, 102
  constant-current, 113
  exciting current
    method of computing, 443-444
    illustrative example
      approximate, 453-455
      exact, 459-461
  half-wave rectifier, 72, 102
  polarized, 102
  use of air gap, 112
Transients
  iron unsaturated
    free response, 340
    magnet with lag coil
      approximate solution, 351-353
      build-down, 348-351
      build-up, 345-348

Transients—*Continued*
  iron unsaturated—*Continued*
    magnet with shunt condenser, 353-356
    magnet with shunt condenser and lag coils, build-down, 356-359
    normal function, 340
    simple magnet
      build-down, 342-344
      build-up, 339-342
    saturation present
      current-time characteristics, 369-372, 382-384
      flux-time characteristic, 366-369, 382-384
      general, 338-339
      magnet with shunt condenser, 393-399
      space-time characteristic, 373-390
Truncated conical plunger magnet
  characteristics, 237
  effect of gap eccentricity, 237, 242
  effect of saturation, 237
  force formula, derivation, 205-206
  force-stroke characteristic, 237
  useful work-stroke characteristic, 237
  weight economy, 237
Tungsten, physical properties, 485
Tungsten steel, 45
  demagnetization curve, 67
  external energy curve, 67
  general data, 70
  magnetization curve, 68
  normal permeability curve, 68
Turn density of coils, 162, 168, 169, 170, 172

U

Underhill, Charles R., 151, 170, 240
Units, system of, 531
Useful work
  computed for different experimental magnets, 239
  general, 80, 228, 400
  long-stroke magnet, 235
Unsymmetrical hysteresis cycles for d-c. magnets, 32-35
Unsymmetrical hysteresis loops in a-c. apparatus, 35

V

Vacuum tubes
  for quick release, 391-392
  operated relay, 521-530
Vanadium, addition to ferrocobalt, 8
Velocity, unit, 532
Voltage, unit, 532
Voltage regulator, 114
Volt-ampere
  consumed in iron core
    approximate method, 454
    exact method, 459-461
  efficiency in a-c. magnets, 430, 431, 461
    example, 436
  excitation of an air gap, 452
  limitation, a-c. magnets, check, 429-431, 437, 461-462
Volume, unit, 532

W

Watson, E. A., 61
Wave form
  effect on depth of penetration, 415-416
  importance in designating incremental permeability, 19-20
  of exciting current of a-c. magnet
    derivation, 443-444
    illustrative example, 459-460
Wave-shape distortion in polarized apparatus, 19-20
Webb, C. E., 7
Wedge-shaped gap and plug, axial force, 210-211
Weight economy, *see particular magnet type*
  as affected by
    index number, 242
    size, 244
    temperature, 243-244
  comparison of various pole-face types, 239-241
Weight limitation in a-c. magnets, 431
Wire
  aluminium, wire table, 157, 158
  areas, 157, 158
  copper, wire table, 157, 158

Wire—*Continued*
  diameters, bare and insulated, 157
  gauge, 156
  square, 162
  sizes, 156
Wire insulation
  asbestos, 160-161
  chromoxide, 160-161
  cotton, 157, 159-160
  double cotton covered, 157
  double enamel covered, 159
  enamel, 157, 159
  glass, 161
  single silk covered, 157
  silk, 157, 159-160
  single cotton covered, 157
  single cotton enamel covered, 157
  single silk enamel covered, 157
Woodruff, L. F., 339
Work
  graphical determination from flux-current loop, 79

Work—*Continued*
  ideal, computed for different experimental magnets, 239
  mechanical, computed for different experimental magnets, 239
  useful
    computed for different experimental magnets, 239
    long-stroke magnet, 235
    optimum from magnet, 251-253
Work available
  analysis from flux-current loop, 73-77
  from electromagnet, 77
  from ideal magnetic cycle, 33-34
Work equation
  for tractive magnet, 77
  use in determining required a-t., 293-295
Work losses in electromagnet, 77-79

Y

Yensen, T. D., 7

RET'D FEB 17 1986

RET'D AUG 25 1986

JUN -2 1995